ZONDERVAN ATLAS OF THE BIBLE

Revised Edition

CARL G. RASMUSSEN

ZONDERVAN

ATLAS

OF BIBLE
THE

Revised Edition

ZONDERVAN
ACADEMIC

ZONDERVAN ACADEMIC

Zondervan Atlas of the Bible
Copyright © 2010 by Carl G. Rasmussen

ISBN 978-0-310-52126-6 (ebook)

Requests for information should be addressed to:
Zondervan, *Grand Rapids, Michigan 49530*

Library of Congress Cataloging-in-Publication Data

Rasmussen, Carl.
    Zondervan atlas of the Bible / Carl G. Rasmussen. — Rev. ed
        p. cm.
    Rev. ed. of: Zondervan NIV atlas of the Bible.
    Includes bibliographical references (p. 263 – 64) and index.
    ISBN 978-0-310-27050-8 (hardcover, printed)
        1. Bible — Geography. 2. Bible — Geography — Maps.
        3. Bible — History of Biblical events. 4 Bible — History of contemporary events. I. Rasmussen, Carl.
    Zondervan NIV atlas of the Bible. II. Title. III. Title: Atlas of the Bible.
    BS630.R37 2009
        220.9'1 — dc22   2009015937

Cover design: Rob Monacelli
Interior design: Kirk DouPonce

Printed in China

20 21 22 23 24 25 26 27 /GPC/ 26 25 24 23 22 21 20 19 18 17 16 15 14 13 12 11 10 9 8 7

This book is dedicated to my wife, Mary, with whom I have spent sixteen exciting years traveling, teaching, and guiding in the lands of the eastern Mediterranean.

CONTENTS

PREFACE

In 1967 my wife and I undertook a journey that changed the course of our lives. In that year we traveled to Jerusalem, where we enrolled as students at the Institute of Holy Land Studies (now the Jerusalem University College). During that time I developed an interest in the historical geography of the Bible, an interest that intensified during my graduate studies in the United States. But it was during my seven-year tenure (1973 – 80) as dean of the Institute of Holy Land Studies that I became keenly aware of the extraordinary usefulness of historical geography as a tool for interpreting Scripture.

After I returned to the United States to assume a teaching position at Bethel University (Minnesota), this conviction deepened as I had the opportunity to teach university and seminary students as well as numerous adult education classes and forums, and I saw these students gain confidence in their understanding of the historical aspects of their faith. Indeed, an understanding of the geographical dimension of history opens up new vistas for students of all texts — both sacred and nonsacred.

Many have aided my work in the historical geography of the Bible, and it is with deep appreciation and thanksgiving that I remember the late Dr. G. Douglas Young and his successors at the Institute of Holy Land Studies — now the Jerusalem University College — for making a study center available for Christian students in Jerusalem. Particular thanks also go to Prof. Anson Rainey, my teacher and later my colleague in Jerusalem, who so generously shared his wealth of knowledge with me as well as with his other students, and to my friend James Monson, whose creativity and expertise have influenced me as well as countless other students and teachers. In the production of the first edition of this book, Dr. Judith Hadley and Wendy (Youngblood) Morrissey read the manuscript in whole or in part and made many helpful suggestions. However, the views presented in this book are my own and in some instances diverge significantly from those of my friends mentioned above.

The writing, and now the revision, of this book has taken place over a number of years, and the staff at Zondervan Publishing House — Stanley Gundry, Ed van der Maas, David Frees, Kim Tanner, Verlyn Verbrugge — have been encouraging, helpful, and patient during the process. The work of Nathan Opsata in helping prepare the indexes is greatly appreciated. Dr Moisés Silva has made many helpful comments that have been incorporated into the Geographical Dictionary. Mark Connally has been especially gracious in allowing me to use a number of his beautiful photographs in this new edition of the *Atlas*. In the first edition of this work Carta Map Company of Jerusalem, headed by Immanuel and Shai Hausman, produced the maps. In this edition International Mapping has produced the all-new maps in cooperation with Zondervan and the author. We have endeavored to make them as accurate and as useful as possible and to effectively illustrate the topography of the lands of the Bible.

After testing a variety of color schemes it was decided it was best to indicate lower elevations in shades of green while higher elevations are represented in degrees of tan and eventually in white. The use of these colors helps to emphasize the topography of the various countries. So the reader of this atlas should be aware that "green" does not indicate vegetation nor does "tan" necessarily indicate lack of vegetation — the variety of colors are used to indicate *elevation*.

In addition, the exact shorelines of bodies of water — such as the Huleh Lake, the Sea of Galilee, and the Dead Sea — and the exact course of rivers such as the Jordan during ancient times are not known with certainty. Indeed, during the 2,200 years of biblical history they must in fact have changed from time to time. If one were to look at these bodies of water on (say) Google Earth, Lake Huleh would not appear at all (the Israelis have drained most of it) and the Dead Sea would have a very odd shape, for it has been significantly contracting during the last 30 years as the Israelis and Jordanians have been diverting the waters of the Jordan and the Yarmuk for their own purposes (drinking and agricultural). Thus, we have tried to estimate what these bodies of water might have looked like prior to significant modern intervention and to produce maps that will assist the modern reader to visualize the lands of the Bible.

It has been a special pleasure to share the joys of the land of the Bible with my wife, Mary, and more recently with three intrepid trekkers, John, Peter, and Andrew. It is because of her encouragement and attention to our family while I was off working on this project in Israel or in the States that I have been able to complete this book. It is to her, with deep gratitude and love, that this book is dedicated, with the hope that the reader will come to share our love for the land, the people, and the God of the Bible.

ABBREVIATIONS

AD	*Anno Domini*, in the year of our Lord (i.e., after the birth of Christ)
ANEP	J. B. Pritchard, ed., *The Ancient Near East in Pictures Relating to the Old Testament*. Second edition. Princeton: Princeton University Press, 1969.
ANET	J. B. Pritchard, ed., *Ancient Near Eastern Texts Relating to the Old Testament*. Third edition. Princeton: Princeton University Press, 1969.
BC	Before Christ
ca.	*circa*, about
ch., chs.	chapter(s)
d.	died
e.g.	for example
EB	Early Bronze Age
ed., eds.	editor(s)
esp.	especially
et al.	and others
etc.	*et cetera*, and so on
F	Fahrenheit
ft.	foot, feet
Gr.	Greek
H.	Horbat
Heb.	Hebrew
in.	inch(es)
J.	Jebel
Ant.	Josephus: *The Antiquities of the Jews*
Life	Josephus: *Life*
War	Josephus: *The Jewish War*
Kh.	Khirbet
KJV	King James Version
L.	Lake
LB	Late Bronze Age
LXX	Septuagint
MB	Middle Bronze Age
mi.	mile(s)
MT	Masoretic Text
Mt(s).	Mountain(s)
N.	Nahar/Nahal
NASB	New American Standard Bible
NEB	New English Bible
NIV	New International Version
NT	New Testament
OT	Old Testament
p., pp.	page(s)
par., pars.	paragraph(s)
R.	River
RSV	Revised Standard Version
sq.	square
T.	Tell (Arabic)/Tel (Hebrew)
V.	Valley
v., vv.	verse(s)
W.	Wadi

INTRODUCTION

When one thinks of the Middle East, many different mental images come to mind. Some envision oil wells dotting the barren landscapes of Saudi Arabia, Iran, and Iraq. Others recall pictures of fighting in the streets of some Middle Eastern capital, or desert battles involving tanks and aircraft. Images of mosques and minarets, of bazaars, of camels and deserts flash into mind. For some, pictures of Jews returning to Palestine, establishing the state of Israel, and making the desert bloom are dominant. Yet, for many others, the Near East remains enigmatic and remote. However, for both Christians and Jews, the Near East is of special interest, since both Christianity and Judaism believe that God has acted in and through the lives of individuals and nations in this area in such a way that those special events and peoples have continuing worldwide significance.

This atlas has been written in the belief that once one has a basic understanding of the geography of the Middle East, one has a much better chance of coming to grips with the flow of historical events that occurred there. This is not to say that the physical environment dictated the events of history; nevertheless, it should be recognized that historical events were oftentimes greatly influenced by the geographical environment in which they occurred. Thus Section 1 of this atlas begins with a geographical description of the Middle East as a whole — studying its major regions. Since most biblical events occurred in the area that is now occupied by the modern states of Israel and Jordan, special attention is given to the geography of those countries.

Section 2 of the atlas is historical in nature and focuses on events beginning in the third millennium BC and continuing up to the fall of Jerusalem in AD 70. Although this book is not intended to be a history of Israel, biblical and extrabiblical texts that lend themselves to geographical illustration have been emphasized. Because of the interplay of time and space in historical events, chronological considerations need to be dealt with, and the rather standard dates found in the *Encyclopedia of Archaeological Excavations in the Holy Land* have been used for archaeological periods and for the kings of Egypt, Assyria, etc. However, dates for the Judean and Israelite kings follow those developed by E. Thiele (see the bibliography). The dates for Israel's early history — the patriarchs, the exodus from Egypt, the conquest of Canaan, and the period of the judges — follow a plain reading of the biblical text and in the mind of this author fit quite well with known extrabiblical chronologies.

The author is well aware of alternate chronological schemes, but an atlas is not the place to discuss all these in detail. A general chrono-

logical chart is located in the appendices, but more detailed charts are presented at the beginning of each chapter to aid in the understanding of that specific period. A separate chapter is devoted to Jerusalem; it reflects an analysis of the most recent topographical, archaeological, and historical studies with regard to this, the most important of biblical cities. The second part of the atlas concludes with an essay on "The Disciplines of Historical Geography."

The final section of the atlas consists of a number of appendices, which include a bibliography of selected important works dealing with the geography and history of the Bible, a glossary of terms, a chronological chart, an index of Scripture references, an index of persons, as well as a geographical dictionary and index. The latter contains notes on biblical site identifications. In general, the reader can consult the standard Bible encyclopedias and dictionaries for more detailed treatments of various geographic names or the works of P. Abel, Y. Aharoni, M. Avi-Yonah, Z. Kallai, A. Rainey, and others. The interested nonspecialist reader is also encouraged to consult popular periodicals such as *Biblical Archaeology Review* and *Near Eastern Archaeology* in order to keep abreast of the most recent archaeological discoveries relating to the history, archaeology, and geography of the Bible.

In the map-making process one has to choose which existing antiquity site is to be identified with a given biblical site so that dots can be placed on the maps. Site identification is not an exact science, and the reader is invited to consult the essay on "The Disciplines of Historical Geography" to become familiar with some of the complexities of this process. In this endeavor we are all dependent on the work of others, and this atlas is no exception. Many have carried out pioneering research in the identification of biblical sites, and the contributions of E. Robinson, Abel, Aharoni, Avi-Yonah, Kallai, and Rainey are of special significance (see the bibliography for their works). The vast majority of site identifications adopted in this atlas are based on their works, but a number of modifications have been made in light of more recent studies.

The spelling of biblical names follows the pattern set in the New International Version. Arabic and Hebrew place names have been transliterated in English characters, and usually these transliterated names sound reasonably close to their Arabic and Hebrew counterparts. Those familiar with either language should not have any difficulty in locating the exact Arabic and Hebrew words if they desire to do so. When quoting from the Bible, I have used the NIV except where specific variances are noted. When extrabiblical texts are cited, reference is made to the standard English translation of these texts, *Ancient Near Eastern Texts Relating to the Old Testament* (= *ANET*). When the reader is referred to the works of the first-century AD Jewish historian Josephus, first the older ("Whiston") reference is given, followed in brackets by the appropriate reference to the now-standard Loeb Classical Series edition.

This atlas with its numerous maps and diagrams contains an immense amount of information and myriads of details. The author, cartographers, and editors have done everything in their power to avoid inaccuracies and inconsistencies and to eliminate errors. However, some may remain, and both the author and the publisher will be grateful for any corrections submitted by readers of this atlas.

I hope that, just as this book has grown out of my classroom and field experiences with university and seminary students as well as with adult study groups, students of the Bible will find it useful both in their personal study of the Bible and in the classroom. In addition, I trust that travelers to the Holy Land will find the geographical section of the book especially helpful as they prepare themselves for a once-in-a-lifetime experience in the land(s) of the Bible, and that the book as a whole will serve them in good stead before, during, and after their travels to the Middle East.

GEOGRAPHICAL
SECTION

INTRODUCTION TO THE MIDDLE EAST AS A WHOLE

The stage on which the major events of Old Testament history took place includes all the major countries shown on page 17, as well as some of the smaller Arab states that are situated in the southern and eastern portions of the Arabian Peninsula. This large land mass is bounded on the west by the Nile River and the Mediterranean Sea, on the north by the Amanus and Ararat Mountains, and on the east by the Zagros Mountains and the Persian Gulf. To the south, the Nafud Desert and the southern tip of Sinai form a rather amorphous boundary. By the time of the New Testament apostles the western horizon of the biblical world had expanded to include Greece, Italy, and even Spain.

Much of the Middle East, in its more limited Old Testament sense, is desert. Large portions of modern-day Syria, Iraq, Jordan, and Saudi Arabia include desert wastes such as the Syrian Desert, the Nafud, the Arabian Desert, and the Ruba al-Khali. These huge deserts cover some 487,000 square miles, or about half of the total area of these countries, and this figure does not even include the desert wastelands of the Negev, Sinai, and Egypt.

▲ *Hill Country of Manasseh during spring*

Besides the huge deserts in the region, the seas and gulfs that help outline the Middle East on the south, east, and west have greatly influenced life in the area. The most important of these bodies of water is the Mediterranean Sea, for it is from it, from the west, that life-giving rains come to the Middle East (except for the monsoon rains in southern Saudi Arabia, which are produced by a different cycle). Much of what has occurred in the Middle East can be summed up as a struggle between the influences of the desert and of the Mediterranean Sea. This is true of its geology, climate, flora, fauna, farming, herding, and the movements of ethnic groups.

For example, during certain periods the dominant ethnic influences have been from the sea — note the arrival and the historical significance of the Philistines, Greeks, and Romans — while during other periods the major ethnic influences have been from the tribes located in or on the fringes of the desert, such as the Amalekites, Moabites, Edomites, Israelites, and Ammonites. The interaction of these diverse groups was sometimes peaceful but often violent. In either case, this interaction was played out on many different levels and can be pictured as a struggle between the desert and the sea, or the desert and the sown.

Climatically, the year in the Middle East can be divided into two major periods: the dry season (the summer months) and the wet season (the winter months). The amount of rainfall the various regions of the Middle East receive during the winter months varies widely, but generally speaking the northern areas receive more rainfall than the southern ones, higher elevations receive more rain than areas of low elevation, and the regions closer to the Mediterranean receive more rain than those distant from the sea. These winter rains nourish the grain crops that grow throughout the area in places where the total rainfall is more than 12 inches annually. In addition, flocks of sheep and goats feed on the winter grasses that cover regions that generally receive more than 8 inches of rain.

Normally, springs, wells, and cisterns supply many of the inhabitants of the Middle East with drinking water throughout the year. The other significant sources for fresh water, apart from rainwater, are the great rivers of the Middle East — the Nile of Egypt and the Tigris and

Black Sea

Skopje **BULGARIA**
MACEDONIA
BANIA

GEORGIA ★ Tbilisi **RUSSIA**

ARMENIA **AZERBAIJAN**
Yerevan

Baku

Ararat
Mt.

*Caspian
Sea*

GREECE

*Aegean
Sea*

Ankara ★

Halys R.

TURKEY

Athens ●

Taurus Mts.

Amanus Mts.

Euphrates R.

Tigris R.

IRAN

Tehran ★

Zagros Mts.

Mediterranean Sea

Nicosia ★
CYPRUS

Beirut ●
LEBANON
Golan Heights
ISRAEL
Jerusalem ★

SYRIA

Damascus ★

Jordan R.

Baghdad ★

IRAQ

Amman ●

LIBYA

Gaza Strip
West Bank
Cairo ★

JORDAN

Sinai

An Nafud Desert

KUWAIT
Kuwait ●

*Persian
Gulf*

EGYPT

Nile R.

*Red
Sea*

SAUDI ARABIA

Arabian Peninsula

BAHRAIN
Manama ●

Doha ●
QATAR

200 km.

200 miles

Riyadh
★

THE ANCIENT NEAR EAST

Black

Sinop ●
Amisus ●

Sea

EUROPE

*Caspian
Sea*

Delphi ●

Hattusa ●

HITTITE – HURRIAN

Mycenae ●
Athens ●
Pylos ● Sparta ●

Kanish ●

Miletus ●

ASIA

MYCENAEAN

Attalia ●

Tarsus ●

Carchemish ●

Haran ●

Tigris R.

Nineveh ●

MINOAN

Rhodes ●

Aleppo ●

Gozan ●

Nuzi ●

Ecbatana ●

Knossos ●

Cyprus

Ugarit ●
Hamath ●

Ebla ●

ASSYRIA

Asshur ●

Crete Phaistos ●

Arvad ●

Euphrates R.

A M O R I T E

Mediterranean Sea

Byblos ●
Sidon ●
Tyre ●

Qatna ●

Tadmor ●

Mari ●

Akkad? ●

Susa ●

Damascus ●
Hazor ●

Babylon ●

BABYLONIA

Cyrene ●

Megiddo ●

Ramoth Gilead ●

AFRICA

Alexandria ●

Zoan ●

Gaza ●

Jerusalem ●

On ●

Memphis ●

EGYPTIAN

Kadesh Barnea ●

Dumah ●

Ur ●

*Persian
Gulf*

Elath ●

Arabian Desert

Tema ●

Nile R.

*Red
Sea*

0   200 km.

0   200 miles

Thebes ●

	Fertile Crescent
——	Land routes
- - -	Sea routes
AMORITE	Cultural spheres

▲ *Roman road in Syria*

Euphrates of modern-day Turkey, Syria, and Iraq. Some of the earliest civilizations developed along the banks of these rivers, where the people could irrigate their crops with river water. Of these great civilizations, the Bible mentions the mighty powers of Assyria and Babylonia, whose heartlands were along the Tigris and Euphrates, and, of course, the perennially powerful Egypt.

One can find the region where the majority of people have lived in the Near East since earliest historical times (ca. 3000 BC) by highlighting on a map the areas watered by the Nile, the Tigris, and the Euphrates, as well as those regions that receive over 12 inches of rainfall annually. This area, in which adequate water supplies make the growing of agricultural products possible, is roughly the shape of a crescent with one point in the Nile River, the other in the Persian Gulf. Its arc passes through Iraq, Syria, Lebanon, and Israel, hovering over the desolate Syrian Desert to the south. In this area, aptly named the "Fertile Crescent," civilizations have risen and fallen throughout the millennia.

Generally, enough wheat and barley were grown in each of the populated areas of the Fertile Crescent to supply the local population, and some countries (e.g., Egypt) were able to export grain to neighboring as well as distant lands during certain periods. Although most of the countries produced sufficient food supplies, many of them lacked other raw materials necessary for daily life. For example, the Mesopotamian region needed timber, building stones, copper, iron, tin, gold, and silver. Egypt, too, lacked local supplies of timber, copper, and iron. Some of these raw materials were available from countries within the Fertile Crescent (e.g., timber from Lebanon and Syria), but other products, including gold, silver, copper, tin, and iron, were often imported from outside the region. Thus, as these raw materials entered the Fertile Crescent and foodstuffs and finished products such as textiles left it, a network of routes developed that connected the various countries with one another.

Although there were many ways to travel from one city to another, travelers tended to follow well-established routes in order to avoid areas that would impede their progress. These areas included swamps, rivers, flooded or muddy terrain, regions that were too sandy or too rocky, places inhabited by hostile tribes or governments, forested regions, and routes that included long, difficult climbs up and down mountains and hills. In addition, long-distance travel over great desert expanses was normally avoided because of the lack of water and the hostility of dangerous tribes.

One of the major international routes ran approximately 1,770 miles from Ur in southern Mesopotamia to Thebes in southern Egypt. Along the way it passed through great urban centers such as Babylon, Mari, Tadmor, Aleppo, Ebla, Damascus, Hazor, and Gaza. It does not appear that this route as a whole had a name, but it was made up of shorter segments that ran from city to city, and in all probability these shorter stretches had special names. For example, the portion of this road that ran eastward from Egypt across northern Sinai into southern Canaan/Philistia was known as the "way of the land of the Philistines" (Exod 13:17 RSV). This name is a typical example of the ancient custom

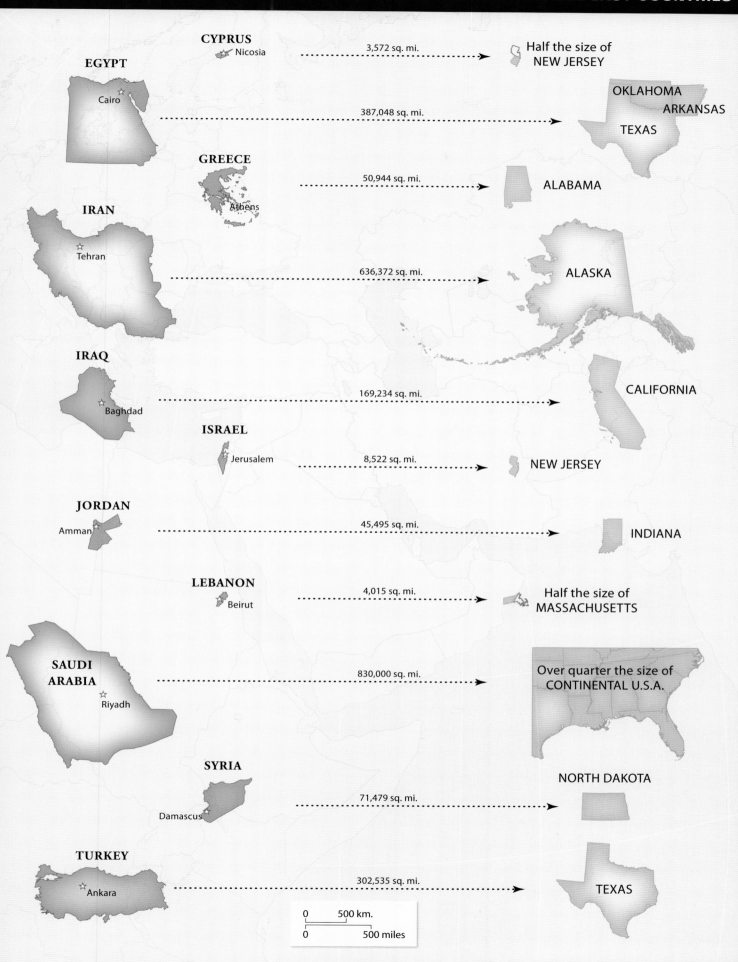

COMPARATIVE SIZES OF MIDDLE EAST COUNTRIES

CYPRUS
Nicosia
· · · · · 3,572 sq. mi. · · · · · →
Half the size of
NEW JERSEY

EGYPT
Cairo
· · · · · 387,048 sq. mi. · · · · · →
OKLAHOMA
ARKANSAS
TEXAS

GREECE
Athens
· · · · · 50,944 sq. mi. · · · · · →
ALABAMA

IRAN
Tehran
· · · · · 636,372 sq. mi. · · · · · →
ALASKA

IRAQ
Baghdad
· · · · · 169,234 sq. mi. · · · · · →
CALIFORNIA

ISRAEL
Jerusalem
· · · · · 8,522 sq. mi. · · · · · →
NEW JERSEY

JORDAN
Amman
· · · · · 45,495 sq. mi. · · · · · →
INDIANA

LEBANON
Beirut
· · · · · 4,015 sq. mi. · · · · · →
Half the size of
MASSACHUSETTS

SAUDI ARABIA
Riyadh
· · · · · 830,000 sq. mi. · · · · · →
Over quarter the size of
CONTINENTAL U.S.A.

SYRIA
Damascus
· · · · · 71,479 sq. mi. · · · · · →
NORTH DAKOTA

TURKEY
Ankara
· · · · · 302,535 sq. mi. · · · · · →
TEXAS

0   500 km.
0   500 miles

of labeling roads as "the way to/of X" (where X = a geographical place name). Other portions of this major international route certainly also had names, but they are rarely preserved in the historical sources.

Although an "international route" may bring to mind images of concrete and asphalt highways crisscrossing a continent or country, it should be remembered that "roads" in the ancient world were, until late in the Roman period (ca. AD 200), usually unpaved dirt paths. These dirt roads were cleared of stones and kept relatively free of weeds and fallen trees, and in some cases they were graded. In the earliest times the most common mode of transportation was walking, with donkeys used as pack animals. Under these conditions, a caravan normally moved at the rate of 2 or 3 miles per hour. Sometime during the second millennium BC, camels began to be used on the desert paths. These animals, which on average could carry 400 pounds of cargo, eventually began to be used on other routes as well. During early times ox-drawn carts were also used for transporting bulky items, but due to the poor condition of the roads the use of carts and carriages for transporting goods and people over long distances did not come into general use until the roads were upgraded during the Roman period.

An international route brought mixed blessings to the inhabitants of the population centers that lay along it. On the one hand, those centers had immediate access to the goods that the traveling merchants were carrying, and the powerful elite could gain added revenue by imposing tolls and by providing services (food, shelter, protection, etc.) to the caravans. On the other hand, the people traveling in these caravans exposed those centers to new external influences — religious, political, economic, etc. — that were not always welcome. In addition, some of the mighty armies of the great powers of antiquity — the Egyptians, Assyrians, Babylonians, Persians, Greeks, and Romans — passed along these same international routes, bringing with them death, destruction, and deportation.

A glance at the map on page 17 shows that the major routes that connected the continents of Europe, Asia, and Africa passed through the regions of Israel and Syria. It was in this area that God placed the descendants of Abraham, that they might live in obedience to his covenant. There they were tested to see if they would keep themselves free from pagan influences, if they would be a light to the nations around them, and if they would trust in God rather than chariots for their ultimate security. They were told that obedience to God's commands would bring blessing and prosperity, while disobedience would bring punishment. Thus the drama of the biblical story develops: How would Israel respond to God's gracious acts and attendant commands?

THE GEOGRAPHY OF ISRAEL AND JORDAN

The Five Major Longitudinal Zones

Introduction to the Five Zones

The land of Israel is situated at the southeastern corner of the Mediterranean at approximately the same latitude as southern Georgia, Dallas, and San Diego. The proximity of Israel to the Mediterranean Sea and the Arabian Desert has greatly influenced her topography, climate, flora, fauna, and human history. Throughout the ages, the desert and the sea have vied with one another for control of the land.

The stability of Israel's permanent western boundary, the Mediterranean Sea, stands in contrast to the fluctuations of her eastern border. At times the edge of the eastern desert served as the boundary, while during other periods Israel's territory ended at the Jordan River. This variability is reflected in the fact that, although the Jordan River formed the traditional eastern boundary of the land of Canaan (e.g., Num 34:12; map p. 106), the Israelite tribes settling east of the Jordan in Gilead considered themselves part of Israel in spite of the fact that the "land of Gilead" was outside the "land of Canaan" (Josh 22; map p. 106).

The classical boundary description of the heartland of Israel was summed up in the phrase "from Dan to Beersheba" (Judg 20:1; 1 Sam 3:20). However, to the south, Judah was allotted territory as far as the Kadesh Barnea/Desert of Zin area, and during periods of strength she extended her rule even farther south, down to Elath on the Red Sea (1 Kings 9:26; 2 Kings 14:22; 2 Chron 26:2). Only on rare occasions was Israel able to extend her rule as far north as Damascus, even though that region was included in the traditional descriptions of the "land of Canaan" (Num 34:7–11; Josh 13:4–5; map p. 106). It seems that Israel was not able to control the Phoenician coast to the northwest, except possibly during the days of David (2 Sam 24:7).

This geographical section will deal with the territory from Mount Hermon in the north to Elath in the south and from the Mediterranean Sea to the Arabian Desert, since most events of biblical history took place within this area. Five major longitudinal zones can be distinguished as one moves from west to east: the coastal plain, the central mountain range, the Rift Valley, the Transjordanian mountains, and the eastern desert. A grasp of this physical stage on which the events of redemptive history occurred can bring those events to life and make it easier to understand and interpret both the records of the events and the message of the prophets and the psalmists who lived there and ministered to God's people.

Coastal Plain

The coastal plain is the westernmost zone that stretches from Rosh HaNiqra in the north to the Nahal Besor, south of Gaza, a distance of approximately 120 miles. Because it is close to the Mediterranean Sea, this zone receives 16 to 25 inches of rain per year; the northern sections receive considerably more rain than the southern. Powerful springs, such as the one at Aphek, provided water, but more commonly the people of the region used wells to tap the water table, which lies just below the surface.

The coastal plain can be divided into four subregions. The Plain of Acco is located in the north, extending from Rosh HaNiqra to Mount Carmel. From the tip of Mount Carmel, the Coast of Dor runs south to the Nahal Tanninim. From there the plain widens and the Sharon Plain extends south to the Nahal Yarkon. And in the south, the Philistine Plain stretches from the Nahal Yarkon to the Nahal Besor. In many of these coastal areas elevations rise from sea level in the west to some 600 feet in the east, before reaching the more pronounced rise in elevation to the central mountain range. The coastal plain consists mainly of low, rolling hills covered with fertile alluvial soils. In the northern sections of the plain several low, narrow kurkar (fossilized dune sandstone) ridges run parallel with the coast close to the shoreline. In the south, narrow strips of sand dunes are more prominent.

Because of the absence of major topographical obstacles, the coastal plain became the most natural route for north–south travel between Babylonia/Assyria/Syria and Egypt. Even so, travelers had to be careful to avoid the sand dunes, large rivers such as the Nahal Yarkon, and the low-lying areas that became swampy during the winter months. They also had to be sure that their route passed near adequate supplies of drinking water and that they chose the most appropriate track through Mount Carmel.

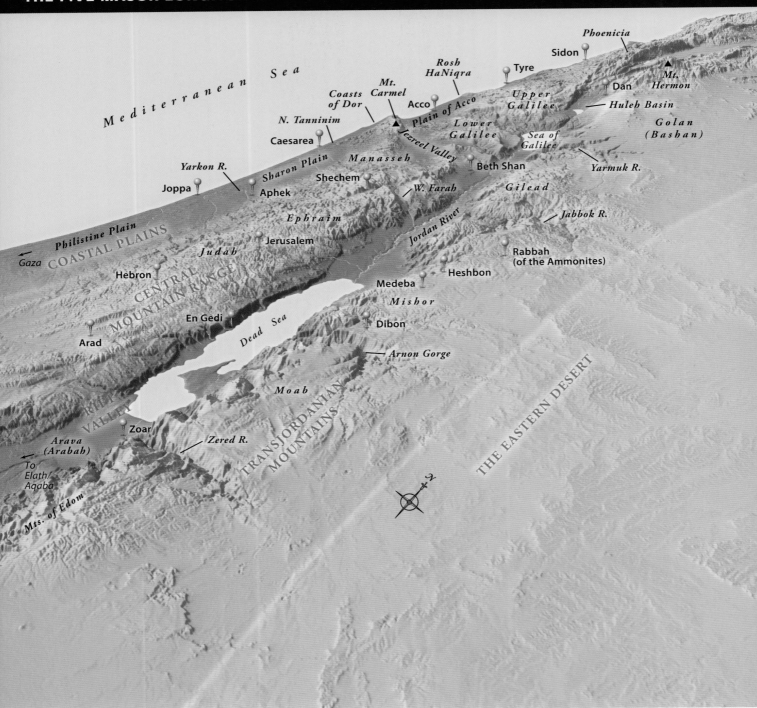

Mediterranean Sea

Phoenicia
Sidon
Tyre
Rosh HaNiqra
Dan
Mt. Hermon
Coasts of Dor
Mt. Carmel
Acco
Plain of Acco
Upper Galilee
Huleh Basin
N. Tanninim
Caesarea
Lower Galilee
Golan (Bashan)
Sea of Galilee
Jezreel Valley
Sharon Plain
Manasseh
Shechem
Beth Shan
Yarmuk R.
Yarkon R.
Aphek
W. Farah
Gilead
Joppa
Ephraim
Jordan River
Jabbok R.
Philistine Plain
Jerusalem
Rabbah (of the Ammonites)
Gaza
COASTAL PLAINS
Judah
Heshbon
Hebron
CENTRAL MOUNTAIN RANGE
Medeba
Mishor
En Gedi
Dead Sea
Dibon
Arad
Arnon Gorge
RIFT VALLEY
Moab
THE EASTERN DESERT
TRANSJORDANIAN MOUNTAINS
Zoar
Arava (Arabah)
Zered R.
To Elath/ Aqaba
Mts. of Edom

One of the most noticeable features of the coastline of Israel is that it is relatively unbroken by any major promontories that could provide natural harbors, except in the Mount Carmel – Acco region. Because it lacked natural harbors, Israel never developed into a seafaring nation as did Phoenicia, its neighbor to the north. Acco, slightly sheltered by a small promontory, was Israel's major port throughout antiquity, while Joppa, which had little natural protection, was of less significance. Eventually some coastal cities also served as harbors, but only Caesarea ever rivaled the Phoenician cities of Tyre, Sidon, and Byblos.

In portions of the coastal plain, grain crops flourished in the winter and spring months while flocks grazed there during the remainder of the year. But to the international powers of antiquity, such as Egypt and Babylonia, the road that passed through the coastal plain was of prime importance for their commercial and military activities. For the local inhabitants this was a mixed blessing; in times of peace they gained income by servicing the caravans, but during times of war the populace suffered as armies swept through the territory, consuming their recently harvested crops and taking their wives and children captive.

Central Mountain Range

The second major longitudinal zone is the central mountain range, which runs from Galilee in the north to the Negev Highlands in the south. This range, rising in places to more than 3,000 feet, is severed in an east–west direction by the Jezreel Valley in the north and the Negev Basin in the south; in these two places east–west traffic can flow with relative ease.

The central mountain range is composed primarily of harder limestones of the Cenomanian-Turonian-Eocene types. These limestones usually erode in such a way that deep V-shaped valleys are formed. Such a valley — usually dry during the summer months but sometimes flowing with water during the winter — is called a *wadi* in Arabic and a *nahal* in Hebrew. The watershed, or central spine of the mountains, runs basically north–south; these deep wadis therefore run off the range roughly to the east and west, draining into the Rift Valley and the Mediterranean Sea.

Travel along the bottoms of these deep wadis is very difficult because of boulders and occasional cliffs. Travel across the wadis is almost impossible; one would have to descend several hundred feet down an irregular slope, cross the wadi, and then climb up several hundred feet on the opposite side. Yet this unusual type of route seems to have been taken by Jonathan and his armor-bearer as they approached the Philistine camp at Micmash (1 Sam 13:23 – 14:14; map p. 132). Because of the difficulty in traveling either in or across the wadis, the roads in the central mountains have generally tended to follow the ridges. Indeed, one of the routes used most often in Bible times was the road that ran along the north–south watershed ridge between Shechem and Hebron, passing Shiloh, Bethel, Ramah, Gibeah, Jerusalem, and Bethlehem along the way (map p. 241).

In the mountain regions between Galilee in the north and Hebron in the south, the western slopes in particular receive considerable amounts of rainfall (20 to 40 in.). The abundant rain, along with the fertile terra rossa and rendzina soil, ensure the fertility of the area. Here — largely on hillside terraces that are partially formed by the natural bedding of the limestone — fields of wheat, groves of olive trees, and vineyards flourish (Deut 8:8; Ps 147:14; Hab 3:17 – 19). Using the numerous stones that they remove from their small fields, the farmers build walls along the terrace edges and then plant their crops in these narrow strips of land, 50 to 100 feet wide. The landscape from Galilee to Judah is characterized by this kind of terrace farming.

▲ *Limestone terracing, terra rosa soil, and fencing and watchtowers typical of the Central Hill Country*

In the hills, because the upper layers of limestone have many cracks and fissures, much of the winter rainwater seeps into the ground. It continues to seep down until it reaches an impermeable layer, where it begins to flow laterally. Eventually the water reaches the ground surface on the edge of a hillside or near the bottom of a valley where it flows out as a spring. At times, settlements developed close to these freshwater springs. But because these small settlements were often located on the wadi's slopes and terraces, where the springs were, they were somewhat difficult to defend against attack from above.

By about 1400 BC, the construction of cisterns solved the problem of being completely dependent on natural water sources such as springs and wells. Hewn out of the limestone and lined with plaster to prevent leakage, the cisterns collected the water from the winter rains for year-round use. Soon these bell-shaped caverns became commonplace in the hill country. Because the opening at the top was small, little sunlight entered the cavity, and the growth of algae in the standing water was thus retarded. Of course, spring water ("living water") and well water were preferred to stagnant cistern water (John 4:1–26). A cracked cistern, from which the water had drained away into the porous limestone, was useless (Jer 2:12–13).

It was in the central mountain range that the Israelites first settled — in Galilee, Manasseh, Ephraim, and Judah. Because the international powers of antiquity were primarily interested in controlling the coastal plain, the location of the Israelite villages and farmsteads in the mountains provided the people with security. Only during periods when they considered their power to be great did the Israelites move out into the coastal plain and attempt to control it (e.g., during the days of Solomon, Uzziah, and Josiah; maps pp. 145, 159); but this almost always resulted in conflict with one or more of the great powers, and in each instance Israel was eventually forced back into her hill-country heartland. The remoteness, security, and provincialism of the central mountains contrast sharply with the openness, insecurity, and more cosmopolitan nature of the coastal plain.

The Rift Valley

The third major zone in Israel is the prominent north–south depression that stretches from Dan to Elath. This depression is part of the Rift Valley system that extends for 3,700 miles from southern Turkey into Africa. North of Israel the system continues in a northeasterly direction into the Lebanese Beqa; to the south it runs through the Red Sea and down into Africa.

In Israel proper, this depression runs from Dan in the north to the southern tip of the Dead Sea, a distance of 150 miles, and then continues south-southwest as far as Elath, a distance of 110 miles. The vari-

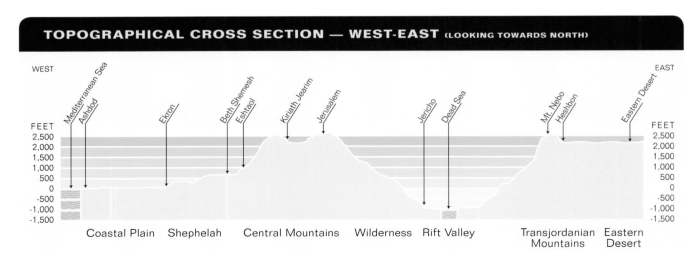

TOPOGRAPHICAL CROSS SECTION — WEST-EAST (LOOKING TOWARDS NORTH)

WEST   EAST

Mediterranean Sea · Ashdod · Ekron · Beth Shemesh · Eshtaol · Kriath Jearim · Jerusalem · Jericho · Dead Sea · Mt. Nebo · Heshbon · Eastern Desert

FEET	FEET
2,500	2,500
2,000	2,000
1,500	1,500
1,000	1,000
500	500
0	0
-500	-500
-1,000	-1,000
-1,500	-1,500

Coastal Plain · Shephelah · Central Mountains · Wilderness · Rift Valley · Transjordanian Mountains · Eastern Desert

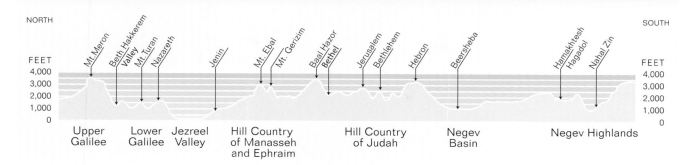

NORTH   SOUTH

Mt. Meron, Beth Hakkerem Valley, Mt. Turan, Nazareth, Jenin, Mt. Ebal, Mt. Gerizim, Baal Hazor, Bethel, Jerusalem, Bethlehem, Hebron, Beersheba, Hamakhtesh Hagadol, Nahal Zin

FEET 4,000 / 3,000 / 2,000 / 1,000 / 0

FEET 4,000 / 3,000 / 2,000 / 1,000 / 0

Upper Galilee   Lower Galilee   Jezreel Valley   Hill Country of Manasseh and Ephraim   Hill Country of Judah   Negev Basin   Negev Highlands

ous sections of the Rift Valley are diverse in character; a considerable amount of rain falls in the northern section (24 in. at Dan), whereas in the south the rainfall is negligible (2 in. at the south end of the Dead Sea). The valley receives runoff water from the mountains to the west and east along its entire length.

The northernmost section of the Rift Valley, called the Huleh Basin, covers an area that measures 20 by 5 miles. The basin has a Mediterranean type of climate and receives about 24 inches of rain each year. The springs at the foot of Mount Hermon, which are fed by the melting snows, form the headwaters of the Jordan River. As the Jordan flows through the basin, it is restrained by a natural basalt dam; its waters collect behind this dam, forming the Huleh Lake, known in antiquity as Lake Semechonitis. Until very recent times, this lake and the marshes it created forced people to live around the edges of the basin. A branch of the International Highway between Damascus and Egypt ran along its western perimeter.

The Jordan, after passing through the Huleh Basin, enters the north end of the Sea of Galilee. This most famous of biblical lakes lies 690 feet below sea level and measures 13 by 7.5 miles. The temperate Mediterranean climate makes the nearby region a desirable place to live. The sea itself is a major source of fish for the inhabitants of the region, and a number of small but fertile plains along its shoreline have been intensively cultivated throughout history.

The Jordan River flows out of the southeast corner of the Sea of Galilee as it begins its descent to the Dead Sea. The actual distance between the Sea of Galilee and the Dead Sea is 65 miles, but the length of the Jordan as it winds its way between these points is 135 miles. Along the way the Jordan receives water from the Yarmuk and Jabbok rivers on the east and the Harod and Farah valleys on the west. The Yarmuk is the major tributary; it doubles the amount of water flowing in the Jordan. Until modern times, when the Israelis and Jordanians began diverting water for commercial purposes, the Jordan averaged 100 feet in width with a depth of 3 to 10 feet, although after heavy rains in the late winter and spring its width could swell to almost a mile in places.

The Rift Valley between the Sea of Galilee and the Dead Sea can be divided into three north–south longitudinal sections. The whole valley, which is bounded on the east and west by the mountains of Transjordan and Cisjordan, is called the *Ghor* in Arabic. The valley varies in width from 3.5 miles at its narrowest point to 14 miles near Jericho; its average width is 7.5 miles. The bed of the Jordan River and its immediate vicinity is called the *Zor* in Arabic. The width of the Zor varies from 600 feet to a mile. This area is lush, teeming with all kinds of plant and animal life. In the Bible it is referred to as the "thickets" (NIV; e.g., Jer 12:5; 49:19; 50:44). On both sides of the Zor is the *Qattara*, a bleak area of chalky, salty marls, where little grows.

South of Beth Shan the climate rapidly changes from a Mediterranean to a steppe/desertlike climate. It is in general difficult to establish permanent settlements in the steppe/desert area, except where springs are available, as at Jericho, or where perennial wadis such as the Jabbok or Farah flow. During the summer months the temperatures are very hot: the average daily high at Jericho during August is 102°F.

The Jordan River completes its course by flowing into the Dead Sea — the lowest spot on the surface of the earth (1,310 ft. below sea level). In the Bible it is called the "Salt Sea" because of its high mineral content. This sea does not have any outlet. Traditionally it measures 47 by 11 miles, but it has shrunk considerably over the last twenty years. Steep cliffs border the sea on the east and the west. Rainfall is sporadic; the southern portion of the sea receives only 2 inches per year. Temperatures are high, especially during the summer months. Only where there are springs or perennial rivers can settlements develop. Because of the steep cliffs and desert conditions, there was never much travel that passed through this area.

South of the Dead Sea, the Rift Valley continues 110 miles to the shores of the Red Sea. This region is called the "Arava" or "Arabah" on modern Israeli maps, although the biblical Arabah was primarily north of the Dead Sea (e.g., Deut 3:17; Josh 11:2; 2 Sam 2:29). The valley rises from 1,310 feet below sea level at the surface of the Dead Sea to 1,165 feet above sea level at a point 50 miles north of Elath before descending to the Red Sea. The whole length of the Arava is a desolate desert region, hemmed in on the east by the towering mountains of Edom and on the west by the Highlands of the Negev. In ancient times the region was practically devoid of settlements, save for an occasional fort built to guard the caravan routes and those near the copper mines at Punon.

Elath marks the extreme southern boundary of modern Israel. At various times the Israelites used the port near Elath or the caravan

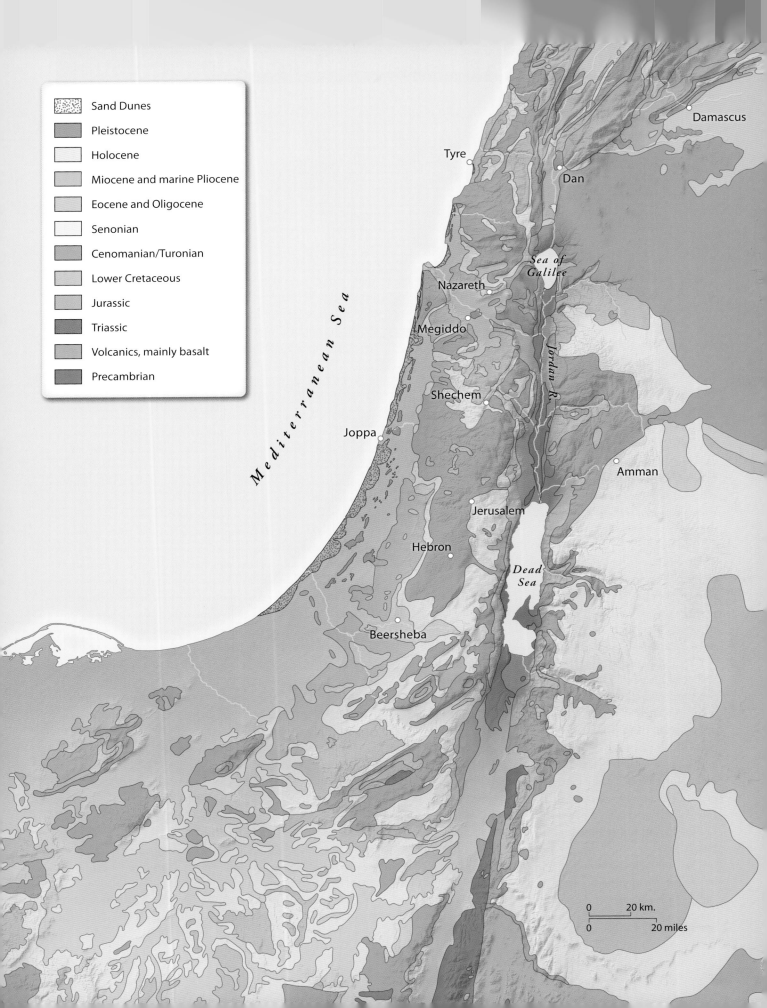

Sand Dunes	
Pleistocene	
Holocene	
Miocene and marine Pliocene	
Eocene and Oligocene	
Senonian	
Cenomanian/Turonian	
Lower Cretaceous	
Jurassic	
Triassic	
Volcanics, mainly basalt	
Precambrian	

Mediterranean Sea

Damascus

Tyre

Dan

Sea of Galilee

Nazareth

Megiddo

Jordan R.

Shechem

Joppa

Amman

Jerusalem

Hebron

Dead Sea

Beersheba

0   20 km.

0   20 miles

route that led to southern Arabia to maintain contact with exotic countries such as Ophir and Sheba. This southern extension of the Rift Valley thus was the gateway to Arabia, Africa, and India.

Transjordanian Mountains

To the east of the Rift Valley rise the mountains of Transjordan, stretching from Mount Hermon in the north to the Gulf of Aqaba/Elath in the south. While the western slopes of these mountains are often quite steep, the eastern slopes descend gradually into the Arabian Desert. The eastern boundary of this region lies approximately along the line of the old Hejaz railroad, which ran between Damascus in Syria and Mecca in Saudi Arabia.

The northernmost section of this region is called the Bashan. It includes the Golan and reaches from the foot of Mount Hermon to the Yarmuk River. Extinct volcanic cones dot the landscape, and the rich soil is derived from volcanic debris. This area is relatively high in elevation and thus receives a considerable amount of rainfall and even some snow. Travel in the Bashan is somewhat difficult during the winter months because of muddy conditions. During certain historical periods one of the branches of the International Highway skirted its northern boundary, just to the south of Mount Hermon, connecting Dan with Damascus.

To the south of the Bashan is the region of Gilead, which stretches from the Yarmuk River in the north to an imaginary east–west line drawn through the north end of the Dead Sea. On the east, the desert is only 25 to 30 miles from the Jordan Valley. The major river, which divides Gilead into two almost equal parts, is the Jabbok. Gilead is composed primarily of the hard Cenomanian limestone, and its western portion is characterized by deep V-shaped valleys, much like the Hill Country of Judah. Topography and sufficient amounts of rainfall make it a good area for wheat, olives, and grapes.

The area to the south of Gilead, reaching to the Zered Valley, was the old tribal territory of the Moabites. It too is bisected by a river, the magnificent Arnon, that flows into the Dead Sea. To the north of the Arnon is a plateau area called the Mishor. This tableland was allotted to the Israelite tribes of Reuben and Gad and became a point of dispute between Moab and Israel.

To the south of the Mishor, between the Arnon and the Zered, was the heartland of Moab. Various types of limestone and, farther east, chalk are predominate in this region, whereas along the western cliffs that face the Dead Sea the lower layers of multicolored Nubian sandstone are exposed. Because some layers of the sandstone are relatively impervious to water, numerous springs dot the landscape. Although some grain crops are grown in this area, agriculturally it is not as productive as Gilead to the north, for as one moves south the amount of rainfall begins to decrease. Mesha, one of the ancient kings of Moab, was famous for the sheep that he raised and supplied to Israel (2 Kings 3:4).

South of the Zered Valley the mountains of Edom extend to Aqaba, some of them reaching elevations of 4,500 feet. Along the western crest of this ridge there is sufficient rainfall for growing wheat and barley, but

▼ *Arnon Gorge east of the Dead Sea*

at lower elevations the amount of rainfall decreases rapidly. In biblical times one of the names for the region was Mount Seir, "The Hairy One," probably because of the scrub forests that covered the mountains. Again, hard limestone and chalk dominate the surface of the landscape, but because the steep western slopes, composed of Nubian sandstone, have a red appearance in the late afternoon sun, the name "Edom" (Heb. for "red") is appropriate for this area. The most famous city of this remote region is Petra, capital city of the Nabateans (ca. 200 BC to AD 200), the people who eventually replaced the Old Testament Edomites.

Roads in the Transjordanian mountains, like those in the central mountain range, avoided the deep valleys wherever possible. The major road was the Transjordanian Highway, which connected Damascus (and hence Assyria and Babylonia) with the countries located in present-day Saudi Arabia. This highway went through such important cities as Ashtaroth, Ramoth Gilead, Rabbah of the Ammonites, and Heshbon. The southern portion of the highway, near Heshbon, was called the "King's Highway" (Num 21:22), although this name was evidently used for another road as well (Num 20:17). In any case, the Transjordanian Highway was second in importance only to the main International Coastal Highway, which connected Gaza and Damascus.

Istanbul, Archaeological Museum

▲ *Canaanite calendar from Gezer describing the seaons of the year — tenth century BC*

The Eastern Desert

The fifth and final longitudinal zone is the great desert expanse east of the Transjordanian mountains. In the north, the great volcanic mountains and the lava flows of the Jebel Druze make this an inhospitable region, although its high elevation ensures adequate rainfall to grow crops. But elsewhere in this zone rainfall is negligible, and thus the barren desert stretches eastward some 400 miles to the Euphrates River.

Out of this desert came many peoples desiring the greener farmland of the Fertile Crescent. At various points in history these groups included the Amorites, the Arameans, the Nabateans, and ultimately the Arabs, all of whom pushed into the settled regions in the continuing struggle between the desert and the sown, between the desert and the Mediterranean Sea.

Weather Patterns

Israel's year is divided into two major seasons: the rainy season (mid-October through April) and the dry season (mid-June through mid-September). These seasons are separated by the transitional months.

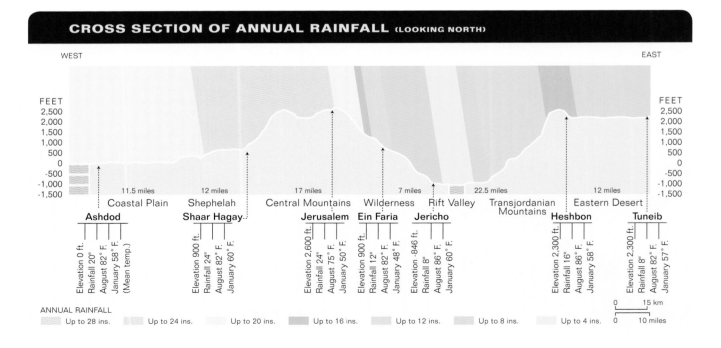

CROSS SECTION OF ANNUAL RAINFALL (LOOKING NORTH)

WEST   EAST

FEET 2,500 2,000 1,500 1,000 500 0 -500 -1,000 -1,500

11.5 miles   12 miles   17 miles   7 miles   22.5 miles   12 miles

Coastal Plain   Shephelah   Central Mountains   Wilderness   Rift Valley   Transjordanian Mountains   Eastern Desert

Ashdod   **Shaar Hagay**   **Jerusalem**   **Ein Faria**   **Jericho**   **Heshbon**   **Tuneib**

Ashdod: Elevation 0 ft. / Rainfall 20" / August 82° F. / January 58° F. (Mean temp.)

Shaar Hagay: Elevation 900 ft. / Rainfall 24" / August 82° F. / January 60° F.

Jerusalem: Elevation 2,600 ft. / Rainfall 24" / August 75° F. / January 50° F.

Ein Faria: Elevation 900 ft. / Rainfall 12" / August 82° F. / January 48° F.

Jericho: Elevation -846 ft. / Rainfall 8" / August 86° F. / January 60° F.

Heshbon: Elevation 2,300 ft. / Rainfall 16" / August 86° F. / January 58° F.

Tuneib: Elevation 2,300 ft. / Rainfall 8" / August 82° F. / January 57° F.

ANNUAL RAINFALL

Up to 28 ins.   Up to 24 ins.   Up to 20 ins.   Up to 16 ins.   Up to 12 ins.   Up to 8 ins.   Up to 4 ins.

0   15 km
0   10 miles

The Dry Season — Summer

In contrast to the variable climatic conditions experienced in many parts of North America, conditions in Israel during the summer months are relatively stable. Warm days and cooler nights are the rule, and it almost never rains. In Jerusalem, for example, the average August daytime high temperature is 86°F, the nighttime average low is 64°F.

Summer days are relatively cloudless; indeed, Israel is one of the sunniest countries in the world. On a typical summer day, temperatures begin to climb immediately after sunrise. Within a short time a cooling sea breeze begins to blow in from the west. After passing through the coastal plain it reaches Jerusalem in the mountains at about noon, and its cooling effect prevents the temperature from rising significantly during the afternoon hours. But the breeze usually does not reach Transjordan until mid to late afternoon, so temperatures there continue to climb through most of the day.

The summer months see grapes, figs, pomegranates, melons, and other crops ripening and being tended by the farmers. The summer dew and deep root systems bring needed moisture to these crops. Most of the fruits are harvested in August and September. During the summer, the shepherds move their flocks of sheep and goats westward, allowing them to feed on the stubble of the wheat and barley fields that were harvested in the spring. Because the soil is dry during the summer months, travel is fairly easy. In biblical times, caravans and armies moved easily through most parts of the country, the armies helping themselves to the plentiful supplies of grain at the expense of the local populace.

The Rainy Season — Winter

The rainy season, extending from mid-October through April, is characterized by occasional rainstorms that roll in off the Mediterranean Sea, normally bringing three days of rain followed by several days of dry weather (although deviations from this norm are frequent). During January the mean daily temperature in Jerusalem is 50°F (see map p. 30 for temperatures in other cities). For this reason, Jerusalem receives snow only once or twice each year, and even then the snow rarely remains on the ground more than a day. However, cold temperatures, combined with wind and rain, make life a bit uncomfortable in the hilly regions — a discomfort the people gladly bear because of the life-giving power of the rains.

During a typical year a farmer plows his field and plants his grain crops after the "autumn rains" of October through December have softened the hard, sun-baked soil. The grain crops grow from December through February, when 75 percent of the rain falls, and they continue to ripen during March and April, as the rains begin to taper off. These "spring rains" are important for producing bumper crops. The Bible actually refers to the three parts of the rainy season in Deuteronomy 11:14: "Then I will send rain [Heb. *matar*; Dec. – Feb.] on your land in its season, both autumn [Heb. *yoreh*; Oct. – Dec.] and

spring rains [Heb. *malqosh*; March – April], so that you may gather in your grain, new wine and oil" (cf. also Jer 5:24; Hos 6:3). Because Israel is situated between arid and wet climatic zones, the amount of rainfall throughout the country varies considerably (Amos 4:6 – 8). The general principles that describe this variation usually work in combination. They are as follows:

- The amount of rainfall *decreases* as one moves from north to south (note the decrease in the amounts for the cities in the Rift Valley from Dan to Elath).
- The amount of rainfall *decreases* as one moves from west to east, away from the Mediterranean Sea (note the sites in the Jezreel and Harod valleys).
- The amount of rainfall *increases* with the elevation (cf. the amounts for cities in the mountains with those of cities at lower elevations).
- The amount of rainfall *is greater* on the windward (Mediterranean) side of the mountains than on the leeward side (cf.

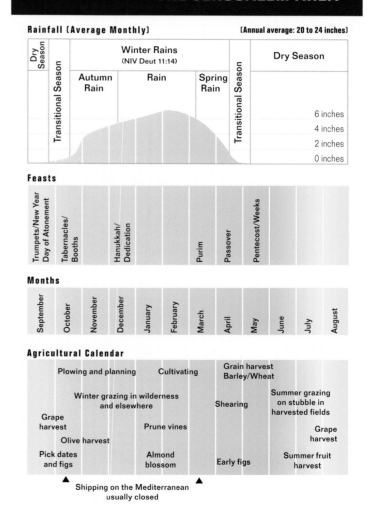

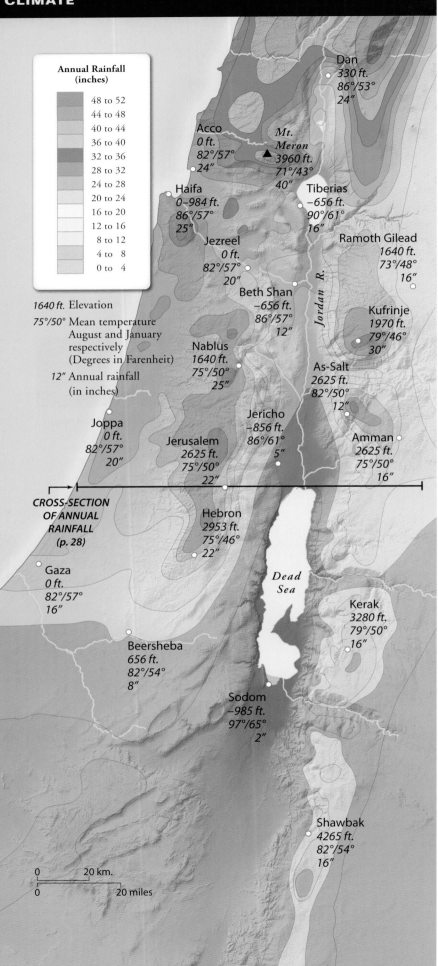

Annual Rainfall (inches)

48 to 52
44 to 48
40 to 44
36 to 40
32 to 36
28 to 32
24 to 28
20 to 24
16 to 20
12 to 16
8 to 12
4 to 8
0 to 4

1640 ft. Elevation

75°/50° Mean temperature
August and January
respectively
(Degrees in Farenheit)

12" Annual rainfall
(in inches)

CROSS-SECTION OF ANNUAL RAINFALL
(p. 28)

Dan
330 ft.
86°/53°
24"

Acco
0 ft.
82°/57°
24"

▲ *Mt. Meron*
3960 ft.
71°/43°
40"

Haifa
0–984 ft.
86°/57°
25"

Tiberias
–656 ft.
90°/61°
16"

Jezreel
0 ft.
82°/57°
20"

Ramoth Gilead
1640 ft.
73°/48°
16"

Beth Shan
–656 ft.
86°/57°
12"

Kufrinje
1970 ft.
79°/46°
30"

Nablus
1640 ft.
75°/50°
25"

As-Salt
2625 ft.
82°/50°
12"

Jordan R.

Joppa
0 ft.
82°/57°
20"

Jericho
–856 ft.
86°/61°
5"

Amman
2625 ft.
75°/50°
16"

Jerusalem
2625 ft.
75°/50°
22"

Gaza
0 ft.
82°/57°
16"

Hebron
2953 ft.
75°/46°
22"

Dead Sea

Kerak
3280 ft.
79°/50°
16"

Beersheba
656 ft.
82°/54°
8"

Sodom
–985 ft.
97°/65°
2"

Shawbak
4265 ft.
82°/54°
16"

0   20 km.

0   20 miles

---

the amounts for Shaar HaGay and Ein Faria, located at the same elevation, on the diagram on p. 28).

Most of the land north of Beersheba receives sufficient rain to grow grain (i.e., more than 12 in. per year). But in those regions where the total annual average is only 12 to 16 inches, the growth of grain crops is by no means assured, for a variation of only 4 to 6 inches can spell disaster. In addition, farmers throughout the country face numerous uncertainties:

- The *beginning* of the rainy season is sometimes delayed until December, resulting in a shorter growing season.
- The *end* of the rainy season can come as early as March, again causing a shortening of the growing season.
- The *total amount* of rainfall in a given locality can, and often does, deviate considerably from the listed mean.
- The *distribution* of rainfall in a given season can vary considerably. For example, in some years, parts of the country can go without rain for four or five consecutive weeks during the months of January and February, usually the rainiest months of the year. A prolonged dry spell such as this can seriously affect crop yield.

One of these negative factors, or several working in combination, can seriously retard the growth of crops. In a good year the farmer can sow and reap "a hundredfold" (Gen 26:12), but a series of drought years can be devastating and, in the past, could drive people from the land (Gen 12:10).

The Israelites knew that it was the Lord God who had his eyes on the land continually, from the beginning of the year to its end, and that their obedience to his commandments would bring blessing while disobedience would usher in drought and disaster (Deut 11:8–17). But given the uncertainties about the amount and distribution of the rainfall, it is no wonder that some Israelites were drawn to participate in the worship of Baal, the Canaanite storm god, who was believed to bring fertility to the land.

The Transitional Seasons

The first transitional season lasts from early May through mid-June. The temperatures gradually rise, and the season is punctuated by a series of hot, dry, dusty days during which the winds blow in from the eastern and southern deserts. On these days — which are called by the names of the winds, *hamsin, sirocco,* or *sharav* — the temperature often rises 25°F above normal, and the relative humidity can drop by as much as 40 percent. *Hamsin* conditions can be very enervating to both humans and beasts, and they completely dry up the beautiful flowers and grasses that covered the landscape during the winter months (Isa 40:7–8). The positive effect of these winds, however, is that the hot, dry weather aids the ripening of the grains by "setting" them before the harvest. It is during this season that first the barley and then the wheat harvest take place.

The second transitional season, from mid-September to mid-October, marks the end of the stable, dry, summer conditions. It is the time of the fruit harvest, and farmers begin to look anxiously for the onset of the rainy season — note the prayers for rain in the Jewish rituals associated with the last day of the Feast of Tabernacles. In the fall, travel on the Mediterranean becomes dangerous (Acts 27:9), and it remains so throughout the winter months.

Major Natural Routes — Roads

Roads and Modes of Travel

The roads that developed in the ancient land of Israel can be divided into three major categories: international routes, interregional routes, and local routes. One of the uses of the international and interregional roads was commercial. The roads were used for the transportation of scarce supplies, such as certain foodstuffs, copper, iron, tin, gold, silver, incense, dyes, and fine pottery (bulkier items such as timber and stones were usually shipped on boats and rafts).

In addition, these roads served as thoroughfares for military expeditions, for itinerant tradesmen such as smiths, for the migration of peoples, for the conveyance of governmental and commercial messages, and for the travel of pilgrims to holy places. Those who controlled the roads, whether brigands or a more permanent central government, could derive considerable income from the traffic on them. The central government could collect tolls from passing caravans, sell food and lodging, and "offer" the services of military escorts, who could be hired by the caravans to "ensure" their safe passage through "dangerous" territory.

Those living along the international routes were exposed to new intellectual, cultural, linguistic, and religious influences, and this inev-

▲ *Jerusalem during a Hamsin*
▼ *Jerusalem three days after the Hamsin*

itably led to a degree of assimilation. For example, the ease of travel in and out of Samaria, when compared with the remoteness of the Hill Country of Judah, helps to explain the openness of the former to non-Israelite religious and cultural influences. This, in turn, eventually led to the deportation of the Northern Kingdom some 130 years before the captivity of Judah (see pp. 150–164). Similarly, during the Persian and Greco-Roman periods, increased exposure to international influences led to a rapidly accelerated process of assimilation.

Besides walking, early modes of transportation included donkeys, solid-wheeled carts, and chariots. Camels eventually began to be used to carry heavy loads, especially in caravans. Horses were used in the

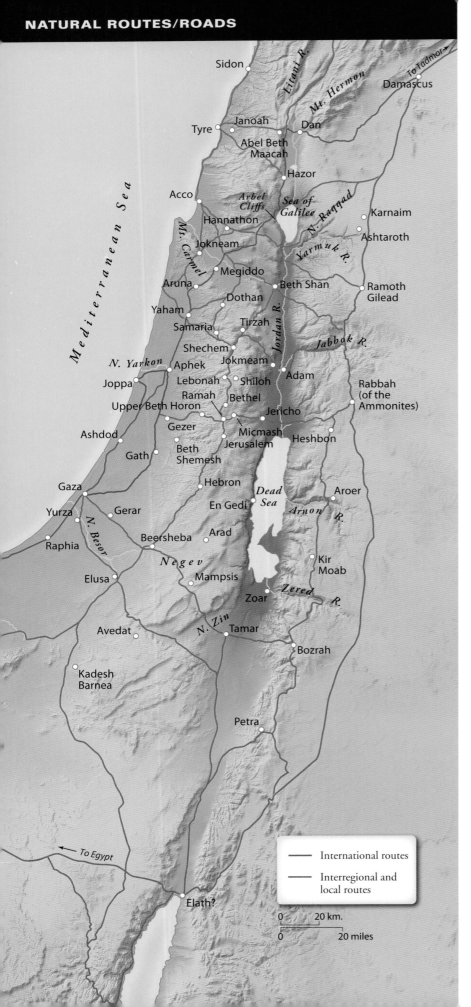

second and first millennia BC to draw chariots and to serve in cavalry units; during the Persian period (538–332 BC) and later, their use for everyday travel became more common. In New Testament times all these means of transportation were used, and the improvement of the road system increased the use of carts and chariots. Travel during the dry summer season was preferred to attempting to negotiate the muddy, rain-soaked terrain in the winter months. The spring and summer seasons were "the time when kings go off to war" (2 Sam 11:1) because the roads were dry and the newly harvested grain was available to feed their troops.

In Old Testament times, tracks between urban areas were prepared but not paved. Basic "road building" operations included the removal of stones from the path, the clearing of trees and bushes, the maintaining of shallow fords in the river beds, and possibly the construction of trails along steep slopes. By New Testament times the Romans had developed advanced road-building techniques, and it is probable that the construction of a rather well-developed road system had already begun in Syria and Judah. Roman road-building techniques included the preparation of the roadbed by leveling the ground and cutting rocks, the use of curbing to mark the edge of the roads, attention to drainage problems, and the laying of paving stones.

The International North–South Route

The most important international route through Israel connected Egypt with its rivals/allies to the north and east (Syrians, Assyrians, Babylonians, Persians, Hittites, Hurrians, etc.). This international route is sometimes called "The Way of the Sea" (cf. Isa 9:1) or the "Via Maris." These are slight misnomers because only portions of this route ran to, or along, the sea coast. It seems that although portions of it were named in ancient times (e.g., the portion leading from Egypt across Sinai to the Philistine coast was called the "way of [to] the land of Philistines," Exod 13:17 RSV), the whole route did not have one special name. It seems best not to use "The Way of the Sea" (= "Via Maris"), for it is not even certain that this term ever referred to a portion of this International Highway (see above, p. 18) — it may in fact have referred to a road that connected Damascus with Tyre, through Upper Galilee.

After crossing northern Sinai, this route proceeded from Gaza through the Philistine and Sharon plains to the area of the city of Yaham, avoiding such obstacles as sand dunes, low-lying muddy spots, the Yarkon River, and the swamps of the Sharon Plain. In order to proceed from Yaham to the Jezreel Valley a traveler had to follow one of the passes that led through Mount Carmel, such as the one that ran from Aruna to Megiddo.

The route northward along the coast led from Megiddo to Hannathon to Acco and from there north to Tyre and Sidon. Several options were available for travel from Megiddo to Damascus, from where one could proceed to Turkey or to the Euphrates River. A major road led southeast from Megiddo to the Beth Shan area. From there it went north to the southern shore of the Sea of Galilee, where it turned northeastward, ascended the valleys of the Yarmuk and Raqqad rivers (or the hills just north of them), and ultimately arrived in the Ashtaroth – Karnaim area, where it joined the Transjordanian Highway. Another route continued eastward from the Beth Shan area, forded the Jordan River, and ascended one of the ridges that led up to the Highlands of Gilead. After arriving at Ramoth Gilead, one could proceed directly northward to Damascus.

Another road took travelers from Megiddo to Hazor by heading northeast from Megiddo, skirting the slopes of Mount Tabor, continuing north of the Arbel Cliffs, and heading north past Kinnereth to arrive at Hazor. During the Roman period and later, a road led east from Hazor, crossed the Jordan River, and then ascended the Golan Heights on the way to Damascus. Alternatively, one could head north from Hazor to Dan and from there head east to Damascus, skirting the southern slopes of Mount Hermon. In addition, an important branch of the International Route led directly north-northeast from Dan up into and through the Lebanese Beqa toward cities such as Lebo Hamath, Kadesh, Hamath, and Aleppo.

The International Transjordanian Route

The other International Route, not quite as important as the preceding, led south from Damascus and traveled the entire length of Transjordan. One branch of this route ran east of the watershed of the Transjordanian mountains. It passed through important cities such as Karnaim, Ashtaroth, Ramoth Gilead, Rabbah of the Ammonites, Heshbon, Aroer, Kir in Moab, and Bozrah. Along this route there were good supplies of water, but one had to cross, often with difficulty, wadis such as the Yarmuk, Jabbok, Arnon, and Zered. The eastern branch of this route ran along the edge of the desert. It avoided the deep wadis, but it did not pass through as many important commercial centers, nor did it have as much water available, and the caravans traveling along it were subject to raids by desert tribes.

These two branches met at Rabbah, the capital of the Ammonites, and then split again as they continued southward. Eventually, in the Bozrah region, the traveler could proceed westward and strike out across the Negev and Sinai toward Egypt or head south-southeast and follow the caravan routes into the Arabian desert. The Transjordanian Highway was especially important for the conveyance of luxury goods such as gold, frankincense, and myrrh from southern Arabia to the Near Eastern commercial center of Damascus.

Interregional and Local Routes

Many of the interregional and local roads are discussed below in the regional sections. However, for biblical studies the interregional route that ran from Beersheba in the south to Shechem in the north — via Hebron, Bethlehem, Jerusalem, Gibeah, Ramah, Bethel/Ai, and Shiloh — is of extreme importance. This route appears again and again in the biblical text. Some call it the "Route of the Patriarchs" because Abraham, Isaac, and Jacob traveled its length, while others refer to it as the "Ridge Route," for in many places it "tiptoes" along the watershed of the Judean and Ephraim mountains. Even when it is not specifically mentioned, it often furnishes the backdrop for many events recorded in the Bible.

Geographical Regions of Israel and Jordan

Bashan

Bashan is a high plateau region located in the northeastern corner of Israel. Measuring some 37 miles east – west by 56 miles north – south, it is bounded on the west by the Rift Valley, on the north by Mount Hermon, and on the east by Mount Bashan, while on the south it eventually merges into Gilead. Running down its center in a north – south direction are several rows of extinct volcanoes, whose conical shapes break up the relatively flat terrain. The name Bashan, which originally seems to have referred to the fertility of the region, is used sixty times in the Old Testament (NIV).

The area north of the Yarmuk River is composed of volcanic soils, rocks, and mountains, while south of the Yarmuk the land consists of softer, chalky limestone formations. The general slope of northern Bashan is from the north-northeast downward to the south-southwest. This northern portion is divided into an eastern and a western section by the Nahal Raqqad. The western section receives close to 40 inches of precipitation per year; this, combined with the fertility of the volcanic soil, makes it a rich agricultural area. In Old Testament times the semi-independent states of Geshur and Maacah were located in the southern and northern portions of this western section respectively (see maps pp. 122, 140).

The fertility of this region was well known, for the biblical writers repeatedly speak of the fatness of the animals that grazed there and of the oak forests that covered portions of the landscape (see, e.g., Ezek 27:6; 39:18). In New Testament times this area west of the Nahal Raqqad was known as Gaulanitis. It became an important producer of wheat in the Roman Empire, in spite of the fact that in certain regions there are basalt boulders that make plowing difficult.

To the east of the Nahal Raqqad rainfall is not as abundant, and although the average of 12 to 24 inches per year occasionally results in bumper crops of wheat, droughts can strike the region in other years. It is probable that in the Old Testament period this territory was called the district of Argob (cf. Deut 3:13 – 14; 1 Kings 4:13); in the New Testament era it was known as Batanea. This area, open to invading armies from the north (see, e.g., Gen 14), is said to have been a region with "sixty large walled cities with bronze gate bars" (1 Kings 4:13).

▲ *Yarmuk River in the southern Bashan*

at the southern foot of Mount Hermon, that Peter made his great confession regarding Jesus' messiahship (Matt 16:13 – 20; Mark 8:27 – 30). Six days later Jesus was transfigured in the presence of Peter, James, and John — probably somewhere on Mount Hermon (Matt 17:1 – 13; Mark 9:2 – 13). To the east of Mount Hermon the prominent lava flow called the *el-Wa'arah* forms the remaining portion of Bashan's northern border. Because it is difficult to travel across the el-Wa'arah the roads heading south from Damascus were diverted to the east or west of the outflow.

On the east, Bashan is separated from the desert by more volcanic flows and by mountains. The northernmost flow, the Leja ("Refuge"), was known in New Testament times as Traconitis, the "Torn Land." This remote and desolate area was under the control of Philip the tetrarch (Luke 3:1). To the east of the Leja is the even more desolate es-Safa, and to the south of the Leja rises the massive Jebel Druze, also known as Mount Bashan and Mount Hauran. Because of its high elevation (ca. 5,800 ft.) it receives good amounts of rain each year, in spite of its easterly location. The roughness of the territory, however, precludes the growing of much wheat; this

On the northern edge of Bashan, Mount Hermon (also called Sirion and Senir, Deut 3:9) rises to an elevation of 9,232 feet, towering over the surrounding area. Its peaks, snow-covered for more than six months of the year and sometimes visible from over 100 miles away, receive more than 60 inches of precipitation annually. This moisture, much of which seeps into the hard limestone foundations of the mountain, reappears as powerful springs that feed the Jordan River as well as the Damascus Oasis. It was near Caesarea Philippi, located

BASHAN

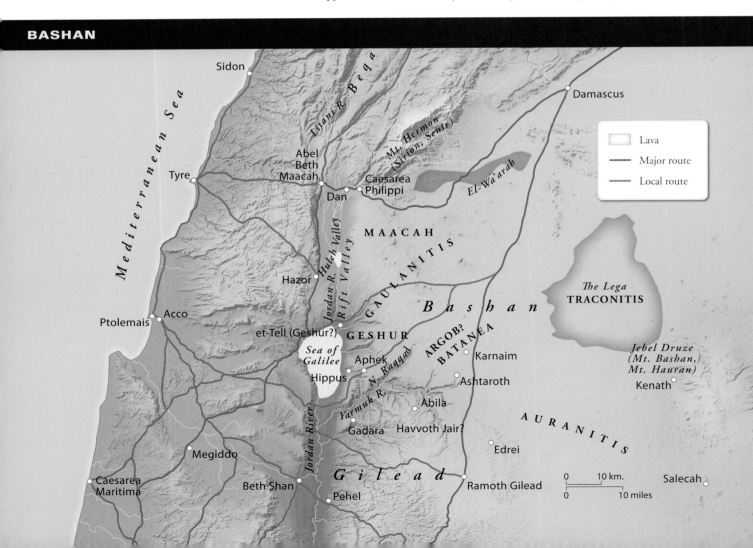

▲ *One of the headwaters of the Jordan River at Banias (ancient Caesarea Philippi)*

volcanic mountain area evidently was primarily used for grazing and for obtaining forest products.

South of the Yarmuk River the plateau region continues for some 18 miles as it gradually blends into the Gilead region. This area was possibly the Old Testament Havvoth Jair, which is variously described as being "in Bashan" (Josh 13:30) and "in Gilead" (1 Kings 4:13). This territory was captured by Israel but remained a source of dispute between Israel and its neighbors to the northeast — Geshur and Aram (Syria) (1 Chron 2:23).

The chief cities of Bashan were Ashtaroth and Edrei, the capital cities of King Og (e.g., Josh 12:4; 13:12). Later in the Old Testament period, nearby Karnaim replaced the formerly prominent Ashtaroth. Salecah seems to have been the easternmost limit of Israelite interest in the region. During the Hellenistic and Roman periods (332 BC – AD 324) the area was heavily populated by Jewish settlers but was also well known for its Greco-Roman cities (including Gadara, Abila, Hippus, Raphana, and Caesarea Philippi).

Not only was Bashan rich in agriculture, but during the Roman period and afterward it also served as a thoroughfare for the main International Highway. One important highway branch, used in both Old and New Testament times, skirted the southern end of the Sea of Galilee and either ascended the western scarp of Bashan, just to the north of the Yarmuk River, or followed the Yarmuk and the Raqqad valleys on up toward Damascus. Aphek was one of the cities guarding this route, and it was here that Israel fought the Arameans (Syrians) on several occasions (1 Kings 20:26, 30; 2 Kings 13:17; see maps pp. 155, 157). This road probably formed part of the Roman "grain route," by which grain from Bashan was shipped via Lower Galilee to the Mediterranean port of Ptolemais (OT Acco).

Other branches of the International Highway led from Hazor and Dan to Damascus, crossing Bashan along the way. In addition, the important Transjordanian Highway headed directly south from Damascus through Bashan or, more specifically, through the district of Argob/Batanea, avoiding the el-Wa'arah, the Leja, and Mount Bashan as it passed by Ashtaroth and Edrei on its southerly course. During most of the year travel through the area was not difficult so long as the caravans avoided the various basalt outflows and boulder-strewn areas. In the winter months, however, the snow and the mud must have slowed the speed of the caravans.

Huleh Valley

The Huleh Valley is the northernmost section of the Rift Valley system in Israel. It stretches from the north shore of the Sea of Galilee up to Dan, a distance of 23 miles. It is bounded on the west by the steep scarp of eastern Upper Galilee, which in places rises 1,600 feet above the valley floor. On the east, the approach to Bashan is much more gradual, while north of the Dan – Abel Beth Maacah region the land rises into the Lebanese Beqa.

The portion of the valley southeast of Hazor is a "plug" of basalt that flowed down from Bashan and filled the Rift Valley all the way to the Sea of Galilee. The waters of the Jordan River collected to the north of this blockage, forming a medium-sized lake. Eventually the river carved its way through this basalt barrier on its way down to the Sea of Galilee. Here the Jordan drops some 900 feet over a distance of only 10 miles, making this the steepest gradient of the river.

Behind this basalt barrier a small lake (Lake Huleh, 5 sq. mi.) and swamp lands (12 sq. mi.) remained until recently; the Israelis drained the area in 1958. Josephus's name for the Lake Huleh was Lake Semechonitis (e.g., *War* 3.10.7 [515]). Cane and papyrus grew in the lake and the swampland. In addition, fowl, animals, and fish abounded in the region.

The southern portion of the Huleh Valley receives about 16 inches of rain each year, the northern portion almost 25 inches. At the foot of Mount Hermon a number of springs contribute to the five rivers that combine to form the headwaters of the Jordan, adding to the lushness of the region.

Because of the lake and swamps, the major settlements of the area — Dan, Abel Beth Maacah, and Hazor — were located along the higher fringes of the valley. This area was exposed to outside influences because the major International Highway from Egypt into the Lebanese Beqa ran alongside of it. In addition, during the intertestamental period and later, a branch of this important route passed just south of the Huleh Basin on its way to Damascus. Thus, during times of invasion from the north or the northeast, this region usually was the first to feel the brunt of the attack (see, e.g., 1 Kings 15; 2 Kings 15; and map p. 157).

GALILEE AND THE HULEH VALLEY

Upper Galilee

The name Galilee is used only six times in the Old Testament but appears sixty-four times in the New Testament, primarily in connection with the ministry of Jesus. The term seems to mean "region" or "district." In the Old Testament period the region was referred to as "Galilee of the Gentiles" (Isa 9:1), but by the New Testament era it had a large Jewish population.

Galilee is divided into two major regions by a fault line that runs from Acco eastward toward the north end of the Sea of Galilee. The Bet Hakkerem Valley marks this fault and serves as the dividing line between the two parts. Upper Galilee, so called because its elevation is much higher than that of Lower Galilee, is located to the north of the Bet Hakkerem Valley. A steep rock scarp rises some 1,500 feet out of the valley, making the transition between the two regions quite dramatic. The eastern boundary of Upper Galilee is clearly defined by the steep north–south scarp that descends rapidly into the Huleh Valley. On the west, Upper Galilee drops off into the Plain of Acco and in one place, at Rosh HaNiqra, actually reaches the Mediterranean Sea. The northern boundary of Upper Galilee is uncertain. Some geographers suggest that the Litani River forms its natural northern limit, while others place the northern limit along the fault line that runs from Rosh HaNiqra northeast toward Tibnine.

The whole region of Upper Galilee, but especially the central southern section, is an uplifted area dissected by numerous fault lines. The eastern portion is composed of hard Eocene limestone interspersed with a few patches of basalt outcroppings; it is drained to the north by the Wadi Doubbé, which eventually joins the Litani River. The western section is composed mainly of hard Cenomanian and Turonian limestones and reaches a height of 3,963 feet in the Mount Meron area; this is the highest elevation in all of Israel, with the exception of Mount Hermon. This western region tilts down toward the northwest, but eventually the streams turn westward and flow into the Mediterranean Sea.

In both the eastern and western sections good topsoil is available, and abundant rains (24 to 40 in. annually) assure the fertility of the region. The natural vegetation of the area most likely consisted of typical Mediterranean scrub forests, but settlers cut down the trees to plant crops. By the beginning of the New Testament era most of the area was under intense cultivation.

Because of the dissected nature of the terrain, the major north–south roads ran around the eastern or western edges of Upper Galilee, bypassing this difficult area. There seems to have been, however, an east–west route that connected Damascus with the port city of Tyre — possibly "the way of the sea" mentioned in Isaiah 9:1. After crossing the northern end of the Huleh Valley, passing cities such as Dan and Abel Beth Maacah, this route ascended the rocky scarp west of Abel Beth Maacah and continued on its way to Tyre along the ridges on the high ground of Upper Galilee, past villages such as Beth Anath and Kanah. Another interregional route led from Abel Beth Maacah up to Kedesh, and from there to Acco or Aczib via Iron and Beth Shemesh. These east–west routes must have been of some importance, for they connected the desert-oasis emporium of Damascus with the Mediterranean Sea.

In the Old Testament period the tribe of Asher settled in western Upper Galilee, while Naphtali was assigned its eastern portion. The northern limits of their settlements do not seem to have extended beyond the northern boundary of Upper Galilee. In New Testament times the area was predominantly Jewish, and most of it was under the control of Herod Antipas. Throughout history the high elevations and the ruggedness of the terrain have caused those living in the area to be somewhat isolated from international influences, but it usually had strong connections with Tyre.

Plain of Acco

The Plain of Acco is situated to the west of Galilee and is bounded on the north by Rosh HaNiqra, on the west by the Mediterranean Sea, and on the south by Mount Carmel. The plain is divided into two parts. North of Acco, the plain is a narrow strip of land (12.5 by 3.7 mi.), filled with alluvial soil that has washed down from the hills of Upper Galilee. The wadis from the hills have cut through the kurkar (fossilized dune sandstone) ridges that run parallel to the coast and thus empty into the sea without difficulty. The shoreline is for the most part made up of a rough, abrasive kurkar platform, making it difficult to land boats in the area.

To the south of Acco, the plain is broader (9.3 by 6.2 mi.), and sand dunes extend inland from the coast. The dunes prevent the wadis from the hills of Lower Galilee from draining directly into the Mediterranean, so the plain area, except for the high ground close to the hills, is somewhat swampy. At the southern extremity of the plain, the Kishon River, coming from the Jezreel Valley, drains into Haifa Bay.

The city of Acco (Ptolemais) is located where the two sections of the Plain of Acco meet. Throughout most of its history Acco/Ptolemais served as the leading port for the whole of the land of Israel and Transjordan (note the roads leading to it; see map p. 36). In addition, a branch of the International Highway passed through the plain up the coast toward the cities of Tyre and Sidon, adding to the flow of traffic through the region. Although this area was assigned to the tribe of Asher, the members of this tribe had great difficulty trying to overcome the Canaanites who lived on the plain (Judg 1:31–32), and the region was always open to outside influences coming from the Mediterranean. As a result, Acco/Ptolemais was either an independent city or under the control of a non-Israelite power during most of its history. It was at Ptolemais that the apostle Paul stopped on his way to Jerusalem at the end of his third missionary journey (Acts 21:7).

Lower Galilee

Lower Galilee, where none of the hills rise above 2,000 feet, lies south of Upper Galilee. It is bounded on the east by the Rift Valley, on the south by the Harod and Jezreel valleys, and on the west by the Plain of Acco.

Lower Galilee can be divided into two rather distinct regions by drawing a north–south line through Mount Tabor. In the western section most of the hills are composed of hard Cenomanian limestone, except in the southwestern corner, where softer Eocene formations

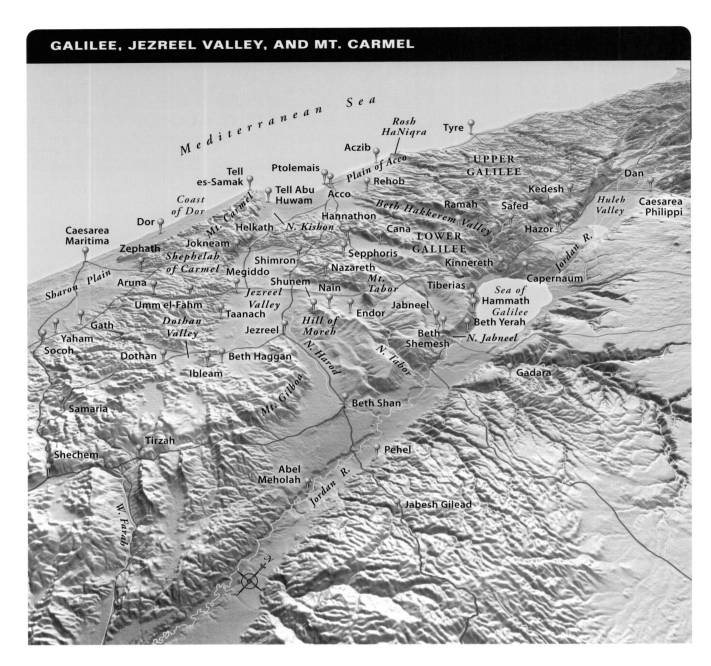

appear. This southwestern corner, with its low rolling hills and valleys, may be the area referred to as the "foothills" (Heb. *Shephelah* = "lowlands" RSV) of the "mountains of Israel" (Josh 11:16); they are structurally similar to the more frequently mentioned "western foothills" of Judah (also called Shephelah).

The major relief lines in western Lower Galilee run in an east–west direction. A traveler going north or south through the area thus encounters a series of medium-sized ridges separated by broad fertile valleys. This topography causes the natural lines of communication to have an east–west rather than a north–south orientation. The area receives approximately 25 inches of rain per year, and the rich alluvial soil adds to the fertility of the region. In ancient times the hills were covered with scrub forests, and the people tended to build their villages and roads on the hillsides in order to leave the valleys open for farming. In the winter, the broad valleys become partially flooded, making travel through them somewhat difficult; but they dry out in the spring and yield a good harvest of wheat and barley.

To readers of the Bible, the most familiar cities of western Lower Galilee are Nazareth and Cana. The former is situated on the southernmost ridge of western Lower Galilee, and from there one could look down on the caravans as they passed along the International Highway through the Jezreel Valley to the south, while on the northern side of the ridge one could see caravans making their way from Bashan to the port of Acco/Ptolemais. In the days of Jesus, Nazareth was a small, out-of-the-way town situated in a chalky basin on the top of this ridge. Its association with Jesus has made it the large city that it is today. Cana was also a rather insignificant city, located on the southern slope of the Yodefat Range; all that remains of it today is the small, barren ruin of Khirbet Qana. It was here that Jesus performed his first recorded miracle: the turning of water into wine (John 2:1 – 11). From extrabiblical sources we know that in New Testament times there were numerous villages in western Lower Galilee, the most important of which was the capital Sepphoris, located 3.5 miles north-northwest of Nazareth.

In eastern Lower Galilee (the area east of the north – south Tabor line) the features of topographical relief run diagonally from northwest to southeast. Here again a series of ridges alternate with a series of valleys (such as the Nahal Jabneel and Nahal Tabor), which drain into the Rift Valley. The northeast slopes of the ridges are fairly steep; the southwest slopes descend more gradually into the valleys. The area is covered primarily with basalt stones and volcanic soil, making the landscape somewhat smoother in appearance than the other areas of Galilee.

In eastern Lower Galilee the southern border reaches all the way to the Harod Valley and includes such famous mountain landmarks as the Hill of Moreh and Mount Tabor. The Hill of Moreh witnessed such events as Gideon's battle with the Midianites (Judg 7), Saul's meeting with the witch of Endor (1 Sam 28), Elisha's raising of the son of the woman of Shunem (2 Kings 4), and Jesus' raising of the son of the widow of Nain (Luke 7). Deborah and Barak met the Canaanite forces at Mount Tabor (Judg 4 – 5), and some consider this to be the site of the transfiguration of Jesus, although Mount Hermon is a more probable location (see above, p. 34).

The International Highway ran through eastern Lower Galilee in a northeasterly direction from Mount Tabor to the Horns of Hattin; from there it descended north of the Arbel Cliffs to the northwestern shore of the Sea of Galilee and continued northward to Hazor. In Roman times the east – west grain road from Bashan, after passing south of the Sea of Galilee, ascended the Nahal Jabneel and then turned westward, continuing past Sepphoris and Hannathon on its way to Acco/Ptolemais.

In Old Testament times, Lower Galilee was only sparsely settled until the arrival of the Israelites (ca. 1400 – 1200 BC). Four of the tribes received their allotments in Lower Galilee: Issachar to the southeast of Mount Tabor, Naphtali to the northeast, Asher to the northwest, and Zebulun to the west (see map p. 116).

Because the terrain of Lower Galilee is much less rugged than that of Upper Galilee, Lower Galilee has been more accessible to outside influences throughout its history. One should also remember that although only a few villages are mentioned by name in the gospel accounts, a good portion of Jesus' early life was spent in the hills and valleys of Lower Galilee.

Jezreel Valley

The term "Valley of Jezreel" is used only three times in Scripture; its Greek form, Esdraelon, occurs only in the extrabiblical literature of the Second Temple period. The valley evidently took its name from the Israelite settlement of Jezreel, situated at its eastern end. During the early years of Israel's residence in the land, most of the cities in the valley were under Canaanite control. An Israelite, when referring to the valley, would call it by the name of the Israelite village of Jezreel rather than by the name of one of the non-Israelite cities, such as Megiddo or Taanach, that were prominent in the valley.

Although there is some dispute as to the proper application of the name Jezreel Valley, the valley proper as understood in this discussion has a triangular, or arrowhead, shape. The points of the triangle are Helkath in the northwest, Mount Tabor in the northeast, and Beth Haggan in the southeast. The northern boundary line, which stretches 17 miles from Helkath to Mount Tabor, is formed by the southernmost ridge of western Lower Galilee, the Nazareth ridge. The southern

▼ *Jezreel Valley from Megiddo looking east at Mount Tabor*

boundary is formed by the Carmel range, which runs 20 miles from Helkath to Beth Haggan. The eastern boundary is less distinct; a good approximation is an imaginary line, 17 miles long, connecting Beth Haggan with Mount Tabor.

The valley itself is composed of rich alluvial soils that have washed down from the surrounding hills. The watershed to the east is the dividing line between the Jezreel and Harod valleys. From the Mount Tabor and Mount Gilboa regions, the Kishon River drains the valley in a northwesterly direction, passing through the narrow exit at Helkath at an elevation of 80 feet. Because of its gentle slope, the valley becomes quite muddy in winter; in ancient times travel across the valley was very difficult during the winter months. Most of the valley surface receives 20 inches of rain annually, while the surrounding hills receive approximately 28 inches. It was probably during a winter rainstorm (with the ensuing muddy conditions) that the Canaanites abandoned their chariots, which were stuck in the mud, and fled on foot before the troops of Deborah and Barak, which were advancing from Mount Tabor (Judg 4:15; 5:20 – 21; Ps 83:9; map p. 126). Although the wet season caused problems for travelers, it enabled the inhabitants of the region to reap in spring and early summer bountiful harvests of wheat and barley, which "God had indeed sown" (in Heb. "Jezreel" means "God sows").

To the east of the watershed, beginning approximately at the city of Jezreel, a smaller, narrower valley, 11 by 3 miles, descends into the Beth Shan region. Some geographers consider this alone to be the biblical Jezreel Valley, but here it will be called the Harod Valley, after the modern name of the *nahal* (Heb. for "valley") that descends along its length. This valley is bounded on the north by the southernmost basalt-soil slope of eastern Lower Galilee and to the south by the steep scarp of Mount Gilboa, which rises some 1,600 feet above the valley floor. Most of the valley is below sea level. Since it cuts through the underground water table, there are numerous springs that make this a fertile region in spite of the fact that Beth Shan receives only 12 inches of rain annually.

Although the Jezreel and Harod valleys are important from an agricultural standpoint, they are of even greater strategic significance. The great north – south International Highway had to pass through the Jezreel Valley, while the easiest east – west route through the whole of Israel crossed both valleys. A traveler heading north along the International Highway, after having traveled through one of the Carmel passes, would be at Jokneam, Megiddo, Taanach, or Beth Haggan. Most commonly he would arrive at the city of Megiddo and from there would cross the Jezreel Valley along the slightly higher ground formed by some basalt outcrops, proceeding in a northeasterly direction toward Mount Tabor. If he was going instead from Megiddo to the ports of Acco, Tyre, or Sidon, he could follow along the foot of the northern scarp of Mount Carmel, proceed through the narrow pass at Helkath (0.3 mi. wide), and then, avoiding the sand dunes, head northward. Another route, possibly used as frequently as the preceding one, led directly north from Megiddo to Shimron. From there it continued through the low hills of western Lower Galilee to Hannathon, and from there westward to the Plain of Acco.

The other major route passing through this area connected Acco, via the cities of Megiddo and Jezreel, with Beth Shan. This east – west connecting route was of extreme importance, for from Beth Shan one could head northeastward to the southern end of the Sea of Galilee and from there ascend into Bashan and proceed on to Damascus, or one could head east from Beth Shan, ford the Jordan River, ascend the hills of Gilead, and eventually join the Transjordanian Highway at Ramoth Gilead.

It is evident that whoever controlled these valleys also controlled the major highways passing through the country. The tribes of Zebulun, Issachar, and Manasseh, which were allotted cities in and around the valley, had great difficulty in overcoming the Canaanites who lived there (Judg 1:27 – 28, 30); Israel probably did not gain actual control of the strategic valley strongholds until the days of David and Solomon (1 Kings 9:15). From the days of the Egyptian pharaoh Thutmose III (1457 BC) until the present, the Jezreel Valley has served as the stage on which the armies of the world have made their entrances and exits. Thus it is altogether fitting that this strategic location should be chosen as the setting of that great eschatological battle, the battle of Armageddon (Gr. form of Heb. *Har Megiddo*, "Mountain of Megiddo"; Rev 16:16).

Mount Carmel

To the south of the Jezreel Valley rise the towering rock scarps of Mount Carmel. The name *Carmel* means "plantation," "garden-land," and/or "garden-growth," and in Scripture the lushness of the vegetation on Mount Carmel became a symbol of great fertility (e.g., Song 7:5). Mount Carmel in a broad sense extends some 30 miles along a line running from the Mediterranean Sea in the northwest to the Dothan Valley in the southeast.

Mount Carmel (map p. 38) actually consists of three distinct geological regions. The northernmost of these is Mount Carmel proper. Composed primarily of hard Cenomanian-Turonian limestone formations with occasional patches of basalt, it rises to a height of 1,790 feet above sea level. Its northwestern tip dips into the Mediterranean Sea, and its western base runs parallel to the seashore for 20 miles. On the north a pronounced scarp rises some 1,500 feet above the floor of the Jezreel Valley. This rugged area, because of its elevation and its proximity to the sea, receives some 32 inches of rain annually, along with a considerable amount of dew; in antiquity it was mostly covered with trees. Because of its massiveness and its steep slopes, particularly on the western and northeastern sides, travel across this portion of the mountain was practically impossible.

The second section of Mount Carmel, southeast of Mount Carmel proper, is called the Manasseh region or the Shephelah (foothill) of Carmel. It is rectangular in shape and is composed of soft, chalky Eocene limestone. Because of its lower elevation (650 to 980 ft.), less rain falls here than on Mount Carmel proper, and the chalky soil is not fertile. Between these two sections is a Senonian chalk valley, or pass, through which traffic could flow from the Sharon Plain to Jokneam.

Farther to the southeast is the third section of Mount Carmel, sometimes called the Umm el-Fahm (Iskandar Uplift) region. This

▲ *Plain of Gennesaret and the Sea of Galilee from the Arbel Cliffs, looking north*

section, also rectangular in shape, is again made up of the harder Cenomanian-Turonian limestones. In some places it reaches elevations of 1,600 feet. The soil is primarily rich terra rossa, and the annual rainfall (20 in.) makes the region quite fertile. Between it and the Manasseh region to the northwest runs a second Senonian chalk pass, this one leading from the Sharon Plain to Megiddo. To the southeast of the Umm el-Fahm block lies another Senonian chalk area, namely the Dothan Valley, which separates this region from the hills of Manasseh; this valley forms the third important pass through the mountain. These strips of Senonian chalk that divide the Carmel Range, like other Senonian outcrops throughout the country, were frequently used as paths for roads because they are usually lower in elevation than the surrounding areas, are relatively free of boulders, and dry out quickly during the winter months.

Throughout antiquity the whole of Mount Carmel was probably covered with thick forests. In the biblical text we learn only of Elijah's and Elisha's associations with the mountain, although the cities bounding it on the northeast — Jokneam, Megiddo, and Taanach — are justly famous as Canaanite centers. Mount Carmel blocks the flow of north – south traffic from the Sharon Plain to the Jezreel Valley or to the Plain of Acco, so travelers, caravans, and armies throughout the ages had to use one of its three passes or attempt to go around the northwestern tip of the mountain. Thus if one could control the passes, one could control the flow of

north – south traffic on the land bridge between the continents of Africa and Asia (recall Josiah's ill-fated attempt to prevent Pharaoh Neco from proceeding north by battling with him at Megiddo [2 Chron 35:22]).

Sea of Galilee

The Sea of Galilee is located in the Rift Valley on the eastern border of Lower Galilee. The surface of the lake is approximately 1,280 feet below the surrounding hills and 690 feet below sea level. The sea measures 13 by 7.5 miles and has about 30 miles of shoreline. It is bounded on the north by the southern edge of the Huleh Valley, on the east by the hills of Bashan, on the west by the steep slopes of the hills of Lower Galilee, and on the south by the continuation of the Rift Valley. The freshwater of the Sea of Galilee has provided the inhabitants of the region with a good supply of fish throughout the ages.

In the Old Testament, the Sea of Galilee is mentioned only five times and is called the "Sea of Kinnereth." Tradition relates the name Kinnereth to the lyre-like shape of the lake, but the lake probably derived its name from the city of Kinnereth (Josh 19:35), located at its northwestern corner. This name, in modified form ("Lake of Gennesaret"), is used once in the gospel accounts. The names used in the New Testament are "Sea of Galilee" (five times), "Sea of Tiberias" after the city of Tiberias (ca. AD 20; John 6:1; 21:1), and simply "the lake" (thirty-one times).

▲ *Plain of Gennesaret looking toward the Arbel Cliffs*

The sources of the lake include the Jordan River, which flows into the lake's northernmost point, and the freshwater and saline springs that are found around and under the lake. In the northwest corner are seven salt springs, called Heptapegon, and along the lake's western shore just south of Tiberias are the hot mineral springs of Hammath (a Hebrew word meaning "hot"). Because of its lower elevation, the immediate vicinity of the lake receives less rain than the surrounding hills; Tiberias, for example, has an annual rainfall of only 16 inches. Yet, because of the rather steep hills that surround the sea, the wind rushing down the slopes can create sudden and violent storms on the lake. There is not much flat farmland along the shore of the lake, except for three small, very fertile plains on the northwestern, northeastern, and southern shores. The northwestern plain was known as the Plain of Gennesaret; Josephus writes of its extraordinary climate and produce, mentioning such diverse crops as walnuts, palm trees, figs, olives, and grapes (*War* 3.10.8 [3.516–21]). The plain in the Bethsaida region northeast of the lake is watered by several perennial streams and is equally fertile.

Although the Old Testament does not record much about the lake, it is known that the cities along its western shore were assigned to the tribe of Naphtali (see map p. 116). By the time of the New Testament the whole area was densely populated. To the east were cities such as Gergesa and Hippus, which were associated with the Greco-Roman Decapolis, while on the western shore the major city was Tiberias, the city Herod Antipas built and made the capital of Galilee. To the south of Tiberias were the hot springs of Hammath, to the north the commercial city of Magdala, or Taricheae (= "salted fish"?; *War* 3.9.7 [445]), where fish were processed for export. Along the shore of the northern third of the lake were cities in which Jesus ministered, including Gennesaret, Dalmanutha, Capernaum, Korazin, and Bethsaida.

Throughout the ages, paths have run around the edge of the lake, bypassing the swampy areas close to where the Jordan enters and exits the lake. Along the northwestern shore the International Highway, after descending from the hills of eastern Lower Galilee north of the Arbel Cliffs to Magdala, crossed the Plain of Gennesaret and then ascended into the Huleh Valley as it proceeded northward to Hazor.

Most of the hills, boulders, and soils surrounding the lake derive from lava outflows. Houses in the area were primarily built of black basalt rock; because they had few windows and the people had only oil lamps for light, living in them must have been rather dismal. This may have been the physical backdrop to the description of the gloomy spiritual state of this region (Isa 9:1). The large basalt boulders in the area could be hand-tooled into quality household grinders, large commercial grinders with their millstones (Mark 9:42), olive presses, and other industrial implements. Thus in New Testament times the area's economic well-being depended on several sources: the presence of the capital of Galilee, Tiberias; the sick who came to the hot springs of Hammath and Hammath Gader for medicinal purposes; the fishing industry; the rich agricultural lands; the trade routes passing through the area; and the manufacture of basalt implements.

Sharon Plain

To the south of Mount Carmel is a section of the coastal plain called the Sharon (map p. 44), which is bounded on the north by the Nahal Tanninim (Crocodile River), on the east by the foothills of Ephraim and Manasseh, on the south by the Yarkon River, and on the west by the Mediterranean Sea. The Sharon measures 30 by 10 miles and is a flat area consisting of alluvial soils washed down from the hills and of deposits of Mousterian red sands. The areas where these latter deposits are found are well known today for the excellent citrus crops that grow there.

Along the western edge of the Sharon Plain three kurkar ridges run north – south, parallel to the seashore. One of these ridges in places forms the shoreline, while the other two are located slightly inland. The ridges block the westerly flow of the wadis, so that in ancient times the runoff water collected behind these ridges, making the Sharon a very swampy area. It was not until Roman times that some outlets were cut through the ridges, and not until the twentieth century were the swamps completely drained.

In Old Testament times the Sharon was not used for agricultural purposes because of the swamps, the dense oak forests, and a thicket of scrub trees; the word Sharon probably indicates a "forest in the plain." The Sharon was used as a place for the grazing of cattle during the Old Testament period (1 Chron 27:29). The swampy conditions also forced the north – south International Highway to run along the eastern edge of the plain as it led northward from Aphek via cities such as Socoh, Yaham, and Gath, to Mount Carmel, avoiding the hills to the east and the swamps to the west. It was not until Herod the Great constructed Caesarea on the coast that any of the major highways ventured out into the plain itself.

To the north of the Sharon Plain is a thin strip of land measuring 20 by 2 miles, tucked in between the western slopes of Mount Carmel and the Mediterranean Sea. This northern extension of the Sharon Plain is sometimes called the Coast of Dor, after the chief

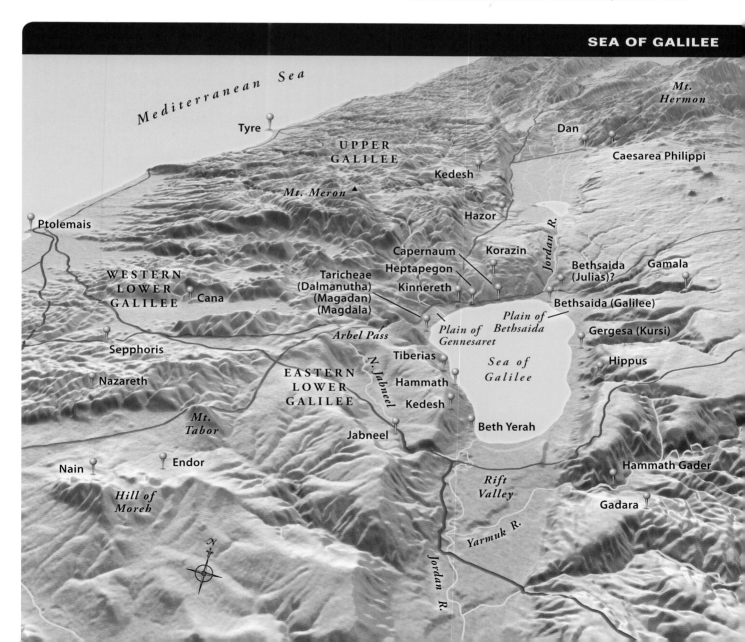

Old Testament city located in the region. This strip of land provided a passageway around the west side of Mount Carmel. But due to the narrowness of the passage and the swamps in the region, this road was not used extensively in ancient times.

The coastline between Mount Carmel and the Yarkon River to the south is relatively smooth, and small natural harbors were found only at Athlit and Dor. Additional anchorages were developed at Crocodilon Polis, Strato's Tower, Apollonia, Tel Mikhal, and sites near the Yarkon. These anchorages could be used by vessels as they plied the coast; around 1100 BC the Tjekker, a tribe of the Sea Peoples, initiated maritime exploits originating from the city of Dor. Not until Herod built the huge

▲ *Kurkar shoreline along the coast of Sharon Plain*

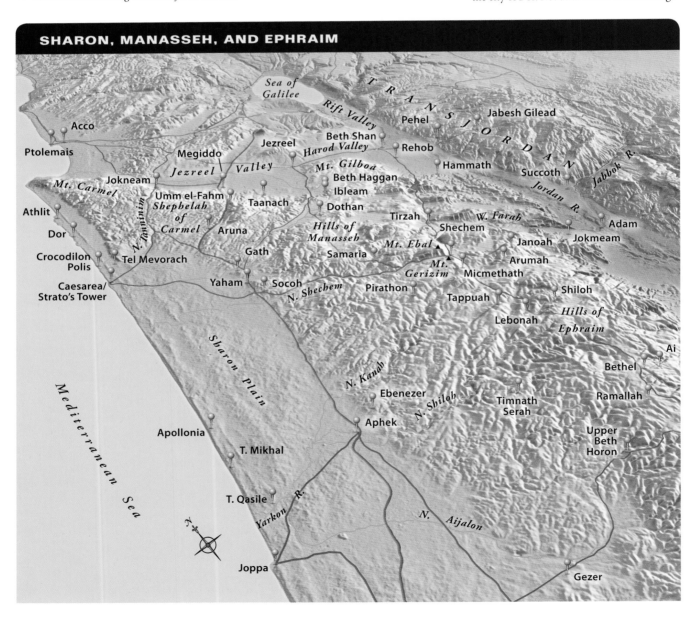

SHARON, MANASSEH, AND EPHRAIM

Sea of Galilee

Rift Valley

TRANSJORDAN

Acco
Pehel
Jabesh Gilead
Ptolemais
Jezreel
Beth Shan
Harod Valley
Rehob
Hammath
Megiddo
Jezreel Valley
Mt. Gilboa
Succoth
Jordan R.
Jabbok R.
Jokneam
Mt. Carmel
Beth Haggan
Ibleam
Athlit
Umm el-Fahm
N. Taninim
Shephelah of Carmel
Taanach
Dothan
Tirzah
W. Farah
Adam
Dor
Aruna
Hills of Manasseh
Shechem
Janoah
Jokmeam
Crocodilon Polis
Gath
Mt. Ebal
Arumah
Tel Mevorach
Samaria
Mt. Gerizim
Micmethath
Caesarea/Strato's Tower
Yaham
Socoh
N. Shechem
Pirathon
Tappuah
Shiloh
Lebonah
Hills of Ephraim
Sharon Plain
N. Kanah
Ai
Mediterranean Sea
Ebenezer
N. Shiloh
Timnath Serah
Bethel
Apollonia
Aphek
Ramallah
T. Mikhal
Upper Beth Horon
T. Qasile
Yarkon R.
N. Aijalon
Joppa
Gezer

▲ *Village of Luban/Lebonah in the Hill Country of Ephraim: Notice the crops in the valley and the terracing of the limestone.*

city of Caesarea (ca. 22 – 10 BC) did a city on this portion of the coastline serve as a major port. At that time some of the swamps were drained and portions of the land were put to agricultural use to serve as a hinterland for this new major urban center. Caesarea was to become the governing center of the land for some six hundred years (ca. AD 6 – 639).

Hill Country of Manasseh

To the east of the Sharon Plain rise the hills of Manasseh and Ephraim. These mountains are part of the massive Cisjordan range, which stretches from Mount Carmel and Mount Gilboa in the north to Beersheba in the south. Going from north to south, one passes through the old tribal territories of Manasseh, Ephraim, Benjamin, and Judah (see map p. 115). The geographical boundaries between these regions are sometimes almost imperceptible; but since each region does have its own distinctive character, they will be treated separately.

The Hill Country of Manasseh stretches from the Dothan Valley in the northwest and Mount Gilboa in the northeast to an imaginary east – west line drawn along the axis of the Kanah Ravine, extending from the Sharon Plain to the Rift Valley. The center of this territory is an uplifted area of Eocene limestone mountains, the southern tip of which culminates in the highest elevations in the whole region of Manasseh: Mount Ebal at 3,084 feet and Mount Gerizim at 2,891 feet. This central area contains some

large valleys, located at higher elevations, which receive 20 to 25 inches of rain annually. At the northeastern edge of the central block is Mount Gilboa, which towers some 1,600 feet over the Harod Valley. Both the amount of rainfall on Mount Gilboa (ca. 16 in. annually) and the number of dew nights fall considerably below those in the rest of the central region (compare David's words at the time of the death of Saul: "O mountains of Gilboa, may you have neither dew nor rain," 2 Sam 1:21).

To the east of the central core is an area composed primarily of Cenomanian limestone, with some outcroppings of Senonian and Eocene chalks closer to the Rift Valley. Near the Rift Valley the annual rainfall decreases (10 to 14 in.), thus allowing only for marginal crops and grazing. The hills of eastern Manasseh are cut through by two parallel fault lines that outline the Wadi Farah. This wadi, from 0.3 to 1.9 miles wide and running in a northwest – southeast direction, begins in the Tirzah region (elevation 660 ft.) and leads down into the Rift Valley near the river ford at Adam (980 ft. below sea level). Because much of its course is below sea level, the wadi cuts through the underground water horizon and contains a number of important springs.

The Wadi Farah has provided a convenient passageway between the Hill Country of Manasseh and the Transjordanian and Rift Valley regions. In addition, there is a thin ribbon-like "moat" of Senonian chalk that runs directly from Shechem toward Beth Shan, geologically separating central and eastern Manasseh. This "moat" provides an additional connecting link leading in and out of the region, as evidenced by the old Roman road that followed this track.

To the west of the uplifted central block are the lower chalk and limestone hills and valleys of western Manasseh. This area is not as rugged as the previous two and has relatively easy access routes into the hills via valleys such as the Nahal Shechem and the Dothan. Although the soil is not the most fertile, it is easily worked, and since the area is on the windward side of the mountains it receives a good amount of rainfall; it was therefore either tree covered or well cultivated throughout antiquity.

Of all the areas in the Cisjordan mountains, the territory of Manasseh is most open to outside influences. Roads lead into the area via the Nahal Shechem, the Dothan Valley, the Wadi Farah, and the Shechem – Beth Shan "moat." Furthermore, one could cross from the Sharon Plain to the Rift Valley by following the zigzag route that led up the Nahal Shechem to Shechem, turning northeastward to Tirzah and from there heading southeastward down the Wadi Farah to the Rift Valley. This route joined the "Watershed" or "Ridge" Route from the Jerusalem region at Shechem. Thus it is no wonder that Shechem, strategically located at this important road junction in a valley between Mount Ebal and Mount Gerizim, was for a long period of time both the major city and the capital of the region.

The move of the capital of the Northern Kingdom to Tirzah (see, e.g., 1 Kings 15:21, 33) seems to reflect a more inward, defensive, conservative approach to international matters. But when the capital was finally moved to Samaria (1 Kings 16), located not far from the Sharon Plain on the western slopes of the mountains, the Northern Kingdom was open to the full force of cultural, religious, and political influences from the west (witness the arrival of Jezebel and her gods; 1 Kings 16:29 – 34). Samaria eventually became a great Greco-Roman city, and in time its name replaced Manasseh as the name of the region.

Hill Country of Ephraim

To the south of Manasseh are the hills that form the mountain heartland allotted to the tribe of Ephraim (see map p. 115). The Hill Country of Ephraim is bounded on the west by the Sharon Plain, on the east by the Rift Valley, on the north by the theoretical east – west line drawn along the Kanah Ravine (see above, pp. 44 – 45), and on the south by a theoretical east – west line through the city of Bethel. From north to south the Hill Country of Ephraim measures 15.5 miles and from east to west some 27 miles. The area is composed entirely of hard Cenomanian and Turonian limestones. These limestones weather into the deep, V-shaped valleys that make the hills of Ephraim a rugged region: almost the entire area consists of very steep-sided, sometimes terraced, rocky hills, separated by dramatic, deep valleys (see above). The main wadi system is the Shiloh, which drains the whole central section of the region toward the Mediterranean Sea.

Elevations in the region range from 1,900 to 3,333 feet; the highest peak being Baal Hazor. The western slopes and central mountains receive abundant amounts of rainfall (20 to 28 in.), while the eastern slopes receive less and less as one approaches the Rift Valley. The soil is fertile terra rossa; terrace farming is common, and vines, wheat, and other crops are grown on the small patches of arable land.

The major road through the area is the "Ridge Route" (see above, p. 32). This road, in the main, sticks close to the watershed and avoids the deep valleys to the east and west as it progresses from Jerusalem, via Bethel and Shiloh, to Shechem. Because the "wall-to-wall" limestone reaches from the coastal plain into the Rift Valley, approaches into the hills from either the west or the east are difficult. This meant that Ephraim was normally spared from direct invasions and that it could serve as a place of retreat during times of political instability. The placement of the tabernacle at Shiloh during the period of the judges may have been motivated by the usefulness of the natural defenses of Ephraim. It is with good reason that one of the Levites was said to live in "a remote area in the hill country of Ephraim" (Judg 19:1), for certainly this was one of the most secluded places in the whole of the Cisjordan mountain chain.

Hill Country of Benjamin

This region approximately coincides with the territory that once belonged to the tribe of Benjamin (see map p. 115). It stretches from Bethel in the north to Jerusalem in the south, a distance of 13.5 miles. Its eastern border is the mountain's edge at the Rift Valley, its western boundary the western slopes of the hills descending into the coastal plain. From east to west it averages about 27 miles in width.

Most of eastern Benjamin is actually the northern extension of the chalky, dry Judean Wilderness (6 to 16 in. of rain annually). It was used primarily as a place of refuge for the outcasts of society and as winter pasturage for sheep and goats. The central and western portions of Benjamin are composed of hard Cenomanian-Turonian limestone; the central section, through which the watershed runs, is a relatively high plateau (2,500 ft.). The "Benjamin Plateau" formed the core of Benjamin's tribal territory. It is circular in shape and is bounded on the south by Gibeah, on the east by Geba, on the north by Mizpah, and on the west by Gibeon. It receives about 25 inches of rain annually. A number of springs are found in this area, as throughout the Cisjordan Mountains, and plastered cisterns were commonly used to collect rainwater.

Settlements sprang up on the tops and slopes of the small limestone hills that dot the plateau. The hard limestone is relatively easy to work and was used in the construction of the walls of houses; the roofs were made of mud and plaster supported by timbers. The plateau's well-cultivated fields are blessed with fertile terra rossa soil. In Bible times, especially at the time of the arrival of the Israelites, the area was probably more densely covered with trees and forests than it is today.

To the south of Gibeah there is a second mountain plateau. It is elliptical in shape and is bounded on the south by Beth Hakkerem and on the west by the Waters of Nephtoah. The city of Jerusalem stands on the eastern edge of this plateau. This "Jerusalem Plateau" was shared by the tribes of Benjamin and Judah. The major internal route running from Hebron to Shechem, the "Ridge Route," ran across these two plateaus, passing close to the villages and sites of Beth Hakkerem, Gibeah, Ramah, Mizpah, and Bethel.

To the east and west of these plateaus are mountain ridges that extend perpendicularly to the central north – south mountain axis. These ridges, especially the western ones, are separated by deeply incised valleys through which travel is next to impossible. North – south travel through the whole region is very difficult if one leaves the plateaus or the watershed. Most east – west routes of any importance ascend into the mountains via one of the limestone ridges. A major interregional east – west road ran from Gezer on the coastal plain, through the Valley of Aijalon, up along the ridge on which Lower and Upper Beth Horon were located, via Gibeon, to Ramah, where it joined the Ridge Route.

From there one could turn south to Jerusalem, north to Bethel, or continue through the wilderness down into the Rift Valley to Jericho. This route from the west was the major road to the capital city of Jerusalem, and throughout the ages those who have controlled Jerusalem have attempted to control this approach (see maps pp. 110 and 153 for several of the many military encounters along this road). The whole Benjamin Plateau was of strategic importance for the control of these internal routes. In fact, the Jerusalem and Benjamin plateaus were some of the "busiest" areas with regard to the flow of biblical events (see maps pp. 132, 154, et al.).

Hill Country of Judah

The tribe of Judah received the largest tribal allotment (Josh 15), which included all of the land stretching south from Jerusalem down to Israel's southern border with Sinai and from the Mediterranean Sea to the Rift Valley. However, the Philistines gained control of the coastal plain area, and the Negev was too dry to support a large population, so the majority of the Judeans settled in the Hill Country of Judah. This hill country is the final continuation of the central mountain spine that begins up at Mount Carmel and Mount Gilboa and disappears at the northern edge of the Negev Basin. The Hill Country of Judah stretches from Jerusalem 40 miles south to Khirbet Ira and is 11 to 14 miles wide; its western border is the Shephelah, its eastern border the Judean Wilderness (see below).

The Hill Country of Judah is composed of hard Cenomanian limestone and thus shares many characteristics with the Hill Country of Ephraim and that of Benjamin (see above, pp. 46 – 47). To the south of the Benjamin and Jerusalem plateaus the mountains begin to rise until they reach an elevation of 3,343 feet just north of the Hebron area, at which point the hills begin their descent to 1,300 feet in the Khirbet Ira region. The central watershed that runs from Jerusalem to Hebron divides into two ridges at Hebron. One of these ridges continues almost due south via Carmel and Maon to Arad in the Negev, while the other proceeds southwest past Adoraim and Debir to Beersheba. The roads that ran along these two ridges provided important links between Judah and the trade routes of the Negev and the Red Sea.

Hebron is at the junction of the main Ridge Route and its two branches. It is the natural hub of the southern region of the Hill Country of Judah. Situated high in the hills, defended on all sides by the typical treacherous wadi systems, it protected Judah from the south. It also served as the spot where the central government in the hill country could make contact with the Bedouin and merchants of the Negev. At

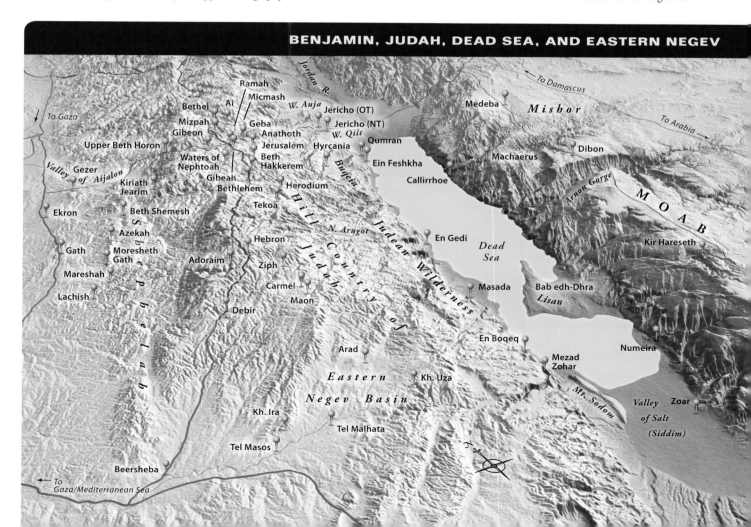

BENJAMIN, JUDAH, DEAD SEA, AND EASTERN NEGEV

▲ *Wadi with water in the Judean Wilderness: Notice the barren limestone slopes.*

Hebron, produce from the hill country — wheat, barley, olives, grapes, pomegranates, and other fruits — could be exchanged for sheep, donkeys, camels, leather goods, and other products brought up from the desert regions to the south. It was to Hebron that the patriarchs gravitated, and it was at Hebron that David first established his kingdom (see, e.g., map pp. 138 – 140).

Because of the high elevation, the amount of rain in the area between Bethlehem and Hebron is high (20 to 28 in.); south of Hebron, however, rainfall begins to drop off rapidly. During the winter months the Hill Country of Judah usually receives at least one snowfall, although the snow generally remains on the ground for only a day or two. Judah's higher elevations make it more difficult to grow olives there than in the Ephraim and Manasseh regions. Vineyards, on the other hand, are much more common in the Hill Country of Judah than in Ephraim and Manasseh. The terra rossa soil is farmed extensively on the small mountain plateaus and in the large valleys such as the Valley of Beracah ("Valley of Blessing"); the terraces on the steep wadi slopes were also cleared and farmed. Although the remote mountain territory between Shechem and Hebron was not of unusual economic, strategic, or political importance to the great powers of antiquity, yet it was there that some of the most significant events of redemptive history occurred. It was there that God settled, protected, and tested his ancient people Israel (see the Historical Section).

Judean Wilderness

The Judean Wilderness (map p. 47) is a rectangular strip approximately 60 miles long and 10 miles wide, sandwiched between the crest of the hills of Judah and the Rift Valley. Its northern limit is close to the Wadi Auja, which enters the Rift Valley 5.5 miles north of Jericho. Its southern border is not well defined but reaches at least to an imaginary east – west line drawn through the southern end of the Dead Sea. The wilderness is primarily composed of Senonian rock formations, chalk, marl, and bands of flint.

The major topographical feature of the region is the steep descent from the Judean mountain ridge (2,500 to 3,000 ft. above sea level) to the surface of the Dead Sea (1,310 ft. below sea level). In some places this means a descent of 4,300 feet over a horizontal distance of only 10 miles. It is not a smooth incline but looks rather like a series of steps descending into the Rift Valley. The final "step" down is the most pronounced and consists of limestone cliffs 350 to 1,300 feet high, stretching along the length of the western shore of the Dead Sea.

As one moves east through the Judean Wilderness, the amount of rainfall drops off drastically. The 14 to 28 inches of rainfall in the Hill Country of Judah decreases to less than 4 inches near the shore of the Dead Sea. More than half of the wilderness area receives less than 8 inches of rain per year. The scarcity of rain and the geology of the area make it difficult to grow crops here. The soils derived from the Senonian rock, chalk, and marl formations are infertile. In addition, when rain falls on these Senonian formations, they develop an impermeable upper layer that prevents the rainwater from soaking into the soil. The water instead rapidly runs off toward the Dead Sea. As a result, the Judean Wilderness is a desolate variegated landscape of plateaus, rounded hills, dramatic scarps, deep canyons, and cliffs. The only places where farming is possible in this inhospitable environment is along the western edge of the wilderness close to the watershed of the Judean mountains and in the Buqeia. Because the wilderness was close to population centers such as Jerusalem and Hebron, all Judeans were well aware of its impact on their lives. It is no wonder that the messages of the prophets Amos and Jeremiah, who came from Tekoa and Anathoth (cities on the edge of the desert) respectively, reverberate with the imagery of the wilderness.

In spite of these adverse conditions, an occasional acacia tree and a few dwarf bushes can be found in the bottom of the wadis. Only during the rainy winter months does some grass cover appear on the plateaus, but even then the meager greenery is basically confined to the northern slopes of the hills. It is during this "wet" season that the shepherds bring their sheep and goats into the wilderness to feed on the sparse vegetation.

The only major springs are located in the northern section of the Judean Wilderness, along the courses of the Wadi Auja and the Wadi Qilt; the central and southern sections are practically devoid of freshwater supplies. Cisterns were hewn throughout the wilderness in an attempt to collect and store as much of the water from the winter rains as possible in order to ensure a minimal supply of water. The Senonian rock formations are well suited for such storage systems. It is no wonder, then, that this rocky, chalky, dry, and desolate region is usually called the "wilderness" (Heb. *midbar*; NIV "desert") in the Bible, or also the "Jeshimon" (the "waste" or the "desert" place; 1 Sam 23:19; 26:1, 3).

Throughout history, this relatively uninhabited wilderness served as a place of retreat and refuge. David fled from Saul into the wilderness (1 Sam 22:1 – 27:6). Herod the Great (37 – 4 BC) built a series of fortresses in the area — including the Herodium, Hyrcania, and Masada — which were to serve as retreats in case of trouble with his Jewish subjects or with his enemy the Egyptian queen Cleopatra (Josephus, *War* 7.8.4 [295 – 303]). The Essenes went out into the solitude of the wilderness, to Qumran, to escape the religious establishment in Jerusalem and to "prepare the way of the Lord." John the Baptizer, while residing in the "Desert [= wilderness] of Judea," called the people of his day to a life of repentance (Matt. 3:1 – 12). Jesus experienced temptations while on a wilderness retreat (4:1 – 11).

The Judean Wilderness also served as the eastern defensive line of Judah. Any invaders from the east — from Moab, Edom, or the Arabian Desert — first had to cross the Dead Sea, scale the imposing cliffs along its western shore, and then march through a waterless wilderness before arriving at major population centers such as Jerusalem and Hebron.

The major roads through the Judean Wilderness all began at the large water sources located in the Rift Valley: the springs of Jericho, Ein Feshkha, and En Gedi. From Jericho at least three roads ascended into the hill country through the northern Judean Wilderness. The northernmost road followed the ridge just to the north of the Wadi Qilt into the Micmash-Ai-Bethel region. It is possible that Joshua led the invading Israelites into the hill country via this route (Josh 7 – 8).

The middle road ascended to Jerusalem on a path slightly to the south of the Wadi Qilt system. This road roughly followed the line of the old border between the tribes of Benjamin and Judah (Josh 15:5 – 8; 18:16 – 19). Today, traces of the old Roman road can be located along its path. In the days of Jesus, Jews traveling between Jerusalem and Perea in Transjordan made frequent use of this route; on occasion it was used by Jesus as he traveled from Jericho to Jerusalem (e.g., Matt 20:17 – 21:3). This road is also the setting of the parable of the Good Samaritan (Luke 10:30).

The third route out of Jericho heads in a southwesterly direction and joins Jericho with Bethlehem.

In the central portion of the Judean Wilderness a road ran northwest from the spring of En Gedi up to the Tekoa-Bethlehem region. At En Gedi it ascended the steep cliffs (660 ft. high) via a narrow, serpentine path that may be the biblical "Pass of Ziz" (2 Chron 20:16). The track then followed the ridge just to the north of the Nahal Arugot up to Tekoa. Remains of Israelite as well as Roman forts along its path have been found. This may have been the route followed by the Moabites, the Ammonites, and the Meunites who invaded Judah during the days of Jehoshaphat (872 – 848 BC; 2 Chron 20:1 – 30; see map p. 159).

In the southern section of the Judean Wilderness an important road led from the southern end of the Dead Sea into the Hebron area. The road ascended the cliffs along the western edge of the Rift Valley near Khirbet Uza, where an Israelite fort guarded the approach. From there the road headed north via Arad and Carmel-Maon-Ziph to Hebron. The queen of Sheba may have used this route as she came from the Red Sea area to visit Solomon (1 Kings 10:1 – 13). In Roman times the road ascended the Rift Valley cliffs via the Nahal Zohar, passing the Roman fort Mezad Zohar in the valley. This route was of great importance for caravans coming from the Red Sea to Jerusalem, as well as for those that transported salt mined in the Mount Sodom region.

Dead Sea/Salt Sea

The largest inland body of water in Israel is the Dead Sea (map p. 47). It is located in one of the deepest sections of the Rift Valley; its surface is some 1,300 feet below the level of the Mediterranean Sea. Because of its position in the Rift Valley, which in the biblical period was known as the Arabah, it is also called the "Sea of the Arabah" (Deut 3:17; 4:49; Josh 3:16; 12:3; 2 Kings 14:25); and because its northern tip is due east of Jerusalem (only 15 mi.), it is sometimes also called the "eastern sea" (Ezek 47:18; Joel 2:20; Zech 14:8).

The Dead/Salt Sea can be divided into two unequal sections by a tongue-shaped peninsula that protrudes into it from its eastern shore (= *Lisan* in Arabic, *Lashon* in Hebrew). The northern portion is 31 miles long, the southern section only 15.5 miles. Its average width is close to 9 miles. The northern section is the deeper of the two; soundings in the northeastern corner indicate that the bottom of the sea lies 1,300 feet below its surface. The southern end is shallow with a maximum depth of only 25 feet.

However, there are considerable variations in the water level of the Dead Sea from year to year. In drought years the level can drop as much as 16 feet, while in especially rainy years its level can rise a corresponding amount, adding 1 mile to the length of the sea. The now-dry basin at the south end of the sea gives evidence of this cycle of contraction and expansion. In recent years the Israelis and the Jordanians have been using the waters of the upper Jordan and Yarmuk rivers for drinking and agricultural purposes, with the result that the amount of water entering the Dead Sea via its main source, the Jordan River, has been considerably reduced, so that the level of the sea has been dropping even more dramatically (about 3 ft. per year). Many recent maps still show a section of water 3 miles wide separating the Lisan from the western shore (as in this atlas); however, today this is practically all dry ground. This area was either dry ground or at least shallow enough to be forded at various times throughout history. It is possible that the invaders from the east during the days of Jehoshaphat (872 – 848 BC) crossed the sea in this area (2 Chron 20:1 – 30), and there is some evidence that it was fordable during Roman times.

Fault escarpments to the west and east of the Dead Sea clearly outline its shore. On the west perpendicular cliffs, some towering 1,300 feet above the sea, rise behind the narrow gravel-and-marl shoreline. In other areas the cliffs descend right into the sea, not even leaving space for a footpath along their base.

Along the eastern shore, the cliffs and mountains rise even more precipitously to elevations of 2,950 feet. These cliffs consist of limestone formations on Nubian sandstone that is exposed in large areas. The mountains of Moab on the east side of the Dead Sea receive much more rainfall than the Judean Wilderness, so that several of the wadis that enter the Dead Sea from the east contain perennial springs. Until modern times there has never been a north – south road along the eastern shore of the Dead Sea because the cliffs, like those along the western shore, often descend directly into the water.

In the south, the eastern shoreline is more expansive. There, jutting out into the sea, is the tongue-shaped Lisan, a very bleak and forbidding region composed of gravel, mud, and marl deposits that have weathered into formations like those of the badlands in the western United States. A basin at the south end of the sea, 13 miles in length, is covered with similar badland formations. This may be the area that in the Bible is called the "Valley of Salt" (2 Sam 8:13; 2 Kings 14:7; 1 Chron 18:12; 2 Chron 25:11) or the "Valley of Siddim" (Gen 14:3, 8, 10).

The region around the Dead Sea is very arid for several reasons: its low elevation, its position in the southern portion of the country on the lee side of the mountains, and its distance from the Mediterranean Sea. At the north end of the sea the rainfall averages 4 inches per year, at the south end only 2 inches. Cloud cover in the area is minimal, and daytime temperatures can be very hot. In the summer, the average daily high temperature is 95 F, in the winter 68 F. Because the sea is bordered by cliffs on the east and west with open areas to the north and south, strong winds can develop during the late afternoon hours, causing whitecaps on the sea.

Because of these hot, dry conditions, there is a large amount of evaporation, which keeps the level of the sea relatively constant. Since there is no outlet and the evaporating water leaves behind most of its minerals, hot sulphur springs, such as those of En Boqeq and Callirrhoe, and the natural salt deposits found in the vicinity also contribute to the exceptionally high mineral content of the sea — over 25 percent as compared to 5 percent for most oceans — that makes the water oily to the touch. The ancients were well aware of the sea's high mineral composition, naming it the "Salt Sea" (Gen 14:3; Num 34:3, 12; Deut 3:17; Josh 3:16; 12:3; 15:2, 5; 18:19). It was not until the modern era, however, that the sodium, magnesium, calcium, and potassium chlorides began to be mined by the Israelis and the Jordanians. The hot sulfur springs, incidentally, were known for their medicinal properties, and it was to Callirrhoe that Herod the Great retired in an attempt to be cured of his fatal illness (Josephus, *War* 1.23.5 [656 – 58]).

Salt was a valuable commodity in ancient times, and there is evidence that Mount Sodom, located at the southwestern corner of the Dead Sea, was mined for this purpose. Rainwater has weathered this "mountain" into fantastic shapes, and from time to time explorers in the region have recognized "Lot's wife" among the salt formations (Gen 19:26)!

Another product that came directly from the sea and from the surrounding area was bitumen, a petroleum product that was used for the caulking of ships and for the making of medicines. The account of the battle between the five kings and the four kings in the days of Lot and Abraham mentions "tar [= bitumen] pits" in the "Valley of Siddim" (the Dead Sea area; Gen 14:10). Josephus later describes how in the Roman period the Dead Sea cast up "black masses of bitumen" that floated on the surface, "their shape and size resembling decapitated bulls," and how the laborers hauled this sticky material into their boats (e.g., Josephus, *War* 4.8.4 [476 – 85]). Because of the presence of bitumen and sulfur springs, the sea was known to some classical writers as "Lake Asphaltitis."

There is little or no vegetation along most of the shoreline of the Dead Sea. Only a few tamarisk trees and other salt-loving plants grow along its banks. However, at oases such as En Gedi and Zoar, some vegetation is cultivated, including such luxury items as date palms and balsam trees. These types of plants flourish in the high-temperature, sweet-water environment of these oases.

Settlement of this desolate area has always been limited. During the middle of the Early Bronze Age (3150 – 2200 BC), large settlements were located at the southeastern end of the sea — at Bab edh-Dhra, Numeira, Zoar, Feifa, and Khanazir (see maps pp. 62, 94). During the period of the Judean monarchy (1000 – 586 BC), small settlements were established along the western shore of the Dead Sea, including such sites as Qumran and En Gedi. During the late Hellenistic and early Roman periods (ca. 152 BC – AD 70), small Jewish settlements were located along the western shore. The Essenes, who have become well known since the discovery of the Dead Sea Scrolls, settled in the Qumran area. Farther south, Jews were living at the oasis of En Gedi, but it was Herod's great fortress of Masada, built on the top of a mesa-like rock, that dominated the southern landscape of the Dead Sea. Although living conditions around the Dead Sea were harsh, it was to places like Qumran, En Gedi, and Masada that the disenchanted or fearful would flee (1 Sam 24:1).

Philistine Plain

South of the Sharon Plain and west of the Shephelah (see below) lies the southernmost section of the coastal plain within Israel proper. Bounded by the Yarkon River on the north, the Shephelah on the east, and the Nahal Besor and Nahal Gerar on the south, the Philistine Plain stretches some 50 miles along the shore of the Mediterranean Sea. It is wider in the south (25 mi.) than in the north (10 mi.).

The coastline itself is uninterrupted by any bays except for one formed by a small promontory near Joppa. The major features of the shoreline are dunes, wider in the south than in the north, and kurkar ridges, which are not as extensively exposed here as they are along the northern coast. Farther inland the landscape is relatively flat. The low,

rolling hills are higher in the southern portion of the plain. In the north, the red sands of the Sharon Plain begin to disappear as they blend with the alluvial soils of the central portion of the Philistine Plain. In the south, the alluvial soils give way to the windblown loess soil that characterizes the Negev region of Israel (see below).

As is typical for the country, precipitation declines as one moves south, from 20 inches annually at Joppa to less than 16 inches at Gaza. Since grain is the major agricultural product grown on the plain, it is with good reason that the Philistines worshiped the grain god Dagon (1 Sam 5:2–5). The southern portion of the plain, near the coast, receives 250 dew nights so that melon crops can be grown there during the rainless summer months.

After the grain crops were harvested in May and June, shepherds, who had been herding their flocks of sheep and goats in the Judean Desert or the Negev, probably brought them to the plain to feed on the stubble. The farmer and the shepherd complemented

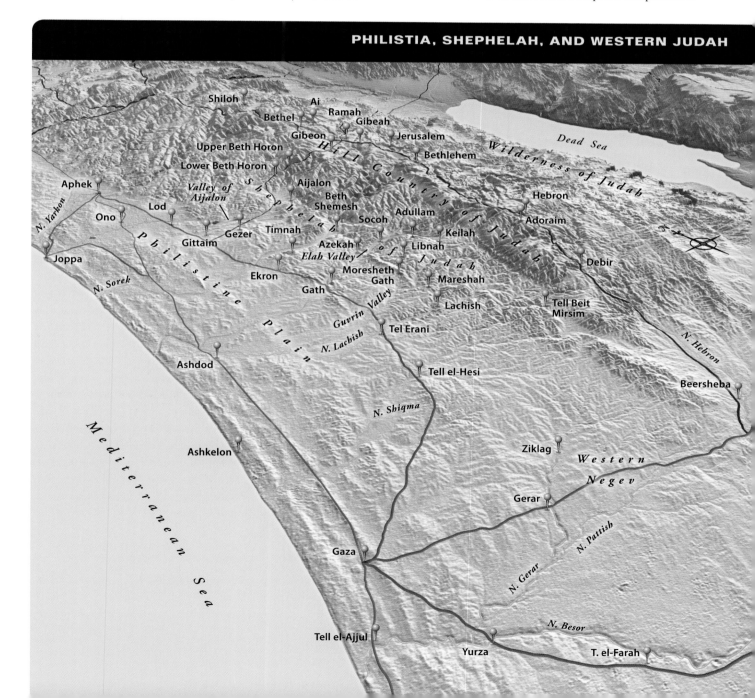

PHILISTIA, SHEPHELAH, AND WESTERN JUDAH

▲ *Nari crust and soft limestone caves in the Shephelah*

each other's lifestyle and may often have been from the same family or tribe.

In the Philistine Plain, drinking water is available from wells dug in the wadi beds and from springs (cf. Abraham and Isaac's activities in this region, Gen 26:15 – 33). Excavations in the region have demonstrated that houses, city walls, and other structures were normally made of sun-dried mud brick, which eventually disintegrated. Subsequent rebuilding on top of the remnants of previous structures has produced numerous "tells" in the area. Travel in the region was relatively easy; the only obstacles were the sand dunes along the coast, the hills of the Shephelah to the east, and the low ground that could become muddy during the wet season. Two branches of the major International Highway ran through the Philistine Plain — one close to the coast via Gaza, Ashkelon, Ashdod, Joppa, and Aphek, the other inland via Tel Erani, Gath, Ekron, Gittaim, and Aphek. Both of these avoided the low, somewhat swampy area between the Nahal Aijalon and the Nahal Yarkon. In addition, the coastal branch, in order to avoid the Nahal Yarkon and the swamps of the Sharon Plain, had to swing eastward to join the inland branch at the strategically placed city of Aphek.

The northern portion of the Philistine Plain, between the Yarkon and the Aijalon, was called the "plain of Ono" (Neh 6:2) or the "Valley of the Craftsmen" (Neh 11:35). The former name was taken from one of the cities of the region; the latter may reflect the ancient memory that the Philistines of the area were blacksmiths, the craftsmen on whom the Israelites were dependent during the reign of Saul (1 Sam 13:19 – 21).

The chief city of the area, next to Aphek, was Joppa, which at various times during the biblical period served as the chief Judean port. It was from here that Solomon brought timbers to Jerusalem (2 Chron 2:16), that Jonah departed from Israel (Jonah 1:3), and that Peter visited the home of Simon the Tanner (Acts 9:43).

Philistia proper occupied the largest portion of the southern coastal plain, from the Nahal Aijalon to the Nahal Besor and Nahal Gerar. This territory with its five Philistine cities (Gaza, Ashkelon, Ashdod, Gath, and Ekron; see maps pp. 51, 130), although assigned to the tribe of Judah (Josh 15:45 – 47), largely remained outside Judean control. With the arrival of the Greeks, the name "Palestine," derived from "Philistine," came into use and eventually began to be used to refer to the whole of the land of Israel.

The whole of the Philistine Plain was open to numerous outside influences. Throughout the ages, from the days of Thutmose III (1457 BC) to those of Napoleon (AD 1799) and Allenby (AD 1918), the armies of the great international powers have marched up and down the coast, using it as a land bridge between Africa and Asia, while others, such as the Philistines and Crusaders, have used it as a beachhead as they attempted to establish themselves in the land.

Shephelah

The Shephelah (map p. 51) is a transitional zone separating the Philistine Plain from the Hill Country of Judah. The Hebrew term *Shephelah* (used twenty times in the OT) is translated "(western) foothills" in the NIV. The term reflects the perspective of an Israelite standing in the Hill Country of Judah looking westward, *down* onto the lower foothills.

The Shephelah is approximately 27 miles long and 10 miles wide. It is composed primarily of soft Eocene chalks and limestones and is pockmarked with numerous caves. Its low rolling hills range from 300 to 1,200 feet in elevation — only about half the height of the "mountains" of the Hill Country of Judah to the east. The hills are covered by nari, a hard, crusty rock. This crusty coating makes the surface of the hills useless for most agricultural purposes save for growing trees and bushes. In Bible times, scrub forests grew on these hills along with sycamore trees, which were cultivated for their fig-like fruit. The area receives adequate amounts of rainfall, so grain and grapes can be grown in the relatively broad, alluvial valleys.

The villages, built with local, mediocre limestones, clung to the hilltops and hillsides so as not to take up valuable farmland in the fertile valleys. The hills in this area do not form continuous ridges like those in the Hill Country of Judah. Thus the roads were laid out along the lower edges of the hills in order to avoid the valley floor, which could become muddy in the rainy season, as well as numerous ascents and descents. Water supplies for the inhabitants of the region included small springs, wells dug into the valley bottoms, and numerous cisterns.

In the northern two-thirds of the Shephelah there are five valleys that figure prominently in sacred history. The Aijalon Valley marks the northernmost limit of the Shephelah. The western approach to this broad chalk valley, now filled with rich alluvial soil, is guarded by the city of Gezer. The major road into the Hill Country of Judah and Benjamin passed eastward through this valley, avoiding the treacherous deep V-shaped valleys in Ephraim to the north as well as the 10-mile-wide band of hills and valleys of the Shephelah to the south. This road then ascended the continuous ridge on which stood Lower and Upper Beth Horon and eventually reached the hill country in the region of Gibeon. It was in the Aijalon Valley that the Lord, on behalf of Joshua and the Israelite army, "hurled large hailstones" down on the fleeing Amorites while the sun and the moon "stood still" (Josh 10:1 – 15; see map p. 110). It was there, in the Valley of Aijalon, that the Philistines, the Egyptians, the Seleucids, the Romans, and many others would assemble to prepare to make assaults into the Hill Country of Judah and Benjamin.

To the south of the Aijalon is the Sorek Valley, guarded on the west by the Philistine city of Timnah and on the east by Israelite Beth Shemesh. Many of the exploits of Samson took place in the Sorek (Judg 13 – 16; see map p. 126). And it was to Beth Shemesh that the Philistines sent the troublesome ark after having kept it in their territory for seven months (1 Sam 5:1 – 7:2; see map p. 131). An east – west road, not as important as the Aijalon – Beth Horon connection, passed through the Sorek to Beth Shemesh before ascending a limestone ridge to Kiriath Jearim on its way into the hills near Jerusalem.

The Elah Valley lies to the south of the Sorek; a road to the Bethlehem area ran through it. This valley, guarded on the west by the Philistine city of Gath and on the east by Azekah and Socoh, formed the stage on which the epic battle between David and Goliath took place (1 Sam 17; see map p. 134).

▼ *Valley of Elah looking east from Kh. Qeiyafa: low rolling hills of the Shephelah*

▲ *Wheat harvest in late spring.*

The Guvrin Valley to the south does not figure so prominently in biblical history, in spite of the fact that an important road to the Hebron area ran through it. The major city in the valley apparently was Mareshah, which became very important during the intertestamental period (then called Marisa). Micah the prophet came from the nearby city of Moresheth (Mic 1:1).

The fifth major valley is today called the Nahal Lachish, after the prominent biblical city Lachish, which lay along its course. This city, first conquered by Joshua (ca. 1400 BC; Josh 10:31 – 33; see map p. 110) and later by the Assyrian Sennacherib (701 BC; see map p. 166), was one of the last of the Judean cities to hold out against the invading Babylonian army (ca. 586 BC; Jer 34:7; see map p. 170). Lachish guarded an important interregional road that led up into the Hebron area via a ridge on which the city of Adoraim was located.

Throughout its history the Shephelah has served as a buffer zone between the provincial peoples living in the hill country and the usually more powerful, internationally oriented peoples of the coastal plain: between Hapiru and Egyptians (ca. 1400 BC), Israelites and Philistines (1200 – 700 BC), Judeans and Assyrians (701 BC), Judeans and Babylonians (586 BC), Arabs and Jews (AD 1948 – 67). Thus this seemingly placid and picturesque agricultural area has in fact served as a bloody battleground as peoples from the coastal plain tried to move into the hills and those living in the hills tried to expand their influence into the coastal plain.

Negev

The Hebrew term *Negev* is used 110 times in the Hebrew Bible, and its basic meaning can be either "dry" or "south." In thirty-eight instances (NIV) it refers to a specific region located in the southern portion of the land of Israel. In the biblical text the Negev, as a regional-geographical term, refers to a limited strip of land extending 10 miles north and 10 miles south of Beersheba and running east to west from the mountain ridge overlooking the Rift Valley to near the dunes along the Mediterranean Sea. The areas to the south of this region were called the "Desert of Zin" and the "Desert of Paran" (NIV). Notice that one could go *up* (= north) from Kadesh Barnea into the "Negev" (Num 13:17, 22).

Topographically, the biblical Negev can be visualized as a double basin in the shape of an hourglass placed on its side, with Beersheba at the "waist" of the glass. To the north, east, and south of both basins is higher ground. The eastern basin, the administrative center of which was Arad, is drained by Nahal Beersheba, which flows west past Beersheba into the western basin, where it joins the Nahal Besor.

The soil of the Negev basins is a fine, windblown (aeolian) soil, called loess. When it rains, the surface of the loess soil becomes relatively impermeable, so that instead of seeping into the ground much of the water rapidly runs off into the wadis, creating miniature "badland" formations. The major water sources of the Negev were wells, which were usually dug in the bed of the wadis, often at the junction of two or more wadis, where water would tend to collect. For example, Beersheba, which can mean the "well of oath" or the "well of seven" (Gen 21:31; 26:33), was located at the junction of the Nahal Hebron and the Nahal Beersheba.

The biblical Negev is a transition zone that receives only 8 to 12 inches of rain annually; however, deviations from this norm are common. If the norm is reached, wheat and barley can be grown, but in only one of every three or four years is an adequate crop harvested. Thus the term *Negev* is a fitting one for this territory in that it is south of the heartland of Israel and considerably drier than most northern regions.

Besides being used for marginal agriculture, the Negev served for the pasturing of flocks of sheep and goats, as well as for the southern defense of Judah. City forts, such as Beersheba and Arad, served as administrative, military, and religious centers. Of these, Beersheba normally was the southernmost outpost of the land of Israel, as reflected in the phrase "from Dan to Beersheba." Caravans traveling from Transjordan, Arabia, and the Gulf of Elath to a Mediterranean port such as Gaza were able to travel with ease across the low, rolling hills of the Negev.

To the south of the Negev is the area known as the Negev Highlands. This

▲ *En Avedat in the Negev Highlands*

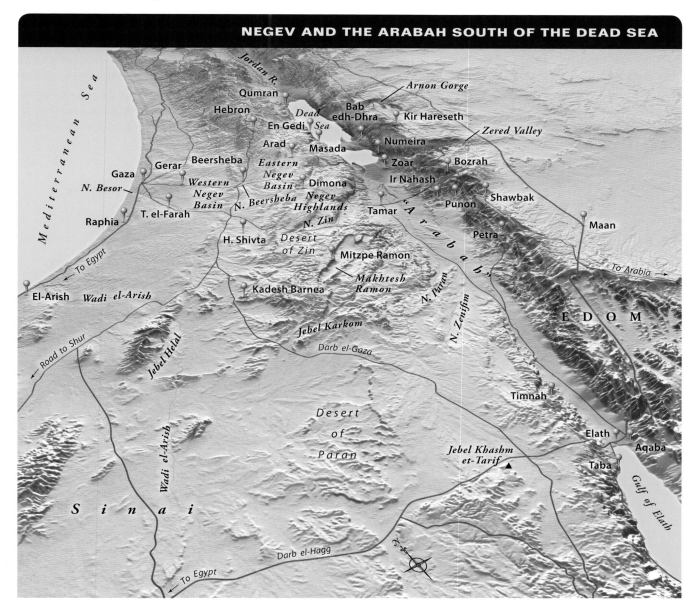

NEGEV AND THE ARABAH SOUTH OF THE DEAD SEA

region, composed of limestone and chalk, contains a series of inland ridges that run parallel to the line of the shore of the Mediterranean Sea — from the southwest to the northeast. Situated on three of these ridges, or anticlines, are three elliptical crater-like depressions called *makhteshim* in Hebrew (sing., *makhtesh*). Travel through any of these craters is difficult, for most of their sides are almost vertical cliffs. The highest elevation reached in the Negev Highlands is 3,395 feet at Har Ramon. The Highlands are divided into northern and southern sections by the large, crescent-shaped Nahal Zin, which flows northeast into the Rift Valley.

The Negev Highlands constitute an arid region that receives only 4 to 8 inches of rain each year. Agricultural products are extremely difficult to grow without costly irrigation techniques, and even grazing is difficult because of the scarcity of grass cover. However, caravans coming from Arabia via Transjordan or the Rift Valley had to cross this territory to take their wares to the port of Gaza. The various ridges and craters lay across the caravan paths and proved to be formidable barriers to travel. The easiest route through the region ran from Tamar in the Rift Valley, up through the Nahal Zin, and from there northwest to Gaza. Roads sometimes ascended the various ridges (especially in the Roman and Byzantine periods), yet the track through the Nahal Zin, the modern "Way of the Sultan," served as the major connecting route.

South of the Negev Highlands rainfall decreases to only 1 or 2 inches annually. The traveler going south encounters broad wadis such as the Paran and the Zenifim. Closer to Elath, sandstone and granite mountains, typical of southern Sinai and Transjordan, appear.

The "Arabah" South of the Dead Sea

The portion of the Rift Valley system that stretches 110 miles from the southern end of the Dead Sea to the tip of the Gulf of Elath on modern Israeli maps is called *Haarava* (map p. 55), "The Arabah." In the English Bible the Hebrew word *arabah* is variously translated as "wasteland," "wilderness," and "desert land" (NIV; e.g., Job 24:5; Isa 35:1; Jer 51:43), but it is also used as a geographical term. In its latter sense, generally with the article ("the Arabah"), it appears twenty-eight times in the NIV and (with the possible exception of 1 Sam 23:24) refers to the area of the Rift Valley between the Sea of Galilee and the southern tip of the Dead Sea, also called the "Sea of the Arabah" (see especially Deut 3:17).

But following the modern Israeli usage, the "Arabah," south of the Dead Sea, is composed of several sections. Extending 25 miles south of the Dead Sea is a severely dissected "badlands" area composed of Lisan marls and salt formations called the Sebkha, which at various times was covered by the Dead Sea (see above, p. 50). Farther south the valley is hemmed in by mountain cliffs on the east and west, but it widens to 15 miles in the Punon region, where it loses some of its "Rift Valley" characteristics. Sixty-two miles south of the Dead Sea it even reaches an elevation of 1,160 feet above sea level before descending to the level of the Red Sea at the Gulf of Elath. Closer to Elath the slopes of

▼ *Southern Rift Valley: looking east across the valley to the Mountains of Edom*

▲ *Jabbok River Valley and the mountains of Gilead*

the mountains east and west of the valley become more prominent, and at one point the valley narrows to a width of only 3.5 miles.

The region of the Arabah receives between 1 and 2 inches of rain each year; thus neither farming nor grazing is possible here. Although the area is basically devoid of vegetation, patches of halophytic (salt-loving) bushes and occasional acacia and broom trees are encountered, especially in the beds of the wadis. In biblical times the Arabah was used mainly as a highway. Since wells and springs are not numerous in this area, those who attempted to control the passage of caravans through the Arabah built small forts at these strategic watering points.

In biblical times the northeast extension of the Red Sea, the Gulf of Elath/Aqaba, was not used as extensively for shipping as one might expect, for the predominantly northern winds in the narrow gulf made the inbound trip difficult. In addition, coral reefs abound, currents are tricky, and freshwater sources along the gulf's shores are rare. However, when ships did use the gulf, they were able to unload at Ezion Geber. The merchandise was then placed on camels or donkeys, and the overland caravans progressed up the Rift Valley to the Tamar region and from there headed northwest across the Negev.

The more popular trade route from the Far East was by sea to southern Arabia, where gold, frankincense, and myrrh could be added to the cargos brought in by ships. From there the caravans proceeded inland in a northwesterly direction via Yemen, Mecca, and Medina up to southern Edom. Caravans heading for Cisjordan may have descended

into the Rift Valley at Elath and then headed north, or, more probably, they continued up the Transjordanian Highway to the Petra or Bozrah regions, where they turned west, descended into the valley, and then continued on their way across the Negev to Gaza. If a caravan was headed for Egypt, it could descend into the Rift Valley at Elath and then continue across the middle of Sinai toward present-day Suez via the Darb el-Hagg, a track formerly used by Egyptians making pilgrimages to Mecca and Medina.

The only other major activity that took place in the Arabah was copper mining in the Timna region near Elath during the Chalcolithic, Late Bronze, and Roman periods and the important copper mines near Punon, located in the Transjordanian mountains 33 miles south of the Dead Sea.

Many have considered the Arabah something of a frontier separating the Negev on the west from Edom on the east, yet there is evidence that at times the Edomites spread from Transjordan into various portions of the Negev (Num 20:14–16; 34:3), as the Nabateans did after them. Indeed, some of the Judean-Edomite conflicts seem to have been over the control of the trade routes through this area.

Gilead

Various forms of the word "Gilead" are used in the Scriptures a total of 129 times (NIV). Although on occasion it is used as a personal or tribal name, it is mainly used as a geographical term referring to the area of the

Bashan

MANASSEH

Sea of
Galilee

Ashtaroth

Kenath

Yarmuk River

Gadara

Edrei

Harod Valley

Bezer

Mts. of Bashan

Beth Shan

Ramoth Gilead

DECAPOLIS

Jordan River

Jebel Umm
ed-Daraj ▲

Tirzah

Jerash

W. Farah

Succoth
(Deir Alla)

Penuel

Shechem

Mahanaim

Gilead

Jabbok River

PEREA

Adam

GAD

ARABAH/GHOR

Auja
Spring

(Rift Valley)

Rabbah

Jericho

Allenby
Bridge

REUBEN

Heshbon

Jerusalem

▲ Mt. Nebo

Medeba

Mishor

Machaerus

Dead
Sea

	Major roads
	Secondary roads

Dibon

Aroer

0   10 km.

0   10 miles

En Gedi

Arnon Gorge

MOAB

country on the east side of the Jordan Valley between the Sea of Galilee on the north and the Dead Sea on the south. More precisely, this territory stretches from the southern edge of Bashan, approximately 18 miles south of the Yarmuk River (see p. 35), for 47 miles to near the latitude of Heshbon and the north end of the Dead Sea. Deuteronomy 3:10 indicates that the plateau (Heb. *mishor*) of Moab, as well as Bashan, was normally thought to be outside of Gilead proper. Northern Gilead is bounded on the east by steppe lands and lava flows and in the southeast by the edge of the Arabian Desert. Although even more precise boundary descriptions are possible, the Israelites often use the term "Gilead" to refer to all Israelite territory east of the Jordan River (e.g., see Josh 22 passim).

The heartland of Gilead (maps pp. 58, 60) is composed of an uplifted dome of hard Cenomanian limestone with stretches of Senonian formations to the north and the southeast. The Jabbok River divides this dome into two portions, the northernmost being the highest, reaching an elevation of 4,091 feet at Jebel Umm ed-Daraj. Because of the high elevations, rainfall in both sections of Gilead is abundant, especially on the western side of the slopes, and olives, grains, and vines flourish. In biblical times the heights were covered with thick forests that ranked with those of Mount Carmel, Bashan, and Lebanon (e.g., Jer 22:6; 50:19). Gilead was famous for a balm extract that was used for medicinal or cosmetic purposes (e.g., Gen 37:25 ; Jer 8:22; 46:11).

The dome area is structurally similar to the Hill Country of Judah, and thus the landscape is dissected by many steep-sided valleys, and plateau areas are not common. The major wadi of the region is the Jabbok, which begins near Rabbah (modern Amman), where it flows north and then heads west to join the Jordan River. Although narrow, the Jabbok is farmed along most of its length, and its descent from the hills of Gilead to Deir Alla (Plain of Succoth) in the Rift Valley is relatively easy. It is possible that the route that ran through this valley to Succoth, crossed the Jordan River, and continued up the Wadi Farah to Shechem, was the route used by Abraham and later by Jacob to enter the land of Canaan (Gen 12:5 – 6; 32:22 – 33:20). The major road through Gilead was a portion of the international north – south Transjordanian Highway that ran northward from Rabbah of the Ammonites past Ramoth Gilead to Ashtaroth in Bashan.

Historically, the northern portion of Gilead was allotted to half of the tribe of Manasseh, while the southern portion was allotted to Gad. The Ammonites seem to have been located on the southeastern fringes of the area at Rabbah, always ready to push into the dome area if possible. During New Testament times the southwestern portion of Gilead was known as part of Perea (see map p. 212), whereas the northern and eastern sections were associated with the cities of the Decapolis. Throughout history the region seems to have served as a place of retreat and refuge. David, for example, fled from Absalom to Mahanaim in Gilead, and Jeroboam I used Peniel (NIV) as his capital for a brief period of time, possibly in an attempt to escape the Egyptian Shishak (1 Kings 12:25; see below, p. 152).

Jordan Valley

Although the Jordan Valley stretches only some 65 miles from the Sea of Galilee to the Dead Sea, the actual length of the meandering river is approximately 135 miles. The width of the river varies considerably

▼ *River Jordan with surrounding "thicket" (=Zor).*

from season to season and in various places, but at the Allenby Bridge, 7.5 miles north of the Dead Sea, its width fluctuates from 70 to 110 feet. Early explorers report that at flood stage certain portions of the river reached a width of almost 1 mile. The depth of the Jordan River at the Allenby Bridge varies from 4 to 13 feet from one season to another. During the winter and spring, especially from January through March, the highest water levels are reached. It has been found that the flooding is primarily due to the increased flow of the Yarmuk River, which drains portions of Bashan and Gilead. It has been observed that the flow of the Yarmuk has been known to increase from 13,200 gal./sec. to 450,000 gal./sec. — a thirty-four-fold increase — in only two days!

In spite of the Jordan's meanderings, its gradient is relatively steep and the current is strong. Travel up and down the river by small boats is almost impossible because of the usually shallow but swift waters and the numerous rapids. The Lynch Expedition encountered some twenty-seven major rapids during their trip along the length of the Jordan in 1848. There is no evidence from biblical times that any bridges were built across the Jordan, but the Madeba Map indicates that by the sixth century AD at least, boats were used to ferry objects across the river. From the mention of "fords" in the Bible, it seems evident that these shallow areas were sought out as the most convenient points of crossing. Over fifty fords have been noted along the Jordan, though this number must have fluctuated considerably through the centuries. Swamps along the banks, swift currents, and muddy or very rocky bottoms can make the crossings treacherous as one wades, swims, or attempts to ride an animal across the stream.

The area of longitudinal strips of land bordering the river on each side is called the *Zor* (photo p. 59) in Arabic and is referred to in the Bible as the "thickets" (NIV; e.g., Jer 49:19; 50:44); the Hebrew *ge'on hayyarden* can be literally translated "the Pride of the Jordan." Each of the two strips of the Zor varies in width from 600 to 5,000 feet, and both were once covered with dense vegetation, including poplars, tamarisks, willows, cane, and reeds. On occasion, during times of flooding, some of the trees were almost completely covered with water. Wild animals — including lions, leopards, and jackals — were commonly found in this dense thicket. Because of the dangers associated with such

GILEAD AND JORDAN VALLEY

▲ *The tableland (Mishor) of Moab*

wildlife, this region was generally avoided by humans and their domesticated animals, except when it was necessary to cross the river.

As one passes through the Zor, away from the river on either side, one comes to another longitudinal zone called the *Qattara* in Arabic. This region is composed of Lisan marl formations of clay and gypsum, which are impermeable to water. Runoff water in this area is considerable, and miniature "badland" formations abound. This type of terrain begins some 17 miles south of the Sea of Galilee and continues all the way down to the Dead Sea. The width of these two strips of the Qattara varies from 0.6 to 2 miles. Because of the infertility of the soil, the region lacks a cover of vegetation, and thus the powers of erosion are magnified. From time to time large blocks of Lisan material can break away, the debris temporarily blocking the river. Earthquakes are common in the region, and on a number of occasions resulting blockages have stopped the flow of the river: in 1547, for two days; in 1906, for ten hours; in 1927, for twenty-seven hours (compare the crossing of the Israelites described in Joshua 3 – 4, esp. 3:16).

The major portion of the Jordan Valley is the area called the *Ghor* in Arabic (the biblical Arabah; see above, p. 25). It is bounded on the east by the hills of Gilead and on the west by the hills of Manasseh and Ephraim. It varies in width from 6 miles in the north to 2.5 – 3 miles in the middle, to 12 miles in the Jericho region. The northern section receives sufficient rainfall to grow crops, but as one progresses southward, the amount of rainfall drops off rapidly to only 4.6 inches at the Allenby Bridge. Where there are springs (such as at Beth Shan, Auja,

and Jericho), or perennial rivers (such as the Yarmuk, Jabbok, Harod, and Wadi Farah), it is possible to irrigate crops. Elsewhere scrub grass grew and was used for grazing flocks from January to March. Temperatures in the valley are warm during the winter and very hot in summer (Jericho's mean low in January is 50 F, its mean high in August is 102° F!). Thus, although people did settle in the northern section of the valley or in the south at an oasis such as Jericho, the valley was not one of the most densely settled areas of the country.

Today the entire region of the Jordan Valley is being transformed into a rich agricultural area as the Israelis and Jordanians divert waters from the Jordan, Harod, Farah, Yarmuk, and Jabbok to irrigate their crops in this natural greenhouse. Thus conditions in the valley are somewhat different today from what they were in the ancient periods described above.

Moab

The name Moab is used most frequently in Scripture as a tribal name referring to the descendants of Lot (Gen 19:36 – 39); less frequently it indicates the territory where they settled. In the latter sense it refers to the region east of the Dead Sea — a region bounded on the north by Gilead, on the south by the Zered Valley, and on the east by the Arabian Desert. It measures some 56 miles from north to south and some 22 miles from east to west. Its eastern boundary is approximately the line of the Hejaz Railway.

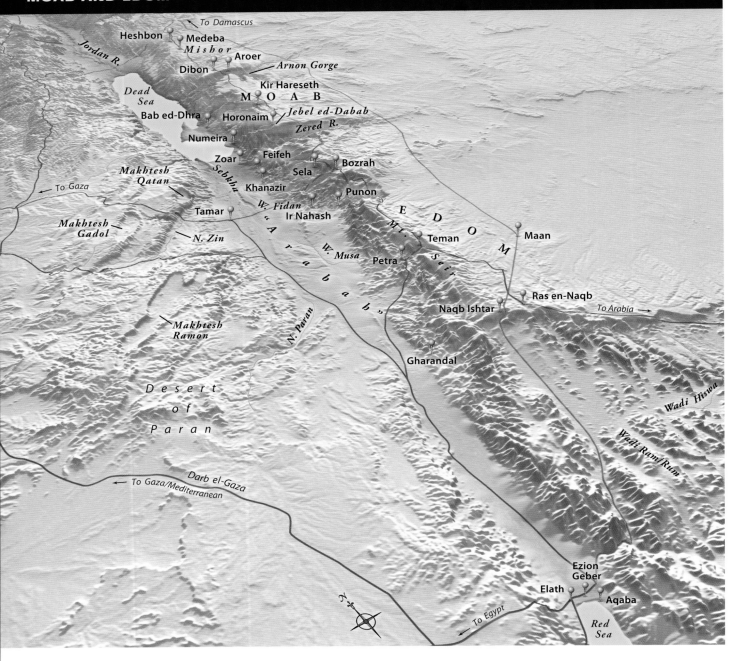

Moab is divided into northern and southern sections by the Arnon Gorge, which in places is 2,300 feet deep and, from rim to rim, almost 3 miles wide. On the south, between the Arnon and the Zered, is the old ancient core of Moab, which in the west consists of hard Cenomanian limestone resting on Nubian sandstone formations; the latter are the beautiful reddish cliffs that line the eastern shore of the Dead Sea. In eastern Moab Senonian chalks and chert prevail. Elevations above 3,600 feet are common, and at Jebel ed-Dabab, on the north bank of the Zered Valley, an elevation of 4,282 feet is reached. Rainfall approaches 16 inches annually at these higher elevations.

The land stretching northward from the Arnon to the Heshbon region is called the "plateau" (NIV; Heb. *mishor*; see, e.g., Deut 3:10; Josh 13:9, 16), and it is considerably lower (2,300 feet) and less dissected than the region to the south of the Arnon. In one passage (1 Chron 5:16) it seems to be called Sharon, which probably means "forest in the plain," reflecting the fact that this area was at least partially forested in biblical times. Because of the lower elevations, rainfall in the Mishor region approaches only 10 inches annually. Throughout Moab the major agricultural crops were wheat and barley, and sheep breeding was a major source of revenue. Mesha, the king of Moab, is reported to have supplied the king of Israel with 100,000 lambs and the wool of 100,000 rams (2 Kings 3:4).

The Israelite tribes of Reuben and Gad were allotted cities in the plateau region (e.g., Num 32; Josh 13:15 – 28), but throughout history this territory was coveted by other nations as well, including the Amorites, the Ammonites from the northeast, and especially the Moabites from the south. It was the Moabites who usually fought with Israel for control of the region during the period of the divided monarchy (931 – 722 BC).

Two branches of the north – south International Transjordanian Highway ran through Moab. One branch ran just east of the western mountain ridge, passing through cities such as Rabbah, Medeba, Dibon, and Kir Hareseth. Along the way it had to descend into the deep gorges of the Arnon and the Zered, making the passage somewhat difficult. The second branch ran along the eastern edge of Moab near the desert frontier. Although geographical obstacles were not as great, caravans taking this route were exposed to raids from the tribes that often swept in off the desert.

▲ *Near Wadi Ram/Rum in southern Edom*

Edom

The name Edom ("red") was used first as an alternate personal name for Esau (Gen 25:30) and then came to be used as a tribal name referring to his descendants (e.g., Gen 36:1 – 17; Num 20:14 – 20). These descendants eventually settled in the mountains east of the Rift Valley in the land between the latitudes of the southern end of the Dead Sea and the northern end of the Gulf of Elath/Aqaba, a distance of 110 miles.

The land of Edom can be divided into three geographical sections. The northern area, which extends from the Zered Valley in the north to Punon, 25 miles to the south, seems to have been the heartland of Edom. The important city of Bozrah was located here. On the west, Nubian sandstone cliffs border the Arabah, while the high ground in the center is composed of hard Cenomanian limestone, which in turn gives way to chalk formations and the desert to the east.

The central section of Edom, stretching 50 miles from Punon in the north to Ras en-Naqb in the south, is a high mountain plateau region where elevations are above 4,900 feet; a high point of 5,696 feet is reached 5 miles northeast of Petra. This high ground stretches 30

▼ *Wadi Ram/Rum in southern Jordan*

miles from north to south and then angles to the southeast into the desert for another 12 miles. Because of the height of this middle area, the rainfall ranges from 6 to 10 inches annually, and so some agriculture is possible. During the summer months, temperatures at these elevations often are moderate; during the winter, snow can remain for days on the high mountain plateau while the lower surrounding areas receive little or no precipitation. To the west of the high plateau, the Nubian sandstone cliffs bordering the Arabah are even more prominent, and nestled in one of the steep-sided valleys is the famous rock-hewn city of Petra, which for a while was the Nabatean capital of the area.

The third region of Edom stretches 3 miles from Naqb Ishtar down to the latitude of the northern end of the Gulf of Elath/Aqaba. At the north end of this region the formations of Nubian sandstone head inland, to the southeast, whereas to the south of Gharandal granite mountains now mark the eastern edge of the Rift Valley. These granite peaks continue down the Gulf of Elath/Aqaba into Saudi Arabia. Between the granite and sandstone areas is the forbidding Wadi Hiswa with its broad alluvial bottom (15 to 20 mi. wide) dotted with massive sandstone mesas. Rainfall in the area is scant — less than 2 inches annually.

Because of both the lack of rainfall and the deep valleys of western Edom, the area was never well populated. In the northern and central sections, especially in the plateau area, the people were able to grow grain crops and raise sheep, goats, and camels. Up until modern times the higher regions of Edom were covered with forests, and this may have been one of the reasons why it was called "the hill country of Seir" (e.g., Gen 36:9). Seir, "the hairy one," may also be a reference to Esau, whose descendants were called Edomites.

The major factor contributing to the economy of Edom seems to have been its control of the caravan routes. The important north – south International Transjordanian Highway ran along the eastern edge of Edom, avoiding its rugged western slope and the high mountain plateaus. Near modern Ma'an this route divided into two branches. One headed southeast into Saudi Arabia, from where luxury items were imported that had originated in Arabia and even in India and Africa. The other branch headed southwest down to Elath at the head of the Red Sea.

East – west routes connecting the Arabah with the Transjordanian Highway included, from north to south, (1) the road that ran up the Wadi Fidan, which passed Punon on its way up to Bozrah, (2) the road up the Wadi Musa, which led to Petra, and (3) a route that led into the mountains from the Arabah oasis of Gharandal. All these routes connected Edom with Egypt and/or the Mediterranean port of Gaza. Thus many of the much-sought-after luxury items had to pass through Edomite territory on their way from southern Arabia to the commercial centers in Egypt and to Damascus and the ports along the Mediterranean Sea. The people living in Edom at times were able to gain great wealth by controlling these caravan routes.

THE GEOGRAPHY OF EGYPT

Located in the northeastern corner of Africa, Egypt has been one of the great power centers of the Near East throughout historical times. The heartland of Egypt is basically a long river oasis situated near the eastern edge of the Sahara Desert. Ninety-five percent of Egypt is stone, sand, and desert, while only 5 percent is rich agricultural land to which the life-giving Nile brings precious water and silt. Here along the Nile, on the "Black Land," lives 95 percent of the population of Egypt. As the Greek historian Herodotus wrote, "Egypt is a gift of the Nile."

The traditional boundaries of ancient Egypt were the Mediterranean Sea on the north, the Red Sea/Gulf of Suez on the east, the first cataract (= rapids) of the Nile near Aswan on the south, and a north–south line of oases that are about 120 miles west of the Nile. At various times Egypt attempted to control or exploit the resources of Sinai to the east, Nubia to the south, and Libya to the west.

The northward-flowing Nile, with its origins in Central Africa (the White Nile) and in the Ethiopian Highlands (the Blue Nile), is the longest river in the world (4,145 mi.). The westerly source, the White Nile, supplies water year-round, while the more easterly Blue Nile provides an abundance of water during the months of June, July, and August — the monsoon season in the Ethiopian Highlands.

Egypt was divided into two major geographical regions. "Upper Egypt," which is upstream (i.e., south), stretches from the first cataract in the south to the beginning of the delta near Cairo, while "Lower Egypt" is the delta proper. In Upper Egypt the arable land along both sides of the Nile varies in width from 6 miles near Aswan to 16 miles near Cairo in the north. In Upper Egypt the land is hemmed in on both sides by limestone cliffs, which in places (e.g., Jebel es-Silsila) approach the riverbed itself. To the west is the foreboding Sahara Desert, the barrenness of which is only occasionally broken by an oasis. On the other side of the Nile, beyond the cliffs, is the Eastern Desert, an upward-sloping plateau of sand and stone culminating in the hills and mountains (4,000 to 7,000 ft.) that overlook the coast of the Red Sea and the Gulf of Suez. Although the Eastern Desert is desolate, at times the Egyptians worked the stone quarries and the gold, tin, and copper mines in the area. Several roads, the most important of which ran through the Wadi Hammamat, crossed the Eastern Desert,

connecting the Nile Valley with the Red Sea and, hence, with eastern Africa, southern Arabia, and even India.

Life in Upper Egypt was regulated by the Nile. Prior to the construction of modern dams, which began in the nineteenth century and culminated with the building of the Aswan Dam in 1970, the Nile rose rather predictably between the fifteenth and eighteenth of June in the Aswan area. Depending on the location, it would usually rise between 15 and 23 feet above its normal level. As it rose, it overflowed its banks and flooded the nearby fields. The muddy floodwaters covered the fields for several months, and as they began to recede in September/October, they leeched out unwanted salts and left behind a fresh layer of fertile silt. The peasants planted their crops in the muddy soil during October/November and harvested from January through March.

Chief among the crops were wheat and barley. During certain periods Egypt served as the "breadbasket" for other parts of the world.

YEARLY CYCLE IN ANCIENT EGYPT

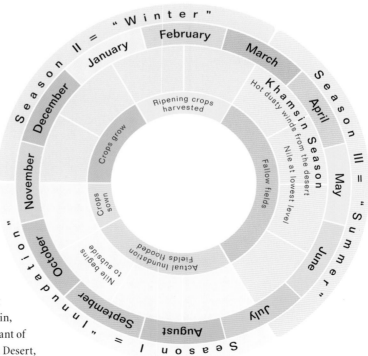

Mediterranean Sea

Damascus

LOWER EGYPT

Alexandria Port Said
Delta
Tel Aviv
Gaza
Raphia
El-Arish
Ismailiya *Wadi el-Arish*
GOSHEN
Giza Cairo Suez
Saqqara Memphis *Jebel et Tih* Elath
Lisht
Faiyum Oasis *Gulf of Suez* *Sinai Peninsula*

Amman
Jerusalem
Beersheba

Timnah
Aqaba
Gulf of Elath/Aqaba

Bahariya Oasis

Farafra Oasis

UPPER EGYPT

El Amarna

Nile R.

Eastern Desert

Abydos
Thebes
Wadi Hammamat

MIDIAN

ARABIA

Ed-Dakhla Oasis

Ed-Kharga Oasis *Jebel es-Silsila*
Aswan
Kurkur Oasis 1st Cataract
Berenike

Medina

Dunqul Oasis

2nd Cataract

Salima Oasis

NUBIA

Red Sea

Mecca

3rd Cataract

Jebel Barkal *Nile R.*
4th Cataract 5th Cataract
CUSH

Sahara Desert

6th Cataract

Atbara R.

White Nile R.
Blue Nile R.

ETHIOPIA

Bab al-Mandab

Ethiopian Highlands

Legend:
— International routes
— Local or regional routes
-- Water routes

0 — 100 km.
0 — 100 miles

In addition, vegetables, dates, and sesame were grown. Fish from the river were abundant, as were fowl — both species that were indigenous to Egypt and species that passed through the area on their biannual migration from Europe to Africa and back. In some areas grapes were cultivated, but the most common beverage was a barley beer rather than wine. Flax was grown and used to produce clothing, ropes, and sails, while papyrus was used for paper production and was exported.

The major center of southern Upper Egypt was Thebes. Throughout the millennia palaces, tombs, and temples were constructed of limestone found in its vicinity. For the most part, the peasants lived in mud

▲ *Temple of Hatshepsut near the Valley of the Kings*

huts that were usually constructed on high ground so as not to be destroyed by the annual inundations. The Nile itself was the major "road," and barges and sailing ships were common modes of transportation. The current easily carried these vessels downstream, and they were able to sail upstream by making use of the prevailing north wind.

Southward travel by boat was interrupted by the first of six cataracts — places where the Nile crossed hard granite ridges that caused treacherous rapids. Located to the south of the first cataract were the lands of Nubia (Egyptian "the land of gold"), Cush, and eventually Ethi-

opia. At various times Egypt was able to exert control up to the second cataract and, on rare occasions, even to the fourth. Conversely, during periods of Egyptian weakness, the peoples from the south expanded northward into Upper Egypt.

The delta region north of modern Cairo is known as "Lower Egypt" (lower = lower in elevation, i.e., downstream). Roughly triangular in shape, it measures 100 miles north – south and 155 miles east – west, with a total area of close to 8,500 square miles. Today, two main branches of the Nile make their way through the delta area to the Mediterranean Sea, but in the days of Herodotus (ca. 450 BC)

▼ *Luxor: Looking west across the Nile at agriculture and the mountains by the Valley of the Kings*

▲ *Canal near the Nile River bringing water to the fields*

there were seven branches of the Nile (as reported in Strabo, *Geography* 17.1.18). This low-lying area is well supplied with fertile silt that has been washed down over the millennia and is crisscrossed by canals. Throughout the area, crops are grown in abundance. Wheat and barley, orchards, vineyards, and fig trees are plentiful. In ancient times the swamps provided some pasturage for cattle, and papyrus, sedge, and lotus plants were abundant. Animals such as the hippopotamus and crocodile were common. Ample supplies of fish, both from the Nile and the Mediterranean, were readily available, as were wild fowl, such as ducks and geese.

Although the coastal area of the delta receives close to 7 inches of rainfall annually, irrigation has always been the main technique used to water crops. The importance of the delta as a source of food supply can be seen in the fact that this area alone comprises 55 percent of the total arable land of Egypt north of the first cataract.

In the delta, settlements have always been located on the high ground between the canals, and although it is generally agreed that the delta was settled from earliest times, it is only from the second millennium BC onward that significant archaeological remains have been found. This state of affairs may be due to (1) silt covering earlier remains, (2) a rising water table, and (3) the effects of salt water on ancient monuments and villages.

Lines of communication are difficult to trace in the delta, but it seems that besides land transport, boat transport on the Nile and the

adjacent canals was very important. For example, when Sinuhe (ca. 1950 BC) returned to Egypt from the land of Canaan, he boarded a ship in the eastern delta and traveled by boat to the town of Lisht (*ANET*, 21). The eastern delta was also the area the Bedouin from Sinai attempted to infiltrate to secure water for their flocks. And from here, the armies of the mighty pharaohs of the Eighteenth and Nineteenth Dynasties (ca. 1500 – 1150 BC) launched their expeditions into Canaan and other countries of Asia. According to the biblical account, it was in the rich and fertile eastern delta, known as the "region of Goshen," that Jacob and his descendants settled and began their sojourn in Egypt (Gen 46 – 50; see Historical Section).

To the east of the delta lies the triangular Sinai peninsula, with an area of 23,600 square miles — smaller than West Virginia. On the north it is bounded by the Mediterranean Sea as it stretches from Port Said on the west to Rafia or Gaza on the east. The western boundary is an imaginary north – south line, 80 miles long, that stretches from Port Said in the north to Suez in the south — basically following the path of the modern Suez Canal, which in turn is situated close to the line of the old easterly branch of the Nile and/ or its major easterly canal. The southwest coast of Sinai borders on the Gulf of Suez, stretching 180 miles from Suez to Sinai's southern tip, Sharm esh-Sheikh. Sinai's southeastern side borders on the Gulf of Elath/Aqaba and is 120 miles long, but its eastern land boundary is more amorphous, although a line drawn from Elath/Aqaba to the

Rafia-Gaza region — 125 miles long — is a reasonable projection of this boundary.

The northern coastline of Sinai is composed primarily of sandy flats and some dunes, the latter extending from 25 to 60 miles inland. The water table is relatively close to the surface, and throughout the ages wells have been sunk in this area to supply the local inhabitants and travelers with fresh water. The major road connecting Asia with Africa ran through this area. Not only did commercial caravans use it, but the great armies of the world have passed this way as well: the ancient Egyptian, Assyrian, Babylonian, Persian, Greek, and Roman armies, as well as modern British, Egyptian, and Israeli forces.

To the south, the dunes eventually give way to a series of mountains. Water supplies can be found in and near these mountains, particularly in the northeast near Kadesh Barnea, where the most powerful spring of the peninsula is located. Fairly close to the line where the coastal region meets the foothills is an important route that connected Beersheba with what is today Ismailia. This may be the biblical "road to Shur" (Gen 16:7), which was used during patriarchal times for travel to Egypt. To the south of these mountains the broad, desolate, lifeless Tih region rises, culminating in the Jebel (Arabic for "Mount") Tih in the south.

Most of northern Sinai is drained by the Wadi el-Arish ("Brook of Egypt"), which enters the Mediterranean Sea at el-Arish. A track called the Darb el-Hagg (the "Way of the Pilgrimage") connects Elath/Aqaba on the east with Suez on the west; it was of importance up through the late 1800s, as Egyptian Muslims used it on their pilgrimages to Mecca. During biblical times, this road connected Egypt with Midian (in Saudi Arabia) and provided the Egyptians with access to the copper mines at Timna, just north of Elath/Aqaba.

The southern tip of Sinai consists of dramatic, jagged granite peaks, some of which reach heights of over 8,600 feet. On the west a broad, flat plain, 10 to 15 miles wide, separates the mountains from the Gulf of Suez, while on the east the mountains often drop precipitously into the Gulf of Elath/Aqaba. Snow sometimes falls in this mountainous granite region, but the total amount of precipitation is minimal. The normal decrease in rainfall from north to south is operative in Sinai as in other regions of the Middle East (e.g., the Mediterranean coastal area receives ca. 8 in. per year, the southern tip of Sinai 0.4 in.). In the southern granite region, however, water collects in the wadis, and some oases are found around both springs and wells. The largest of these oases in southern Sinai is the Feiran Oasis, located at the foot of Jebel Sirbal.

Because of its rainfall deficiency and correspondingly rugged terrain, Sinai has never boasted a large population. A census in 1937 listed 18,000 inhabitants, of which only 2,500 lived in the south. Sinai's major

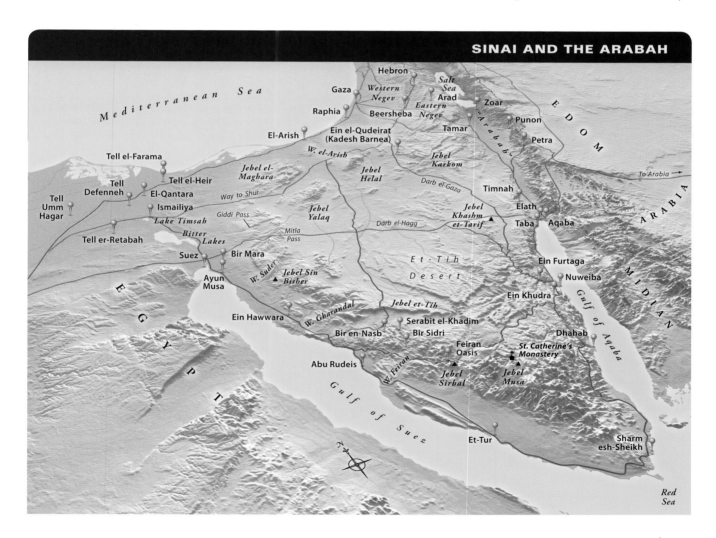

importance has been as a land bridge between Africa and Asia. In modern times its mineral wealth has begun to be exploited, but in ancient times the Egyptians were primarily interested in mining the turquoise deposits in the Wadi Maghara and Serabit el-Khadim regions — for turquoise was in great demand for the production of rings, beads, necklaces, amulets, and scarabs. Although some deposits of copper are located in Sinai, it appears that the mines at Timna, just north of the northern tip of the Gulf of Elath/Aqaba, were the major sources of copper during certain periods. For students of the Bible, the events surrounding the giving of the Law at Mount Sinai are of prime importance (see the Historical Section).

The recorded history of Egypt began around 3100 BC, when Upper and Lower Egypt were united into one country. As early as the third millennium BC the Egyptians had divided Upper Egypt into twenty-two nomes, or districts, and by the late first millennium BC twenty delta nomes were added to the total. During periods when the central government was relatively weak, the rulers of the nomes — the nomarchs — were often powerful.

Historians, following the lead of Manetho (an Egyptian priest who wrote ca. 282–245 BC), divide the line of kings into thirty or thirty-one "dynasties." Modern historians usually begin the First Dynasty at about 3100 BC and end the series with the Ptolemaic Dynasty (ca. 30 BC). In addition, Egyptologists and historians combine these dynasties into more comprehensive periods or eras:

"Kingdom Periods" (Old, Middle, New) were periods of strength and stability while "Intermediate Periods" (First, Second, Third) were periods of disorder and political chaos.

Dynasties	Periods	Approximate Dates (BC)
1st–2nd	Early Dynastic	3050–2700
3rd–6th	Old Kingdom	2691–2176
7th–10th	First Intermediate	2176–2023
11th–14th	Middle Kingdom	2116–1638
15th–17th	Second Intermediate	1638–1540
18th–20th	New Kingdom	1540–1070
21st–25th	Third Intermediate	1070–664
26th	Saite Renaissance	664–525
27th–31st	Late Dynastic	525–330
	Conquest of Alexander	332
	Macedonian Domination	332–304
	Ptolemaic Dynasty	304–30
	Roman Conquest	30

▼ *Nile Valley near Luxor: Note Late Kingdom temple, lower right, the irrigated fields, the Nile (flowing from right [south] to left, and Luxor and desert on the far east side of the Nile*

One of the first systems of writing was developed in Egypt. Although its origin is shrouded in the mist of antiquity, by the time of the First Dynasty (ca. 3100 BC) writing had been established in the country. For over three thousand years (until the fourth century AD), hieroglyphic writing was used in Egypt. The earliest form of the script made use of ideograms, or sense-signs, which represented a specific thing or idea, or something suggested by the thing (e.g., a sun dish ☉ represented the sun; an outline of hills ⏜, hilly or foreign lands; and writing and related words were represented by scribal tools — palette, water pot, and brush case). In addition, sound signs, or phonograms, soon came into use; here the pictures represented a sound or series of sounds (e.g., "r" from mouth ⬯, "pr" from the ideogram house ⬜). Altogether, a scribe needed to master some 700 characters.

The Egyptians referred to their language as the "words of the god" and considered it a gift of the god Thoth. The Egyptian priests were among the few who knew how to read and write (carve) the language when the Greeks took control of Egypt (ca. 332 BC), and thus the language came to be called hieroglyphics (Gr. *hieros*, "sacred," and *glyphein*, "to carve"). As early as the time of the Middle Kingdom, Egyptian scribes were using reed pens and papyrus to write a modified form of hieroglyphics called hieratic (Gr. "priestly") writing. Later, during the Twenty-Fifth Dynasty (ca. 700 BC), demotic (Gr. "popular"), a more rapid form of hieratic writing, came into use. The final phase of ancient Egyptian is Coptic, the script known from the third century AD onward. It makes use of the Greek alphabet plus seven special characters. This language is still read and spoken, particularly in the liturgy of the Coptic Church.

A wealth of written material has been preserved from ancient Egypt, for almost every vertical surface of ancient temples, tombs, and buildings was covered with hieroglyphic inscriptions. In addition, the use of papyrus was common, not only for literary texts, but also for economic, administrative, and personal documents. Because of Egypt's dry climate (only 1 in. of rain per year from Cairo south), many monumental texts and papyrus documents have been preserved. Those that are important for understanding biblical history and geography will be discussed at the appropriate places in the Historical Section.

THE GEOGRAPHY OF SYRIA AND LEBANON

The area shown below comprises primarily the modern states of Syria and Lebanon. It was not only an important area in its own right but also served as a crossroads that connected Babylonia and Assyria with Anatolia (modern Turkey) to the northwest, with the Mediterranean to the west, and with Palestine and Egypt to the southwest.

The region is bounded on the west by the Mediterranean Sea, on the north by the Amanus and Malatya mountains, on the east by a north – south line drawn through Jebel Sinjar, on the south by the Syrian Desert, and eventually on the southwest by Damascus and the Litani River. Often difficult to rule due to pressures from hostile desert and mountain tribes, it was not until the first millennium BC that the area was subdued and unified by the great empires of Assyria, Babylonia, Persia, and especially the Seleucids. It is difficult to find a single ancient name that refers to the whole region, although the area west and south of the Euphrates was called Amurru (the "West Land") and

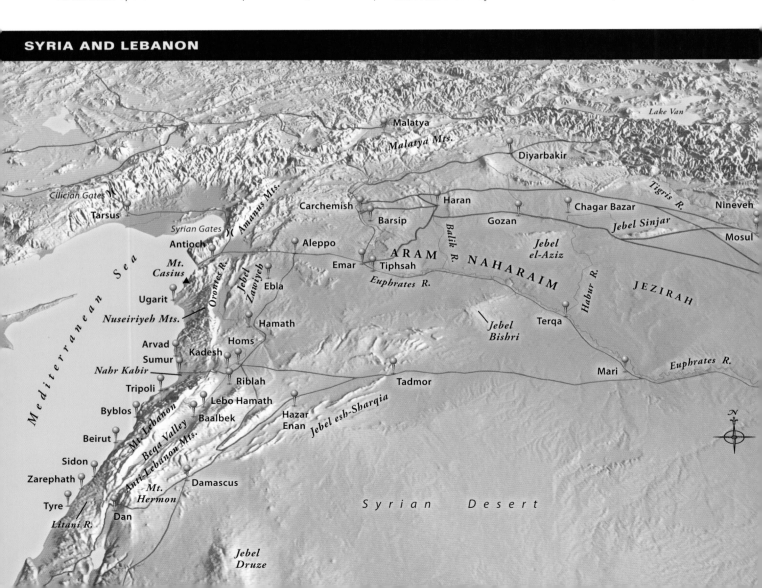

SYRIA AND LEBANON

Lake Van

Malatya

Malatya Mts.

Diyarbakir

Tigris R.

Cilician Gates

Tarsus

Amanus Mts.

Carchemish

Haran

Chagar Bazar

Nineveh

Syrian Gates

Barsip

Gozan

Jebel Sinjar

Antioch

Aleppo

ARAM NAHARAIM

Jebel el-Aziz

Mosul

Mt. Casius

Emar

Tiphsah

JEZIRAH

Orontes R.

Jebel Zawiyeh

Ebla

Euphrates R.

Ugarit

Habur R.

Nuseiriyeh Mts.

Hamath

Jebel Bishri

Terqa

Arvad

Homs

Kadesh

Sumur

Nahr Kabir

Riblah

Tadmor

Mari

Euphrates R.

Tripoli

Lebo Hamath

Byblos

Baalbek

Hazar Enan

Jebel esh-Sharqia

Beirut

Mt. Lebanon

Beqa Valley

Anti-Lebanon Mts.

Sidon

Zarephath

Damascus

Mt. Hermon

Syrian Desert

Tyre

Dan

Litani R.

Mediterranean Sea

N

Jebel Druze

▲ *Cedar of Lebanon on snow-covered mountain slopes*

Mark Connally

the "land beyond the River [Euphrates]" during the second and first millennia BC and was also known as "Aram," after the "Arameans" who lived in the area.

Many varieties of landscape and lifestyle are found in this region. Grain crops can be grown north of an arc that runs from Damascus in the southwest, via Homs and Tiphsah, to Jebel Sinjar in the northeast. The land to the north and west of this rough line receives at least 10 inches of rain annually, while rainfall drops off rapidly to the south of this arc. There lie the expanses of the Syrian steppe/desert, where, along its northern and western fringes, nomads wander with their herds in search of winter grasses.

To the north and east of the Euphrates is a steppe area known today as the Jezirah, which in actuality is a westward extension of Assyria. This area is drained by the Habur and Balik rivers. The land between the Habur and the Euphrates may very well be the original "Aram Naharaim" (Gen 24:10; "the land between the two rivers"; Gr. *mesopotamia*). North of an east – west line drawn through Jebel Sinjar and Jebel el-Aziz, rainfall is adequate to grow grain crops in winter (over 12 in. annually), and springs are plentiful. South of that line rainfall drops off, and the flat steppeland is mainly given over to grazing. Most of the major cities were either along or just north of this line. They include, from east to west, Chagar Bazar, Haran, and Carchemish. The roads connecting Assyria, and even Babylonia, with Carchemish ran through this area, skirting Jebel Sinjar to the north or south. Carchemish was one of the key cities along this route, since from there caravans and/or armies could head northwest into Anatolia, westward to the Mediterranean, or southward into Syria, Israel, and Egypt.

To the south and west of the Euphrates is the area of Syria proper. This area can be conveniently divided into northern and southern portions by the "Homs-Palmyra [Tadmor] Depression [or corridor]." This depression stretches westward from the bend of the Euphrates just southeast of Mari, through Tadmor and Homs, to the Nahr Kabir, which flows into the Mediterranean. When political conditions were relatively stable, and especially after the camel began to be used as a pack animal, this caravan route across the steppe/desert from Mari to Tadmor (ca. 145 mi.) became important. From Mari, bulk goods, especially lumber from Jebel Bishri and the Anti-Lebanon and Lebanon ranges, could be shipped down the Euphrates River. From Tadmor, caravans could continue almost due west along the depression, via Homs, to Mediterranean seaports near the Kabir River.

In addition, a very important route ran southwest, along the edge of the desert, connecting Tadmor with Damascus, a distance of about 130 miles, and hence with Israel, Egypt, and south Arabia. Tadmor, later called Palmyra, was one of the great caravan cities of antiquity. The Bible mentions how Solomon built "Tadmor in the desert" as he attempted to gain control of the trade routes of the Levant (2 Chron 8:4). To the south of Tadmor and to the east of Jebel esh-Sharqia lies the tableland of the Syrian Desert, broken by rugged escarpments and bounded on the southwest by great, almost impassible basalt outflows. This area, with less than 8 inches of rainfall, supports only pastoral nomads, who barely subsist here.

Situated in the southwest corner of this area is the magnificent oasis of Damascus. Receiving only 8 inches of rain annually, the oasis is fed by the Barada River flowing down from the Anti-Lebanon mountains. Damascus was a key to Israel, for nearly all traffic entering or exiting Israel from the north had to pass through it. Because of this, the control of Damascus has been much disputed throughout history, yet rarely has Damascus been able to extend its control far in any direction, for it is hemmed in by mountains to the west and north and by desert and basalt outflows to the east and south. Here Israel's archenemies, the Arameans of Damascus, were headquartered.

Most of the area sandwiched between Damascus and the Mediterranean lies within the modern state of Lebanon. The southern boundary of this region is slightly to the south of the Litani River, and its northern boundary is at the Nahr Kabir at the western end of the Homs-Tadmor corridor. This river is actually the modern boundary between Lebanon and Syria. The region itself can be divided into four major longitudinal zones that run from the southwest to the northeast.

The "coastal plain" is at best very narrow, 5 to 10 miles wide, and in many places spurs of the Lebanon Mountains reach the shores of the Mediterranean, interrupting north–south travel along the coast. This area came to be known as Phoenicia. Its orientation has always been westward, toward the Mediterranean Sea. Great ports such as

Tyre, Sidon, Beirut, Byblos, Tripoli, and Sumur alternately cooperated or vied with each other for control of shipping on the Mediterranean. The agricultural hinterland of each of these cities was restricted, often necessitating trade with Israel and especially with Egypt for needed foodstuffs. In return they supplied much desired building timber to these and other countries throughout the millennia.

Byblos, in fact, became a great distribution center, connecting Egypt with the remainder of the Mediterranean world. At certain times (e.g., during the Late Bronze Age, 1550 – 1200 BC), Byblos almost had the status of an Egyptian colony, as evidenced in the written documents from that period and by the many Egyptian artifacts found in excavations at Byblos (ancient Gebal). From Byblos, papyrus that had originated in Egypt was shipped all over the Mediterranean, and it should be remembered that the Phoenicians may have played a role in the transmission of the alphabetic script westward to Greece at the end of the second millennium BC.

Immediately inland from the coast are the majestic Mountains of Lebanon, rising in places to heights of over 10,000 feet. These mountains, composed of hard Cenomanian limestone, sandstone, and some Jurassic rock, receive over 40 inches of precipitation annually. For about six months of winter they are snow covered, and this whiteness may have given rise to the name "Lebanon," which is related to a Hebrew root meaning "white." Both the eastern and western slopes were tree covered in antiquity, and it was here that the prized "cedars of Lebanon" grew. The trunks of these stately trees, along with those

▼ *The island city of Arvad. Tyre was also an island city like this until the days of Alexander the Great (ca. 332 BC) when he linked it to the land.*

of cypresses and pines, were cut and used wherever long beams were necessary — for planks and masts of boats, for columns and beams in palaces and temples, for paneling, etc. Deep V-shaped wadis dissect the western slopes of the mountains, and travel is difficult.

On the east, a steep escarpment drops off into the third longitudinal region, the Beqa ("Valley"). The long and narrow Beqa Valley stretches some 85 miles up to the area just south of Homs. On the west towers the escarpment of the Mountains of Lebanon, while on the east the Anti-Lebanon range rises. Swamps and lakes that form near Kadesh block the north end of the Beqa; the southern portion ends in a jumble of ridges and valleys. Most of the Beqa is above 3,000 feet, but since it is on the lee side of the Lebanon range and slightly removed from the sea, it is in a partial rain shadow. Baalbek receives 16 inches of precipitation annually, while the mountains to the west receive 40 inches. In Beirut, on the coast, the average is 34 inches. In the Beqa, gardens flourish, as do olive and fruit trees, vines, and grains. Located near Baalbek is the north – south watershed of the valley; the Orontes drains the valley to the northeast, and the Litani drains it to the southwest. The latter river eventually turns due west and, after passing through a deep, steep, narrow gorge, flows into the Mediterranean 5 miles north of Tyre.

East – west travel through the Beqa was not common, for the Lebanon and Anti-Lebanon ranges are very high and do not contain convenient passes. Southwest to northeast travel is also difficult due to ridges that intersect the valley and, in addition, the entrances at both ends of the valley are difficult — hills to the south, swamps to the north. Thus, the main International Route passed to the east of the Anti-Lebanon Mountains through Damascus.

The Anti-Lebanon mountain range stretches from Mount Hermon (9,232 ft.) in the south approximately 85 miles to an area overlooking the Homs-Tadmor corridor in the north. In most places precipitation reaches 40 inches annually, and a thick forest covered the rugged mountains during much of antiquity. From the slopes of the Anti-Lebanon Mountains the Barada River flows eastward, sustaining the oasis city of Damascus.

North of the Homs-Tadmor corridor, the major lines of relief are basically north – south. The narrow coastline, stretching 105 miles from the Nahr Kabir north to the Amanus Mountains, boasts a series of ports, including Arvad (an island anchorage), Tripoli, and Ugarit. Both Arvad and Tripoli served as ports for the western end of the route that ran through the Homs-Tadmor corridor, while Ugarit was connected via a pass to the northeast through the Nuseiriyeh Mountains with Aleppo. It was here at Ugarit that precious documents from the fourteenth century BC, written in cuneiform script, were found. Documents written in the Ugaritic language furnish us with a vast literature that sheds welcome light on Canaanite religious beliefs and practices, some of which are referred to in the Old Testament.

Close to the north end of the coastal plain, Mount Casius (5,771 ft.) forms a prominent landmark along the shore of the Mediterranean. Just north of this, the plain of Antioch/Alalakh provides a swampy but adequate connecting route from the Mediterranean to Aleppo on the east. The steep scarp of the Amanus Mountains (ca. 6,000 to 7,000 ft.) rises to the north of the plain, and through these mountains a pass leads to the Cilician Plain and on to Anatolia.

East of the coast, behind Arvad and Ugarit, rise the Nuseiriyeh Mountains (4,000 to 5,380 ft.) along a north – south axis. Both the gentle western and steep eastern slopes were forested in antiquity, forming an impassible barrier for east – west traffic.

To the east of the Nuseiriyeh Mountains is the low-lying, marshy Ghab Valley. The Orontes River meanders through this swampy area, flowing from south to north before swinging around north of Mount Casius and exiting into the sea via the Antioch Valley.

To the east of the Ghab rises Jebel Zawiyeh (3,080 ft.) with gentle eastern slopes that eventually blend into the steppeland that runs to the Euphrates. In the north, this plateau was dominated by the city of Halab (Aleppo), while the southern portion of the plateau was often controlled by Hamath on the Orontes, a city/country located just north of the traditional boundaries of the land of Canaan (see map p. 106).

THE GEOGRAPHY OF MESOPOTAMIA

The Tigris and Euphrates rivers dominate life at the eastern end of the Fertile Crescent. This large region is bounded on the east by the Zagros Mountains that separate Mesopotamia from Iran. The Zagros and Malatya mountains enclose the region on the north, while the northern reaches of the Syro-Arabian Desert help define the habitable area on the west and south. Most of this area, which in historical studies is often called "Mesopotamia," lies in what today is Iraq, while the rest is located in Syria, Turkey, and Iran.

The name Mesopotamia is derived from the Greek and means "land between the rivers." Originally it may have referred to the land between the Euphrates and the Habur rivers, for it is used in the Greek translation of the Old Testament (Septuagint) to refer to Aram Naharaim (NIV; lit. "Aram of the two rivers"), which was located near Nahor (Gen 24:10). Both Polybius (second century BC) and Strabo (first century AD) use "Mesopotamia" to refer to the area between the Euphrates and the Tigris. Today it is used by extension to refer to the land between and beside these two great rivers.

The Euphrates and Tigris both have their origins in the mountains of Armenia and are fed by the melting snows and local rains of this rugged northern area, which receives 20 to 40 inches of precipitation annually. Although the source of the Tigris is within a few miles of where the Euphrates passes, the two rivers diverge and follow different paths. The Euphrates, also known as "the River" in the Bible, is over 1,780 miles long. It begins in Turkey and, after flowing through the mountains of Armenia, it heads due south and joins the North Syrian Plain (see map p. 72) near Carchemish. There it is only some 100 miles inland from the Mediterranean Sea. It then turns southeast and continues its gradual descent to the Persian Gulf.

After the Euphrates passes through the Syrian plain, limestone escarpments hem it in on both sides from Abu Kamal to Hit, leaving only a narrow alluvial valley for farming. Since this area between Abu Kamal and Hit receives only 4 to 8 inches of rain each year, agriculture is primarily confined to the narrow river valley, although some grain crops are grown on the steppes where flocks of sheep and goats feed on the winter grass cover.

The Tigris, which begins in Lake Hazar southeast of Elazig, only 2 or 3 miles from the Euphrates, is 1,150 miles long. After leaving the mountains of Armenia, the Tigris flows south and is fed from the east first by the Greater and then by the Lesser Zab rivers. These two rivers in fact double the amount of water flowing in the Tigris. This area is the old heartland of Assyria, and great cities (including Nineveh and Asshur) were once located along the banks of the Tigris. Today this highland region is largely deforested and heavily eroded, but during the winter grain crops are grown. Winter rains — 10 to 20 inches annually — are the primary source of moisture for the crops. In some periods, irrigation canals have been used as well, although this involved somewhat advanced technology and large amounts of capital, for the Tigris is lower than the surrounding tableland. From the Greater Zab south to near Samarra there are few canals, scant rainfall (ca. 8 in. annually), and little habitation.

Near Samarra, however, the landscape changes. Canals begin to branch off the Tigris to the east and to the southwest, and the "delta" or plain region of the Tigris and Euphrates begins. This plain is triangular in shape and is marked on the north by a line drawn from Samarra southwest to Hit, on the southwest by a line from Hit to Basra, and on the east by a line from Basra to Samarra. Near Baghdad the Tigris and Euphrates are only 18 miles apart, although they diverge again before actually joining at Qurna, northwest of Basra. The low-lying landscape between the two rivers is flat, expansive, and treeless. The silt, deposited over the millennia, is 15 to 25 feet deep.

Since this area receives only 4 to 8 inches of rain each year, agriculture is dependent on irrigation techniques. Although the courses of the Tigris and Euphrates have in this region changed from time to time, the channels of both rivers are basically above the surface of the surrounding plain. If the rivers overflow their embankments in the spring, vast stretches of the plain are inundated. These inundations are problematic, for the amount of water brought down each year is erratic. This irregularity can mean devastating floods in some years and disastrous droughts in others. Even when the flooding is neither too much nor too little, it comes too late for the winter crops and too early for the summer ones. Thus the residents of the area from time immemorial have tried to harness the rivers by diverting floodwaters to low-lying areas upstream (e.g., into the Tharthar depression) to avoid flooding downstream.

Because in this area the Tigris and Euphrates are higher than the surrounding countryside, canals and channels crisscross the plain, bringing irrigation water to the crops during the fall, winter, and spring months. Since the water is slow moving and the plain is flat, these canals and channels often silted full, and it was necessary to organize labor gangs to clean them out. To coerce residents to show up and actually do the work, a rather strong local government was needed, and many city-states sprang up. At various times one of these city-states would gain supremacy over the others, and during that period a region was often named after its ruling city. Thus the geographical, political, and chronological terminology has often been derived from the names of cities (e.g., Sumer, Akkad, Ur, and Babylon).

After repeated cleanings, the banks of the canals and channels became so high that it was difficult to clean out the waterways. New waterways, parallel to the old ones, were dug. Eventually, low-lying fields lacking proper drainage were enclosed by the canals and channels. Water that entered these fields for the most part evaporated, leaving behind mineral deposits. In addition, since the water table in the plain was high, water and salts percolated out of the ground to the surface of the fields and added to the problem of salinization of the soil. As early as the third and second millennia BC good agricultural land was lost because of this. First wheat could no longer be grown in the salty soil; then barley, which can tolerate more salt, could no longer be

grown; and eventually, even date palms stopped producing fruit. When the land could no longer support barley or date palms, the people would move on to more fertile fields to raise their crops, leaving behind the useless land. Thus, the plain between the Tigris and Euphrates, rather than being a uniformly lush region, is an area where fertile irrigated fields alternate with infertile wastelands.

The southern part of the plain, where the Tigris and the Euphrates come close together, turns into marshland and eventually into swamp. In ancient times the Tigris and Euphrates entered the Persian Gulf separately; today they join at Qurna and form the Shatt al-Arab, which flows into the gulf. The area is hot, humid, and mosquito-infested. The Ma'dan, the swamp Arabs, inhabit the area today, living in reed huts constructed on high ground or on floating bogs. Their reed huts, along with their reed canoes, are reminiscent of the life of ancient inhabitants of this area. The area is certainly not prime agricultural land, and the natives subsisted by hunting wild fowl and wild boars, by fishing, and by tending date palms. Today rice is also grown. Archaeological remains are scanty in this region.

Internally, Mesopotamia produced enough foodstuffs to feed its population, although at times grain had to be shipped from the

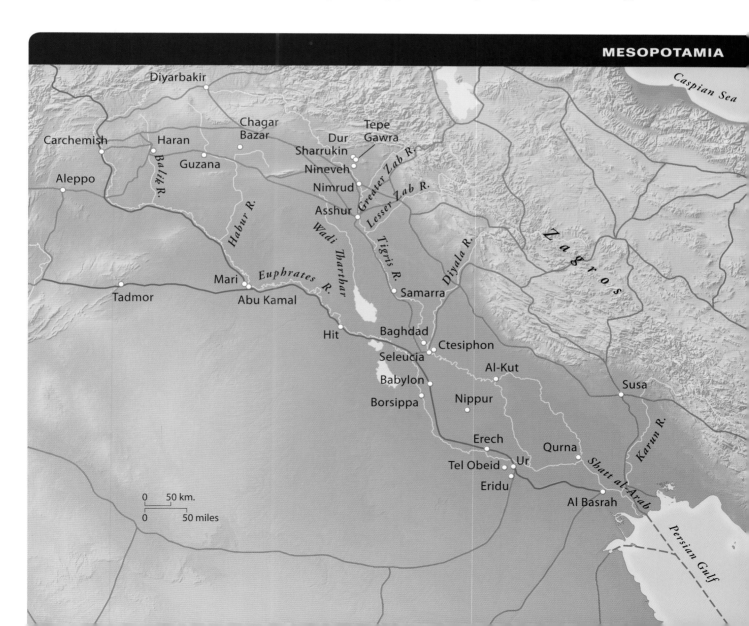

MESOPOTAMIA

southern plain to Assyria. Land transport was mainly by foot or donkey, although the camel was probably introduced sometime in the second millennium BC. Land transport in the southern plain was impeded by the need to cross the Tigris and Euphrates rivers, as well as numerous canals and channels. In addition, the plain was often covered by mud and/or inundated by floodwater during the winter and spring.

The main thoroughfares for travelers going from north to south were the rivers and canals. Bulk goods (such as timber and stone) were transported down the Tigris and Euphrates on rafts, although the Euphrates was a bit easier to navigate. These rafts were basically wooden platforms placed over inflated animal skins and could carry up to 35 tons of cargo. After the trip down the river was completed and the cargo had been delivered, the wooden frames were sold and the skins packed onto donkeys for the return trek northward. Since in antiquity bridges were almost unknown, people often made use of these rafts (today called kalaks) to cross the rivers and canals. Guffahs — large circular baskets covered with bitumen that could carry up to twenty people — were also used for this purpose.

Because Mesopotamia lacked many raw materials, it was necessary to import them from near and far. Although the sources varied from period to period, tin was imported from Iran, Afghanistan, and the Caucasus regions; silver from the Taurus Mountains; common tim-ber from the Zagros Mountains; prized cedar wood from the Lebanon and Amanus mountains; and copper from many areas to the north-east, northwest, and even from sources reached via the Persian Gulf. In addition, luxury items were imported from India (spices and cloth) and south Arabia (frankincense and myrrh), both by overland transport and by sea.

One of the main routes of international as well as local significance that passed through northern Mesopotamia began at Nineveh and ran westward via Shubat-Enlil (= Chagar Bazar?), Guzana, and Haran to Carchemish (see also map p. 72). From Carchemish, roads led northwest into Anatolia, westward to the Mediterranean, and southward toward Damascus, Israel, and Egypt. Also at Nineveh routes led northwest, via Diyarbakir (in Turkey), directly into Anatolia; eastward through difficult passes in the Zagros into Persia; and of course southward, via the Tigris, to the delta region.

In addition, an important route led northwest along the Euphrates, from Sippar in the delta toward Mari. From Mari it continued along the Euphrates to Carchemish, while another branch headed westward across the steppe/desert to Tadmor. From the delta, routes led east and southeast through the Zagros to the Persian plateau and other points eastward. In the first millennium BC, routes also led southwest through the Arabian desert toward sources of frankincense and myrrh in southern Arabia.

▼ *The Euphrates at Dura Europos*

During peaceful times, caravans would ply these and other routes, while in less stable eras the great armies of Assyria, Babylonia, Persia, etc., set out on campaigns of conquest and devastation along these same paths.

Although remains of human occupation in Mesopotamia date back at least to the Neolithic period (ca. 8000 to 4000 BC), Mesopotamia, like Egypt, entered the light of history at the beginning of the Early Bronze Age (ca. 3150 BC).

The civilizations that sprang up along the two great rivers were constantly pressured by tribes located in the mountains to the north, in the Zagros to the east, and in Elam to the southeast, as well as by Bedouin marauders coming from the Syro-Arabian Desert. However, in many periods the Mesopotamian rulers were able to hold these enemies at bay, sometimes even gaining territory at their expense. Of particular note for biblical studies are the westward expansions of the Assyrians, Neo-Babylonians, and Persians during the first millennium BC.

Since life in Mesopotamia can be characterized as a "mud culture" — crops were grown in mud; houses, palaces, temples, ziggurats, etc., were built of mud — it was only natural that mud or clay should often be used as a medium for communication. By 3100 BC cuneiform (wedge-shaped) writing had developed in Mesopotamia. This script was used to write a number of different languages (including Sumerian, Akkadian, Assyrian, Hittite, Babylonian, Persian, Ugaritic, and Eblaite) in a way similar to the use of Latin characters in writing English, German, and French. Cuneiform, which at first was composed of pictographs, quickly developed into stylized wedges, impressed on clay tablets by reeds. Many of the languages that adopted the cuneiform script used syllabic rather than alphabetic characters, along with determinatives — signs placed before and after words to indicate the class in which the word belonged (e.g., wooden objects, cities, mountains, etc.). Eventually, over 500 signs were used.

ANNUAL CYCLE IN SOUTHERN MESOPOTAMIA

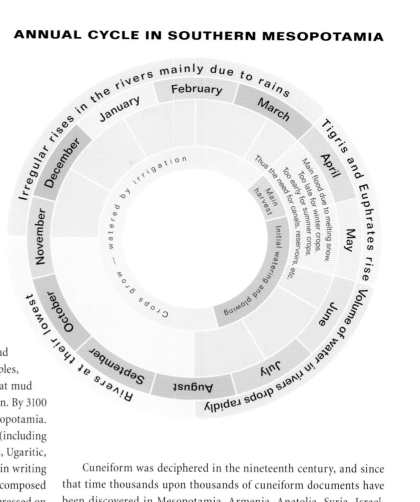

Cuneiform was deciphered in the nineteenth century, and since that time thousands upon thousands of cuneiform documents have been discovered in Mesopotamia, Armenia, Anatolia, Syria, Israel, and even in Egypt. This script was eventually discontinued in the first century AD, after having been in use for over 3,000 years. From these millennia a rich and varied literature has been preserved, and relevant texts will be discussed in the Historical Section.

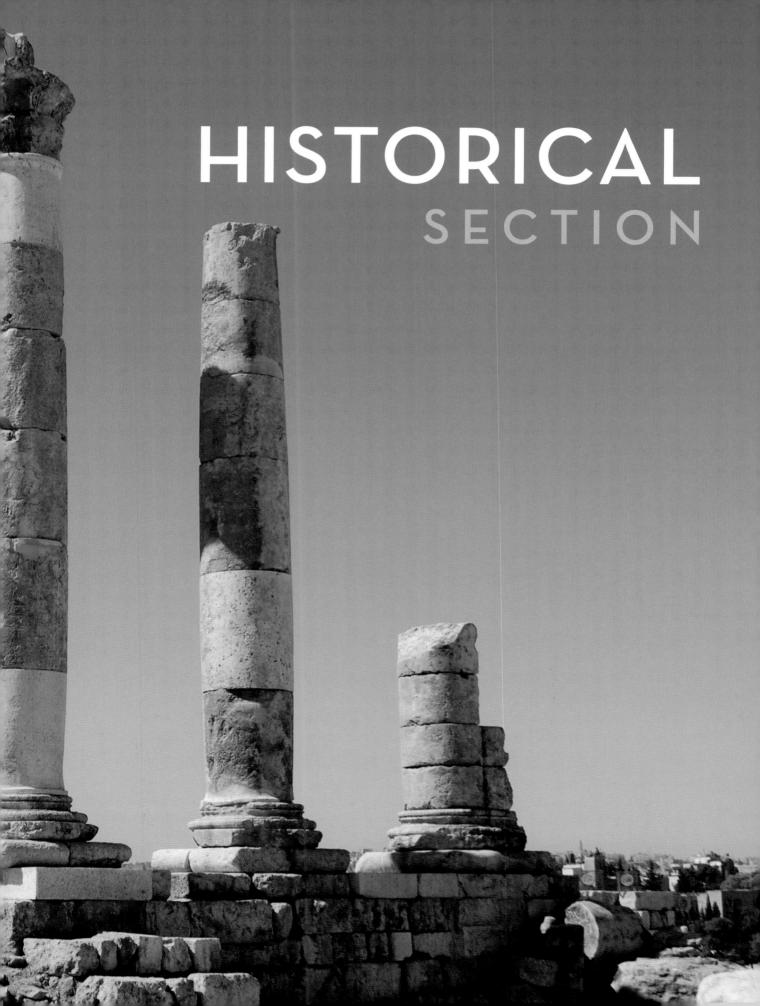

HISTORICAL
SECTION

THE PRE-PATRIARCHAL PERIOD

Garden of Eden (Genesis 1 – 3)

The book of Genesis, using very terse yet picturesque language, recounts the history of the universe and the world from the time of creation until the call of Abram/Abraham and Jacob's descent into Egypt. The second section of the creation account (Gen 2:4 – 25) focuses on the creation of the first humans. The Bible places these sinless people in a perfect environment called the garden of Eden.

The name Eden, if derived from Hebrew, means "delight" and is rendered "paradise" in the Septuagint (the Greek translation of the Old Testament, abbreviated LXX). However, some scholars have suggested that Eden is related to the Sumerian/Akkadian word *edin(u)*, meaning "plain," and that this is a description of the location of Eden. It is, of course, possible that the author may have had both of these ideas in mind when using the term. In any case, it seems that Eden was the name of a locality, and the special garden mentioned was

planted in the eastern portion of it (Gen 2:8). It is described as containing trees and an abundance of water. We can infer that it was in a warm climate — note the mention of fig trees and the lack of clothing (3:7). The major river that watered Eden is said to have separated into four headwaters, called Pishon, Gihon, Tigris, and Euphrates (2:10 – 14). The latter two rivers are well known, but the identities of the first two are not, and their identifications usually depend on where one places the garden.

The first of two major views places Eden in eastern Turkey, in Armenia, near the headwaters of the Tigris and the Euphrates. If this is correct, then the Pishon and the Gihon might be identified with other rivers in the area, such as the Araxes and the Murat, or they may now be nonexistent. This view has the advantage of locating Eden at a high altitude, from which the river(s) flowed down, and of uniting the symbolism of Eden (the place of the beginning of human history) with the Ararat region (the place where the second beginning, with Noah and his descendants, took place). However, this region does not match the warm conditions implied in the biblical account.

The second view places Eden in southern Mesopotamia (modern Iraq). This view has the advantage of locating the garden in a warmer climate in the vicinity of the two identifiable rivers, the Tigris and the Euphrates. The Pishon and the Gihon could then have been tributaries of the Tigris and/or Euphrates, though the text seems to indicate that one river was divided into four branches — reminiscent of a delta region — rather than four tributaries coming together to form one river. Alternatively, the Pishon and Gihon could have been canals that branched off of the Tigris and the Euphrates, but their paths through Havilah and Cush seem to imply considerable length.

The identification of Havilah and Cush is also difficult. Havilah (Gen 2:11 – 12) may have been a country/region located in Sinai and/or Arabia (25:18), a locality in which there were supplies of gold, aromatic resins (RSV "bdellium"; Heb. *b^edolaḥ*), and onyx. If this was the case, it is interesting to note that the Israelites wandered in this barren area for forty years, observing firsthand the disastrous effects of the curse and receiving manna from heaven — which looked like resin (Heb. *b^edolaḥ*; Num 11:7). The "land of Cush" usually refers to the land south of Egypt but north of Ethiopia, although it might possibly also refer to a portion of Arabia or, according to some, a territory in the mountains east of the Tigris River. Thus, even though the exact location of Eden continues to elude modern interpreters, its theological and spiritual significance has certainly been appreciated by both ancients and moderns alike.

The Table of Nations (Genesis 10)

The biblical text describes the progress of sin in the world after the fall of the first humans and their expulsion from the garden of Eden. The sinfulness of humankind was judged climactically in the flood (Gen 6 – 9), after which a new beginning for humanity was initiated through the line of the only righteous man, Noah. Genesis 10 describes how the nations of the then-known world were derived from Noah's three sons — Japheth, Ham, and Shem. Their seventy descendants mentioned in this list are considered to have been the ancestral heads (= eponyms) of the clans and nations that bore their names (v. 32). The list is divided into three major sections, beginning with those peoples farthest removed from Israel's horizon and moving toward her nearest neighbors.

First, the fourteen descendants of Japheth are listed (vv. 2 – 5); special attention is given to the "sons" of Javan, with whom the Israelites came

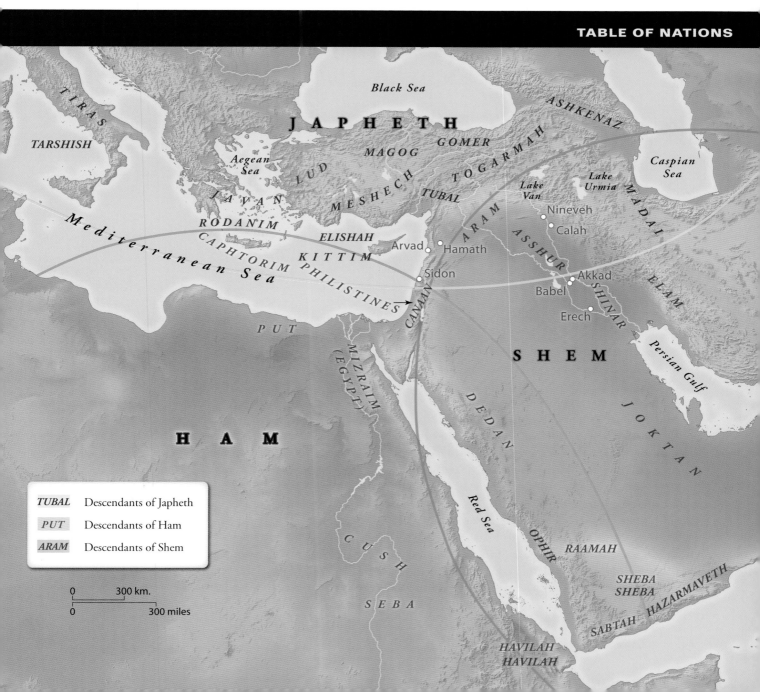

TUBAL	Descendants of Japheth
PUT	Descendants of Ham
ARAM	Descendants of Shem

0   300 km.

0   300 miles

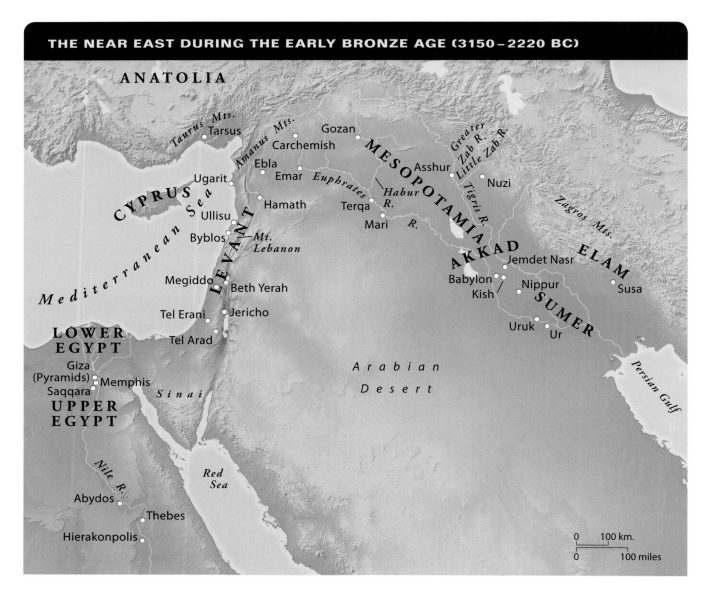

most often into contact. Although not all of the peoples/nations can be certainly identified, commonly accepted identifications include Gomer = Cimmerians, Madai = Medes, Javan = Ionians, Ashkenaz = Scythians, Elishah = Alashiya/Cyprus, and Rodanim = Rhodes. The general area where these peoples lived was to the west and north of the land of Israel.

The second portion of Genesis 10 (vv. 6–20) lists the thirty-one descendants of Ham. Generally accepted identifications for the descendants of Ham include Cush = Ethiopia/Nubia, Mizraim = Egypt, Put = Libya, Canaan = Canaan, Sheba = Saba (in southern Arabia), Dedad = Dedan (the el-Ula oasis in Arabia), Caphtorites = Cretans, Arkites = Arqad (in Lebanon), Arvadites = Arvad, Hamathites = city of Hamath, in addition to peoples well known from the Old Testament: Philistines, Sidonites, Hittites, Jebusites, Amorites, Girgashites, Hivites, and others. In general, the descendants of Ham settled in and around the land of Israel but also to the southeast and in Africa. The list gives special attention to the descendants of Canaan (vv. 15–19), with whom the Israelites came into very close contact, and to the cities associated with the warrior Nimrod (vv. 8–12). The first four of the cities mentioned in

connection with him were important political/cultural centers in Shinar, in southern Mesopotamia, while the latter four (v. 11) were located along the upper Tigris River and are mentioned in connection with Nimrod's northern migration.

The third and final section of the list contains the twenty-six descendants of Shem (vv. 22–31). As is typical in Genesis, this, the most important line, is dealt with last — Abram/Abraham being a descendant of Shem. The people mentioned in this list mainly settled to the northeast and southeast of Israel. Commonly accepted identifications include Elam = Elam, Asshur = Asshur/Assyria, Arphaxad = Chaldeans/Kasdim(?), and Aram = Arameans. The list gives special attention to the descendants of Joktan, people who evidently were the eponymous heads of various Arabian tribes.

Although modern linguists sometimes speak of Hamitic and Semitic languages — making a seemingly vague reference to this text — these language-classification schemes are modern in origin, and it is not necessary to try to explain the list in Genesis 10 in terms of related languages. Because some of the peoples mentioned in the list appear for the first time in extrabiblical documents of the first millennium BC, some scholars have concluded that the list must date from that period. However, it is possible that these "late-appearing" peoples were known as tribes that existed in the second half of the second millennium BC but happened not to be noted in extrabiblical documents until later periods. Simply put, the purpose of the list in Genesis 10 is to describe how the descendants of Noah were fruitful and filled the earth (9:1; cf. 1:22, 28).

Mesopotamia — Early Bronze Age

It was during the Early Bronze Age (ca. 3150 – 2200 BC) that some of the countries of the Near East entered into the light of history. In Mesopotamia, the beginning of this era, called the Jemdet Nasr Period (3100 – 2900 BC), forms the transition from the Protoliterate to the Early Dynastic Period. During this period in Mesopotamia writing began, and large cities, containing examples of monumental architecture (such as temples, palaces, and fortifications), were first built on the plains near the Tigris and Euphrates rivers. This urban revolution was hastened by the use of newly developed copper tools and weapons, while the interaction of the people of the cities with those of the surrounding countrysides led to an increase in the population. This was the era, according to the Sumerian King List, during which the long-lived members of the preflood dynasties ruled.

The urban revolution continued during the Early Dynastic Period (2900 – 2300 BC). The leadership of the country passed between great cities such as Kish, Uruk, and Ur. It seems that there was no strong central government controlling the whole area; rather, individual city-states ruled over limited territories, with first one and then another gaining dominance. These city-states seem to have had a unified culture, religion, and language, known today as Sumerian. The religious center of this civilization was the city of Nippur. It is to this period that much of the Akkadian epic literature owes its origin, although the now-extant copies of this literature come from later times.

The Early Dynastic Period came to an end when a Semite — Sargon of Akkad, from the area north of Sumer — took control of the old city-states. Sargon had already conquered territory to the northeast and the west; according to the texts, he eventually pushed into Anatolia to Purush-khanda. Thus it was Sargon who established the first empire in the Mesopotamian region. His grandson, Naram-Sin, seems to have been especially active in the west, campaigning as far as the Amanus, Taurus, and Lebanon mountains, even reaching the Mediterranean coast near Ullisu. During this period of great prosperity the arts flourished, literature developed, and many foreign contacts were made. The Sargonic Empire (2300 – 2100 BC) came to an end because of internal and external pressures, but the idea of a large, unified empire became a recurring concept that continued to express itself at various periods in Mesopotamian history.

Egypt during the Early Bronze Age

At the end of the fourth millennium BC, Egypt too was moving out of the prehistoric period into the full light of history. During the Early Bronze Age there were two major eras in Egyptian history: the Early Dynastic Period and the Old Kingdom. Just prior to, or at the beginning of, the Early Dynastic Period (1st and 2nd Dynasties; 3050 – 2700 BC), Lower Egypt (the delta region) and Upper Egypt (the Nile Valley south of Cairo) were united into a single state. Memphis was established as the capital and remained as such throughout the Old Kingdom Period (2691 – 2176 BC).

▼ *Giza, Egypt: Pyramids built by rulers of the Fourth Dynasty (ca. 2600 BC), hundreds of years before Abraham*

Mark Connally

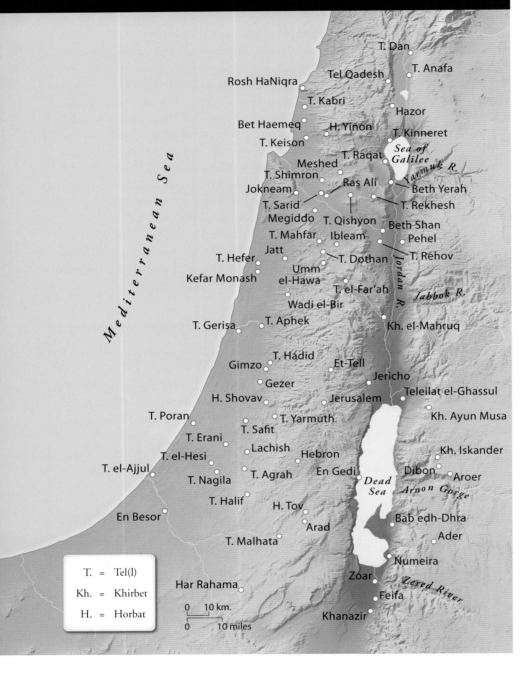

T. Dan
Tel Qadesh
T. Anafa
Rosh HaNiqra
T. Kabri
Hazor
Bet Haemeq
H. Yinon
T. Kinneret
T. Keison
T. Raqat
Sea of Galilee
Meshed
Yarmuk R.
T. Shimron
Ras Ali
Jokneam
Beth Yerah
T. Sarid
T. Rekhesh
Megiddo
T. Qishyon
Beth Shan
T. Mahfar
Ibleam
Pehel
Jatt
Jordan R.
T. Hefer
T. Dothan
T. Rehov
Umm el-Hawa
T. el-Far'ah
Jabbok R.
Kefar Monash
Wadi el-Bir
T. Gerisa
T. Aphek
Kh. el-Mahruq
T. Hadid
Gimzo
Et-Tell
Gezer
Jericho
H. Shovav
Jerusalem
Teleilat el-Ghassul
T. Poran
T. Yarmuth
Kh. Ayun Musa
T. Safit
T. Erani
Lachish
Kh. Iskander
T. el-Hesi
Hebron
Dibon
T. el-Ajjul
En Gedi
Aroer
T. Nagila
T. Agrah
Dead Sea
Arnon Gorge
T. Halif
H. Tov
En Besor
Arad
Bab edh-Dhra
T. Malhata
Ader
Numeira
Har Rahama
Zoar
Zered River
Feifa
Khanazir

Mediterranean Sea

T. = Tel(l)
Kh. = Khirbet
H. = Horbat

0   10 km.
0   10 miles

The Third Dynasty through the Sixth Dynasty is collectively known as the Old Kingdom (2691 – 2176 BC). It was during this period that many aspects of Egyptian life reached their classical form: the traditional southern boundary at the first cataract was established, the symbols of hieroglyphic writing were stabilized, traditional administrative structures were put into place, and artistic compositions assumed their stylized forms. By the end of the period, the traditional twenty-two nomes (districts) of Upper Egypt were well established, although the twenty-nome structure of the Delta was not finalized until later.

The Old Kingdom is known as the age of the pyramid builders, the first of whom was Djoser. This king from the Third Dynasty (2692 – 2617 BC) is known not only for centralizing the government in Memphis but especially for his stepped pyramid, which he constructed at Saqqara, two miles west of Memphis. Thirty-four pyramids, out of a total of forty-seven, were built during the Old Kingdom period. The pyramids are thought to symbolize the power of the king as well as to emphasize his association with the sun god Re. During this period the kings themselves were considered to be gods, and the pyramids served not only as burial places but also as religious centers where the cult of the deceased god-king was practiced. The Fourth, Fifth, and Sixth Dynasties are considered to have been the high point of the Old Kingdom, and the pyramids of three of the kings of the Fourth Dynasty — those of Cheops, Chephren, and Mycerinus, located at Giza — are the most famous of the pyramids of Egypt.

Although the Old Kingdom is well known because of the pyramids, not many documents have been preserved from which a history of the period can be reconstructed. It is known that there were extensive contacts with Nubia to the south, from which luxury items such as gold, ivory, and ebony were obtained. To the north, Egyptian gold work has been found in Turkey, while stone vases of Chephren and Pepi I have been found at Tell Mardikh (Ebla), and additional Egyptian artifacts at Byblos. These data, along with the tomb inscription of Uni (see below), indicate important Egyptian contacts with the Levant (= the whole eastern shore of the Mediterranean Sea from Turkey in the north to Sinai in the south).

The first king of the First Dynasty (3050 – 2900 BC) was Narmer. The pallet of Narmer shows him wearing the crowns of both Upper and Lower Egypt, indicating his authority over both regions. It seems likely that Egypt had some contacts with Palestine during his reign, because pottery inscribed with Narmer's name/symbol has been found at Tel Erani, Tel Arad, Tel Maahaz, Tel Malhata, En Besor, Lod, and Tel Halif. It also seems that Egypt had contacts with Mesopotamia, since cylinder seals of the Protoliterate Period have been found in Egyptian cemeteries and Mesopotamian-style niched facades were used on Egyptian tombs. Artistic motifs (such as the intertwined necks of animals) and the early use of the pictographic writing system also seem to indicate Mesopotamian influence upon Egypt.

The stability of the country began to crumble during the rule of the fifteen kings of the Seventh and Eighth Dynasties. During this time the power of the nomarchs and other officials was growing at the expense of the king. The collapse of the Old Kingdom was probably due to internal factors, such as the economic burdens brought about by the high maintenance costs of the pyramid cults, by weak kings, and possibly by a series of below-normal inundations of the Nile that led to famine. These factors, along with the incursions of Asiatic peoples into the delta region, seem to have led the country into the turbulent days of the First Intermediate Period.

The Southern Levant during the Early Bronze Age

Although human settlement in the southern Levant has had a long history, it was not until the country passed from the Chalcolithic Period (fourth millennium BC) to the Early Bronze Age (3150 – 2200 BC) that the transition from small, scattered settlements into large urban centers took place. In a report on Early Bronze sites in modern Israel (those located north of an imaginary line running from Beersheba to Arad), it has been noted that there are a total of 260 sites, of which 47 are over 10 acres in size. It has been estimated that the average urban population during this period could have been in the neighborhood of 150,000. Not until 3,000 years later, during the Roman and Byzantine periods (63 BC – AD 640), would the number of large settlements in Israel again approach the magnitude of those found in this early period. Evidently the population group was Semitic, for in later written documents the names of many of the towns that they founded are clearly Semitic (e.g., Megiddo, Beth-Yerah, Jericho, Aphek, etc.).

Most of the large urban centers — Megiddo, Beth-Yerah, Arad, et-Tell, Tel Yarmuth, etc. — were protected by strong city walls, often massive in size (15 to 25 ft. thick). Some of the walls had protruding semicircular towers built into them to aid in the defense of the city (e.g., Arad, et-Tell). At Arad it is estimated that forty such towers were placed at intervals along the 3,840-foot wall.

Typical houses from this period were rectangular in shape with the entry placed close to the center of one of the long walls — hence the name "broad house." From street level, one would usually descend a few steps into a house. Archaeologists have discovered that the door socket, on which the door swung, was normally placed on the left side of the entry. Low benches lined the walls of the houses, and often small mortars used for the grinding of grain were found in or on the floors. On occasion one or more stone slabs, located along the long central axis of the house, have been found; these were evidently used to

▲ *Megiddo: Sacrificial altar (25 feet in diameter) from the Early Bronze Age had gone out of use by the time Abraham entered Canaan*

support pillars, which in turn supported the roof of the building. The clay model of an Early Bronze house, found at Arad, indicates that the buildings were flat-roofed and windowless.

At Megiddo, et-Tell, Bab edh-Dhra, Arad, and elsewhere, large public buildings, all in the shape of "broad houses," have been found, which appear to have served as temples. At Megiddo four such buildings were discovered in the sacred area. Three of them, similar in plan and measurements, were built in close proximity to the large (25 ft. in diameter) circular altar.

Alongside the large urban centers one would expect correspondingly large cemeteries, but this is not usually the case. However, large burial grounds have been found at Bab edh-Dhra and Feifa, Early Bronze cities located at the southeastern end of the Dead Sea. At Bab edh-Dhra, it is estimated that some 20,000 tombs contain the remains of 500,000 people — and the cemetery at Feifa is almost as large! Since the number of burials seems to be larger than the number of people that could have lived in these cities, and since some of these are secondary burials (the bones were brought from elsewhere and reburied here), it

▼ *Arad: Outline of a portion of the 25-acre Early Bronze Age (2800 BC) found at Arad. Note city wall, semicircular towers, streets, and buildings*

▲ *Arad: Foundations of a ten-foot-thick city wall and semicircular tower from the Early Bronze Age*

the reign of Pepi I (Sixth Dynasty; *ANET*, 227–28). Although the "land of the Sand-Dwellers" in other contexts could refer to Sinai, it seems probable in this case that Canaan, to the northeast of Sinai, is indicated, since (fortified) enclosures, fig trees, and vines are mentioned in connection with the land. Indeed, the reference in the text to a prominent mountain height situated close to the sea and called the "Antelope-Nose" could refer to the promontory of Mount Carmel that juts out into the Mediterranean.

Another source of historical data that could pertain to Canaan during this period is the Ebla archives, which were discovered in the late 1970s. Early reports indicated that Canaanite cities such as Hazor, Megiddo, Gaza, Urusalima, Salem, Lachish, Dor, Ashtaroth, and Joppa, as well as Sodom, Gomorrah, Admah, Zeboiim, and Bela were mentioned in the texts. Since then, the claim that the last three names appear in the texts has been withdrawn, and the mention of some of the other geographical names has been called into question. However, if the names of any Canaanite places do in fact appear in the texts, their appearance will shed welcome light on the history of this otherwise relatively dim period.

appears that these cemeteries may have been used as common burial sites for many other Early Bronze urban centers.

In contrast to Egypt, Ebla, and Mesopotamia, no archives dating to the Early Bronze Age have yet been found in the southern Levant. As far as the written record goes, we are thus limited to such historical references as are found in the literature of the neighboring countries. At Abydos in Egypt, the tomb inscription of Uni describes how he led five campaigns to the "land of the Sand-Dwellers" during

▼ *Jericho: Neolithic tower 25 feet high discovered by Kathleen Kenyon*

SYRIA/MESOPOTAMIA

UR III		ISIN LARSA		OLD BABYLONIAN	

Abraham

Entrance into Canaan

Jacob to Haran

Jacob to Canaan

Hammurabi destroys Mari

| 2100 BC | 2000 BC | 1900 BC | 1800 BC | 1700 BC | 1600 BC |

CANAAN

MIDDLE BRONZE I		MIDDLE BRONZE II A		MIDDLE BRONZE II B	

Abraham

Isaac

Jacob

Jacob and the family to Egypt

Expulsion of the Hyksos

| 2100 BC | 2000 BC | 1900 BC | 1800 BC | 1700 BC | 1600 BC |

EGYPT

1ST INTERMEDIATE PERIOD		MIDDLE KINGDOM PERIOD		2ND INTERMEDIATE PERIOD	
Dynasties VII-XI		Dynasties XI-XII		XIII Dynasty (Memphis)	

XI Dynasty (Thebes)

XII Dynasty (Lisht)

XIV Dynasty (Xois)

Sinuhe to Canaan

Asiatics to Egypt

XV-XVI (Avaris)

XVII (Thebes)

| 2100 BC | 2000 BC | 1900 BC | 1800 BC | 1700 BC | 1600 BC |

THE PATRIARCHS AND THE EGYPTIAN SOJOURN

The end of the third millennium BC (ca. 2200 – 2000 BC) marks the beginning of the era of the biblical patriarchs: Abraham, Isaac, Jacob, and Joseph. Genesis 12 – 50 records Abram's emigration from Ur of the Chaldeans, in southern Mesopotamia, to Canaan (ca. 2091 BC) and the events surrounding the lives of Abraham, Isaac, and Jacob in and around the land of Canaan. This section of Scripture concludes by describing the circumstances surrounding the migration of Jacob and his family to Egypt about 1876 BC, 215 years after Abram had entered Canaan, which marks the beginning of the period of the sojourn of the Israelites in Egypt.

The internal chronological markers found in Exodus and Genesis indicate that Abram was born in 2166 BC. Abram's time was an era of relative peace and prosperity in southern Mesopotamia, during

which his home city of Ur controlled most of the other city-states in the region. This era, known as the Ur III Period (ca. 2130 – 2022 BC), is well known from the thousands of cuneiform documents that have been found in Ur, Lagash, Umma, Drehem, Nippur, and other cities. During this period many of the old Sumerian epics and myths, which had originated in the Early Dynastic Period (see above), were put into their final form. These documents show that a resurgent Sumerian

▲ *Et-Tell: Looking east at the ruins of the Early Bronze Age site of Patriarchal Ai. Possibly Abraham pitched his tent near where this photo was taken (Gen 12:8).*

Possible routes of Abram's journey

0 ___ 100 km.
0 ___ 100 miles

Although the dating of the patriarchs is somewhat problematic, the Bible presents some rather straightforward "relative dates" (e.g., Abram was 75 years old when he departed from Haran, Gen 12:4), from which some "absolute dates" (specific years, such as 2091 BC) can be derived. The keystone in the attempt to convert "relative" to "absolute" dates is the passage found in 1 Kings 6:1:

> In the four hundred and eightieth year after the Israelites had come out of Egypt, in the fourth year of Solomon's reign over Israel, in the month of Ziv, the second month, he began to build the temple of the LORD.

It is generally agreed that Solomon began his reign in 970 BC (an "absolute date"), which means that his fourth year (a "relative date") would have been 966 BC. Since this, according to the text, was 480 years after the exodus, the exodus must have taken place ca. 1446 BC. The biblical text indicates that the exodus took place 430 years after Jacob and his family entered Egypt (Ex 12:40), which implies that this descent into Egypt occurred about 1876 BC. From data given in Genesis (12:14; 21:5; 23:1; 25:7, 26; 35:28–29; 47:9) one can work back to the date of the birth of Abram (2166 BC) and determine that the probable date of his entrance into Canaan was 2091 BC. These and other "absolute dates" derived from the internal biblical chronology will be used in this atlas, and the dates so generated will—where possible—be correlated with extrabiblical data in an attempt to help define the geographical, archaeological, and cultural environment in which biblical events occurred.

culture was again flourishing in the region, albeit for the final time. Thousands of economic, legal, and judicial texts witness to the complex and pervasive roles of the palace and the temple in the everyday life of the people. The rich archaeological finds from Ur indicate that it was a thriving commercial center during this entire period. It was from such a culturally advanced city that Abram began his earthly pilgrimage of faith (Gen 11:31; Acts 7:4).

When Abram was about seventy years old (ca. 2096 BC), he, along with his wife, Sarai; his nephew Lot; his father, Terah; and a group of household retainers; left Ur of the Chaldees for the "land of Canaan" (Gen 11:31). There were two major routes that Abram might have taken on his journey from Ur to Haran. One headed north, through what would later become Assyria, before it turned westward through the Jezirah to Haran (a total distance of ca. 670 mi.; see p. 78). The other route proceeded northwest, alongside the Euphrates River, to the city of Mari, and from there continued to Haran via the Euphrates and Balik rivers (a total distance of ca. 630 mi.). It is estimated that the trip would have taken at least thirty-one actual days of travel at the pace of 20 miles per day, and this figure does not include needed days of rest.

Haran ("roadway"), also known from extrabiblical texts, was an important caravan city around 2000 BC. It is not known how long Abram stayed in Haran, but it appears that he settled there for at least a year, for his father Terah died there. The roots of Abraham in the Haran/Aram area were such that the Israelites would later refer to their ancestor as a "wandering Aramean" (Deut 26:5).

At age seventy-five (ca. 2091 BC; Gen 12:4), Abram departed from Haran, setting out for "the land of Canaan" (12:5). Abram's route probably took him south through Damascus into Transjordan, to the area of Ramoth Gilead. From there he would have descended into the Jordan Valley, possibly via the Jabbok River to Succoth, where he would have crossed the Jordan River into Canaan. He probably entered the Hill Country of Manasseh via the relatively broad and gentle Wadi Farah, which took him up into the mountains just a few miles north of Shechem. All totaled, this trip covered about 400 mi. and took at least twenty actual days of travel. This basic route to and from Haran is of some importance, for it was probably later used by Abraham's servant as he went to secure a bride for Isaac (Gen 24), by Jacob as he fled from his brother Esau to his uncle Laban who lived in Paddan Aram (Gen 27–29), and again by Jacob on his return to Canaan (see esp. Gen 31:19–33:20).

Shechem, Abram's first stop in the land of Canaan, was located near the watershed between the twin mountains of Ebal and Gerizim. Roads from all directions feed into this high, plateau-like area, and from early times Shechem served as the major urban center in the Hill Country of Ephraim and Manasseh. Although Shechem has been excavated, it is not necessary to try to find any particular archaeological remains from this period (ca. 2091 BC), for the biblical evidence does not demand that an actual city was located there in Abram's day. It was at Shechem, at the site of the "great tree of Moreh," that the Lord appeared to Abram and promised, "To your offspring I will give this land." In response to this promise Abram built an altar and worshiped God there (Gen 12:7).

From Shechem Abram traveled south, through the Hill Country of Ephraim, to a mountain situated east of Bethel and west of Ai. There he pitched his tent and constructed another altar (Gen 12:8). The topographic details given in this passage fit very well with the usual identification of Bethel with modern Beitin and Ai with et-Tell. At the time of Abram's arrival, the large Early

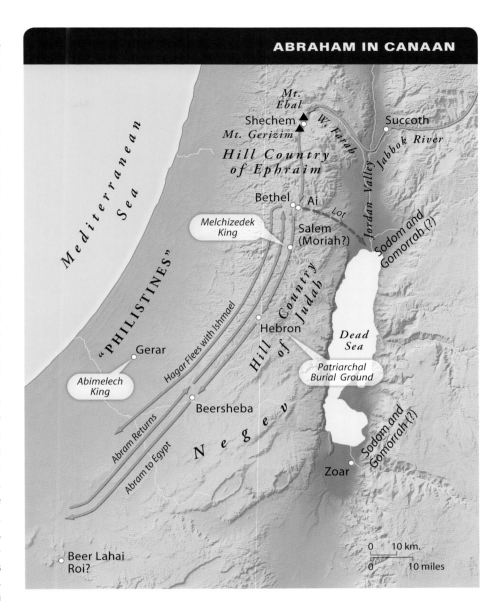

ABRAHAM IN CANAAN

▼ *Negev: Reconstructed campsite from the MB I Period. Probably a circular tent was erected and supported by the partially standing center pole.*

▲ Golan: A dolmen marking a burial site from the Middle Bronze I Period (ca. 2200 – 2000 BC); the era when Abram entered Canaan

Bronze Age city of Ai, which covered 27 acres, had been in ruins for hundreds of years. Abram continued his journey southward through the Hill Country of Judah to the Negev (Gen 12:9). The route he traveled from Shechem to Bethel/Ai, to Hebron, and on to Beersheba/Negev, was subsequently used frequently throughout sacred history. This route is often called the "Ridge Route" (see p. 33) but could also be called the "Route of the Patriarchs" during this period.

The country in which Abram now lived was no longer characterized by the large Early Bronze urban centers, for these had collapsed by 2200 BC, if not earlier. The reason for this collapse is not completely clear, but possibilities include Egyptian invasions (Pepi I?), internal conflicts, attacks by the MB I peoples, or, more probably, climatic changes — or a combination of any of the above.

In any case, the characteristics of the MB I Period (2200 – 2000 BC), during which Abraham lived, are well known. Instead of living in large fortified cities, people of the MB I Period lived in tents (tents are mentioned twenty-two times in the patriarchal narratives) and huts. In the Negev, where numerous settlements have been found and excavated, the typical settlement consisted of a cluster of small, flimsy, circular or rectangular installations grouped around a central courtyard. The largest of these settlements are less than 3 acres in size and unwalled. In addition, numerous intermediate and single-unit settlements have been found scattered over the Negev Highlands, the Uvdah Valley, and southern Transjordan.

Although remnants of this civilization have been found at tells in the northern parts of the country (e.g., Bethel, Tell Beit Mirsim, Jericho, Megiddo), no fortifications and no public buildings have been discovered in Palestine, perhaps indicating that these settlements were poor and temporary. Evidently the Hill Country of Judah was not heavily settled during the MB I period although numerous burials have been found there.

The burial customs of the MB I people were unique. Their tombs were usually shaft tombs, oftentimes hewn into the hard limestone hills of Palestine. A vertical shaft led to one or more chambers, but in contrast to the EB practice, one burial per chamber was the rule. Tens, or even hundreds, of these tombs are often found together (e.g., near what are called today Ein Samiya, Elazar, Efrat, Jebel Qa'aqir). In the Golan region, Transjordan, and elsewhere, fields containing hundreds of dolmens have been found. These low, table-like structures were built out of three or four large rocks and were sometimes used to mark shallow grave sites. In other instances, especially in the Negev, piles of stones called tumuli mark the burial sites.

Soon after Abram entered Canaan, the land experienced one of its occasional droughts, which led to insufficient supplies of food and fodder. Abram, knowing that Egypt normally had sufficient supplies of food, crossed the northern Sinai peninsula, possibly following the way of Shur, to Egypt, where he found sustenance for his family (Gen 12:10 – 20).

The Egypt that Abram entered had been experiencing political and social upheavals for the previous one hundred years. This turbulent era, known as the First Intermediate Period, followed the collapse of the stable Old Kingdom (Pyramid Age; see above, pp. 85 – 87). According to the description of this period found in the Admonitions of Ipuwer (ANET, 441 – 44), it was a time of great social unrest — the wealthy were consigned to perform menial tasks while the poor became their masters, the tombs of the kings were plundered, droughts (low inundations of the Nile) occurred, and death and destruction were everywhere.

Although monumental remains are few, literary works such as the Instruction for King Meri-ka-Re (ANET, 414 – 18) aid in understanding this era. This First Intermediate Period came to an end around 2025 BC, when Mentuhotpe II of the Eleventh Dynasty reunited the north with the south, ushering in the Middle Kingdom.

Abram's return to Canaan from Egypt marked the beginning of nearly two hundred years of patriarchal residence in the land. From the biblical account it appears that most of this time was spent in the region of the Negev, with occasional trips to the Hill Country of Judah and Ephraim. The patriarchs were occupied with raising sheep and goats and growing grain crops (see, e.g., Gen 13:2, 5–7; 24:35; 26:12). This meant that from season to season they moved from one pastureland to another, but in all probability they returned to previous camps each succeeding year. Wells, probably dug in and near the junctions of wadis, provided water for their families and flocks. The digging of these wells was difficult, and thus the control of these sources of life-giving water was often disputed. Both Abraham and Isaac had conflicts with the king of Gerar over the control of wells located between Gerar and Beersheba (Gen 21:25; 26:12–33).

The attempt of Abram and his clan to live in the Bethel/Ai region in the Hill Country of Ephraim (Gen 13) seems to have been somewhat problematic. This may have been due to the fact that there was not enough grazing land in the area for his large herds because of extensive forests, or possibly because the Canaanites and Perizzites were hostile to him (13:7). In any case, Lot, Abram's nephew, chose to live in one of the cities located in the Jordan Valley while Abram remained in the hill country and Negev regions. The text indicates that the Jordan Valley was intensively culti-

vated at that time, in all probability being irrigated by ditches that conducted water from the powerful springs to the fields in the area.

Although for a time Abram remained in the hill country, it does not seem that he or the other patriarchs spent a great deal of time in the region. Places like Shechem and Bethel/Ai were visited, altars built, Yahweh worshiped, and divine revelations received, and on one of the mountains in the "region of Moriah" — probably one of the mountains in the Jerusalem area — Isaac was "bound" to be sacrificed (Gen 22:9; cf. 2 Chron 3:1). Yet it appears that the major patriarchal encampment was in the (western) Negev.

The exception to this state of affairs was the patriarchal settlement close to Hebron in the hill country. This association with Hebron was to be expected, however, for throughout the ages Hebron was the contact point between the peoples of the Negev and the peoples of the hill country — the place where goods and services were bought and sold. At Hebron, Abraham purchased the cave of Machpelah (Gen 23), where ultimately Abraham and Sarah, Isaac and Rebekah, and Jacob and Leah were buried. Today, in the center of modern Hebron, the traditional

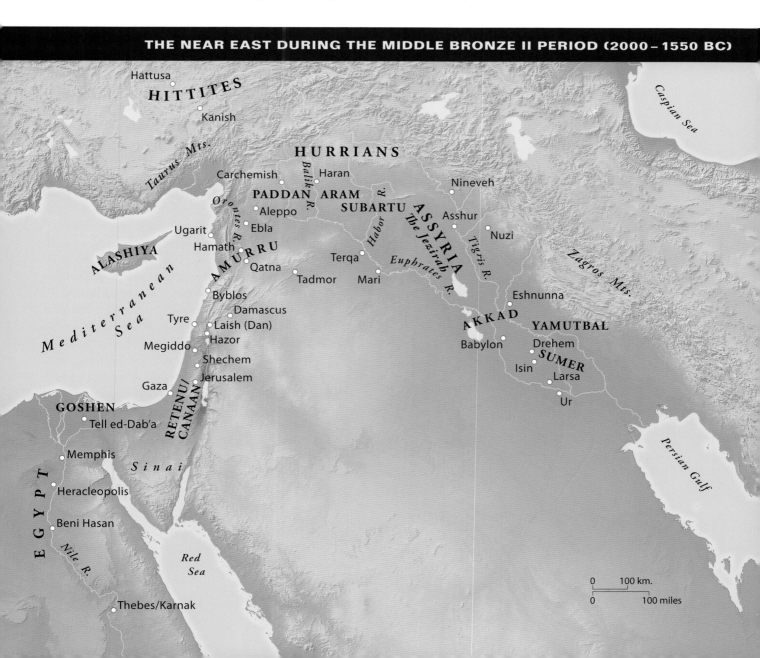

THE NEAR EAST DURING THE MIDDLE BRONZE II PERIOD (2000–1550 BC)

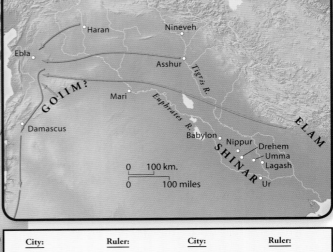

Legend:
- △ Selected Early Bronze Age cities (3150–2200 B.C.)
- △ Middle Bronze I encampments
- △ Middle Bronze I burial grounds
- → Invasion of Kedorlaomer and kings allied with him

City:	Ruler:	City:	Ruler:
Sodom	Bera	Goiim	Tidal
Gomorrah	Birsha	Ellasar	Arioch
Admah	Shinab	Shinar	Amraphel
Zeboiim	Shemeber	Elam	Kedorlaomer
Bela (Zoar)	-----------		
Salem	Melchizedek		

burial sites of the patriarchs are marked by a mosque and synagogue that are located in an old Crusader church that in turn was built within an enclosure wall that dates to the Herodian era (ca. 37–4 BC). Genesis 23:10 suggests that an actual city existed in Hebron during the days of Abraham, for it was in the city gate that he negotiated the purchase of the cave of Machpelah. Evidently the city of Abraham's time was located at Jebel er-Rumeidah, where Bronze Age remains have been found.

Although Lot's choice of living in the well-watered Jordan Valley seemed logical from a human standpoint, he, in fact, faced a number of difficulties and twice had to be saved from death through the intervention of his uncle Abraham. For example, Genesis 14 relates how four kings from the north invaded the area and made war on the five kings of the plain who had revolted against their rule. The kings of the north came from Shinar, Ellasar, Elam, and Goiim. Evidently their major objective was the conquest of the cities of Sodom, Gomorrah, Admah, Zeboiim, and Bela (i.e., Zoar).

The four invading kings headed south along the Transjordanian Highway, defeating the Rephaite and Zuzite tribes in Gilead, the Emites in Moab, and the Horites in Mount Seir (Gen 14:5–6). The kings evidently continued south to El Paran before swinging northwestward into the Negev Highlands to En Mishpat, where they defeated the Amalekites (vv. 6–7). Turning eastward, the four kings attacked and defeated the Amorites of Hazazon Tamar before initiating their major encounter with the five kings of the plain in the Valley of Siddim. This battle may have taken place in the salt badlands south of the

Dead Sea or possibly in the area now covered by the salt pans at the southern end of the Dead Sea. The defeated kings of the plain were taken captive and, along with the booty, were carried off northwards as the invading kings returned to their homelands. Abram, together with his allies, pursued them as far as Dan and Hobah, rescuing Lot and the kings with whom he had been captured. Upon Abram's return he met first with the king of Sodom in the "Valley of Shaveh (that is, the King's Valley)" and then with "Melchizedek king of Salem" (v. 18; Salem = Jerusalem [Ps 72:6]).

The second time Lot was delivered is recorded in Genesis 18 – 19, in the account of the destruction of Sodom and Gomorrah. These two cities, along with Zoar (the place to which Lot and his family fled to escape the destruction), were already mentioned in Genesis 14. According to the internal biblical chronology followed here, the destruction of Sodom and Gomorrah took place about 2067 BC (i.e., toward the end of MB I [2200 – 2000 BC]). To date, no MB I sites with which any of the five cities could be identified have been discovered near the south end of the Dead Sea. Some have proposed that the cities are now buried under the south end of the Dead Sea; while this is possible, the recently receding waters of the sea have not revealed evidence of any ancient cities having been located there.

However, recent surveys and excavations along the Transjordanian foothills east and southeast of the Dead Sea have located five sites that date to the EB period (3150 – 2200 BC): Bab edh-Dhra, Numeira, Zoar, Feifa, and Khanazir (see pp. 87 – 88). Some have wondered if these five sites could be the remains of the five cities mentioned in Genesis 14:2. Given the current absolute dates attached to the various archaeological periods, this identification is difficult to maintain, for the only period during which these sites were all settled was the EB III period (ca. 2650 – 2350 BC), at least three hundred years prior to the date of the events mentioned in the Bible. At this time, no sure identification of the five cities has been made.

Approximately one hundred years of Abraham's life were spent in the land of Canaan. At the time of his death (ca. 1991 BC) the country was moving from the "nomadic" MB I period into another "urban" age. The culture that was established at the beginning of the MB II period (2000 – 1550 BC) actually continued well into the Late Bronze Age (1550 – 1200 BC). During the first part of the MB II period (MB IIA, 2000 – 1750 BC), Isaac and Jacob were active in the land, until 1876 BC, when Jacob went down to Egypt. During the MB IIA period, new urban centers were built on the ruins of EB cities or over the temporary encampments of the peoples of the MB I period. Cities were now constructed with

monumental city walls, gates, palaces, and temples. Bronze replaced copper as the choice metal for making agricultural implements and weapons. New pottery forms were introduced, and evidence shows that contacts with Egypt now began to increase. Statues, scarabs, and other artifacts of Egyptian origin from the days of the Twelfth Dynasty are found at major sites in the Levant.

In Egypt, the period of the Twelfth Dynasty is known as the Middle Kingdom. It was a time of great prosperity in a reunified Egypt. Pyramids were again being built (possibly as many as nine of them), administrative and bureaucratic structures were in place and functioning, and the arts and letters flourished; indeed, the Middle Kingdom is considered to be the "classical" period of Egyptian literature. Egypt, like Palestine, was turning from the inward confusion of the First Intermediate Period to a time of stability and even outward expansion. In

MIDDLE BRONZE II SITES IN THE SOUTHERN LEVANT

Dan
Tel Qadesh
Abdon   T.
Aczib   er-Ruweisa
Hazor
Nahariya   T. Kabri
T. Bira
Acco
T. Kinnereth
T. Kison   *Sea of Galilee*
Mediterranean Sea   T. Shimron   *Yarmuk R.*
T. Qashish
T. Yin'am
Jokneam
Dor   Megiddo   T. Rekhesh
T. Mevorach   Taanach   Beth Shan
Jenin
T. Zeror   Dothan   Pehel
T. Hefer   Gath   T. Rehov   *Jordan River*
T. Poleg   T. el-Far'ah   *Jabbok R.*
T. Mikhal   Shechem   (Tirzah)
Aphek
Joppa   T. Gerisa   Shiloh   Kh. Marjameh
Bethel   (Ein Samiya)
Yavneh Yam   Gezer   T. es-Sultan
T. Mor   Gibeon   (Jericho)
T. Miqne   Jerusalem
Ashdod   Beth Shemesh
Ashkelon   T. es-Safi
Beth Zur
T. el-Hesi   Lachish
T. el-Ajjul   Hebron   *Dead Sea*
T. Nagila   Tell Beit Mirsim   *Arnon Gorge*
T. Ridan   T. Haror
T. Jemmeh   T. Masos
T. el-Far'ah (S)   T. Malhata
Zered River

T. = Tel(l)
Kh. = Khirbet

0   10 km.
0   10 miles

▲ *Gezer: Standing stones from ca. 1600 BC — the time of the Israelite sojourn in Egypt — possibly testifying to a covenant/treaty between local Canaanite tribes*

▼ *Dan: Well-preserved gate from ca. 1800 BC — the time of the patriarchs*

tained and supported Egyptian messengers who passed through his country, and how he protected the frontier of his overlord's territory. It even gives a detailed description of the produce of the land of Retenu — figs, grapes, wine, honey, olives, fruit, barley, emmer-wheat, and cattle — a list that is strikingly similar to that found in Deuteronomy 8:8 with reference to the land of Canaan. In addition, Sinuhe's daily fare included bread, wine, cooked meat, roasted fowl, and milk. Although Sinuhe may have lived in a more "urban" context than his contemporaries Isaac and Jacob, it can be assumed that the patriarchs also enjoyed foodstuffs similar to those described in the Story of Sinuhe.

In addition to the Story of Sinuhe, a group of texts called the Execration Texts shed welcome light on life in the Levant during the MB IIA period (2000 – 1750 BC). These texts, discovered in Egypt, are divided into two groups. The earliest, written on bowls, consist of curses mentioning the names and locations of the persons execrated (cursed). Composed in the late twentieth or early nineteenth century BC, the texts date to a time just prior to Jacob's descent into Egypt. About twenty towns and regions in the Levant are mentioned, including Jerusalem, Ashkelon, Beth Shan, Rehob, and Byblos. Since oftentimes several rulers are connected with one city or region, it seems that the common form of government at that time was the oligarchy.

The later Execration Texts, written on figurines that evidently represented captives, date to the late nineteenth century BC and thus to a time shortly after Jacob's descent into Egypt. These later texts mention some sixty-four geographical names from the Levant, and usually only one ruler is associated with each geographical name. The personal names from both groups of texts that can be analyzed are Amorite (West Semitic). Thus it seems that during the MB IIA period (2000 – 1750 BC) the ruling class in the Levant was of Semitic stock, as were the masses.

The Beni Hasan inscription (*ANET*, 229; *ANEP*, 2 – 3) also helps illuminate life in the Levant during the Patriarchal Period. It was found on the wall of the tomb of Khum-hotep III and dates to the sixth year of the king Sesostris II (ca. 1872 BC) — approximately the same time as Jacob's descent into Egypt. The painting represents thirty-seven Asiatics bringing stibium, a substance used in the production of eye paint, to Egypt. The colorful dress of both the men and the women is well represented, as are their weapons, tools, and instruments. On the backs of the donkeys are bellows, or possibly copper ox-hide ingots, indicating that the company may also have included a group of traveling smiths. (It should be noted that at approximately this same time Joseph was sold to a cara-

Upper Egypt fortresses were built south of the second cataract to control the movement of the Nubians. Luxury goods, either originating in or transiting through Nubia, were shipped north. Commercial contacts with the Levant, especially with Byblos, were common, although the archaeological and literary evidence is not strong enough to indicate that Egypt "controlled" Canaan.

Egyptian contact with the Levant during this period is further reflected in the Story of Sinuhe (*ANET*, 18 – 23), a tale of an Egyptian who, upon the death of Amenemhet I (ca. 1944 BC), fled from Egypt to the Levant. He first journeyed to the city of Byblos, on the coast of the Mediterranean Sea, but then turned inland toward Qedem, apparently located near the border of the eastern desert. He settled in the land of Araru (possibly in the region of Gilead or Bashan), where he was given land by a ruler of Upper Retenu, and there he lived until he returned home to Egypt to die. The story describes how Sinuhe enter-

van of Ishmaelites/Midianites who were transporting spices, balm, and myrrh from Gilead to Egypt [Gen 37:25, 36].)

The picture is also a reminder that there were always Asiatics attempting to enter Egypt via her northeastern frontier for commercial purposes (as here), to avoid famine (as did Abram), or to find water and pasturage for their flocks. To keep these Asiatics out, Amenemhet I (1973 – 1944 BC) dug a canal and built a corresponding "wall," not only to protect this northeastern frontier, but also to provide water for the flocks of these peoples so that they would not have to enter Egypt to secure water.

According to the biblical data, Jacob and his family moved to Egypt in 1876 BC, which would place this event early in the reign of Sesostris II (1878 – 1872 BC). Unfortunately, there are no Egyptian documents that make direct reference to the 430-year Israelite sojourn in Egypt, but the general features of Israel's life in Egypt can nevertheless still be outlined. According to Genesis 47:4, Jacob's family settled in the eastern delta of the Nile, in the agriculturally rich land of Goshen, where they evidently remained during their long stay in Egypt (cf. Exod 8:22 and 9:26). Although political stability was maintained in Egypt during the Twelfth Dynasty, conditions began to deteriorate during the Thirteenth and Fourteenth Dynasties.

During the Thirteenth Dynasty, more and more Asiatics infiltrated the eastern delta until they became powerful enough to establish what are now known as the Fifteenth and Sixteenth Dynasties. This was the era of the Hyksos domination of the eastern delta; during this time their control occasionally even extended upstream along the Nile. These Asiatic rulers of Egypt were called "rulers of foreign countries" by the native Egyptians. The Hyksos kings were evidently Amorites (West Semites) who had moved into the eastern delta from Asia. They had come from the Levant, where numerous city-states had been established by their compatriots.

During the second part of the Middle Bronze period (MB IIB) in Canaan, old cities were rebuilt and a number of new urban centers were established. These city-states were defended by glacis (sloping ramps built of packed earth, stone, and plaster) and dry moats. Temple and palace architecture is well-known from the excavations at Hazor, Megiddo, Shechem, and other sites, indicating a high cultural level. The small finds of gold, silver, ivory, alabaster, etc. also indicate a great measure of prosperity in Canaan at this time.

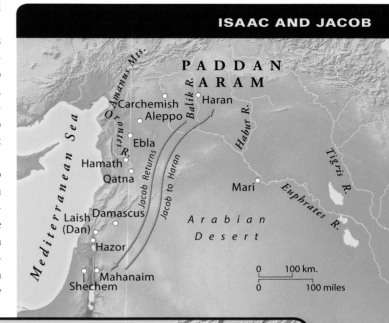

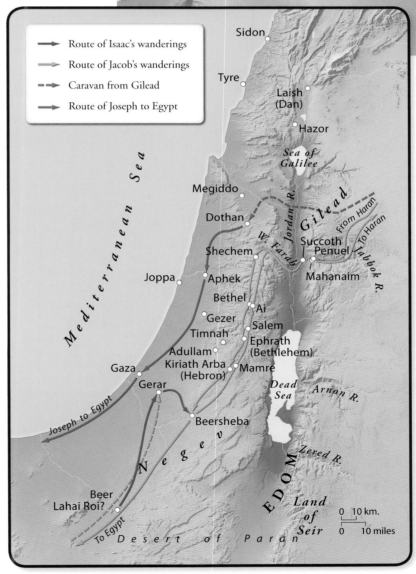

Militarily, the most important development was the introduction of the horse-drawn battle chariot. The charioteers, called *maryannu*, comprised a new social class that had to be reckoned with. These charioteers probably formed the power base on which the various city-states maintained their existence. Besides typical cities built on tells, large enclosures were built in Canaan, Syria, and the eastern Nile delta. The purpose of these enclosures is disputed and may have changed over time, but initially some may have provided protected storage areas for chariots, horses, and related gear. The wide geographical distribution of similar architectural remains, pottery, defensive works, and large enclosures indicates continuing contacts between the regions of the Levant and Egypt.

During the MB II period there were also extensive contacts between Canaan and northern and southern Mesopotamia. The northern Mesopotamian connections are well-documented by information contained in the cuneiform tablets that were discovered at the city of Mari on the bank of the Euphrates River. These 20,000 tablets, dating to the late nineteenth and early eighteenth centuries BC, shed welcome light on many aspects of life in the ancient Near East during this period. In particular, conditions in the Euphrates – Balik – Habur River region are described in detail — the very region where Jacob lived with his uncle Laban for some twenty years (= Paddan Aram). One Mari tablet actually refers to the "men of *Ki-na-aḫ-um*" — which is one of the earliest usages of "Canaan" as a geographical term.

In addition, Canaanite cities (including Hazor and Laish [Dan]) are mentioned in the Mari texts. Of these, pride of place goes to Hazor, which to date has been found in seven Mari documents. It is evident that Hazor was a very important city, for its ambassadors were found in royal courts all over the Near East. In addition, large quantities of tin were shipped to Hazor, which indicates that it was a major producer of bronze (one part tin, ten parts copper). The prominence of Hazor during this period is also seen in its sheer size — 200 acres — and from the rich finds discovered at the site. It is probable that the statement "Hazor had been the head of all these kingdoms" (Josh 11:10) refers back to this period.

The other major Palestinian city mentioned in the Mari texts is Laish (= biblical Leshem = Dan; Josh 19:47; Judg 18:29). It is mentioned along with Hazor as a recipient of deliveries of tin. It should be noted that Wari-taldu, the ruler of Laish (Dan), bears a Hurrian, not an Amorite, name — one of several indications of the growing importance of a Hurrian element in the Levant. In fact, many of the rulers of city-states in the late MB II and Late Bronze (LB) ages, as well as some of the later kings of the Hyksos dynasty in Egypt, bore Hurrian names. The Hurrian element in the Levant became so important that at times Egyptians referred to the country as "Hurru

▼ *Jabbok River: Looking at the Jabbok Valley near Penuel (brown mound below the horizon in upper right of image) near where Jacob met Esau on his return from Haran*

Mark Connally

Land." Thus it seems that the Hurrian element in the population, which was growing during the MB II period, became prominent during the LB Age.

A good portion of Israel's stay in Egypt took place during the turbulent Second Intermediate Period, and it may very well be that the "new king, who did not know about Joseph" (Exod 1:8) was a Hyksos. This would mean that the oppression of the Israelites began during the Hyksos rule. Additional support for this thesis is found in the statement that "the Israelites have become much too numerous for us" (1:9), which would fit much better in the mouth of a Hyksos king — limited in number as the Asiatics were — than in the mouth of a native Egyptian ruler.

The end of the MB II period (1550 BC) is not marked by any great cultural break in the Levant but rather by a historical event — the expulsion of the Hyksos from Egypt. The first king of the Eighteenth Dynasty, Ahmose, drove the Hyksos from the country. Egyptian texts describe battles in the eastern Nile delta, the siege of the Hyksos capital of Avaris (= Tell ed-Dab'a), and the driving of the Hyksos rulers out of Egypt and back into Canaan (*ANET*, 230 – 34, 553 – 55). Ahmose besieged the Hyksos city of Sharuhen for three years and, after conquering it, established an Egyptian base there at the southwest entrance to the land of Canaan. This was the area from which later Egyptian kings of the powerful New Kingdom launched their conquests of Canaan, increasingly bringing the land into the Egyptian sphere of influence.

At the end of the MB II period, Israel found herself in bondage in Egypt. She had increased in number, yet she did not possess the land promised to Abraham.

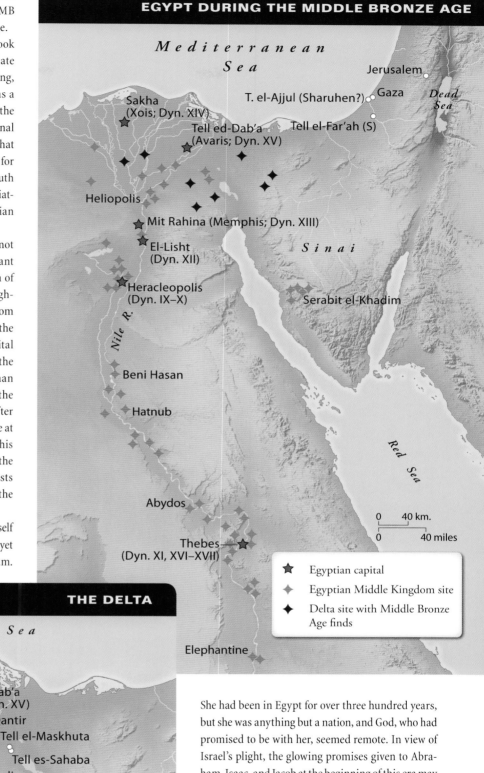

EGYPT DURING THE MIDDLE BRONZE AGE

Mediterranean Sea

Jerusalem

T. el-Ajjul (Sharuhen?) Gaza *Dead Sea*

Sakha
(Xois; Dyn. XIV)

Tell ed-Dab'a
(Avaris; Dyn. XV) Tell el-Far'ah (S)

Heliopolis

Mit Rahina (Memphis; Dyn. XIII)

Sinai

El-Lisht
(Dyn. XII)

Heracleopolis
(Dyn. IX–X)

Serabit el-Khadim

Nile R.

Beni Hasan

Hatnub

Red Sea

Abydos

0   40 km.

0   40 miles

Thebes
(Dyn. XI, XVI–XVII)

★ Egyptian capital

◆ Egyptian Middle Kingdom site

◆ Delta site with Middle Bronze Age finds

Elephantine

THE DELTA

Mediterranean Sea

Sakha
(Xois; Dyn. XIV)

Tell ed-Dab'a
(Avaris; Dyn. XV)

Kom
el-Hisn Farasha Qantir

Tell Basta Tell el-Maskhuta

Tell Atrib

Ghita Tell es-Sahaba

Nile R.

Tell el-Yahudiya
Heliopolis

Saqqara Mit Rahina
Dahshur (Memphis; Dyn. XIII)

0   40 km.

0   40 miles

She had been in Egypt for over three hundred years, but she was anything but a nation, and God, who had promised to be with her, seemed remote. In view of Israel's plight, the glowing promises given to Abraham, Isaac, and Jacob at the beginning of this era may have seemed to many to be incapable of fulfillment; yet it would be during the next phase of Near Eastern history, in the face of the most powerful nation on earth, that God would act decisively for his people.

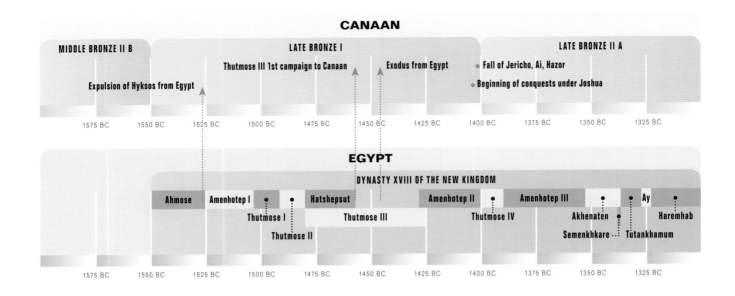

CANAAN

MIDDLE BRONZE II B	LATE BRONZE I	LATE BRONZE II A
	Thutmose III 1st campaign to Canaan	• Fall of Jericho, Ai, Hazor
Expulsion of Hyksos from Egypt	Exodus from Egypt	• Beginning of conquests under Joshua

1575 BC	1550 BC	1525 BC	1500 BC	1475 BC	1450 BC	1425 BC	1400 BC	1375 BC	1350 BC	1325 BC

EGYPT

DYNASTY XVIII OF THE NEW KINGDOM

Ahmose	Amenhotep I		Hatshepsut		Amenhotep II		Amenhotep III		Ay
	Thutmose I		Thutmose III		Thutmose IV		Akhenaten		Haremhab
	Thutmose II						Semenkhkare	Tutankhamum	

1575 BC	1550 BC	1525 BC	1500 BC	1475 BC	1450 BC	1425 BC	1400 BC	1375 BC	1350 BC	1325 BC

EXODUS AND CONQUEST

The Exodus from Egypt

One of the most important configurations of events in the Old Testament centers on the exodus from Egypt, the revelation of God's law at Sinai, and the establishment of Israel in the Promised Land. Since the exodus, the wilderness wanderings, and the initial conquest of Canaan (Exod 1 – Josh 11) form a continuous narrative covering approximately forty-five years, these events will be treated together in this chapter, while the settlement of the Israelites in the land (Josh 12 – Judges 21) will be discussed in the next chapter.

According to the chronology adhered to in this atlas, the exodus and conquest occurred at the end of the archaeological period known as the Late Bronze I Age (= LB I; 1550 – 1400 BC). In Egypt, the Israelites were suffering from oppression at the beginning of the New Kingdom, an Egyptian era that included the Eighteenth, Nineteenth, and Twentieth Dynasties (1540 – 1070 BC). In particular, the first half of the Eighteenth Dynasty coincided with the biblical events under consideration.

The first king of the Eighteenth Dynasty, Ahmose (1540 – 1515 BC), not only unified Egypt but also expelled the Hyksos from the eastern delta. In addition, he pushed Egypt's northeastern frontier eastward across Sinai to Sharuhen and its southern boundary southward, up the Nile to the second cataract. Subsequently Thutmose I (1494 – 1482 BC) conducted a military campaign through Canaan into Syria, even reaching the Euphrates River (*ANET*, 234, 239 – 40)! This was actually as far north as Egypt would expand during this period, for later kings would fight in Canaan and Syria, but not beyond the Euphrates.

▲ *Karnak: Thutmose III slaying his Canaanite enemies held in his left hand. With upraised hands they plead for mercy. Below Thutmose are three rows of cartouches of captive enemies.*

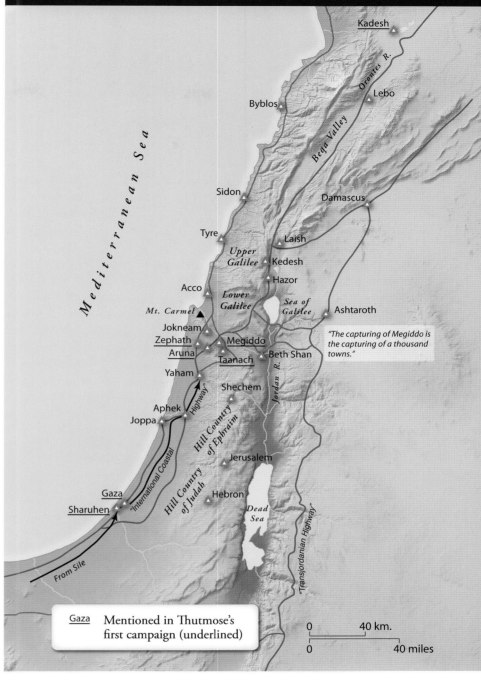

Kadesh

Orontes R.

Byblos

Lebo

Beqa Valley

Mediterranean Sea

Sidon

Damascus

Tyre

Laish

Upper Galilee

Kedesh

Acco

Hazor

Lower Galilee

Sea of Galilee

Ashtaroth

Mt. Carmel ▲

"The capturing of Megiddo is the capturing of a thousand towns."

Jokneam

Zephath

Megiddo

Aruna

Taanach

Beth Shan

Yaham

Jordan R.

"Highway"

Shechem

Aphek

Joppa

Hill Country of Ephraim

Jerusalem

Hill Country of Judah

Gaza

Hebron

Dead Sea

Sharuhen

"International Coastal"

"Transjordanian Highway"

From Sile

Gaza — Mentioned in Thutmose's first campaign (underlined)

0   40 km.

0   40 miles

At about this same time a new political power was emerging in northern Mesopotamia — the kingdom of Mitanni. For the next one hundred years, Egypt and Mitanni would vie for control of the northern Levant. By the end of the reign of Thutmose I, Egypt had conquered territory stretching from Syria in the northeast to the fourth cataract of the Nile in the south. Egypt's territory had never before been so extensive, and it would never be again. From Nubia in the south, gold and luxury goods flowed northward, while Egypt imported cedar and other important forest and agricultural products from Canaan and Lebanon to the northeast.

During the reigns of Thutmose II and Queen Hatshepsut (1479 – 1457 BC), Egypt seems to have lost ground in Asia, for in 1457 BC, at the beginning of Thutmose III's sole reign (1457 – 1425 BC), he departed on the first of some seventeen campaigns into the Levant. His first expedition was by far his most important and in it he was able to secure control of much of the southern Levant.

The first phase of his campaign took him from Sile (near the present-day Suez Canal) across northern Sinai to Gaza — covering a distance of 130 miles in nine or ten days. From Gaza he headed northward to Yaham. This distance of 75 miles was covered in eleven or twelve days, a bit slower than his trek across northern Sinai, yet fast enough to indicate that he met with little resistance along the way. Since the enemy was headquartered at Megiddo, Thutmose held a war council at Yaham to consider which of the three passes he would take across Mount Carmel: the Zephath-Jokneam, the Aruna-Megiddo, or the Taanach pass. In spite of advice to the contrary, he chose the narrower, more difficult, and more direct route from Aruna to Megiddo.

Fortunately for the invading Egyptians, the Canaanites had split their chariot forces, stationing them northwest and southeast of Megiddo. Thus the Egyptians were able to emerge from the pass and set up camp near Megiddo unmolested.

The next day the Egyptians engaged the Canaanites in battle. The Canaanites were thoroughly routed, and the survivors fled to Megiddo, where they were pulled up on ropes over the city wall into the fortress. After a siege of seven months the city was conquered, and considerable booty, including 924 chariots, was taken. The leader of the Canaanite coalition seems to have been the king of Kadesh (i.e., Kadesh on the Orontes), who had been joined by city-state rulers of Canaan, as well as by forces from Naharin (Mitanni), Hurru, and Kode. By capturing Megiddo, Thutmose III was able to defeat and capture the leaders of the revolt and was, in effect, able to take control of most of the southern Levant without having to lay siege to each individual city. No wonder a scribe wrote, "The capturing of Megiddo is [like] the capturing of a thousand towns" (*ANET*, 237). Mopping-up operations continued during the reign of Thutmose III, and it was not until his sixth campaign that Kadesh on the Orontes was actually taken (*ANET*, 239).

Three copies of the topographical text that commemorates Thutmose's victory have been discovered (*ANET*, 242 – 43). Two of these texts, probably composed soon after his first campaign, list 119 towns

▲ *Karnak: Three cartouches, among many, representing captured enemies of Thutmose III. From right to left the cities of Kadesh (on the Orontes River), Megiddo (in Canaan), and Haszi (in the Beqa Valley) are represented.*

and regions that were now under his control, while the third lists 231! These rosters are the largest listings of geographical names from the Levant, of which approximately half have been identified with some certainty. From the lists it appears that Egypt was primarily interested in controlling the International Coastal Highway and the Transjordanian Highway, as well as the latter's connecting routes to the Mediterranean via the Jezreel Valley and/or Galilee. Conspicuous by their absence are cities in the Hill Country of Judah, Ephraim, and Manasseh — evidently indicating that these regions were sparsely settled at this time and that the centers that did exist did not pose any immediate threat to Egyptian interests in the area. These were the areas where the Israelites would soon begin to settle.

Thutmose III's successor, Amenhotep II (1427 – 1401 BC), conducted three campaigns into the Levant. The descriptions of these campaigns (*ANET*, 245 – 48) suggest that, although early in his reign the battles were fought in the northern Levant, by his ninth year he was fighting battles in the region of the Sea of Galilee — a clear indication that he had been losing ground in the area. Indeed, after his ninth year, Egyptian military activity in the Levant was limited until the days of Seti I (1294 – 1279 BC). It seems that Egypt had become prosperous and stable and had come to terms with its northern enemy, the Mitanni. This new alliance may have been forged to meet the threat common to both, namely, the growing power of the Hittites in Anatolia and in Syria.

The oppression of the Israelites that had begun during the days of the Hyksos (1638 – 1540 BC) continued and probably intensified during the early rule of the Eighteenth Dynasty (Exod 1:13 – 22). Although not all aspects of Israel's oppression are known in detail, the text indicates that "they built Pithom and Rameses as store cities for Pharaoh" (v. 11). The city of Rameses was evidently in or near the land of Goshen, for at the time of the exodus it was the starting point of the Israelite departure from Egypt (Exod 12:37; Num 33:3, 5). The site of Rameses has been well identified with Tell ed-Dab'a that is located near modern Qantir. This huge mound has extensive Hyksos remains and was formerly called Avaris, the capital of the Hyksos.

▼ *Nile Delta: The fertile "land of Goshen" where Israel lived in northeast Egypt*

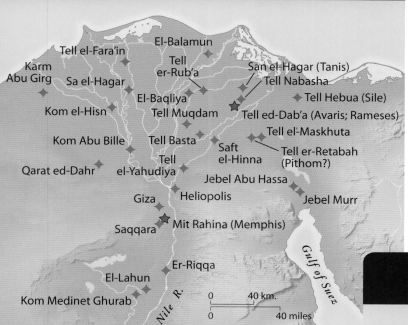

Israelites journeyed to Succoth (possibly located in the Wadi Tumilat at Tell el-Maskhuta). It is well to note that for fear of their becoming discouraged because of war, "God did not lead them by the way of the land of the Philistines" (Exod 13:17 NASB). This well-known route across northern Sinai from Sile to Gaza was the one Thutmose III and Amenhotep II had used so effectively on their frequent campaigns to Canaan, and it must have been well fortified by Egyptian troops. Thus a northern route for the exodus seems excluded. Since the Israelites were led by the "way of the wilderness to the Red Sea" (v. 18), it appears that they were heading southeast toward modern Suez (Exod 13:20 – 14:9). The location of Etham ("fort" in Egyptian), Migdol ("fort"

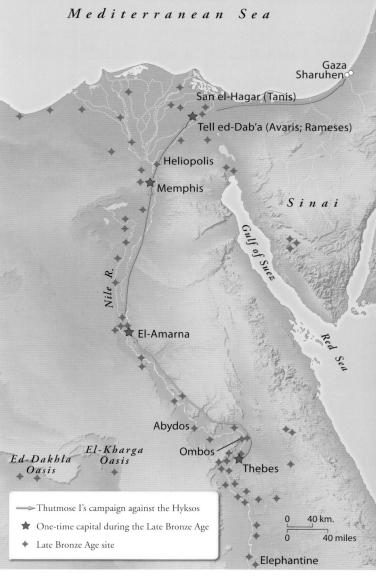

EGYPT IN THE LATE SECOND MILLENNIUM BC

Thutmose I's campaign against the Hyksos

★ One-time capital during the Late Bronze Age

✦ Late Bronze Age site

In addition, at nearby Ezbet Helmi two palaces from the Eighteenth Dynasty have been excavated — they would have been in use at the time of the exodus. Although the use of the geographical name "Rameses" is not attested this early for the site, its appearance in the Bible may be due to the work of a copyist who "updated" the text for his readers. The other city mentioned in the Bible, Pithom, can be identified with Tell er-Retabah or posibly Tell el-Maskhuta.

Although a considerable amount of geographical data is presented in connection with the exodus and the trek to the land of Canaan, the exact identification of many places and regions mentioned remain unknown. The major reason for this is the lack of continuity among the populace of the desert-wilderness regions of the Sinai peninsula, the Negev, and parts of southern Transjordan. Without this continuity of population, the preservation of ancient place names is almost impossible. The other difficulty is that archaeologists have not discovered any artifactual remains that can be attributed to the Israelites in those regions through which they traveled. This, however, is somewhat expected, for a nomadic people, living in tents and using animal skins instead of pottery for containers, would leave few permanent remains behind.

Thus, scholars are divided in their opinions as to the location even of major landmarks such as the Red Sea and Mount Sinai. Indeed there are at least ten different proposals for the location of the Red Sea or Reed Sea — including three lakes near the Mediterranean Sea, five lakes along the line of the present-day Suez Canal, as well as the Gulf of Suez and the Gulf of Elath. In addition, there are at least twelve different candidates for Mount Sinai: five in the southern part of the peninsula, four in the north, one in the center, one in Midian (Saudi Arabia), and another in Edom (southern Transjordan).

In spite of these uncertainties, a few suggestions can be made regarding the exodus and wanderings. After leaving Rameses, the

in Semitic), Baal Zephon, and Pi Hahiroth are problematic. It has been suggested that Hahiroth refers to the low ground between Jebel Geneife and the Bitter Lakes, but this is not certain. Etham and Migdol could be any one of a number of Egyptian forts located near the present-day Suez Canal.

On the next stage of their journey the Israelites crossed the Red Sea. Since the Hebrew text literally means the "Reed Sea," many scholars look for a location in the lake/marsh areas that used to exist in the region through which the Suez Canal now passes. A location near the junction of the Great and Little Bitter Lakes is as plausible as any. According to nineteenth-century travelers, the water at that spot was not very deep, and they even mention that at times the depth of the water decreased when the wind shifted. It is interesting to note that according to the text the "Lord drove the sea back with a strong east wind" (Exod 14:21).

The location of Marah, where the water was bitter (15:23), and Elim, where there were twelve springs and seventy palm trees (v. 27), depends on where one locates Mount Sinai. If Jebel Sin Bisher is the location of Mount Sinai, then the identifications of Marah and Elim

with Bir Mara ("bitter well" in Arabic) and Ayun Musa ("the spring of Moses") are plausible. If the more traditional site of Sinai at Jebel Musa is maintained, then identifications of Marah and Elim with Ein Hawwara and Gharandal are also possible.

The identification of Mount Sinai (Horeb) with Jebel Musa ("Mount Moses") is based on Christian tradition dating back to the fourth century AD, about 1,750 years after the event. There, during the Byzantine period (AD 324 – 640), the desert monastery of St. Catherine was established. Although the Greek Orthodox monks today point out the very site of the giving of the Law, the place where the golden calf was erected, the plain where the Israelites camped, the site of the burning bush, etc., the suggested identification of Mount Sinai with Jebel Sin Bisher deserves careful attention.

Indeed, its location agrees with some of the biblical data. For example, it is located approximately three days' journey from Egypt (Exod 3:18; 5:3; 8:27) at a desert junction where there are fair supplies of water; possibly the Amalekites fought with Israel for control of this junction and the water sources (Exod 17). It is close to Egypt on the road that led directly from Midian to Egypt (the modern Darb el-Hagg,

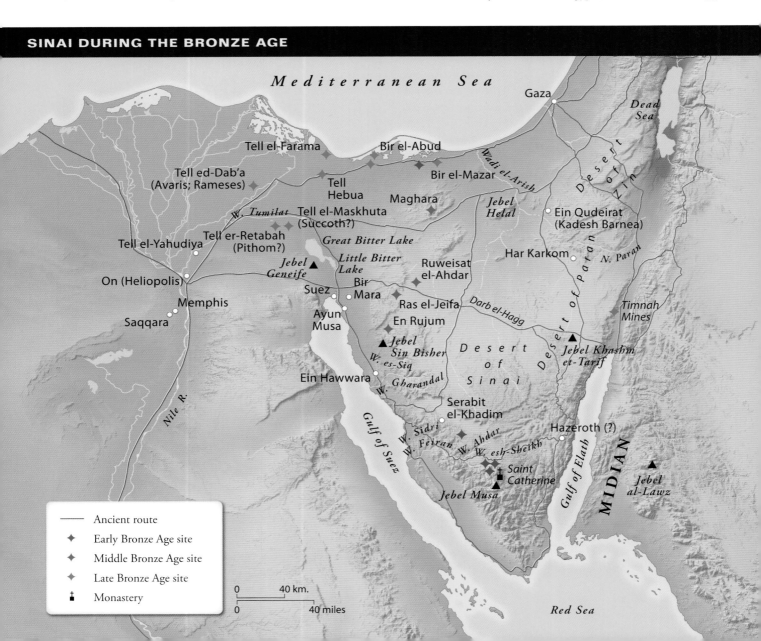

SINAI DURING THE BRONZE AGE

Legend:
— Ancient route
✦ Early Bronze Age site
✦ Middle Bronze Age site
✦ Late Bronze Age site
‡ Monastery

0   40 km.
0   40 miles

▲ *Sinai: Mountains surrounding Jebel Musa —
the traditional site of Mt. Sinai*

although historical geographers — in the absence of
better candidates — sometimes attach these names
to the larger oases along the route.

From the Kadesh region twelve men were sent to
"explore the land of Canaan" (Num 13:2). They went
up through the Negev into the hill country, traveling as far north as Rehob and Lebo Hamath (13:21).
It is clear that the "land of Canaan" is presented as a
geopolitical entity with definable boundaries. These

**CANAAN DURING THE LATE
BRONZE AGE (1550 – 1200 BC)**

Laish
(Dan)

Kedesh

Beth Anath

Kabri   Hazor

Acco   Kinnereth

Tel Shiqmona   Tell Abu   Hannathon   *Sea of
                Huwam   Galilee*

T. Regev   Shimron

Tel Qashish   Anaharath

Dor   Jokneam   Kishion

Megiddo

Tel Mevorach   Taanach   Beth Shan

Tel Zeror   Gina

                Rehob   Pehel

T. Hefer   Dothan

                            Tell
                            es-Saidiya

T. el-Far'ah

Tel Michal   Shechem

Tel Gerisa   Succoth

Joppa   Aphek   Shiloh

Azor

Bethel   Rabbah

Tel   Gezer

Tel Mor   Miqne   Jericho

Ashdod   (Ekron) Timnah   Gibeon

          Gath   Jerusalem

Ashkelon   Beth
                Shemesh

Tel Zippor   Gedor

Tell el-Hesi   Lachish   Beth Zur

Tell   Tel Eton   Hebron   Aroer

el-Ajjul   Gaza

        Tell Beit   *Dead
        Mirsim   Kh. Rabud   Sea*

Deir   Tel
el-Balah Gamma   Tel
                  Sera

        Tell el-Farah (S)

                            Ader

0   10 km.

0   10 miles

Mediterranean Sea

Jordan R.

p. 69), and thus it would make a plausible location for the burning bush incident. Moses could
have been bringing Jethro's sheep along this road
in order to use the water and pasturage found on
the eastern edge of the Nile delta (see above, p. 97,
for the canal) when the Lord appeared to him in
the burning bush. This is said to have taken place
near the mountain where he would later worship him (3:1). Since it is reasonable to assume
that Moses used the way of the wilderness on his
return to Egypt, the meeting of Moses and Aaron
at the "mountain of God" could well have been at
this spot (Exod 4:27).

After camping at Mount Sinai for about a
year, the Israelites set out for the Kadesh Barnea
region. This trek normally took eleven days (Deut
1:2), which fits the identification of Mount Sinai
with Jebel Sin Bisher better than the identification with Jebel Musa. On their way they passed
from the Desert of Sinai through the Desert of
Paran to the Desert of Zin near Kadesh (Num
10:12; 33:36). These "deserts," or better, "wil-
dernesses," are difficult to locate with accuracy,
although the Desert of Paran seems to have been
large, covering most of the southern and central
sections of the Sinai peninsula, while the Desert
of Zin was evidently located to the north of this.

Along the way to Kadesh, several judgments
of God befell the Israelites — at Taberah (Num
11:1 – 3), Kibroth Hattaavah (vv. 4 – 35), and
Hazeroth (11:35 – 12:16). None of these places
has been identified with certainty, for the place
names have not been preserved in the area,

Map labels:
Mt. Hor, Zedad, Hazar Enan, Byblos, Lebo Hamath, Mediterranean Sea, Eastern Desert, Damascus, Kadesh, Hazor, Acco, Sea of Kinnereth, Hannathon, Ashtaroth, Jebel Druze, Ain, Megiddo, Jordan R., Shechem, Land of Gilead, Jerusalem, Salt Sea (Dead Sea), Gaza, Hebron, Gerar, Arad, Negev, Zoar, "Wadi of Egypt", Mt. Hor, Desert of Zin, Tamar, Wadi el-Arish, Scorpion Pass, Azmon, Kadesh Barnea, Hazar Addar, EDOM

Legend:
Land of Canaan
Route travelled by the 12 scouts

0   40 km.
0   40 miles

boundaries, which are described in Numbers 34:1 – 12 and Ezekiel 47:13 – 20, can be traced with some degree of accuracy. The southern boundary stretched from the southeastern corner of the Salt Sea westward, ran south of Kadesh Barnea to the "Wadi of Egypt" (= W. el – Arish), and ended "at the [Mediterranean] Sea" (Num 34:3 – 5; Josh 15:2 – 4; Ezek 47:19).

The Mediterranean Sea formed the western boundary, but the northern one is a bit more difficult to define accurately (Num 34:6 – 9; Ezek 47:15 – 17, 20). The key point necessary to determine its proper line is "Lebo Hamath" (= "the entrance to Hamath [a country in Syria]"). This frequently mentioned northern point has been well identified with Lebweh, located in the Beqa Valley. Geographical references that occur in the texts both before and after Lebo Hamath were located near a general east-west line that ran through Lebweh. The boundary is described as running out into the eastern desert, where it swings to the south to include the Jebel Druze region, before turning westward to the hills at the southeast corner of the Sea of Kinnereth (Num 34:11). From there, the border followed the Jordan River south to the Salt Sea, back to its starting point.

Since the account of the exploration of Canaan gives prominence to the Hebron region and the nearby Valley of Eshcol (Num 13:22 – 23), it is probable that these Israelite scouts traveled the old patriarchal road, the Ridge Route. Although other cities are not mentioned by name, they are described as being "fortified and very large" (v. 28). The land itself is described as being very fruitful, as evidenced by the large cluster of grapes, the pomegranates, and the figs that the scouts brought back to camp with them (vv. 23 – 24). In addition, the land was described as flowing "with milk and honey" (v. 27), symbols of bounty and abundance.

Because of the people's disobedience (failing to enter the land at the Lord's command), they were consigned to forty years of "wandering" in the "desert." After an abortive attempt was made to enter the land of Canaan (Num 14:39 – 45), the Israelites began their bleak wilderness experience. It seems that they spent much of their time in the desolate region between Kadesh and Ezion Geber (near/on the Red Sea [33:36]), in all probability camping near Kadesh, in the western Negev Highlands as well as in eastern Sinai.

At the end of this period the Israelites again found themselves in the Kadesh Barnea area. It was there that Miriam, Moses' sister, died (Num 20:1). It was there that Moses was disobedient to God's command, striking the rock twice when he should have spoken to it (vv. 2 – 13), and

Land of Canaan

The concept of the "land of Canaan" is found throughout Scripture (over 88 times, from Gen 12:5 through Ezek 47:15 – 20). The term "Canaan" is attested as far back as the Ebla tablets (ca. 2400 BC) where reference is made to "Dagan [a deity] of Canaan." Later, one Mari tablet (ca. 1800 BC) actually refers to the "men of Ki-na-ah-um" (which is one of the earliest usages of "Canaan" as a geographical term), and one of the cuneiform tablets found at el-Amarna in Egypt (ca. 1400 BC) indicates that Hannathon was in the "land of Canaan." Thus at the time of the exodus and conquest, the "land of Canaan" was a definite geopolitical entity, known both to the Egyptians and to the Israelites, with specific boundaries (see p. 105 – 106 and compare later map on p. 178).

▲ *Sinai/Negev Wilderness in which Israel spent 40 years*

it was from Kadesh that Moses requested the king of Edom to allow the Israelites to pass through his land along the "King's Highway" (v. 17; in this passage possibly to be identified with today's Darb es-Sultan; see p. 56). Since this request was made from Kadesh, it is evident that the Edomites had extended their control from the Transjordanian Mountains westward, across the Arabah Valley into the Negev Highlands. In spite of Israel's promise to stay on the highway and to purchase water (vv. 17–19), their request was refused. Setting out from Kadesh, the Israelites arrived at Mount Hor, where Aaron died and Eleazar was appointed in his place (vv. 22–29). Although Mount Hor cannot be identified with certainty, the suggestion of Imaret el-Khureisheh, near Kadesh on the border of Edom, is much more plausible than the traditional spot, Jebel Nebi Harun, located in Transjordan.

After these events the Israelites attempted to enter Canaan from the south (Num 21:1–3). At first the king of Arad, who lived in the Negev, defeated them but then, after praying to God, the Israelites were victorious in their second encounter. In spite of this victory they decided not to enter Canaan from this direction; instead, they followed a circuitous path to the south and east. Their major route seems to have taken them south along the "way of the Red Sea" (= Darb el-Gaza) to Elath and Ezion Geber, then northward around (i.e., east of) the lands of Edom and Moab (Judg 11:18), traveling on what was called the "King's Highway" (Num 21:22; here the eastern branch of the Transjordanian Highway).

From the "Desert of Kedemoth" (Deut 2:26–30) messengers were sent to Sihon, the king of the Amorites (who lived in Heshbon), requesting permission to pass, from east to west, through his territory to the Jordan River. Sihon, who refused this request, marched to Jahaz, where he fought against and was defeated by the Israelites (Num 21:23–24). Having subdued Sihon, the children of Israel took possession of the land that he had controlled, from the Arnon River in the south to the Jabbok in the north (Num 21:24–25; Deut 2:36; Judg 11:22).

Some commentators believe that another "wave" of Israelites took a more westerly route, actually passing through Edomite and Moabite territory. The validity of accepting the use of such a route depends on the certainty of the identification of certain towns — including Zalmonah, Punon, Oboth, and Iye Abarim (Num 21:1, 11; 33:41–44). But since none of these has been identified with complete certainty (even Punon's location can be challenged), it seems best to hold to a single, more easterly route rather than supposing a second route directly through Edom and Moab, for the text does not give any indication of direct conflict with them as the result of an invasion.

Ever since Nelson Glueck's exploration of southern Transjordan (1932 to 1947), many have believed that an "occupation gap" in the area lasted from the nineteenth to the thirteenth centuries BC. If this were true, then the Israelites could not have encountered a sedentary population in the area (as the biblical account seems to imply). However,

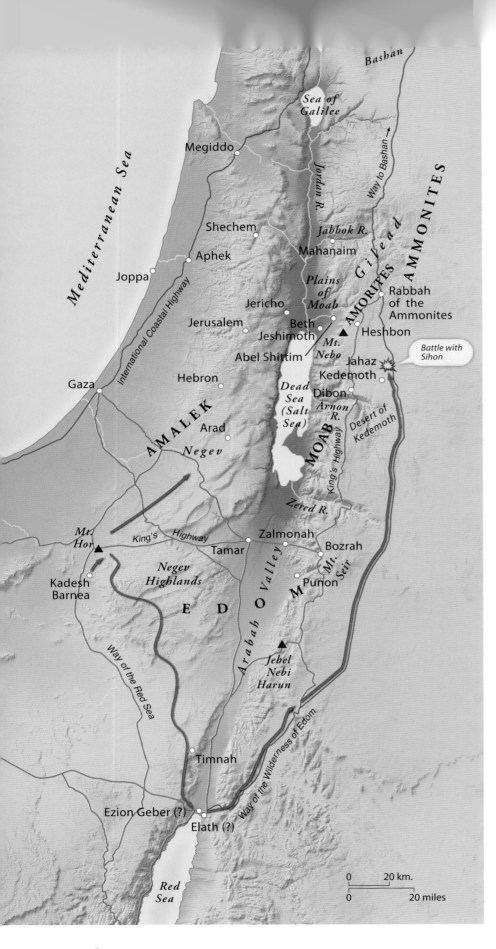

recent excavations and surface surveys have shown that this "occupation gap" really did not exist, for some MB II (2000 – 1550 BC) and LB (1550 – 1200 BC) sites have been located. To be sure, the number and size of MB II and LB remains are not as extensive as those during some of the previous and later periods, but the "occupation gap" has been filled in recent years, and it is no longer necessary to doubt Israel's encounters with the kings of Edom and Moab and with Sihon, the king of the Amorites.

After securing Sihon's former territory, the Israelites marched north on the "way to Bashan." Og, the king of Bashan, who lived in Ashtaroth (Deut 1:4), came out and battled Israel at Edrei. The forces of this "giant" (3:11) were defeated, and his territory from the Jabbok River to Mount Hermon — including sixty cities as well as Gilead and Bashan — came under Israelite control (Num 21:33 – 35; Deut 3:4 – 11).

Numbers 21 through Deuteronomy 34 describe the events that occurred during Israel's encampment, east of Jericho, in the Plains of Moab between Beth Jeshimoth and Abel Shittim (Num 22:1; 33:49). There Balaam blessed, rather than cursed the Israelites (Num 22 – 25) and there Moses preached his last sermons before ascending Mount Nebo to die (Deut 34).

The Conquest of Canaan

After the death of Moses, Joshua led the Israelites into Canaan and directed the initial conquests in the land (Josh 1 – 12). Although the account is rather detailed, it should be remembered that in all probability, not every military event that took place under Joshua's leadership was recorded; rather, a number of key events were selected. The initial phase of the conquest could have begun as early as 1406 BC, forty years after the exit from Egypt. This initial phase probably did not take longer than seven years, although Joshua apparently lived until 1375 BC.

Under Joshua's leadership Israel crossed the Jordan River at one of the fords located east of Jericho. The text indicates that this event took place in the spring of the year, for the Jordan was overflowing its banks after the winter rains and the (barley?) harvest was taking place (Josh 3:15); also, the Passover (March – April) was celebrated soon afterward at "Gilgal on the plains of Jericho" (5:10). Whereas the waters of the Red Sea had been "divided," the waters of the Jordan "piled up in a heap a great distance away, at a town called Adam in the vicinity of Zarethan, while the water flowing down to the Sea of the Arabah (the Salt Sea) was completely cut off" (Josh 3:16; see p. 61). The Israelites set up camp at Gilgal, where they circumcised the new generation and celebrated the Passover, and where, after the people tasted the produce of the land, the manna ceased.

The first city to be captured by the Israelites was Jericho (Josh 6). This city has been well identified with Tell es-Sultan, a 10-acre mound situated beside a powerful spring in an otherwise arid region (6 in. of rain annually). At most 2,000 people lived at Jericho. Although the name Jericho has been preserved in the name of the village er-Riha, and the geographical location of Tell es-Sultan matches that of Jericho as described in the Bible, some scholars do not believe that the archaeological profile of the site corresponds to the history of Jericho as found in the sources.

If one holds to a late date for the conquest (i.e., ca. 1250 – 1230 BC), then according to the archaeological record no one was living at Jericho at the time when Joshua supposedly conquered it. However, if one holds to an early date for the conquest (i.e., ca. 1406 BC), then the archaeological profile fits much better. To be sure, archaeological remains from LB I (1550 – 1400 BC) are not plentiful on the tell but pottery from nearby tombs indicates that people were living there at the time of the conquest. In addition, it is interesting to note that scarabs of the early kings of the Eighteenth Egyptian Dynasty up through the rule of Amenhotep III have been found at Jericho. Since the scarabs of later Egyptian kings are not found at Jericho, it is not unreasonable to conclude that this is because Jericho ceased to be a viable city during the reign of Amenhotep III (1391 – 1353 BC) — possibly because of the Israelite conquest around 1406 BC. The much-sought-after walls that collapsed in the attack (Josh 6:20) were probably the walls of the MB II

CONQUEST OF GILEAD AND BASHAN

Damascus

Mt. Hermon

Laish (Dan)

Bashan

Hazor

Sea of Kinnereth

Karnaim
Ashtaroth
Og, King of Bashan

Edrei

Beth Shan

Ramoth Gilead

Gilead

Jebel Druze

Salecah

Israelites

Jordan R.

Jabbok R.

Mahanaim

Approximate territory of Og of Bashan under Israelite control

0   10 km.
0   10 miles

city, found by Kenyon, that were used by people living at Jericho during LB I — up until the conquest of the city by Joshua.

The conquest of Ai (Josh 7 and 8) presents the historical geographer with serious problems. Biblical Ai is normally identified with the site named et-Tell. This identification is based on several factors, including Ai's geographical relationship to Bethel, which is said to lie to the west of it (Gen 12:8; Josh 8:9, 12). Thus the usual identification

▼ *Jericho oasis from the mound of Jericho*

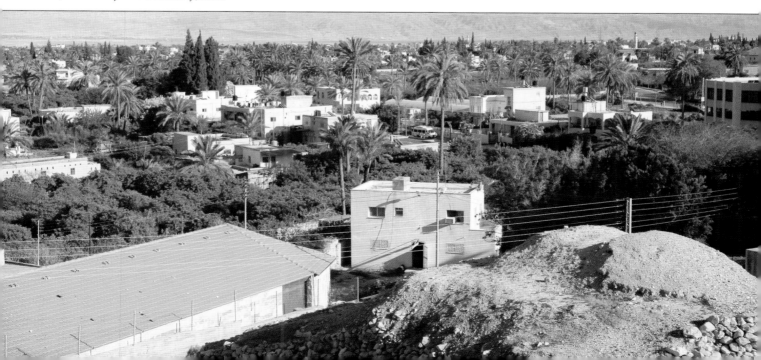

Jericho: Middle Bronze "revetment wall" upon the top of which one of the city walls stood

However, the archaeological picture of et-Tell does not agree at all with the historical data found in the Bible. According to the excavators, the large Early Bronze city was destroyed ca. 2400 BC, and the site was not inhabited again until ca. 1200 BC. Thus, no matter which date of the conquest is espoused, whether 1406 or 1250 BC, et-Tell was apparently unoccupied at the time Ai was supposed to have been conquered by Joshua.

In fact, there were probably three different places called "Ai" in the Bible. The Ai mentioned in the patriarchal narratives (ca. 2000 BC; Gen 12:8; 13:3) is to be identified with the just-mentioned remains at et-Tell — the EB ruins would have been very impressive at this time.

of Bethel with modern Beitin lends support to the identification of Ai with et-Tell. In addition, other geographical data in Joshua 7 – 8 concerning Ai — a place for the ambush party to the west of the city (8:13), a valley to the north (8:13), and a "place overlooking the Arabah" to the east (8:14) — all fit well with the et-Tell identification.

The Ai destroyed by Joshua (ca. 1400 BC, Josh 7 – 8) is probably to be located at the 2-acre fortified site called Khirbet el-Maqatir. This site is located only 0.6 mi. to the west of et-Tell and 1 mi. south-southeast of Beitin (= Bethel). Khirbet el-Maqatir has been excavated and a LB I walled fort has been found there. Its archaeological profile

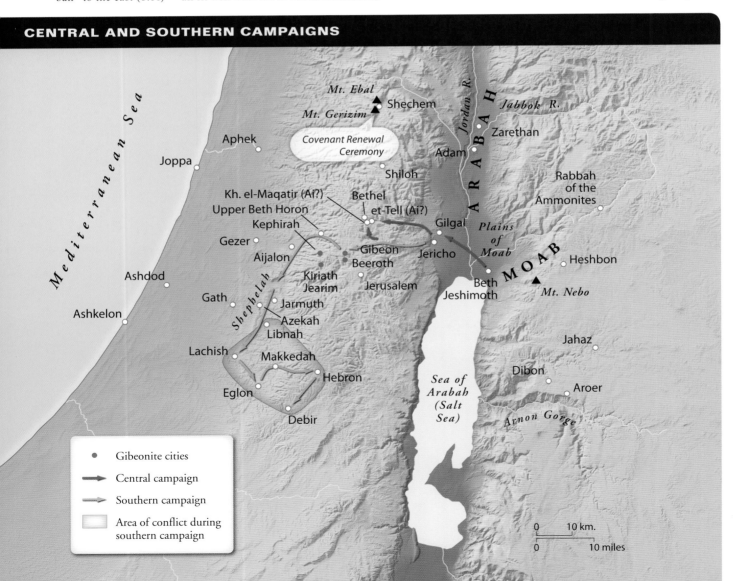

CENTRAL AND SOUTHERN CAMPAIGNS

Mediterranean Sea

Mt. Ebal
Mt. Gerizim
Shechem
Covenant Renewal Ceremony
Aphek
Joppa
Shiloh
Adam
Zarethan
ARABAH
Jabbok R.
Jordan R.
Rabbah of the Ammonites
Kh. el-Maqatir (Ai?)
Upper Beth Horon
Kephirah
Gezer
Aijalon
Bethel
et-Tell (Ai?)
Gibeon
Beeroth
Gilgal
Jericho
Plains of Moab
Heshbon
MOAB
Kiriath Jearim
Jerusalem
Shephelah
Gath
Ashdod
Jarmuth
Azekah
Libnah
Beth Jeshimoth
Mt. Nebo
Ashkelon
Lachish
Makkedah
Hebron
Eglon
Debir
Sea of Arabah (Salt Sea)
Dibon
Aroer
Jahaz
Arnon Gorge

Legend:
- • Gibeonite cities
- → Central campaign
- → Southern campaign
- ☐ Area of conflict during southern campaign

0   10 km.
0   10 miles

▲ *Gibeon: The chief of the four Gibeonite cities that made a treaty with Israel (Josh 9 – 10)*

and location, slightly "east of Bethel" (= Beitin), but mainly south, fits comfortably with the geographical data from the book of Joshua noted above.

The Ai of Ezra 2:28 and Nehemiah 7:32 (sixth and fifth centuries BC) is to be located at Khirbet Haiyan (about 1 mi. southeast of et-Tell).

After the conquest of Ai, the Israelites moved northward to the Shechem area. Nearby, on Mount Ebal, Joshua built an altar to God, sacrificed, and wrote on stones (monumental, standing stones?) a copy of the Mosaic law (Josh 8:30 – 35). There on Mount Ebal and Mount Gerizim the Israelites read the curses and blessings of the law (Josh 8:33 – 34; Deut 27:11 – 14) as the covenant between the people and Yahweh was renewed.

When the people of the land of Canaan heard of the Israelite victories at Jericho and Ai, they banded together for mutual defense (Josh 9:1 – 2). However, a group of Hivites living in the cities of Gibeon, Kephirah, Beeroth, and Kiriath Jearim (vv. 7, 17) chose to make a treaty with Israel. At the time when they made the treaty, the Israelites thought that the Hivites were "from a distant country" (v. 6). But Israel soon discovered that the Gibeonites lived in the heart of the land. However, Israel still observed the terms of the treaty and did not destroy the Gibeonites, although they consigned them to "serve as woodcutters and water carriers" (v. 23).

When the king of Jerusalem heard of the Gibeonite-Israelite treaty, he was alarmed: a major neighboring city (10:2), Gibeon, had gone over to the side of the invading Israelites, and it, as well as its three allied cities, sat astride the two major roads that led from Jerusalem to the coast. Thus Jerusalem's lifelines to the coast and to her Egyptian ally were cut off, and she could become easy prey to the Israelites.

Jerusalem countered by assembling a coalition consisting of the kings of Hebron, Jarmuth, Lachish, and Eglon (Josh 10:5). As these rulers moved against the city of Gibeon, probably to reopen the road to the coast, the Gibeonites appealed to Joshua for assistance on the basis of their treaty. Joshua responded by marching all night from Gilgal up into the hill country to relieve the siege of Gibeon. The coalition was defeated, and the kings and their armies fled westward, down the descent of Beth Horon, heading for the safety of their cities located in the Shephelah — to Jarmuth, Lachish, and Eglon (Josh 10). Their line of retreat led them through the Valley of Aijalon, past Azekah, to Makkedah. The Israelites, with divine assistance — hailstones and a prolonged day — defeated these armies, and eventually the kings themselves were executed at Makkedah (vv. 22 – 27). Cities captured as a result of this campaign included Libnah, Lachish, Eglon, and Makkedah in the Shephelah, as well as Hebron and Debir, both located in the southern portion of the hill country. Thus, what began as a rescue mission ended with the conquest of southern Canaan — the second major campaign led by Joshua.

The final phase of the conquest of Canaan — the northern campaign — occurred when Jabin, the king of Hazor, put together a coalition that included (among others) the kings of Madon, Shimron, and Acshaph (Josh 11:1 – 3). These kings and their armies, horses, and chariots encamped at the "Waters of Merom" (vv. 4 – 5). Although

▲ *Hazor: Canaanite worship center with numerous "standing stones."*

THE NORTHERN CAMPAIGN

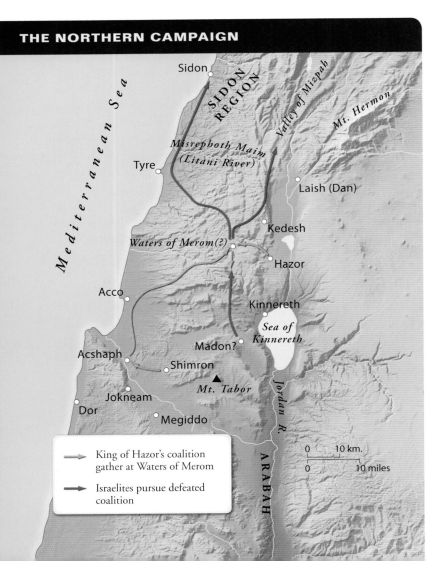

King of Hazor's coalition gather at Waters of Merom

Israelites pursue defeated coalition

details are not given, the Israelite attack was successful, and as the defeated kings retreated, the Israelites pursued them to the Sidon region, to Misrephoth Maim, identified with the Litani River, and to the "Valley of Mizpah on the east" (v. 8), probably a portion of the Rift Valley north of modern Metulla. Jabin's city, Hazor, which "had been the head of these kingdoms," was burned (vv. 10, 13). The Israelites, who evidently had no immediate use for them, hamstrung the horses and burned the chariots (v. 9). However, it is evident that the Israelites did not follow up their victory by establishing a settlement at Hazor, for archaeology shows that Canaanites reoccupied the city and lived there until its conquest by Deborah and Barak (Judg 4–5), around 1200 BC.

Thus, by the end of the fifteenth century BC the initial stages of the conquest of the land of Canaan had been completed. Yet the biblical writers were well aware that within Canaan there were still large portions of country controlled by non-Israelites (e.g., Josh 13:1–7). The apportionment of the land, the settlement of the Israelites in it, and the attempt to deal with the non-Israelite population groups would preoccupy the Israelites for the next four hundred years.

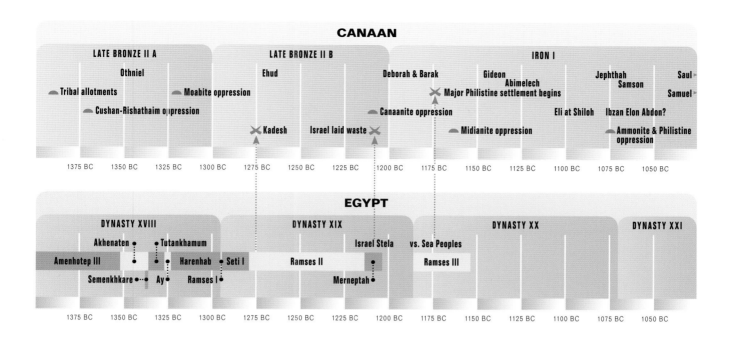

SETTLEMENT IN THE LAND OF CANAAN

Allotment of the Land

Soon after the initial conquest of the land of Canaan had been completed, the various Israelite tribes began to receive territory. The basic record of these allotments is found in Joshua 13 – 21. Although the actual apportionment took place during the days of Joshua and Eleazar the son of Aaron (early fourteenth century BC; Josh 14:1; 19:51), it would seem that later copyists and editors of the book of Joshua may have "updated" the lists of cities mentioned as belonging to the various tribes.

The tribal territories are described in three basic ways. In some instances the boundary of a given tribe is described as a line running from point A to point B to point C, etc., in a "dot-to-dot-to-dot" fashion (e.g., Judah's boundary in 15:1 – 12). The second method is to simply list cities belonging to a given tribe, as in the cases of Judah (15:21 – 63) and Benjamin (18:21 – 28). The third method is to note the cities on the extremities of a tribal territory, as in the case of Reuben (13:16 – 17; "from Aroer … to Heshbon").

Judah (Josh 15; Judg 1:8 – 18)

The first tribe to be allocated territory in the land of Canaan was the tribe of Judah. The author of Joshua was particularly interested in this tribe, for all of chapter 15 (sixty-three verses) describes its allotment. The first twelve verses give a "dot-to-dot-to-dot" description of Judah's boundaries. Its southern border (15:1 – 4) was identical with that of the land of Canaan (Num 34:3 – 5), while its northern border coincided first with the southern boundary of the tribe of Benjamin (Josh 15:5 – 10; 18:14 – 19) and then, following the path of the Nahal Sorek westward to the Mediterranean Sea (15:10 – 11), with the southern boundary of the tribe of Dan.

▼ *Hill Country of Judah where Israel settled. Terrace farming, rich soil, stone buildings.*

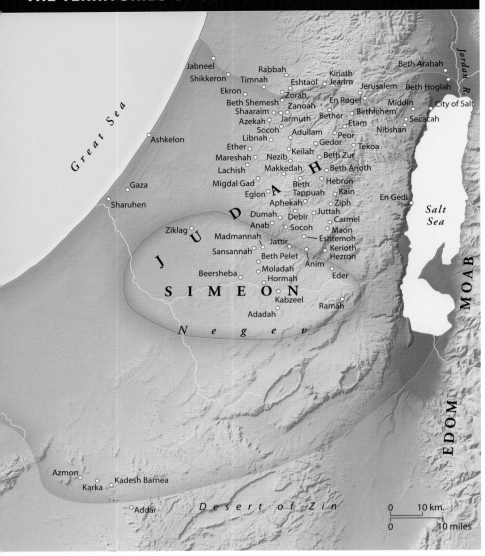

Jabneel
Shikkeron
Rabbah
Timnah
Ekron
Eshtaol
Kiriath Jearim
Beth Arabah
Jerusalem
Beth Hoglah
Zorah
Beth Shemesh
Zanoah
En Rogel
Middin
City of Salt
Shaaraim
Jarmuth
Bether
Bethlehem
Secacah
Azekah
Socoh
Adullam
Etam
Nibshan
Libnah
Gedor
Peor
Ether
Keilah
Tekoa
Mareshah
Nezib
Beth Zur
Lachish
Makkedah
Beth Anoth
Migdal Gad
Beth
Hebron
Eglon
Tappuah
Kain
Aphekah
Ziph
En Gedi
Dumah
Debir
Juttah
Anab
Socoh
Carmel
Ziklag
Madmannah
Jattir
Maon
Eshtemoh
Sansannah
Beth Pelet
Kerioth
Hezron
Beersheba
Moladah
Anim
Hormah
Eder
Kabzeel
Ramah
Adadah

J U D A H

J U

S I M E O N

N e g e v

Great Sea
Ashkelon
Gaza
Sharuhen

Salt Sea

Jordan R.

MOAB

EDOM

Azmon
Karka
Kadesh Barnea
Addar

D e s e r t   o f   Z i n

0   10 km.
0   10 miles

A description of the granting of Hebron and Debir to Caleb and Othniel is followed by a long list of 132 cities. This list is divided into four major geographical areas: the Negev (Josh 15:20 – 32), the Shephelah (vv. 33 – 47), the hill country (vv. 48 – 60), and the Eastern Wilderness (vv. 61 – 62). These in turn are broken down into even smaller districts (notice the district totals in vv. 32, 36, 41, etc.) for a total of ten districts within the tribe — actually eleven, since an additional district, which included Bethlehem in the Hill Country, is found in the Greek translations of Joshua (15:59b) and probably should be included in the total.

Most historical geographers believe that originally the city list functioned as an administrative document (for the collection of taxes, etc.) and propose adding an additional district to make a total of twelve, one for each month of the year. The date of the city list is difficult to determine, but it has been suggested that it is not from Joshua's time (ca. 1400 BC), since archaeological surveys and excavations have turned up little evidence that many of the cities in the list existed at that time. Later dates proposed for the list range from the days of Solomon (970 – 931 BC) to the period of the Babylonian exile

(after 586 BC), with perhaps the most probable of these dates being in the reign of King Josiah (640 – 609 BC).

This city list describes the essential heartland of Judah. Throughout much of its history, Judah was confined to the heights of the mountains, but during periods of strength, expansion westward into the Shephelah and southward into the Negev occurred. Only on rare occasions did Judah in fact control the Philistine Plain. The rugged mountainous terrain was useful in providing Judah with a measure of defense against the hostile powers that were usually located on the coastal plain to the west.

Simeon (Josh 19:1 – 9; 1 Chron 4:24 – 43)

Although the allotment of land to the tribe of Simeon is described later in Joshua, the tribe received territory inside the allotment of Judah. Of the seventeen cities mentioned as being assigned to Simeon (Josh 19:2 – 7), fifteen were previously mentioned in the city list of Judah — most in the Negev district but a few in the Shephelah. Simeon's primary location was in the western Negev, between Beersheba and the Mediterranean Sea. It should be noted that Simeon did not lose its tribal identity by being absorbed into the tribe of Judah, for it was still a distinct and growing tribe during the days of the Judean king Hezekiah (715 – 686 BC; 1 Chron 4:41 – 43). Indeed, the Chronicler describes its expansion northwestward toward Gedor (LXX, "Gerar") and eastward toward the Hill Country of Seir (1 Chron 4:39). Since Simeon's territory received only 10 inches of rain each year, the tribe specialized in keeping flocks, although in most years some grain crops could be grown.

Ephraim (Josh 16; Judg 1:29)

The second distribution was made to the tribe of Ephraim, the first son of Joseph. Only incomplete portions of its boundary description are given. Its southern boundary matched the northern one of Benjamin (Josh 16:1 – 5; 18:12 – 13), while on the west it theoretically was blocked from the sea by the tribe of Dan (19:40 – 48). Its northwestern boundary along the Kanah Ravine was well defined (16:8; 17:7 – 10), while on the northeast and east fewer border points are provided. For most of its history, Ephraim was confined to the rugged mountainous area. Deep V-shaped valleys provided it with security and inaccessible areas (see pp.46). Since no city list for the tribe of Ephraim is provided, relatively little is known about its population centers. In addition, some Ephraimites settled in the territory of Manasseh located to the north (16:9).

Manasseh (Josh 17; Judg 1:27 – 28)

The third lot fell to the second son of Joseph, Manasseh. While a portion of the tribe settled in the land of Gilead (see below), the other part settled in Canaan. Since no list of settlements is provided for Manasseh, there is somewhat of a "blank" spot on the map in comparison to tribes bordering it on the north and south. In addition, only the southwestern boundary is described in some detail (Josh 17:7 – 11), so that its remaining borders are difficult to define with precision.

Manasseh stretched from the sea to the Jordan River; the southern boundary was conterminous with Ephraim, while on the north, from west to east, Manasseh bordered on Asher, Zebulun, and Issachar. The boundary with these northern tribes evidently passed through the Jezreel and Harod valleys, which were divided among them. But cities such as Beth Shan, Ibleam, Dor, Endor, Taanach, and Megiddo, while assigned to Manasseh (v. 11), were difficult to capture because of the strength of the Canaanites in the plain, who possessed iron chariots (v. 16) — a fact well known from the annals of Thutmose III (p. 101). It probably was not until the reigns of David and Solomon four hundred years later (i.e., 1010 – 931 BC) that the Israelites were able to take control of these powerful cities. Since the choice land on the plains was controlled by the Canaanites, the sons of Joseph (Ephraim and Manasseh) settled in the heavily forested hill country, cutting down trees to secure needed farmland (vv. 15 – 18).

After the assignments to Judah, Ephraim, and Manasseh had been completed, the focus of activity shifted from Gilgal to Shiloh. At Shiloh, in the Hill Country of Ephraim, the tabernacle was set up. From there, Joshua sent out three men from each of the remaining tribes to prepare a written survey of the remaining portions of Canaan (Josh 18:4, 9). Based on this survey, the remaining allotments were made.

Benjamin (Josh 18:11 – 28; Judg 1:21)

Benjamin's territory fell between the two powerful tribes of Ephraim and Judah. Its allotment is described in two ways: first by a boundary list and then by a list of cities in Benjamin's territory. Its northern boundary coincided with that of Ephraim and its southern with that of Judah. On the east it was bounded by the Jordan River, and on the west its boundary, instead of extending into the Aijalon Valley, confined Benjamin to the hills overlooking the valley. Of particular interest — as evidenced by the minute details — are the notations of its southern boundary in the Jerusalem area (Josh 18:16 – 17), where

it is explicitly noted that Jerusalem was located in Benjamin, not in Judah, despite the fact that later a Judahite (David) would capture the city and make it his personal possession.

Benjamin's city list (Josh 18:21 – 28) consists of two sections of twelve and fourteen cities respectively. It seems that the list is divided into eastern and western groupings of cities — approximately along the line of the watershed in the mountains.

The strategic importance of Benjamin cannot be overemphasized. One of the main approach roads from the coastal plain into the hill country ran through its western portion (see p. 47). On the east, several roads led down into the Rift Valley and joined at the oasis of Jericho and from there proceeded eastward across the fords of the Jordan into Transjordan. Thus, throughout history, Benjamin was one of the busiest tribal areas, for invading international powers often entered the hills via the roads from the east or the west, and the northern and southern Israelite tribes occasionally met in battle in the territory of the Benjamites as they sought to expand their influence.

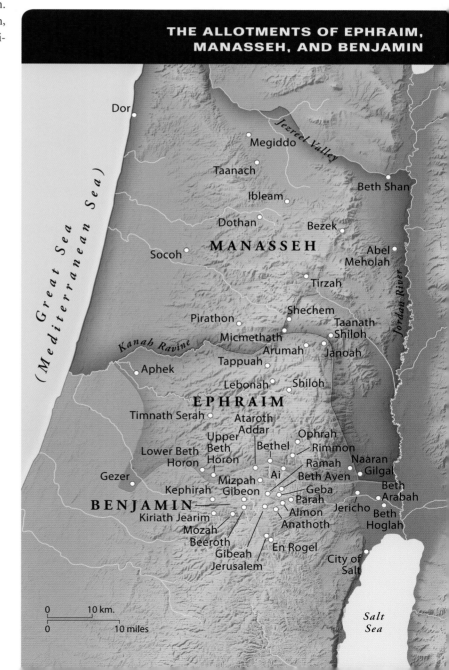

THE ALLOTMENTS OF EPHRAIM, MANASSEH, AND BENJAMIN

Zebulun (Josh 19:10 – 16; Judg 1:30)

A correct understanding of the boundaries of the tribe of Zebulun is a key in attempting to understand the other tribal allotments in Galilee, for Issachar, Naphtali, and Asher bordered it on the southeast, northeast, and northwest respectively. First the southern boundary of Zebulun is given. Sarid, a city near the center of the southern border, is the starting point, and the boundary is first traced westward from Sarid to Dabbesheth and then eastward from Sarid to Daberath, Japhia, and Gath Hepher. The boundary passed through Gath Hepher on the east side and circled around so that its northern portion touched Rimmon and passed through the modern Bet Netofa Valley to Hannathon. From there it headed southwest and south, following the Valley of Iphtah El (Wadi el-Malik) for a short distance and then the Wadi Musrarah to the Kishon.

Fourteen cities are mentioned as being included in Zebulun. The greater part of Zebulun's allotment was confined to the high ground overlooking the Jezreel Valley to the south, yet on occasion, Zebulunite territory stretched across the valley to include a city such as Jokneam (cf. Josh 21:34). However, the tribe of Zebulun was not able to drive out the Canaanites who lived in Kitron and Nahalol (the Nahalal of Josh 19:15; see Judg 1:30).

Zebulun was situated near several major trade routes that ran through the Jezreel Valley, the east – west route that ran through Hannathon to Acco, and the route that connected Megiddo with Acco via Shimron and Hannathon. Because of its proximity to these routes, the Zebulunites were open to considerable outside influences and probably maintained commercial relations with Mediterranean port cities (Deut 33:18 – 19).

Issachar (Josh 19:17 – 23)

The towns given to Issachar were located in the valleys and on the basalt heights of eastern Lower Galilee as well as in the eastern portion of the Jezreel Valley. Because of the basalt rocks and the lack of water sources, the heights were never densely settled; the major settlements were located in the valleys. The description of the allotment of Issachar is composed of a city list, along with brief boundary descriptions (Josh 19:22). Its northern line seems to have run from the hills overlooking the Jabneel Valley westward past Mount Tabor to Kesulloth. On the west, its boundary ran from Kesulloth to the city of Jezreel, indicating that only the eastern portion of the Jezreel Valley was within its territory. On the south, its boundary headed southeast from Jezreel, following the Harod Valley to the Jordan, its eastern limit.

Issachar's location was such that the major international highways ran close to its southern and western borders, and because of the relatively gentle terrain it would have been easy for armies on the march to invade. Indeed, major cities such as Beth Shan and Anaharath probably remained under non-Israelite control until the period of David and Solomon.

Asher (Josh 19:24 – 31; Judg 1:31)

The tribe of Asher was apportioned territory in the northwestern corner of Israel. It was bounded on the south and east by Manasseh, Zebulun, and Naphtali and on the north by Phoenicia. Its territory stretched from Mount Carmel in the south to the Litani River on the north and theoretically included the coastal plain and the western hills of Upper and Lower Galilee. However, according to Judges 1:31, Asher was not able to take control of seven important cities located on the coastal plain: Acco, Sidon, Ahlab, Aczib, Helbah, Aphek, and

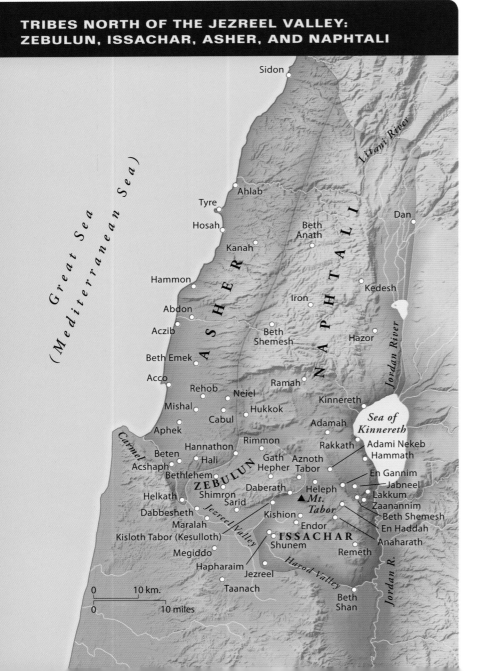

TRIBES NORTH OF THE JEZREEL VALLEY: ZEBULUN, ISSACHAR, ASHER, AND NAPHTALI

▲ *Rebuilt four-room house. A typical Israelite dwelling down to the Babylonian exile (586 BC). Three parallel rooms and one long cross room.*

Rehob. Thus the Asherites were said to have "lived among the Canaanite inhabitants of the land." It probably was not until the days of David and Solomon that the plain came under Israelite control (1 Kings 4:16) and then only for a brief time, for it was soon traded to Hiram, king of Tyre, in exchange for financial benefits (9:10 – 14).

Strategically and economically, Asher's territory was important. The one natural harbor of the whole country, Acco, was located there — but rarely controlled by Israel. The international north – south route along the coast ran through Asher, as did east – west routes that connected the port of Acco with Transjordan and even Damascus. During the Old Testament period, the maritime powers of Tyre and Sidon took great interest in controlling the area, to the economic detriment of Israel.

Naphtali (Josh 19:32 – 39; Judg 1:33)

The tribe of Naphtali was bounded on the south by Issachar and on the west by Zebulun and Asher. Its northern boundary is not recorded, but it probably extended as far as the Israelites settled — that is, south of an east – west line that coincided with the east – west portion of the Litani River. Its eastern boundary seems to have been the Jordan River and the Sea of Galilee.

The outline of Naphtali's territory is first sketched by a boundary description, which is then supplemented by a city list. In the former, the report of its southern boundary with Issachar is most detailed. The starting point in the description is Heleph, from which the boundary pro-

ceeded eastward. It is clear that Naphtali controlled the Jabneel Valley, while Issachar was situated on the heights to the south of it. Then the border proceeded westward from Heleph to the Mount Tabor area and then northward along its common line with Asher. It is interesting to note how the prominent hill of Mount Tabor served as the meeting point for the boundaries of three tribes — Issachar, Naphtali, and Zebulun.

The city list of Naphtali proceeds geographically from south to north. The cities mentioned prior to Ramah were located in Lower Galilee, while those after it were in Upper Galilee (see p. 37). Since a fair portion of Galilee was unoccupied prior to the Israelites' arrival, Naphtali was probably able to settle in the hills relatively early in the period of the judges. However, it was unable, save for relatively brief periods of time, to take control of the larger Canaanite centers, such as Beth Shemesh and Beth Anath (Judg 1:33), which sat astride an important east – west road in Upper Galilee.

Dan (Josh 19:40 – 48; Judg 1:34 – 35; 17 – 18)

The tribe of Dan was allotted territory to the west of Benjamin, and Dan's eastern boundary, on the western slopes of the mountains overlooking the coastal plain, coincided with the western boundary of Benjamin. Its southern boundary was identical to Judah's northern one and followed the Sorek Valley out to the Mediterranean Sea (Josh 15:10 – 11). Dan's northern boundary is not outlined, but from the cities mentioned and from comparison with the cities allotted to Ephraim (e.g., Gezer

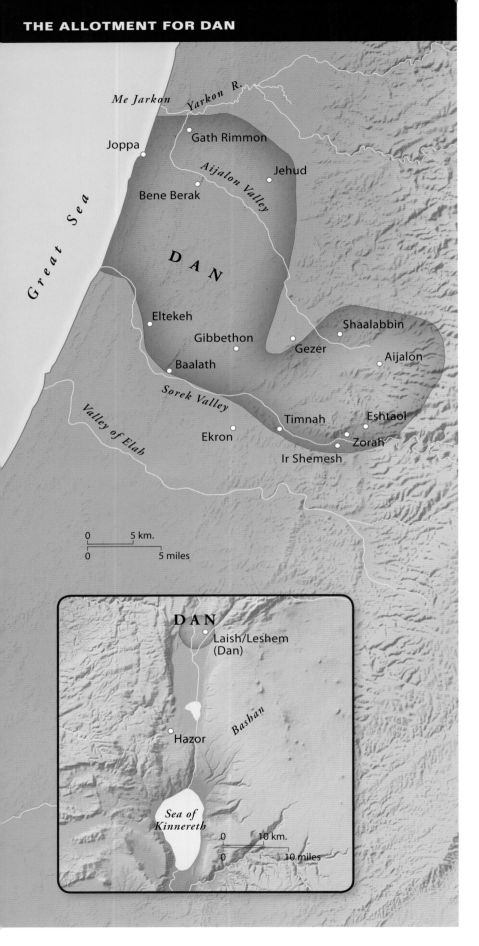

[16:3, 10]), its irregular shape can be surmised, and it appears to have stretched to "Me Jarkon" (i.e., to the waters of the Jarkon [River], 19:46). Concerning several of the cities mentioned there is some uncertainty as to whether they belonged to Judah or Dan (e.g., Zorah [15:33; 19:41], Eshtaol [15:33; 19:41], Ir Shemesh [probably = Beth Shemesh; 15:10; 19:41], and Ekron [15:45; 19:43]).

Dan's territory sat astride the Aijalon Valley, through which the major approach road into the Hill Country of Ephraim, Benjamin, and Judah ran. In addition, the main international north – south route ran through Dan's western extension. Thus the great powers of antiquity were interested in controlling this territory, as were the powerful local Canaanite/Amorite chiefs, with the result that the Danites were not able to expand westward but were rather confined to the western slopes of the mountains in the Beth Shemesh, Zorah, Eshtaol, and Kiriath Jearim region (see Judg 1:34 – 35; 13 – 16). Not until the days of David and Solomon did Israel take control of Danite territory, and then only for a brief period of time (see 1 Kings 4:9; 2 Chron 2:16).

Due to Amorite pressures, some of the tribe of Dan moved northward to Laish/Leshem, which they captured and renamed "Dan" (Judg 17 – 18). They thought that they had moved from an area of insecurity to one of security and prosperity (18:7, 27 – 28); yet later in biblical history this was the area that invaders from the north (the Arameans, Assyrians, and others) attacked first.

Reuben, Gad, and Manasseh (Josh 13:8 – 33; Num 32)

After the conquest of the kingdoms of Sihon and Og but before crossing the Jordan River into Canaan, the tribes of Reuben, Gad, and part of Manasseh asked Moses to apportion to them the territory east of the Jordan, because of its suitability for pasturing their large herds and flocks (Num 32:1). After securing a pledge from them that the adult males would cross into Canaan and participate in the wars of conquest, Moses apportioned them territory according to their wish. Two descriptions of these allotments are given in the Bible: Numbers 32:33 – 42 and Joshua 13:8 – 33. The longer list in Joshua begins by mentioning cities at the extremities of the tribes and then enumerates the cities belonging to each.

The tribe of Reuben (Josh 13:15 – 23; Num 32:37 – 38) received most of the territory stretching from the Arnon Gorge in the south to the city of Heshbon in the north (the Moabite tableland, or *Mishor* [see p. 62]), although at some point in time it seems that Gadites settled in some of their cities. The territory allotted to Gad stretched from Heshbon in the south to Mahanaim on the Jabbok River in the north and from just west of Rabbah (of the Ammonites) to the Jordan River on the west. In addition it seems that Gad received the whole of the Jordan Valley east of the river, all the way up to the Sea of Kinnereth (Josh 13:27). Finally, certain clans from the tribe of Manasseh settled the territory north of the Jabbok (i.e., northern Gilead and the Bashan [see pp. 33 – 35, 57 – 59], possibly north to Mount Hermon [Num 32:39 – 42; Josh 13:29 – 31]).

These tribes, although settled in the land of Gilead, attempted to maintain relations with the other Israelite tribes in the land of Canaan (Josh 22). Throughout the remaining portion of Old Testament history, these tribes felt continuous pressure from a number of sources: from the Moabites south of the Arnon Gorge, who laid claim to Reubenite and Gadite territory north of the Arnon; from the Ammonites on the east, who made similar claims; from desert tribes, such as Midianites and Ishmaelites; from the Arameans of Damascus, who desired to control the north – south Transjordanian Highway that led to Arabia; and also from the Assyrians, who pressed similar claims. Despite these and other pressures, an Israelite/Jewish presence was maintained in Transjordan through the end of the biblical period and beyond.

Levitical Cities (Josh 21; 1 Chron 6:54 – 81)

As part of the process of settling in the land of Canaan, Joshua and Eleazar the priest assigned towns and their pasturelands to the Levites. Not settled exclusively by Levites, forty-eight cities were assigned to the three Levitical clans, and approximately four such cities were designated from each of the tribes. A good number of the cities were old Canaanite centers, some of which did not come under Israelite control until the days of David and Solomon. By the time of David, the Levites were involved not only with matters pertaining to God but also with "the affairs of the king" (1 Chron 26:29 – 32).

Thus in their towns, scattered throughout Israel, the Levites probably were spreading a godly influence through their teaching activities, but they also seem to have been promoting loyalty to the

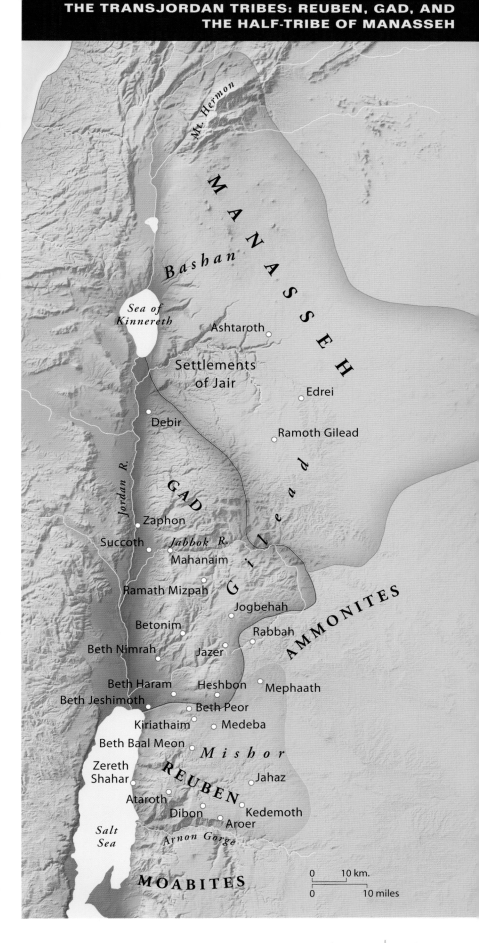

ASHER

NAPHTALI

Kedesh

Abdon

MANASSEH

Acco   Rehob

Mishal

Golan

Rimmon

Ashtaroth

ZEBULUN

Kartan

En Gannim

Sea of
Kinnereth

Yarmuk R.

Helkath

Daberath

Jokneam   Kishion

ISSACHAR

Jarmuth

Taanach

MANASSEH

Ramoth
Gilead

Great Sea
(Mediterranean Sea)

Shechem

Jabbok R.

Mahanaim

Gath
Rimmon

GAD

Joppa

EPHRAIM

Shiloh

Beth
Horon

Jazer

DAN

BENJAMIN

Mephaath

Eltekeh   Gezer

Geba

Heshbon

Gibbethon

Aijalon   Gibeon

Almon

Anathoth

Bezer

Beth
Shemesh

Jerusalem

REUBEN

Libnah

Jahaz

JUDAH

Hebron

Salt Sea
(Dead
Sea)

Kedemoth

Gaza

Debir   Juttah

Arnon Gorge

Jattir   Eshtemoa

Beersheba

SIMEON

Hebron   Cities of refuge (underlined)

Levitical cities:

△   Towns received by Kohathite clans

△   Towns received by Gershonite clans

▲   Towns received by Merarite clans

Zered River

0   10 km.

0   10 miles

Davidic dynasty. Because of their attachment to the temple in Jerusalem and its chief patron, the Davidic king, it is no wonder that many Levites moved south to Judah and Jerusalem at the time of the revolt of the north in the days of Jeroboam I (931 BC; 2 Chron 11:13 – 17).

Cities of Refuge (Josh 20; Deut 4:41 – 43)

As part of the allotment process, six cities were set aside as places to which a person who committed manslaughter could flee. There the case would be tried (Josh 20:4, 6), and if it was indeed judged as manslaughter rather than murder, the slayer was required to remain in that city until the death of the high priest (Josh 20:6; Num 35:9 – 34; Deut 4:41 – 43; 19:1 – 14). Three of the cities were located to the west of the Jordan River (Kedesh, Shechem, and Hebron) and three to the east (Golan, Ramoth, and Bezer). All of these cities were Levitical cities, where presumably the law of God was well known due to the teaching activity of the Levites. Both sides of the Jordan had northern cities (Kedesh and Golan), central cities (Shechem and Ramoth in Gilead), and southern cities (Hebron and Bezer), so that the accused could readily find refuge from the hand of the "avenger of blood" (Josh 20:3).

Thus, in theory at least, by the end of the period of initial conquest, the land that God had promised to Abraham, Isaac, and Jacob had been assigned to their descendants, and religious and political institutions were in the process of being established. Yet, as the period of the book of Judges will show, complete possession of the land had not yet been attained. That remained for future generations to accomplish.

The Period of the Judges

At the end of the initial stage of the conquest of Canaan, which had been led by Joshua (ca. 1390 BC), the Israelites were well aware that not all of the land had been conquered, despite the fact that, theoretically at least, the land had been allotted to the various tribes. The southwestern portion of the land, which was then counted as Canaanite and which would later become Philistine, was not yet under Israelite control (Josh 13:2 – 3; Judg 3:3; and the LXX of Judg 1:18). In fact, the whole of the coastal plain and its vicinity — including the cities of Gezer (Judg 1:29), Aijalon, Shaalbim (v. 35), and Dor (v. 27) — remained Canaanite. The Jezreel Valley and its southeastern extension to Beth Shan remained in Canaanite and/or Egyptian hands (vv. 27 – 28), while the Mediterranean coastline north of Mount Carmel was under Sidonian control (Josh 13:4; Judg 1:31 – 32; 3:3). Indeed, the whole of the land of Canaan north of Dan up to Lebo Hamath remained unconquered (Josh 13:4 – 5; Judg 3:3). In addition, there were foreign enclaves that remained independent, such as the city of Jebus (Jerusalem; Josh 15:63; Judg 1:21; 19:11) and the regions of the Geshurites and Maacahites (Josh 13:13). Most of these areas came under Israelite control by the days of David and Solomon (e.g., see 1 Kings 4:7 – 19).

It is thus evident that the initial areas of Israelite settlement included (1) the hill countries of Judah, Benjamin, Ephraim, and Manasseh; (2) portions of Upper and Lower Galilee; and (3) Gilead

and the tableland of Moab. The Israelites were in effect moving into areas that had not previously been densely populated, save for a few scattered urban centers such as Hebron and Shechem. The process of Israelite settlement must have proceeded in a somewhat peaceful fashion over the next 340 years, as farmsteads were established, scrub forests cleared, terraces built, and crops (especially grapes, olives, figs, almonds, and wheat) planted.

In addition, the recent invention of rock-hewn, plaster-lined cisterns allowed the people to settle in areas distant from springs, thus opening up new territorial vistas for the Israelites. It appears that throughout the period of the judges (ca. 1370 – 1050 BC) Israel was able to remain in the mountains, somewhat aloof and removed from the threats of Egyptian kings (e.g., Seti I and Ramses II; map p. 124), who in the main were concerned with controlling the international routes through the country.

Israel was, however, threatened by lesser peoples, both near and far. In the process of settling down, some Israelites began to worship pagan deities, including Baal (who was thought to be responsible for the fertility of the land) and his consort Ashtoreth (goddess of war and fertility), as well as Asherah. As punishment for these sins, Yahweh sent foreigners who oppressed the Israelites. Eventually Israel cried out in repentance, and God sent "judges" who delivered Israel from her oppressors and ushered in periods of "rest."

This cycle of lapse into sin, divine punishment, repentance, deliverance, and rest (Judg 2:10 – 19) is illustrated in the accounts of the six major judges (those whose exploits are recounted at length) and of the six minor judges. Since the judges operated in various parts of the country, it is not necessary to assume that their periods of leadership were successive; in some cases their activities may have been concurrent. For example, the Ammonite oppression in Transjordan and the Philistine oppression from the west may have occurred at the same time, so that the judgeships of Samson and Jephthah may have coincided, with Jephthah active in the east and Samson in the west.

Although the literary presentation of the judges seems to be roughly in chronological order, in all probability the events recorded in Judges 17 – 21 took place early in the period (i.e., mid-fourteenth century BC). The Danite move from the region of the valleys of Aijalon and Sorek (Josh 19:47; Judg 17 – 18; and see 1:34 – 35) has already been mentioned (see p. 118), but it is well to note that the move was not attributed to Philistine pressure (that would come later), and that not all Danites moved from their tribal allotment. Later Samson, a Danite, would deal with the Philistines.

The final episode in the book of Judges — that of the Levite and his concubine (chs. 19 – 21) — is set in the days of Phinehas, son of Eleazar (20:28), early in the fourteenth century BC. It illustrates the moral corruption of the inhabitants of Gibeah and, evidently, of many in Benjamin as well, for the latter came to the defense of the former. All the Israelites from "Dan to Beersheba and from the land of Gilead [still distinguished from Canaan proper]" (20:1) assembled for battle against the Benjamites. The result of the various military encounters left only six hundred Benjamite men alive; from them the tribe eventually regained its numerical strength.

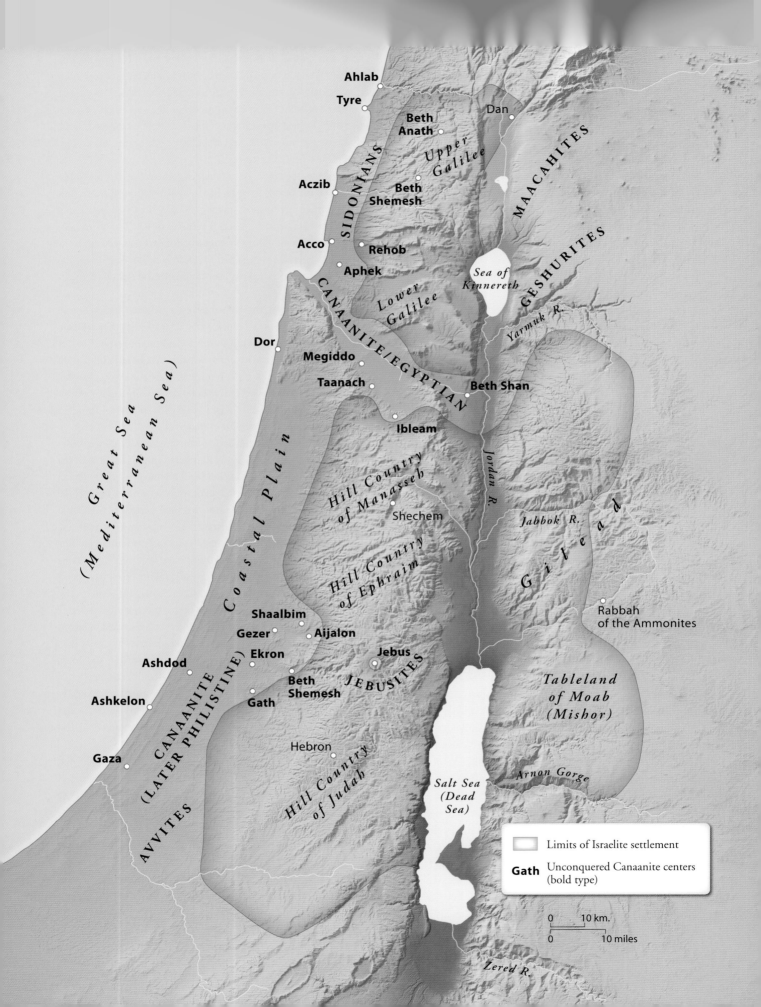

Ahlab

Tyre

Beth Anath

Dan

Upper Galilee

MAACAHITES

Aczib

Beth Shemesh

SIDONIANS

GESHURITES

Acco

Rehob

Aphek

Sea of Kinnereth

CANAANITE/EGYPTIAN

Lower Galilee

Yarmuk R.

Dor

Megiddo

Taanach

Beth Shan

Ibleam

Jordan R.

Jabbok R.

Great Sea (Mediterranean Sea)

Hill Country of Manasseh

Shechem

G i l e a d

Hill Country of Ephraim

Rabbah of the Ammonites

Shaalbim

Gezer

Aijalon

Ekron

Jebus

Tableland of Moab (Mishor)

Ashdod

Beth Shemesh

JEBUSITES

Coastal Plain

Ashkelon

Gath

CANAANITE (LATER PHILISTINE)

Hebron

Gaza

Hill Country of Judah

Salt Sea (Dead Sea)

Arnon Gorge

AVVITES

Limits of Israelite settlement

Gath Unconquered Canaanite centers (bold type)

0   10 km.

0   10 miles

Zered R.

The account contains a number of interesting geographical features. For example, Benjamite connections with Jabesh Gilead (Judg 21:12) were established through intermarriage. The Ridge Route from Bethlehem to Ramah is described (19:10–15); note the south–north sequence of Bethlehem, Jebus (Jerusalem), Gibeah, and Ramah, as well as the fact that one needed to turn off the north–south Ridge Route and go east in order to get to Jerusalem (vv. 11–12). In addition, the description of the location of Shiloh — to the north of Bethel, and east of the road that goes from Bethel to Shechem, and to the south of Lebonah (21:19) — is the most detailed description of the location of any city mentioned in the Bible.

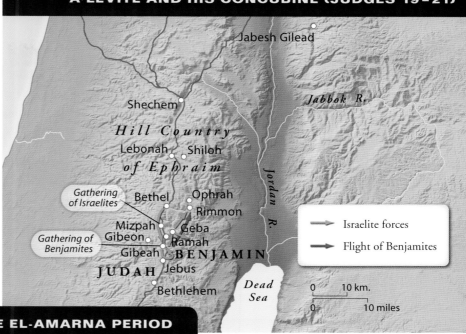

A LEVITE AND HIS CONCUBINE (JUDGES 19–21)

CANAAN IN THE EL-AMARNA PERIOD

Throughout the period of the judges, Egyptian sources shed welcome flashes of historical light on conditions in Canaan. In the late nineteenth century, Egyptian peasants discovered a large number of cuneiform tablets at el-Amarna, a site now known to have been the capital of the "heretic" king Akhenaton. These tablets, of which almost 400 have survived, date to the reigns of the Egyptian kings Amenhotep III (1391–1353 BC) and Akhenaton (1353–1337 BC). These clay tablets were written in Akkadian, the diplomatic language of the day.

Although the content of these texts is varied, from the historical and geographical point of view the letters are of extreme interest. Correspondence between the king of Egypt and his contemporaries in Babylonia, Assyria, Mitanni, Alashia, and Hatti shed light on the international scene. In one of the letters, for example, the king of Babylonia complains that his caravan has been attacked at Hannathon in the land of Canaan and that since the land of Canaan is under Egyptian control, the king of Egypt is responsible for its fate.

Most of the letters, however, are records of the correspondence between the numerous city-state rulers in the Levant and the king of Egypt. This correspondence demonstrates that the Egyptians were treating Canaan with benign neglect yet still maintained nominal control over the area. Not that Canaan was in total chaos, but its numerous rulers were constantly forming alliances — first with one neighbor and then with another — in an effort to gain the upper hand in the country.

One of the most prominent of these manipulators was Labayu of Shechem. From his base in the hill country, he was reaching out in all directions like an octopus, attempting to take control of the trade routes that ran through the Sharon Plain, the Jezreel Valley, and the region of Pehel in Transjordan. Kings such as Biridiya of Megiddo complained of his antics, requesting the intervention of the Egyptians, while Labayu consistently swore his (feigned?) allegiance to the Egyptians. Eventually Labayu's enemies in Canaan murdered him, but soon his sons were following in his footsteps.

Jerusalem, the other prominent hill country center, is also represented in the correspondence. The ruler of Jerusalem, Abdi-heba, was attempting to keep the Habiru at bay and was constantly interacting with the rulers of cities in the Shephelah (e.g., Gezer, Gath, Keilah, and Lachish). The web of intrigue there was similar in scope to that noted in the north.

When these letters were first discovered, some scholars thought that they were firsthand accounts of the Israelite (Hebrew) conquest of Canaan. However, the names of the kings of Jerusalem (Adoni-Zedek), Lachish (Japhia), and Gezer (Horam) that are mentioned in the Bible (Josh 10:3, 33) do not correspond to the names of the kings mentioned in the el-Amarna letters, namely, Abdi-heda (Jerusalem), Zimredda and others (Lachish), and Milkilu and Yapai (Gezer). According to the chronology adopted here, the events chronicled in these letters occurred soon after the initial conquests under Joshua (ca. 1406 – 1400 BC).

One of the most interesting groups mentioned in the el-Amarna letters are the Habiru, who were usually involved in anti-Egyptian activities and who threatened the stability of many of the city-states. It is possible that the term Habiru is related to the word Hebrew, and it may be that in some instances the activities of the Israelites (Hebrews), as they continued their conflict with the Canaanites and as they began to settle the hill country, are alluded to in the Amarna letters as activities of the Habiru. However, it does not appear that all Habiru were

EGYPTIAN CAMPAIGNS IN CANAAN

CYPRUS

HITTITES

Kadesh

Orontes R.

Lebo

Damascus

Great Sea
(Mediterranean Sea)

Tyre

Hazor

Sea of
Galilee

Yanoam

Megiddo

CANAAN

Beth Shan
Pehel
Rehob
Hammath

Jordan R.

Gezer

Ashkelon

Jerusalem

Gaza

Hebron

Dead
Sea

| 0 | 40 km. |
| 0 | 40 miles |

Zoan

Sile

EGYPT

→ Seti I's campaign in Canaan

→ Ramses II's expedition to Kedesh

→ Merneptah's campaign to Canaan

Hebrews, for the Habiru were active not only in the south but also in the northern Levant, where the Israelites were not active, nor were all Hebrews Habiru. Thus, it does not seem possible to maintain a one-to-one correspondence between the Habiru and the Hebrews.

In any case, it is interesting to note that cities such as Jericho, Bethel, Gibeon, and Hebron — cities in the Rift Valley and hill country that were conquered early by the Israelites — are not mentioned in the el-Amarna correspondence. Thus, although the letters do not seem to offer any one-to-one correspondence with known biblical events or personages, they do bear witness to the chaotic political conditions in Canaan and to Egyptian and Canaanite priorities along the coastal plain and in the valleys during the early fourteenth century BC — the very time when Israel was beginning to settle in the Hill Country of Judah and Ephraim.

The first of the six major judges was Othniel (Judg 3:7 – 11). During his days Cushan-Rishathaim (= "Cushan, the doubly wicked"?), king of Aram Naharaim, oppressed Israel. Aram Naharaim is usually identified with northwestern Mesopotamia, but this seems quite distant from Judah, where the oppression probably took place. It has been suggested that "Aram" is corrupt for an original "Edom," for the Hebrew writing of the two names is almost identical except for two Hebrew letters (*r* and *d*) that could easily be confused due to similar appearance. This view would make the nearby Edomites the oppressors, a more logical enemy than distant Mesopotamia. Yet the appearance of "Naharaim" is still a problem.

The second of the major Judges was Ehud who was one of the surviving members of the tribe of Benjamin (Judg 3:15; and see above, p. 121). Eglon, the king of Moab, along with his Transjordanian neighbors, the Ammonites, and also the Amalekites, oppressed Israel from their headquarters in the "City of Palms" (Jericho, according to 2 Chron 28:15). Ehud,

Istanbul, Museum of the Ancient Orient

▲ *Cuneiform copy of the treaty of Kadesh between Rameses II and Hatussilis III, the Hittite King.*

after delivering tribute to Eglon, killed him and escaped to Seirah (the forest) in the Hill Country of Ephraim (Judg 3:26 – 27). There he mustered the Israelite forces from the tribes of Benjamin and Ephraim and led them down into the Rift Valley to the fords of the Jordan River, where they were able to decimate the retreating Moabites and their allies. At least a portion of the land then enjoyed "peace" for eighty years (vv. 28 – 30).

Although the Bible does not mention it, the Egyptians were active in the land of Canaan at the end of the fourteenth and in the thirteenth centuries BC. For example, Seti I (1318 – 1304 BC) suppressed a revolt in the Beth Shan area that had been led by the kings of the cities of Hammath and Pehel. Besides reasserting Egyptian control in the Beth Shan region, he subdued the Habiru in Mount Yarmuta (i.e., in territory that had been assigned to the tribe of Issachar). He also subdued Yenoam in the Bashan and moved northward, conquering Hazor (lower city Level II) before moving farther northward along the Phoenician coast.

His successor, Rameses II (1279 – 1213 BC), battled the Hittites at Kadesh on the Orontes in his fifth year (1275 BC), indicating that

▼ *Ramasseum, Egypt: Rameses II in chariot conquering his enemies*

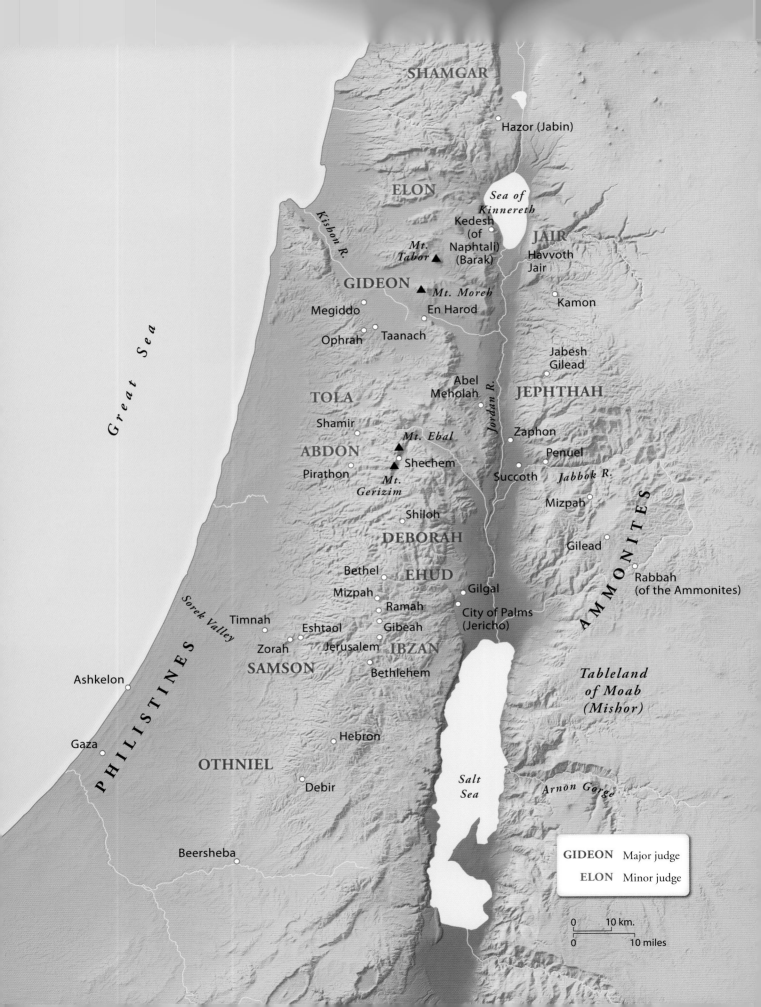

SHAMGAR

Hazor (Jabin)

ELON

Sea of Kinnereth

Kishon R.

Kedesh (of Naphtali) (Barak)

JAIR

Havvoth Jair

Mt. Tabor ▲

GIDEON

▲ *Mt. Moreh*

Megiddo

En Harod

Kamon

Ophrah

Taanach

Jabesh Gilead

Great Sea

TOLA

Abel Meholah

JEPHTHAH

Jordan R.

Shamir

Zaphon

ABDON

▲ *Mt. Ebal*

Penuel

Shechem

Pirathon

Mt. Gerizim

Succoth

Jabbok R.

Shiloh

Mizpah

AMMONITES

DEBORAH

Gilead

Bethel

EHUD

Mizpah

Gilgal

Rabbah (of the Ammonites)

Ramah

City of Palms (Jericho)

Sorek Valley

Timnah

Gibeah

Eshtaol

Jerusalem

IBZAN

Zorah

SAMSON

Bethlehem

Tableland of Moab (Mishor)

Ashkelon

Hebron

OTHNIEL

PHILISTINES

Salt Sea

Arnon Gorge

Gaza

Debir

Beersheba

| GIDEON | Major judge |
| ELON | Minor judge |

0   10 km.

0   10 miles

Canaan to the south was firmly under Egyptian control. Indeed, later the Hittites and Rameses II made a treaty that fixed their common boundary just to the south of Kadesh (ca. 1258 BC). In addition, "A Satirical Letter" (*ANET*, 476 – 78), dating to the thirteenth century BC, also gives clear evidence of Egyptian control of the coastal plains and valleys of Canaan, noting that Egyptian chariots and messengers moved freely through the country.

Although Merneptah (1213 – 1203 BC), the successor of Rameses II, was not as powerful as his forerunner, his stele, among other items, mentions his victory over Israel:

> Desolation is for Tehenu (Libyans); Hatti is pacified;
> Plundered is the Canaan with every evil;
> Carried off is Ashkelon; seized upon is Gezer;
> Yenoam is made as that which does not exist;
> Israel is laid waste, his seed is not;
> Hurru is become a widow for Egypt!
> All lands together, they are pacified;
> Everyone who was restless, he has been bound
> by the king … Mer-ne-Ptah. … (*ANET*, 378)

This is the first extrabiblical reference to "Israel," and it attests to Israelite presence in Canaan by at least 1209 BC.

The question as to how all of this Egyptian activity could be going on in Canaan without being mentioned in the Bible is a relevant one. The most plausible solution is that the Egyptians were interested in the plains and open valleys, the places where the international routes ran, while the Israelites were striving to secure a foothold in the more rugged mountainous areas. Thus the encounters between the two peoples seem to have been minimal.

The first of the minor judges, Shamgar, son of Anath, is mentioned in Judges 3:31 and 5:6. Since he was active against an early group of the Philistines, it seems reasonable to locate his activities in or near the Philistine Plain. Since he is called the "son of Anath" (a Canaanite goddess), it has been suggested that he came from the city of Beth Anath in Naphtali (Josh 19:38; Judg 1:33), but this would place his home far from the scene of his activity. Another possibility is that he was from Beth Anoth, a city of Judah (Josh 15:59), which would place his home much closer to the assumed site of his exploits.

During the first half of the twelfth century BC, the Canaanites were again asserting themselves in northern Israel. Led by Jabin, the king of Hazor, and his commander Sisera, they oppressed northern Israel for some twenty years. Under the leadership of Barak of Kedesh in Naphtali and of the prophetess Deborah, whose headquarters were at the palm tree of Deborah between Ramah and Bethel in the Hill Country of Ephraim (4:4 – 6), the Israelites mustered their forces on the slopes of Mount Tabor at the northeastern end of the Jezreel Valley while the Canaanites gathered near Megiddo in the Jezreel Valley. There on the wooded slopes of Tabor the Israelites were relatively safe from the operations of the nine hundred chariots controlled by Sisera (vv. 3 – 13).

At the appropriate moment, Barak led his forces down the mountain to battle the Canaanites in the plain below. Due to a sudden downpour and the flooding of the Kishon River, the Canaanite chariots were rendered ineffective, and Israel scored a decisive victory (Judg 4:14 – 16; 5:20 – 21). Sisera, possibly fleeing back toward Hazor, was killed by the wily Jael, who drove a tent peg through his head while he slept. Instead of the Canaanites dividing the silver and plunder of the Israelites near Taanach, by the waters of Megiddo (5:19; possibly their major military

▼ *Karnak, Egypt: Part of a relief of Pharaoh Merneptah conquering his foes. This may actually be an image of ancient Israelites between the legs of Merneptah's horse from 1209 BC! Note headdresses and pointed beards!*

▲ *View east toward Mt. Tabor where Deborah and Barak defeated the Canaanites. Hills of Galilee on the left, Jezreel Valley in lower right.*

camp), it was Israel that was victorious, and the northern portion of the land experienced relative peace for forty years (v. 31).

The exploits of Gideon, the judge who led Israel against the Midianites, their allies the Amalekites, and other eastern peoples, are recorded in Judges 6 – 8. Each year these peoples invaded Israel in the late spring/early summer, confiscating the newly harvested crops and grazing their livestock and camels in the fields. Gideon, from the tribe of Manasseh and the village of Ophrah, which overlooked the Jezreel Valley from the south, responded to this call. Mustering the tribes of Manasseh, Asher, Zebulun, and Naphtali at the spring of Harod, Gideon prepared for battle by culling his troops down to a select 300. Dividing this corps into three units, he surprised the Midianites — whose main camp was evidently on the northern slope of Mount Moreh at Endor (Ps 83:10) — with a night attack. The frightened Midianites fled in a southeasterly direction toward the Jordan River, Beth Shittah, Zererah, Abel Meholah, and Tabbath.

The Ephraimites joined the battle by seizing control of the fords of the Jordan and capturing the Midianite leaders Oreb and Zeeb. Gideon and his troops continued pursuing the Midianites eastward, passing Succoth and Peniel and heading up the Jabbok River into Gilead. Following the "route of the nomads east of Nobah and Jogbehah" (Judg 8:11), Gideon overtook the unsuspecting army of the eastern peoples at Karkor, a desert oasis in the Wadi Sirhan. After all but annihilating the opposing forces, Gideon and his men captured considerable booty. Subsequently, the land had peace for forty years.

Although Gideon's response to the request to rule over Israel was "no" ("The Lord will rule over you" [Judg 8:23]), one of his sons, Abimelech,

whose mother was from Shechem, had different ambitions. By slaughtering seventy of his brothers at Gideon's home in Ophrah, Abimelech was able to set himself up as king in Shechem (Judg 9). Only Jotham, the youngest son of Gideon, survived the slaughter, and in a fable Jotham predicted how the people of Shechem and Abimelech would eventually destroy each other (vv. 7 – 20). This, in fact, occurred after only three years of Abimelech's rule. Gaal, Abimelech's adversary who had replaced him as the leader in Shechem, and his forces were killed, and when his allies, the citizens of Shechem, took refuge in the temple of El/Baal-Berith (v. 46), Abimelech and his troops took wood from Mount Zalmon (Jebel el-Kabir [?] to the northeast of Shechem) and burned the tower down.

Shortly thereafter, as Abimelech was laying siege to Thebez, he approached too closely to a tower on the wall, and a woman dropped an upper millstone on his head, cracking his skull (vv. 50 – 55; 2 Sam 11:21). It seems likely that Abimelech had been attempting to assert his kingship partly on the basis of Shechem's traditional role as the dominating city in the region. Shechem's dominance is well illustrated in the el-Amarna letters, which describe Labayu's antics in the region (see above, p. 124).

The date and extent of the activities of the two minor judges Tola (Judg 10:1 – 2) and Jair (vv. 3 – 5) are not known with certainty. Although Tola was from the tribe of Issachar, he lived and was evidently active in Shamir in the Hill Country of Ephraim. Jair, however, was from the tribe of Manasseh and lived in Transjordan, in Gilead.

At the beginning of the eleventh century BC, the Israelites were pressed on the east by the Ammonites and on the west by the Phi-

listines. As was to be expected, Ammonite pressure was greatest in Gilead, but it was also exerted on Judah, Benjamin, and Ephraim (Judg 10:9). The major dispute was over the control of the tableland of Moab and southern Gilead. This land was claimed at various times by the Moabites and Amorites (Num 21:26) and in this case by the Israelites and Ammonites. The elders of Gilead chose Jephthah to head their forces. Jephthah's argument that Yahweh had given the land to Israel, that Israel had taken it from the Amorites, and that Israel had lived there for three hundred years, was not accepted by the Ammonites (Judg 11:14–27). The ensuing battle with the Ammonites most likely took place south of Mizpah of Gilead (11:29–33), and the Ammonites were soundly defeated. Twenty of their towns were devastated, and the territory from Aroer to the vicinity of Minnith and as far as Abel Keramim was secured (vv. 32–33).

In the aftermath of the Gileadite victory, the Ephraimites, who had been offended because they had not been invited to participate in the battle, crossed the Jordan River to Zaphon (possibly now the home and headquarters of Jephthah [?]; Judg 12:1) to fight with the Gileadites. In the battle that followed, the Ephraimites were routed, and fleeing survivors were executed as they attempted to cross the Jordan westward back to Ephraim (vv. 4–7). The story of Jephthah, like that of Gideon, illustrates how the tribes of Israel experienced pressure from their neighbors who lived in or on the fringes of the eastern desert.

Not much is known about the three minor judges who are mentioned next. Ibzan (Judg 12:8–10) was from Bethlehem, which may have been either in Zebulun (Josh 19:15) or in Judah (cf. Josephus, *Ant.* 5.7.13 [271]). Elon (Judg 12:11–12) the Zebulunite was from Aijalon, a city of Zebulun, and Abdon was from Pirathon (vv. 13–15).

Close to the beginning of the twelfth century BC a new political force began to make itself felt in the land of Canaan. Called in Egyptian sources the Sea Peoples, these newcomers included the Tjekker, Danuna, Sherden, Tursha, Lukka, and Peleset (Philistines). The beginning of their migration in the thirteenth century BC brought to an end the Minoan civilization in Crete and the strong Hittite state in Anatolia, and they attacked northern Syria, including places such as Alalakh and Ugarit.

The Egyptian king Ramses III defeated these invaders (ca. 1174 BC) both on land and at sea. Subsequently the Sea Peoples began to settle along the coast of Canaan, primarily in seaport towns. From the biblical standpoint, the most interesting of these peoples were the Philistines, who settled in Gaza, Ashkelon, Ashdod, Gath, and Ekron. Although they brought with them elements of their Aegean cultural heritage — distinctive language, dress, pottery, weapons, etc. — they seem to have quickly assimilated major features of Canaanite culture, including its religion (note Canaanite deities such as Dagon, Ashtoreth, and Baal-Zebub), pottery (their distinctive pottery disappeared by

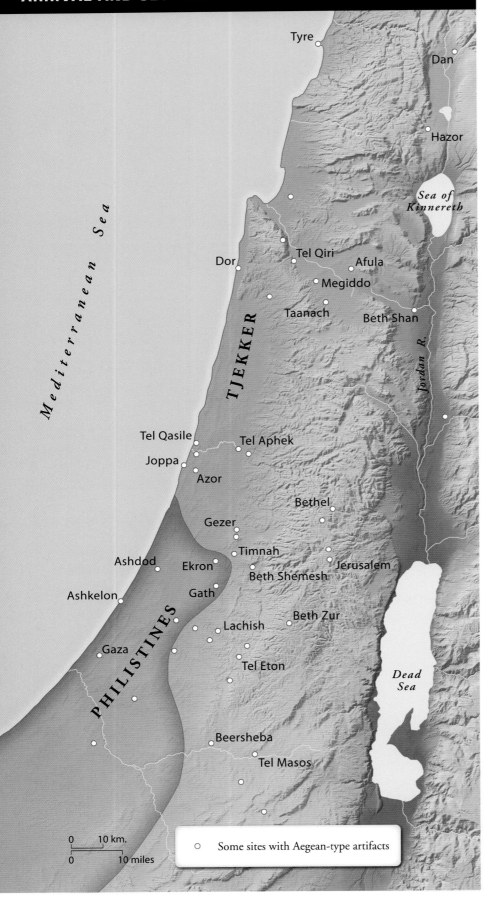

Map labels: Tyre, Dan, Hazor, Mediterranean Sea, Sea of Kinnereth, Dor, Tel Qiri, Afula, Megiddo, TJEKKER, Taanach, Beth Shan, Jordan R., Tel Qasile, Tel Aphek, Joppa, Azor, Bethel, Gezer, Timnah, Jerusalem, Ashdod, Ekron, Beth Shemesh, Ashkelon, Gath, Beth Zur, PHILISTINES, Lachish, Gaza, Tel Eton, Dead Sea, Beersheba, Tel Masos

Scale: 0 — 10 km. / 0 — 10 miles

○ Some sites with Aegean-type artifacts

1000 BC), and probably a Semitic language. They may have brought with them the technology for smelting and forging metal products, including those made of iron, and this seems to have given them a technological and military edge over the other peoples of Canaan (see, e.g., 1 Sam 13:19 – 22).

During the first half of the eleventh century BC the Philistines began to exert serious pressure on the Israelites. The natural place for this conflict between the people of the coast and the people of the hill country was in the buffer zone between them, the Shephelah. There, in its northern region, in the Valley of Sorek, the Nazirite judge Samson rose to meet the threat. The son of Danite parents from Zorah, Samson's base of operation was in Mahaneh (the camp) of Dan, which was located in the vicinity of Zorah and Eshtaol (Judg 13:25).

His earliest encounter with the Philistines was when he married a Philistine woman who lived in Timnah, which was located in the Sorek Valley west of Beth Shemesh. Later, as an offended husband, he burned the fields of grain, the vineyards, and the olive groves that the Philistines maintained in the Sorek Valley. Although his exploits had taken him to Ashkelon (Judg 14:19) and would eventually lead him to Gaza and Hebron (16:1 – 3), it was an affair with "a woman in the Valley of Sorek whose name was Delilah" that led to Samson's heroic but tragic death at Gaza (vv. 4 – 31).

With the death of Samson, the last of the heroes of the book of Judges passed from the scene. For 350 years (ca. 1400 – 1050 BC) the Israelites had been settling the land; yet even at the end of this long period, large pockets of non-Israelite populations still existed in the country. King Saul would provide protection from some of these enemies, especially the Ammonites and Philistines (whom Jephthah and Samson had not silenced completely), but it would be King David who would eventually bring them to submission.

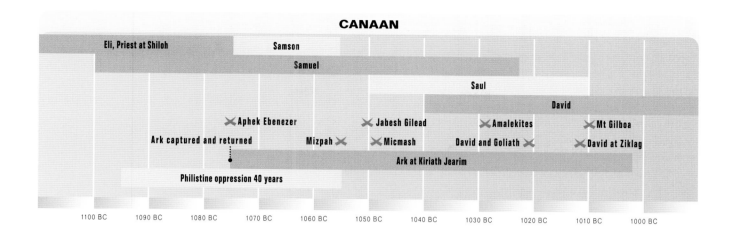

Eli, Priest at Shiloh   Samson

Samuel

Saul

David

✕ Aphek Ebenezer   ✕ Jabesh Gilead   ✕ Amalekites   ✕ Mt Gilboa

Ark captured and returned   Mizpah ✕   ✕ Micmash   David and Goliath ✕   ✕ David at Ziklag

Ark at Kiriath Jearim

Philistine oppression 40 years

1100 BC   1090 BC   1080 BC   1070 BC   1060 BC   1050 BC   1040 BC   1030 BC   1020 BC   1010 BC   1000 BC

TRANSITION TO THE MONARCHY: SAMUEL AND SAUL

During the whole of the eleventh century BC, Israel was in the process of shifting from the rule of judges to that of a monarch. Also, the Philistines were growing in strength and it was almost inevitable that they would come into conflict with Israel.

At the beginning of the century (ca. 1100 BC), Samuel was born to parents who lived in Ramathaim (1 Sam 1:1), very likely the same place as the Ramah where Samuel later made his home as an adult. During his childhood years he served the high priest Eli at the tabernacle at Shiloh (1 Sam 3). Probably one reason the tabernacle was located there was that the rugged terrain of the Hill Country of Ephraim that surrounded Shiloh provided natural topographical defenses, making it difficult for Israel's enemies to reach it.

When Samuel was approximately twenty-five (ca. 1075 BC, about the same time Samson became active against the Philistines in the Sorek Valley), the Philistines mustered their forces at the city of Aphek, to prepare for an invasion into the hill country (1 Sam 4:1). To meet this threat, the Israelites set up camp near Ebenezer. After an initial defeat, the Israelites believed that if the visual symbol of the presence of God would go with them into battle, they would be victorious. Thus they brought the ark of the covenant from Shiloh to Ebenezer. Yet when the time came for the battle, the Philistines soundly defeated the Israelites and even captured the sacred ark! Upon hearing news of this disaster, particularly of the loss of the ark, Eli the high priest collapsed and died (vv. 12 – 18). Although the Bible does not mention it, from the archaeological evidence it seems that Shiloh was destroyed at this time, in all probability by the Philistines.

The ark of the covenant was first taken to Ashdod and placed in the temple of Dagon. But the statue of Dagon repeatedly fell over in obeisance to the ark, and a plague broke out in the city, so the ark was sent to Gath (5:1 – 8). There a similar plague broke out, and the troublesome ark was transferred to Ekron. For a third time a plague broke out in the city where the ark resided.

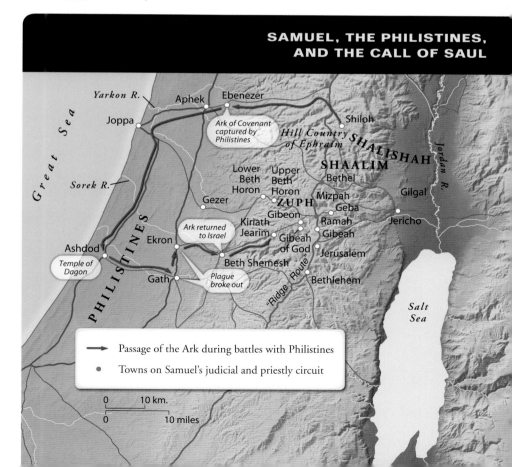

SAMUEL, THE PHILISTINES, AND THE CALL OF SAUL

The rulers of the Philistines decided that a test should be made to determine if the ark really was the cause of the plagues: the ark was placed on a cart, which was to be drawn by two cows that had young calves and that had never been yoked. If the cart headed toward the nearby Israelite city of Beth Shemesh, then the Philistines would know that Yahweh had brought this disaster (1 Sam 6:1 – 9). And indeed, as the five rulers of the Philistines watched from the ridge that overlooks the Sorek Valley, the cows pulling the cart headed eastward up the valley toward Beth Shemesh.

The people of Beth Shemesh, who were harvesting wheat in the valley (which means that it must have been sometime around May), rejoiced to see the ark and offered sacrifices in celebration of its safe return (vv. 10 – 18). But some of the people of Beth Shemesh looked into the ark — which only the high priest was allowed to do, and then only once a year, on the Day of Atonement — and God struck them down for their sin. Because of this the ark was transferred to Kiriath Jearim, one of the four Gibeonite cities, where it remained until David transported it to Jerusalem (ca. 1003 BC; 2 Sam 6).

During his adult life, Samuel made his home in Ramah, which was strategically located at the important junction of the west – east connecting route (Gezer – Beth Horon – Ramah – Jericho) and the north – south Ridge Route. His annual duties, which were probably judicial and priestly, took him on a circuit that led from Ramah to Bethel, to Gilgal, and to Mizpah (1 Sam 7:16). It is interesting to note that all of these villages were situated on or east of the central watershed, away from areas to the west that were under Philistine control.

Even after the return of the ark, Philistine pressure on Israel continued. About 1055 BC Samuel gathered the Israelite tribes at Mizpah to pray for them and to offer sacrifices (1 Sam 7:5). This assembly may have convened soon after Samson's death at Gaza, and it may have been called to deal with anticipated Philistine retaliation. The ensuing battle between Israel and the Philistines occurred somewhere on the western side of the Benjamin Plateau. God intervened on Israel's behalf, sending a violent thunderstorm that panicked the Philistines. Israel, seizing the opportunity, pursued them westward toward the coastal plain, as far as Beth Car (identification unknown). In order to commemorate God's gracious assistance in Israel's victory over the Philistines, Samuel set up a memorial stone called Ebenezer ("stone of help") between Mizpah and Shen. As a result of this victory, Israel regained territory in the Ekron and Gath areas (vv. 11 – 14), but the lull in the Philistine threat would last only a few years.

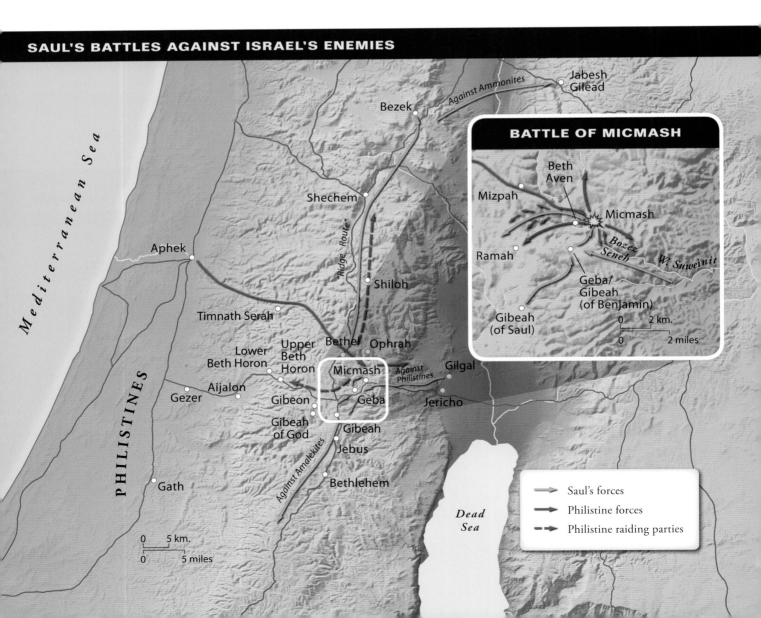

SAUL'S BATTLES AGAINST ISRAEL'S ENEMIES

BATTLE OF MICMASH

While Samuel was ministering in the Benjamin region, two of his sons were serving as judges at the southern extremity of Israel, in Beersheba (1 Sam 8:1–3). Because of the corruption of these sons, and possibly because of new Philistine inroads into western Benjamin, the elders of Israel approached Samuel, requesting that he appoint a king over them (vv. 4–5). At this point in the narrative (1 Sam 9), Saul of Benjamin is introduced. Looking for lost donkeys in the Hill Country of Ephraim, in the area of Shalisha and in the districts of Shaalim and Zuph, he finally approached Samuel in Ramah for guidance. After Saul was informed that the lost donkeys had been found, Samuel privately anointed him king (10:1). To help confirm in Saul's mind that he was God's choice, Samuel predicted certain events that would occur near Rachel's tomb at Zelzah on the border of Benjamin, at the great tree of Tabor, and at Gibeah of God (vv. 2–8).

These events indeed occurred, and at Gibeah of God (Nebi Samwil) Saul met "a procession of prophets [and] the Spirit of God came upon him in power" (1 Sam 10:10). Since Saul's family was from nearby Gibeon (1 Chron 9:35–40), it is not surprising that as Saul descended from Gibeah of God, his uncle approached him with questions regarding what Samuel had said to him (1 Sam 10:15; Gibeah of God and Gibeon are only 1.5 mi. apart). Soon after Israel assembled at Mizpah and there Saul was chosen by lot to be king (vv. 17–27).

After the selection process Saul returned to his residence in Gibeah (Tell el-Ful). Although this site had lain abandoned since its destruction after the affair of the Levite and his concubine (Judges 20:33–45), Saul evidently built a fortress at this centrally located site and renamed it Gibeah or Gibeah of Saul (to avoid the ban on rebuilding a wicked city that had been destroyed? Deut 13:13–18); the earlier name Geba/Gibeah of Benjamin had been transferred to the nearby site of Jaba. Here at Tell el-Ful (Gibeah of Saul), archaeologists have uncovered the remains of a fortress dating to the time of Saul.

Saul soon had the opportunity to exhibit his leadership qualities by mustering Israelite and Judean forces for the purpose of delivering the people of Jabesh Gilead from their Ammonite oppressors (1 Sam 11:1–13). Saul may have been eager to do this, for many Benjamites of Saul's day were probably descendants of women whose ancestral homes were in Jabesh Gilead (Judg 21:6–12). Leaving Gibeah of Saul, he proceeded northward to Bezek, where he prepared the forces of Israel and Judah for battle (1 Sam 11:6–8). After crossing the Jordan River, he was able to end the siege of Jabesh and defeat the Ammonites. Convinced of Saul's prowess, Israel confirmed him as king at the old cultic center of Gilgal (vv. 14–15), and from that point on his kingship was not doubted by the populace at large.

Soon after this confirmation ceremony (possibly ca. 1048 BC) Saul and his son Jonathan mustered small numbers of Israelite forces in Micmash, in the Hill Country of Bethel, and in Gibeah of Benjamin (1 Sam 13:2). These localities of Israelite control were again in central and eastern Benjamin, for it is probable that the Philistines controlled western Benjamin by means of garrisons at Gibeah of God (10:5) and Geba (probably = el-Jib [Gibeon] in this instance; Gibeon; 13:3). With these garrisons, the Philistines' line of communication to the coastal plain, via the Beth Horon road, was secure. They were also in a posi-

tion to make further inroads into the plateau area of Benjamin and to sever Israelite-Judean connections along the north–south Ridge Route. Militarily, the Israelites were at a distinct disadvantage, for they were dependent on the Philistines for the manufacture and repair of their copper (and iron?) agricultural and military implements (vv. 19–22).

Saul's son Jonathan met the Philistine threat head-on by assaulting their garrison at Geba (= Gibeon; 1 Sam 13:3). Since by this defeat the Philistines had lost their toehold in the hill country, and since the Israelites now controlled the Beth Horon approach, the Philistines evidently regrouped, possibly in the Aphek area, and then reentered the hill country via the more northerly route that led through Timnath Serah. This brought them into the northwestern and northeastern sections of the Benjamin plateau. Bringing with them chariots, horsemen, and foot soldiers, the Philistines set up camp at Micmash (v. 5). From there they sent out raiding parties to the north (Ophrah), west (Beth Horon), and east ("the borderland overlooking the Valley of Zeboim," vv. 17–18).

After assembling troops at Gilgal in the Jordan Valley near Jericho (prudently well away from the Philistine menace!), Saul and Jonathan moved into the hill country, establishing bases south of the Philistines at Gibeah/Geba of Benjamin (1 Sam 13:15–16) and at "Gibeah [of Saul]" (14:2). Then Jonathan and his armor-bearer headed north from Geba to Micmash, crossing the wadi between the two villages (the modern Wadi Suweinit), in the process climbing down and up the cliffs (Seneh and Bozez) that lined the chasm (v. 4). At Micmash, Jonathan and his armor-bearer subdued the guards of the Philistine camp (v. 14), and as the earth quaked, the Philistines panicked and ran (v. 15).

Saul's watchmen, who had been observing from Gibeah of Benjamin (v. 16), informed him of the panic, and Saul led the remaining Israelite forces into the fray. Israelites who had been hiding in fear in the Hill Country of Ephraim to the north (v. 22) also joined the battle in the Beth Aven area, and they helped to drive the Philistines from the hill country back to the Aijalon region (v. 31).

Later in his reign (ca. 1025 BC), Saul had to deal with the Amalekites, who were apparently making an incursion into southern Israel. Saul mustered his troops at Telaim (cf. Josh 15:24, but identification unknown). Upon approaching the city of Amalek (possibly Tel Masos), Saul instructed the Kenites — descendants/relatives of Moses' father-in-law, who at times dwelt in Judean territory (Judg 1:16; 1 Sam 27:10), but some of whom were now associated with the Amalekites — to separate themselves from the enemy (1 Sam 15:5–6). This accomplished, Saul attacked the Amalekites and slaughtered them from "Havilah to Shur, to the east of Egypt" (v. 7). To commemorate this victory, Saul set up a monument in "Carmel [of Judah]" (v. 12), which was probably one of the southern Judean cities that had been harassed by the Amalekites.

Besides noting Saul's victories over the Philistines and Amalekites, the biblical text mentions his victories over Moab, Edom, and the Ammonites to the east, and over the kings of Zobah to the north (1 Sam 14:47). Although these enemies were evidently kept at bay, it was not until the reign of David that Israel took control of these kingdoms, and then for only a brief period of time.

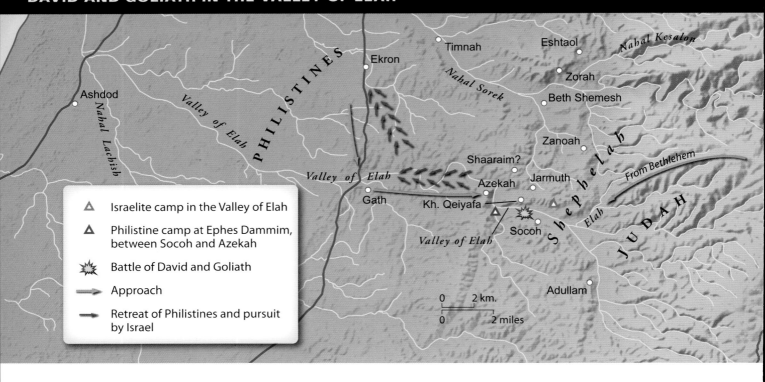

Map legend:
- △ Israelite camp in the Valley of Elah
- △ Philistine camp at Ephes Dammim, between Socoh and Azekah
- 💥 Battle of David and Goliath
- → Approach
- → Retreat of Philistines and pursuit by Israel

It was soon after the battle with the Amalekites that Samuel anointed David (1 Sam 16) and that David began to serve at Saul's court as a musician. The Philistines and Israelites were still vying for power, but now — probably to the relief of the Israelites — their battles were being fought in the Shephelah, the buffer zone between the Philistines on the west and the Israelite settlements in the mountains on the east. It is in this context that the battle of David and Goliath (1 Sam 17) in the Valley of Elah should be placed. The Philistines, moving eastward from Ekron and Gath, camped at Ephes Dammim between Socoh and Azekah (17:1). The Israelites, defending the approaches to the hill country, camped on the north side of the valley (vv. 2 – 3), probably east of the Philistine camp.

▼ *Valley of Elah from Kh. Qeiyafa. Looking west with Azekah on the left (south) side of the image. David fought Goliath near here (1 Sam 17).*

The encounter between David and Goliath took place in the broad valley itself, from which David took five smooth stones for his sling. Emboldened by David's example, Saul's troops successfully attacked the Philistines. The latter at first fled northward, on the Shaaraim road in the valley east of Azekah, and then north of Azekah; they turned west and followed the valley to the security of their cities of Gath and Ekron (v. 52). This battle was probably one of a number that occurred between the Israelites and the Philistines in the Shephelah. For example, in the Shephelah David defended the inhabitants of Keilah against the Philistines (23:1 – 13). Thus the account of David and Goliath not only provides geographical details concerning the Valley of Elah region but also illustrates the fact that the Shephelah served as a military buffer zone between the inhabitants of the coastal plain and those of the mountains to the east.

Following this stunning victory, chapters 18 through 21 of 1 Samuel describe David's growing popularity among the people and Saul's growing jealousy. These chapters are punctuated with descriptions of David's narrow escapes from Saul, most of the action occurring in the Benjamin area. For example, after Michal (David's wife) helped him escape from Saul (1 Sam 19:11 – 17), David fled to nearby Naioth, the Bedouin encampment that was close to Ramah. Saul was overcome with a prophetic spirit as he approached Naioth, thus setting the stage for David's escape (vv. 18 – 24).

After a number of close encounters, David decided it was necessary to leave the Benjamin Plateau in order to avoid Saul's wrath. With this decision, David began his flight from Saul. After receiving meager provisions as well as the sword of Goliath from the priest who served at the tabernacle at Nob (1 Sam 21:1–7), David fled into the hands of his archenemy, Achish, the king of Gath (vv. 8–15) — possibly believing that Achish would welcome any enemy of Saul. However, Achish viewed David's arrival as an opportunity to rid himself of David, who had to flee for his life. At the cave of Adullam in the Shephelah (Josh 15:35) David gathered a fighting force of four hundred men.

After he sent his parents to Moab — Davidic connections with Moab via Boaz and his Moabite wife, Ruth, are well attested (Ruth 4:9–22; 1 Chron 2:12–15) — David's movements are described in detail in 1 Samuel 22–27. He was sometimes west of the Judean mountain ridge in the forest of Hereth (22:5), sometimes delivering Keilah from the Philistines (23:1–13). But most of the time he was east of the watershed, either in or near the Judean Desert. In this connection the cities of Ziph, Carmel, and Maon are mentioned, as are the deserts of

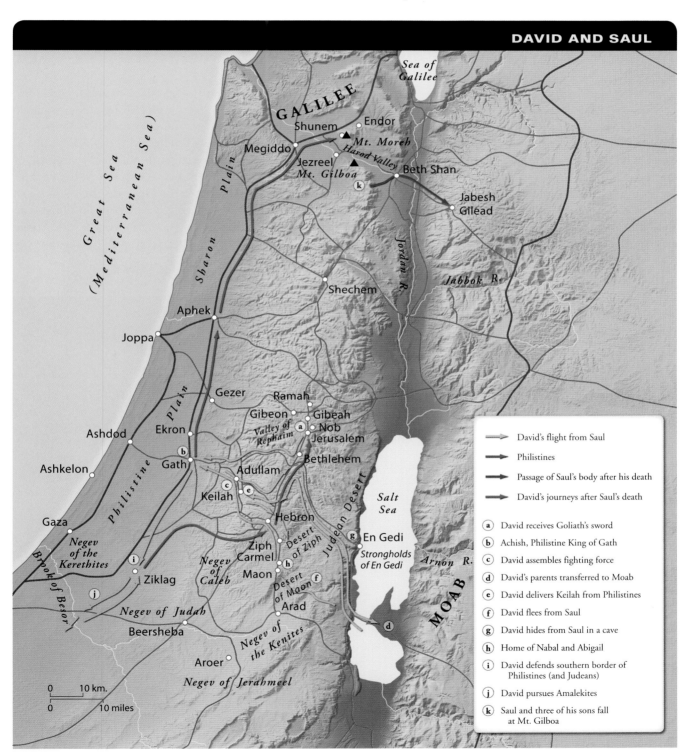

DAVID AND SAUL

Legend:
→ David's flight from Saul
→ Philistines
→ Passage of Saul's body after his death
→ David's journeys after Saul's death

(a) David receives Goliath's sword
(b) Achish, Philistine King of Gath
(c) David assembles fighting force
(d) David's parents transferred to Moab
(e) David delivers Keilah from Philistines
(f) David flees from Saul
(g) David hides from Saul in a cave
(h) Home of Nabal and Abigail
(i) David defends southern border of Philistines (and Judeans)
(j) David pursues Amalekites
(k) Saul and three of his sons fall at Mt. Gilboa

Ziph, Maon, and En Gedi as well as the Arabah and Jeshimon to the east and south.

Saul pursued David out into these deserts but was continually foiled in his attempts to capture him. It is probable that most of the local inhabitants in some way supported David — a fellow member of the tribe of Judah — and that David in turn "protected" the Judeans from their enemies. Exceptions to this assumed support included the inhabitants of Ziph, who at times attempted to deliver David into Saul's hands (1 Sam 23:19 – 29; 26:1 – 2), as did the insolent Nabal (a resident of Maon and Carmel), whose beautiful wife Abigail married David after Nabal's untimely death (1 Sam 25). Since it would be difficult to support a band of six hundred men and their families in the Judean Desert, David's flight to En Gedi and the nearby Crags of the Wild Goats (23:29 – 24:2) was to be expected, for the freshwater spring at En Gedi is the largest on the western shore of the Salt Sea. Even though Saul's pursuit of David actually led him to the very cave where he was hiding, David still was able to escape (24:3 – 21).

Eventually David must have realized that he would either have to kill Saul in self-defense or else be killed by him. To avoid this, David again sought asylum with Achish, the king of Gath. A year or so had passed since their previous encounter (1 Sam 21:10 – 15), and by this time Achish was well aware that David was indeed a true enemy of the Israelite king. Achish, planning to make use of David's troops and military prowess, stationed him at Ziklag. There, on his southern border, David was to protect Achish against raiders from the south.

David, in fact, conducted raids on the Geshurites, the Girzites, and the Amalekites, all of whom dwelt in northern Sinai between Ziklag and Shur (1 Sam 27:8). Leaving no survivors, David was free to perpetrate a lie by telling Achish, who lived in Gath (24 mi. to the north), that he had made raids on the Negev of Judah, the Negev of Jerahmeel, and the Negev of the Kenites (27:10), all of which are sub-districts of the biblical Negev. In this way David was leading Achish to believe that Judean hostility toward David was growing, when in fact Judean appreciation for David was increasing because he was defending them from these desert bands!

After David had served sixteen months at Ziklag (1 Sam 27:7), the final confrontation between Saul and the Philistines began to unfold. It is difficult to pinpoint the exact cause of the battle. Possibly Israel, by camping at the spring by Jezreel in the north (29:1), was attempting to cut the line of communication between the Philistines and their allies garrisoned at Beth Shan. The Philistines responded to this threat by assembling their forces at Aphek, the point where the Philistine and Sharon plains meet, and from there they marched north and eastward, setting up camp opposite Israel on the southern slope of Mount Moreh at Shunem. By establishing their camp at Shunem they effectively cut

▼ *Oasis of En Gedi on the western shore of the Dead Sea. Here David hid from Saul (1 Sam 24).*

▲ *David worked for the Philistines in this area of the western Negev/Besor Ravine (1 Sam 27; 30). The light brown soil is "loess soil."*

the north – south lines of communication between Israelite settlements in Galilee and Manasseh.

As the Philistine and Israelite forces faced each other across the Harod Valley, David, whom the Philistines did not allow to take part in the battle (1 Sam 29), returned to Ziklag, only to find that the Amalekites had made a raid on the Negev of the Kerethites, the Negev of Caleb, Judean territory, and Ziklag, and that they had taken captive the women and children that had remained behind. David and his six hundred men pursued the retreating Amalekites southwest to the Wadi Besor. After crossing the wadi with a small force he overtook the Amalekites, slaughtering them and rescuing the captive women and children. The spoil captured from the Amalekites was not kept by David and his men but was sent to the elders of Judah and distributed among cities located in the southern Hill Country of Judah and in the Negev (30:26 – 31).

Meanwhile, back in the Jezreel-Shunem region, the Israelites were engaged in a life-and-death struggle with the Philistines. Some of the fleeing Israelites went up Mount Gilboa, on the south side of the valley, possibly thinking that the mountain would offer protection from the pursuing Philistines, whose chariotry could not operate effectively in the rocky and forested mountainous terrain. It was there on Mount Gilboa that Saul and Jonathan died as a result of the battle (1 Sam 31). As part of the Philistine victory celebration the weapons of Saul and Jonathan were placed in the temple of the Ashtoreths and their headless bodies were fastened to the wall of Beth Shan in public display.

The men of Jabesh Gilead, the town that Saul had delivered from the Ammonites and with which he may have had family ties (see above, p. 133), removed the bodies from the wall, and buried the first king of Israel, along with his son Jonathan, under a tamarisk tree at Jabesh. With the burial of Saul and Jonathan the transition period between the period of the judges and that of the monarchy had come to an end. Within a few years the idea of a dynastic monarchy would be firmly established, at least in the minds of the Judeans.

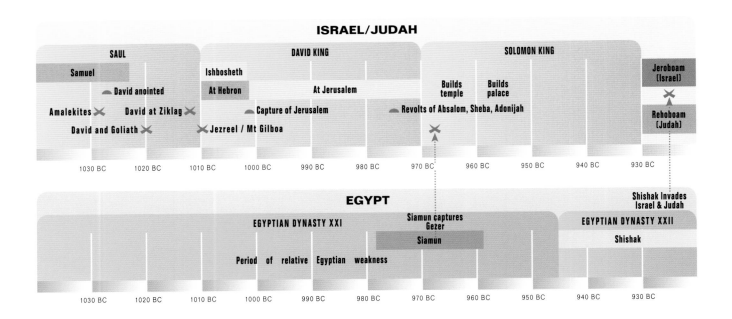

ISRAEL/JUDAH										
SAUL		DAVID KING			SOLOMON KING					Jeroboam (Israel)
Samuel		Ishbosheth				Builds temple	Builds palace			✕
⌐ David anointed		At Hebron	At Jerusalem							Rehoboam (Judah)
Amalekites ✕	David at Ziklag ✕	⌐ Capture of Jerusalem				⌐ Revolts of Absalom, Sheba, Adonijah				
David and Goliath ✕		✕ Jezreel / Mt Gilboa				✕				
1030 BC	1020 BC	1010 BC	1000 BC	990 BC	980 BC	970 BC	960 BC	950 BC	940 BC	930 BC

EGYPT										Shishak Invades Israel & Judah
EGYPTIAN DYNASTY XXI						Siamun captures Gezer			EGYPTIAN DYNASTY XXII	
						Siamun			Shishak	
Period of relative Egyptian weakness										
1030 BC	1020 BC	1010 BC	1000 BC	990 BC	980 BC	970 BC	960 BC	950 BC	940 BC	930 BC

THE UNITED MONARCHY: DAVID AND SOLOMON

A fter the death of Saul it was probably somewhat unclear in the minds of the common people whether a king should rule over them. One of the surviving sons of Saul, Ish-Bosheth, was established as "king over Gilead, Ashuri [= Asher?], and Jezreel, and also over Ephraim, Benjamin and all Israel" (2 Sam 2:9). The real power behind the throne, however, was Abner, the former commander of Saul's army. Ish-Bosheth ruled from the Transjordanian city of Mahanaim (a town discreetly removed from the clutches of the recently victorious Philistines).

In the south David moved from Philistine Ziklag to Hebron, where he was anointed king of Judah (2 Sam 2:1 – 7, 11). It should be remembered that David had already ingratiated himself to the Judeans by protecting their southern flank from the Amalekites, Geshurites, and Girzites (1 Sam 27:8) and by sending gifts to eleven of their cities and to two local tribes (30:26 – 31). Thus it is no wonder that they accepted one of their own as king rather than the weak Benjamite Ish-Bosheth.

For a period of time there was war between the house of Saul and that of David, as exemplified by the battle of selected men from both armies at the pool of Gibeon (2 Sam 2:12 – 32). Abner, realizing the weak position of Ish-Bosheth, proposed to David that he would deliver "all Israel" to him in order to make him king of the entire country, but before the arrangements could be completed, Joab intervened and killed Abner in cold blood (3:22 – 29).

Ish-Bosheth was murdered by two men from Beeroth, one of the old Gibeonite cities, possibly in revenge for Saul's extermination of many of their kinsmen (2 Sam 4:1 – 12; compare 21:1 – 14). At that point the elders of Israel came to David at Hebron, made a covenant with him, and anointed him king over Israel (5:1 – 3; 1 Chron 11:1 – 3).

▼ *Gihon Spring: The spring chamber of the chief water source of Jerusalem*

▲ *Jerusalem: Ancient tunnel leading from where water was gathered to a point inside the city. Possibly Joab and his men gained access to the city via this tunnel (2 Sam 5:8; 1 Chron 11:6).*

The Philistines, who up to that time may have considered David a client king, realized that a united Israel posed a serious threat to their control of the country. They responded by twice trying to divide the country into two parts by establishing themselves in the hill country in the Valley of Rephaim, southwest of Jerusalem. Twice they were defeated, and the last time David drove them out of the hill country, via the Geba/Gibeon – Beth Horon road, back to Gezer and the coastal plain (2 Sam 5:17 – 25; 1 Chron 14:8 – 17).

After ruling in Hebron for seven years and six months (ca. 1010 – 1002 BC), David captured Jebus (Jerusalem) and made it his capital (2 Sam 5:6 – 10; 1 Chron 11:4 – 9; see pp. 240 – 253 for a detailed description of the city). The move of the capital from Hebron to Jerusalem was very important. David's conquest of the non-Israelite city made it the personal possession of him and his descendants, so that neither Judah nor Israel could lay claim to it. Neither group would be offended by being governed from it, since it was situated on the boundary between the northern and southern tribes and was thus in a neutral location. By bringing the ark of the covenant to Jerusalem (2 Sam 6; 1 Chron 13) David made it the religious as well as the political center of a united Israel. From a defensive standpoint its mountainous location, removed from the coastal plain, meant that any enemies who wished to attack Jerusalem had to make the difficult ascent into the mountains with their chariots, siege machinery, etc., before they could even begin the attack on the city. In addition, its local topography (see pp. 241 – 242) made it easily defensible on the west, south, and east.

It is to be assumed that David, after establishing his capital in Jerusalem, consolidated his kingdom internally before beginning his wars of expansion. This means that he must have taken control of the old Canaanite centers that were located in the Jezreel and Harod valleys — Megiddo, Taanach, Ibleam, and Beth Shan — along with the Canaanite centers located on the Plain of Acco and in Galilee (Judg 1:21 – 35). To the southwest David was able to subdue the Philistines, capturing the important city of Gath (1 Chron 18:1, but compare 2 Sam 8:1).

The exact sequence of David's wars of expansion, as described in 2 Samuel 8 – 12, is difficult to determine with precision, since the textual arrangement of the events is not necessarily their chronological order. David's first war in the Transjordanian region was with the Ammonites. Their king, Nahash, who had been a friend of David, died, and his son Hanun became king in his place. David sent emissaries to Hanun to congratulate him on his rise to the kingship, but instead of receiving the messengers graciously they were rebuffed (2 Sam 10:1 – 5; 1 Chron 19:1 – 5). The Ammonites, realizing that their insult to David would lead to war, prepared for it by hiring mercenary troops from their neighbors to the north. Twenty thousand Aramean foot soldiers were supplied by the federated countries of Beth Rehob and Zobah, along with a thousand from the king of Maacah, a country located in the northeastern part of the Huleh Valley, while twelve thousand were supplied from the area of Tob (2 Sam 10:6). In addition, numerous chariots and charioteers were hired from Aram Naharaim (1 Chron 19:6 – 7).

These forces, combined with the Ammonite army, assembled at Medeba to fight the Israelite forces led by Joab. Joab found himself trapped between the Ammonites, stationed in and near the city of Medeba, and

HAMATH

Cun

Lebo
Hamath

Gebal

ZOBAH

Berothai

BETH REHOB

Damascus

ARAMEANS

Sidon

Abel
Beth
Maacah

Dan

MAACAH

Tyre

GESHUR

Aczib

Hazor

Helam

Acco

Geshur

Plain of Acco

GALILEE

*Sea of
Galilee*

TOB

Tob

*Jezreel
Valley*

Megiddo

*Harod
Valley*

Ramoth
Gilead

Taanach

Beth
Shan

Ibleam

Jabesh
Gilead

Shechem

Jordan R.

Mahanaim

Joppa

Upper
Beth
Horon

David makes Jebus
his political and
religious capital

Rabbah of
the Ammonites

Gezer

Gibeon

AMMON

Baalah of Judah/
Kiriath Jearim

Jericho

Jerusalem

Medeba

Gath

Baal
Perazim

Gaza

PHILISTINES

Joab battles Ammonites
and allies from the north

Hebron

*Dead
Sea*

Aroer

Ziklag

Arnon Gorge

Beersheba

Kir Moab

Mediterranean Sea

MOAB

0   20 km.

0   20 miles

Brook of Egypt

*Valley
of Salt*

AMALEKITES

EDOM

Israelites

Arameans

Edomites

Ammonites

Subdued by David

▲ *Geshur: Iron Age gate at et-Tell northeast of Sea of Galilee. This is probably Geshur, where Absalom was from. Note the cult center on the right side of the entrance (stairs, basin, and standing stone).*

the Arameans, operating in the open countryside. By splitting their forces, Joab and his brother Abishai were able to drive off the Aramean and Ammonite forces (2 Sam 10:6 – 14; 1 Chron 19:6 – 15). It is likely that Israel continued to press its advantage in the area by conquering the territory of Moab (2 Sam 8:2; 1 Chron 18:2).

The Arameans, whose forces were still intact following the battle of Medeba, regrouped at Helam, 35 miles east of the Sea of Galilee. Hadadezer, the king of Zobah and Rehob, supplemented his forces by bringing in additional Arameans from "beyond the River" (= Euphrates). In the ensuing battle, David and his troops emerged victorious (2 Sam 10:15 – 19, but compare 1 Chron 19:16 – 19). It is probable that nearby Tob and Maacah passed into Davidic control at that time.

After the battle of Helam the Israelite forces returned to Rabbah to deal decisively with the Ammonites. While the army under the leadership of Joab was laying siege to the city, David was back in Jerusalem having his sinful affair with Bathsheba (2 Sam 11). David then had Bathsheba's husband, Uriah the Hittite, stationed close to the city wall during the siege of Rabbah, so that Uriah was "killed in the battle." After David repented of his sins (12:1 – 24), the royal citadel and water supply of Rabbah were captured and David took the crown of the Ammonite king and placed it on his head (vv. 26 – 31; 1 Chron 20:1 – 3). Some of the Ammonites were executed for their participation in the revolt, and from that time on, David probably ruled over the Ammonites via an appointed governor.

The decisive blow against the Arameans came at a time when Hadadezer had gone north to quell a revolt (2 Sam 8:3). In this northern battle David defeated the Arameans of Damascus who had come to the aid of Hadadezer (vv. 5 – 6; 1 Chron 18:5 – 6). Eventually, the Aramean states of Beth Rehob and Zobah, along with their cities of Tebah, Berothai, and Cun, submitted to David's control (2 Sam 8:3 – 4, 7 – 8; 1 Chron 18:3 – 4, 7 – 8).

Finally, the army of David defeated the Edomites in the Valley of Salt, located to the south of the Dead Sea (2 Sam 8:13 – 14; 1 Chron 18:12 – 13). Although many Edomites were killed, Hadad, who was of royal Edomite descent, fled to Egypt via Midian and Paran and later returned to lead his people in revolt against Solomon (1 Kings 11:14 – 22).

By the end of the period of expansion David had gained control of many of the neighboring states. To the far north, Tou (or Toi), the king of Hamath, acknowledged David's supremacy by sending his son Joram (Hadoram) along with "gifts" of silver, gold, and bronze to Jerusalem. Certainly the Aramean states of Zobah, Beth Rehob, and Maacah now came under Israelite control. A garrison was placed in Damascus and tribute extracted from the populace (2 Sam 8:5 – 6; 1 Chron 18:5 – 6). It is probable that the state of Geshur, located just east of the Sea of Galilee, remained semi-independent although its king, Talmai, formed an alliance with David. This alliance was probably sealed by the marriage of Talmai's daughter to David; the offspring of their union was Absalom (2 Sam 3:3). It is thus no wonder that Absalom, after killing his half-brother Amnon (who had raped Absalom's sister Tamar), fled for refuge to Geshur, the homeland of his mother (13:37 – 39).

East of the Jordan the Israelites solidified their control of the land of Tob (in Gilead) and the portion of Moab north of the Arnon Gorge, so that the territory that had been assigned to the tribes of Reuben, Gad, and Manasseh was now firmly in Israelite hands. David assumed royal prerogatives over the Ammonites (2 Sam 12:30). He placed garrisons in Edomite territory, and the Moabites sent him tribute (2 Sam 8:2, 14; 1 Chron 18:13). To the south and the southwest the Amalekites and Philistines were subdued (2 Sam 8:12; 1 Chron 18:11). Thus all the countries surrounding Israel were now firmly under Davidic control.

Although the exact boundaries of the Davidic empire are not given, the Chronicler, describing the bringing up of the ark from Kiriath Jearim to Jerusalem, notes that "David assembled all the Israelites, from the Shihor River in Egypt to Lebo Hamath" for the event (1 Chron 13:5). Since the Shihor is normally considered one of the eastern branches or canals of the Nile, located in the vicinity of the present-day Suez Canal, this implies that David controlled northern Sinai all the way to the eastern Nile delta. Never before and never again would Israelite control extend this far. The northern most point, Lebo Hamath, was the northern boundary of the land of Canaan. Thus, while David and Solomon reigned, the Israelites did, in fact, rule over most of the land of Canaan that had been promised them as an inheritance four hundred years earlier (Num 34).

As far as the core of David's kingdom (Israel proper) is concerned, east of the Dead Sea it stretched from Aroer in the south to Dan in the north; west of the Jordan Valley it reached from the Sidon and Tyre region in the north to Beersheba in the Negev of Judah, as evidenced by the extreme limits of the territory in which Joab conducted a census for David (2 Sam 24:1 – 8). By David's day the traditional boundaries of the area of major Israelite settlement, "from Dan to Beersheba," were firmly established (v. 2).

Within these limits David appointed men who supervised the crown's property. These administrators were in charge of the agricultural workers; the vineyards and their produce; the oil and sycamore trees in the Shephelah; the cattle grazing in the Sharon Plain and in the valley; the camels, donkeys, and flocks that belonged to David; and the storehouses scattered throughout the country (1 Chron 27:25 – 31). Since there is no evidence that David taxed his subjects, it seems reasonable to suppose that the farmlands of the Canaanite cities that had recently passed into Davidic hands were now considered crown property and that the produce of these lands was used to support the monarchy.

Toward the end of David's reign the question of his successor to the throne of Israel and Judah became acute. His oldest living son, Absalom, decided to take matters into his own hands (2 Sam 14 – 19). After having been proclaimed king by "all Israel" at the Judean tribal center of Hebron, Absalom with his followers marched north to Jerusalem. David and his followers fled eastward from Jerusalem, crossed the Kidron Valley, ascended the Mount of Olives, and proceeded to the fords of the Jordan River via Bahurim. After crossing the Jordan, David found refuge in the city of Mahanaim and received support from Shobi (an Ammonite from Rabbah), from Makir

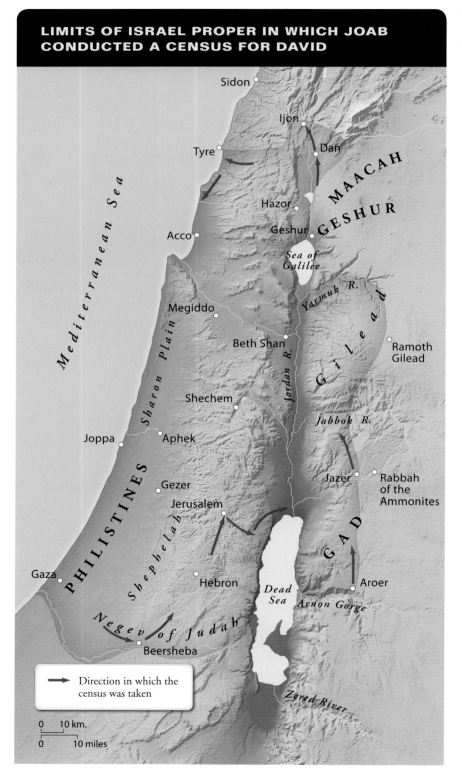

LIMITS OF ISRAEL PROPER IN WHICH JOAB CONDUCTED A CENSUS FOR DAVID

→ Direction in which the census was taken

0   10 km.
0   10 miles

(of Lo Debar), and from Barzillai (the Gileadite from Rogelim; 2 Sam 17:24 – 29). The battle between the forces of David, led by Joab, and the forces of Absalom, led by Amasa, took place in Transjordan in the "forest of Ephraim" (18:6). It was there that Joab put an end to the revolt by killing the instigator, Absalom.

Upon David's return to Jerusalem he had to deal with a second revolt, this one led by Sheba, a Benjamite (2 Sam 20:1 – 22). It is clear that a considerable element of the northern tribes was in sympathy with Sheba, while the tribe of Judah backed David. As usual, it was Joab who dealt decisively with the matter, pursuing Sheba to the northern Israelite city of Abel Beth Maacah, where Sheba was killed.

As the elderly David languished in Jerusalem, his (apparently) eldest son, Adonijah, decided to secure the kingship for himself (1 Kings 1:5 – 27). With the assistance of Joab, the commander of the army, and Abiathar the priest, he assembled many of the notables of the kingdom near En Rogel, a small spring located in the Kidron Valley just south of Jerusalem (vv. 5 – 10). Upon hearing of this, Nathan the prophet and Bathsheba approached David in order to confirm Solomon as king. Nathan, along with the priest Zadok and Benaiah, the commander of the Kerethites and Pelethites, anointed Solomon as king at the Gihon Spring (vv. 11 – 48). Soon after the line of succession was clarified, David died, and his son Solomon inherited the vast kingdom.

According to the Bible, most of the years of Solomon's rule (970 – 930 BC) were taken up with building projects, commercial ventures, and the continued maintenance of the kingdom he had inherited. During at least a portion of Solomon's rule the kingdom extended from Tiphsah on the Euphrates River in the northeast to the border of Egypt on the southwest (1 Kings 4:21 – 24; 2 Chron 9:26). The one military expedition attributed to Solomon was his conquest of Hamath Zobah (2 Chron 8:3), by which he evidently extended his kingdom from its previous border at Lebo Hamath northeastward up to the Euphrates. Solomon consolidated his hold on this region by building the city of Tadmor, as well as store cities in the country of Hamath (v. 4). In all probability these served as military and administrative outposts by means of which he controlled and serviced the caravans passing through the area. In addition, some of these cities may have served as centers for the breeding of horses and

for the buying and selling of both horses and chariots to the neo-Hittite states in the region (1 Kings 10:26 – 29; 2 Chron 1:14 – 17).

Solomon also moved to solidify his control of the trade routes that passed through Israel proper. Along the major International Highway he fortified the strategic cities of Hazor, Megiddo, and Gezer (1 Kings 9:15). All three of these cities have been excavated, and similar, six-chambered Solomonic gates have been discovered at each of them. Gezer

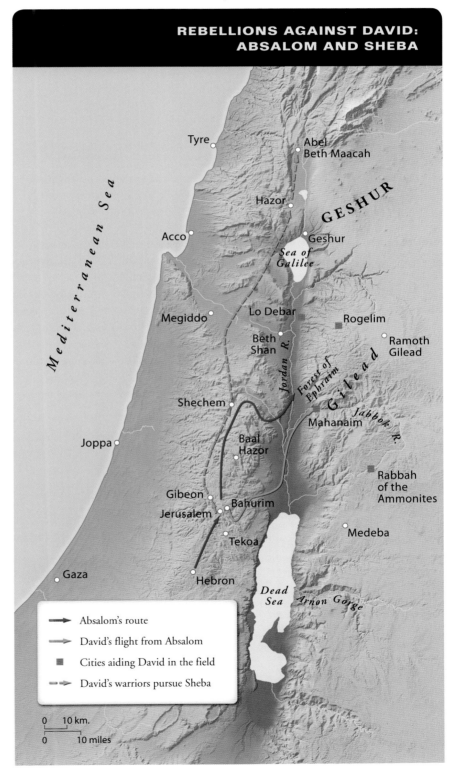

REBELLIONS AGAINST DAVID:
ABSALOM AND SHEBA

▲ *Gezer: Six-chamber gate at Gezer — probably associated with the Solomonic building program (1 Kings 9:15). View from inside the city looking out — note the drain in the center of the gate over which the road ran.*

is of particular interest, for it guarded the road that led up to Jerusalem from the west. Along this western approach to Jerusalem, Solomon also fortified Lower and Upper Beth Horon and, on a road slightly to the south, Baalath (= Kiriath Jearim; 9:17 – 18; 2 Chron 8:5 – 6). To the south and east of Israel, he built Tamar in the wilderness (Heb. and NASB; the NIV reads "Tadmor" [1 Kings 9:18]). In addition, store cities and cities for his chariots and horses were set up around the country.

Archaeological excavations and surveys have shown that a string of over forty forts, scattered over the Negev Highlands, were constructed during this period. These small forts were usually built with a double wall (casemate), had a gate, and in some cases were guarded

▼ *Hazor: Double (casemate) wall attached to the six-chambered (Solomonic) gate*

by attached towers. Located nearby were four-room houses along with farmsteads consisting of terraces, dams, and cisterns. It is obvious that people had not moved into this desolate region because of its agricultural richness, for it receives only 4 to 8 inches of rain each year. Rather, it is to be supposed that the soldiers who were stationed at the forts lived in these houses and that their families constructed the agricultural installations in an attempt to supplement their government rations.

Presumably, at least early in his reign, Solomon controlled the Transjordanian states of Ammon, Moab, and Edom, for it is likely that his marriages to women of these countries sealed alliances made with their leading families (1 Kings 11:1; cf. 3:1). Certainly Solomon controlled Gilead, where Israelites were settled, and, for a period of time, Damascus as well.

Thus it is evident that Solomon controlled the major north – south route that led from Arabia to Damascus through Transjordan, the overland routes that connected Arabia with Egypt and the Mediterranean Sea via the Negev, the overland route that connected Egypt with Mesopotamia, and many of the routes in the north that connected Syria and northwestern Mesopotamia with the Mediterranean Sea. It is probable that by providing these overland caravans with food, water, and protection, and by collecting tolls from them, Solomon became very wealthy. Indeed, the visit of the Queen of Sheba to Jerusalem is commonly interpreted as reflecting her interest in maintaining good relations with Solomon so as to expedite the passage of her caravans through his territory (1 Kings 10:1 – 13; note especially her camel caravan, which brought such luxury items as spices, gold, and precious stones). In addition, the Bible notes that Solomon received revenues from merchants and traders as well as from the kings of Arabia and the governors of the land (2 Chron 9:14).

His middleman status is nicely illustrated by the fact that he both imported and exported horses and chariots. Horses were imported from Egypt and Kue, while quality (ceremonial?) chariots were built in Egypt and then exported to the "kings of the Hittites and of the Arameans" (1 Kings 10:28 – 29; 2 Chron 1:16 – 17).

Mediterranean Sea

Damascus

Tyre

Dan

Cedar wood

Hazor

Acco

Geshur

Cabul

Sea of Galilee

Megiddo

Beth Shan

Ramoth Gilead

Gilead

Shechem

Jordan R.

Succoth

Zarethan

AMMON

Tel Qasile

Rabbah of the Ammonites

Joppa

Lower Beth Horon

Upper Beth Horon

Gibeon

Heshbon

Gezer

Jerusalem

Beth Shemesh

Baalath

Gaza

Hebron

Dead Sea

PHILISTINES

Arad

Aroer

Beersheba

MOAB

Kir Moab

Fortresses in the Negev

Tamar

Bozrah

Kadesh Barnea

Kh. en-Nahas

EDOM

Timnah Copper Mines

Solomon's Kingdom

Shipping route

Fortified by Solomon

0   20 km.

0   20 miles

Ezion Geber

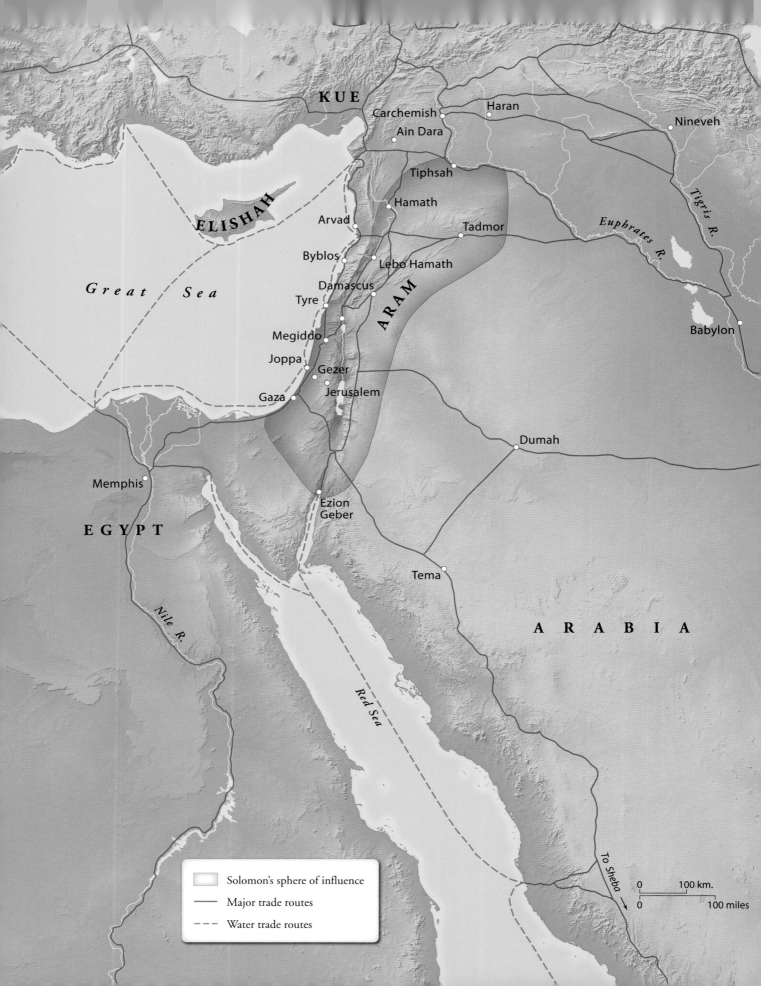

KUE

Carchemish

Haran

Nineveh

Ain Dara

Tiphsah

Tigris R.

Hamath

Euphrates R.

ELISHAH

Arvad

Tadmor

Byblos

Lebo Hamath

Great   Sea

Damascus

ARAM

Tyre

Babylon

Megiddo

Joppa

Gezer

Gaza

Jerusalem

Dumah

Memphis

Ezion
Geber

EGYPT

Tema

Nile R.

ARABIA

Red Sea

To Sheba

	Solomon's sphere of influence
——	Major trade routes
– – –	Water trade routes

0   100 km.

0   100 miles

The other sources of revenue frequently mentioned in connection with Solomon are his copper mines and the export of agricultural produce. It was at one time thought that a major source of Solomon's revenue came from exporting copper mined at Timnah near the Red Sea. However, there is no textual support for this theory, and recent archaeological investigations have shown that the Timnah mines were not used during the Solomonic period and thus could not have been a source of revenue, although recent studies have shown that copper was mined and processed in the vicinity of Khirbet en-Nahas (near the exit of Wadi Feinan) in the tenth century BC. As for the export of agricultural produce, it should be noted that shipments

Mark Connally

▲ *Ain Dara, Syria: View of an Iron Age temple that has strong design parallels with the Solomonic temple in Jersualem.*

of wheat, barley, wine, and oil were made to the workers of Hiram of Tyre, who were cutting trees to be used in Solomon's building projects (2 Chron 2:10, 15), and that Solomon also exported wheat and olive oil to Hiram on a yearly basis (1 Kings 5:10). But all of this was in exchange for cedar and pine logs and in itself could not have been considered a primary source of surplus revenue.

Almost half of Solomon's rule was taken up with his two great building projects in Jerusalem. The temple, which was begun in the fourth year of his reign (966 BC), took seven years to build, while his palace took thirteen years. The limestone used in the two buildings was cut from the Judean hills. The timber — choice cedar and pine logs — was supplied by Hiram of Tyre. After being cut in the mountains of Lebanon, the logs were floated down to Joppa (1 Kings 5:1 – 12; 2 Chron 2:3 – 16), and from there they were transported up to Jerusalem, probably via the Gezer – Beth Horon road. From the biblical description of these two buildings it seems probable that their basic outlines were somewhat similar to temple and palace designs found in the neo-Hittite states to the north. In any case, the Solomonic age was remembered as a time when luxury items of gold and silver and cedar abounded in the capital (1 Kings 10:14 – 27; 2 Chron 1:15; 9:13 – 24).

In spite of all the wealth flowing into his kingdom, it seems that Solomon had to resort to a program

of taxing some of his subjects in order to maintain his large court. He did this by dividing all of the territory of Israel north of Jerusalem into twelve administrative districts. Evidently Judah, the tribe from which David and Solomon came, was exempt from this burden, and the hard feelings that resulted from this probably fueled the north's eventual desire for independence. The twelve districts each had to provide food for the king and his household for one month each year (1 Kings 4:7, 22 – 23, 27 – 28).

In some cases the districts seem to have coincided with old tribal territories, such as Ephraim (1 Kings 4:8), Dan (v. 9), Naphtali (v. 15), Asher

▼ *Ain Dara, Syria: Three foot(!) long "footprints of the deity" carved on the paving stones of the entrance to the temple at Ain Dara*

Mark Connally

TWELVE ADMINISTRATIVE DIVISIONS IN SOLOMON'S KINGDOM

ARAM

DAMASCUS

Tyre

Dan

Baana, son of Hushai in Asher and Aloth-Zebulun

Ahimaaz in Naphtali

Bashan

Hazor

GESHUR

Geshur

Acco

LAND OF CABUL

Cabul

Sea of Galilee

Ben-Abinadab in Naphoth Dor

Ben Geber in Ramoth Gilead, Argob in Bashan and its sixty large, walled cities

Mediterranean Sea

Dor

Jehoshaphat in Issachar

Megiddo

Beth Shan

Taanach

Ramoth Gilead

Baana in Taanach, Megiddo, Beth Shan

Arubboth

Jordan R.

Hepher

Socoh

Abel Meholah

Ben-Hesed in Arubboth, Socoh and all the land of Hepher

Mahanaim

AMMON

Shechem

Ben-Hur in the hill country of Ephraim

Jokmeam

Ahinadab in Mahanaim

Joppa

Rabbah of the Ammonites

Ben-Deker in Makaz, Shaalbim, Beth Shemesh and Elon Bethhanan

Shaalbim

Aijalon

Shimei in Benjamin

Heshbon

Beth Shemesh

Jerusalem

Geber, son of Uri, in Gilead

JUDAH

Dead Sea

Aroer

0   10 km.
0   10 miles

(v. 16), Aloth (= Zebulun; v. 16), Issachar (v. 17), and Benjamin (v. 18). Other districts appear to have been composed of cities that had formerly been major Canaanite centers but that now received a status of their own: the Sharon Plain and western Manasseh (v. 10); the Plain of Dor (v. 11); and the Jezreel, Harod, and Jordan valleys (v. 12). The delimitation of the Transjordanian districts is problematic, but those districts may have roughly coincided with the territories allotted to Reuben, Gad, and the half tribe of Manasseh. It has been suggested that, since portions of territory once allotted to Ephraim and Manasseh — the coastal

Mark Connally

▲ *Ain Dara, Syria: A winged creature (cherub?) guarding the entrance to the temple at Ain Dara*

plain and the Jezreel, Harod, and Jordan Valleys — now were considered independent districts, part of the Solomonic agenda was to neutralize the strength of these two powerful northern tribes. In all probability this administrative list dates to the second half of Solomon's reign, for two of the administrators were Solomon's sons-in-law (vv. 11, 15).

Although many Bible readers consider Solomon's reign to be one of unmatched wealth and power, there are some indications from the Bible that the empire was showing signs of weakness. This can be seen in Solomon's relationship with Egypt. On the one hand, the fact that Pharaoh gave his daughter to Solomon to seal a political alliance between the two states is evidence that Solomon was strong and had to be reckoned with. Reference is made to Pharaoh's daughter five different times in the Bible in an attempt to aggrandize the power, wealth, and prestige of Solomon.

But on the other hand, it must be remembered that this pharaoh, probably Siamun (978 – 959 BC) of the Twenty-First Dynasty, also captured Gezer (only 30 mi. from Jerusalem!), burned it, and then gave it as a dowry present to his daughter (1 Kings 9:16). Since Solomon fortified Gezer (v. 15), and since the transport of shipments of cedar from Joppa to Jerusalem had to pass by Gezer, this expedition of Siamun must have taken place early in Solomon's reign, probably before his fourth year (966 BC), the year in which the temple was begun (6:1). What was Siamun doing in the heart of the Solomonic empire? Reestablishing Egyptian sovereignty over Philistia? Driving the Israelites out of Philistia? Attempting to invade Israel, but, not succeeding, settling for a treaty with Solomon? It is difficult to be dogmatic on this point, but

it is clear that the king of Egypt felt himself to be sufficiently strong to make this bold move soon after David's death (970 BC).

From the north, Hiram of Tyre supplied Solomon with lumber and gold (1 Kings 9:11 – 14), workmen (5:18; 7:13 – 14), and ships and sailors for maritime ventures on the Red Sea (9:27 – 28). In return, Solomon provided Hiram with agricultural products, including wheat and oil (5:11). The motive behind Solomon's gift to Hiram of twenty cities from the tribe of Asher in the Plain of Acco area (9:11 – 14) is much disputed. It is possible that Solomon gave up these cities because of his unpaid debt to Hiram.

Toward the end of his reign, Solomon faced a number of serious problems. Internally, he had to face the dissatisfaction of some of the leaders of the northern tribes, who objected to his building projects and probably also to his taxation and conscription policies (1 Kings 11:27). This dissatisfaction can be seen in Jeroboam's rebellion against Solomon, although this abortive attempt failed and Jeroboam had to flee to Egypt, where he remained until Solomon's death (vv. 26 – 40). Externally, Hadad the Edomite became Solomon's adversary to the southeast (vv. 14 – 22), while to the northeast Rezon of Damascus became a leader of rebels who took control of and settled in Damascus (vv. 23 – 25). In addition, Solomon began to worship some of the foreign deities he had introduced into Jerusalem in deference to his foreign wives (vv. 1 – 8). It is thus evident that both internal and external factors were weakening the empire, and it is no wonder that it collapsed almost immediately upon the death of Solomon.

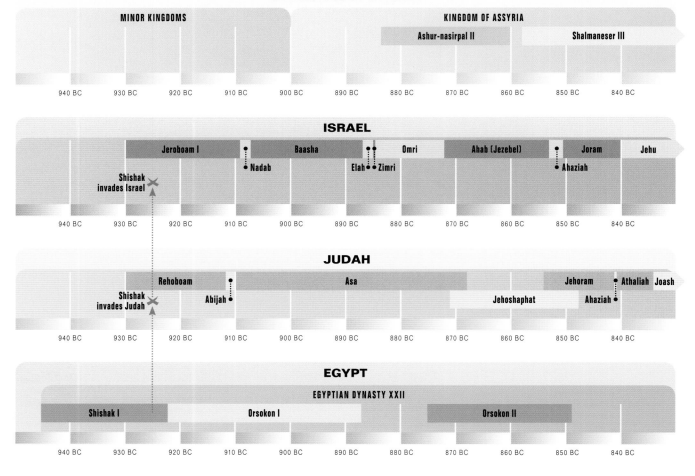

SYRIA/MESOPOTAMIA

| MINOR KINGDOMS | | | | | KINGDOM OF ASSYRIA | | | | |
| | | | | | Ashur-nasirpal II | | | Shalmaneser III | |

940 BC   930 BC   920 BC   910 BC   900 BC   890 BC   880 BC   870 BC   860 BC   850 BC   840 BC

ISRAEL

Jeroboam I — Nadab — Baasha — Elah — Zimri — Omri — Ahab (Jezebel) — Ahaziah — Joram — Jehu

Shishak invades Israel

940 BC   930 BC   920 BC   910 BC   900 BC   890 BC   880 BC   870 BC   860 BC   850 BC   840 BC

JUDAH

Rehoboam — Abijah — Asa — Jehoshaphat — Jehoram — Ahaziah — Athaliah — Joash

Shishak invades Judah

940 BC   930 BC   920 BC   910 BC   900 BC   890 BC   880 BC   870 BC   860 BC   850 BC   840 BC

EGYPT

EGYPTIAN DYNASTY XXII

Shishak I — Orsokon I — Orsokon II

940 BC   930 BC   920 BC   910 BC   900 BC   890 BC   880 BC   870 BC   860 BC   850 BC   840 BC

THE DIVIDED KINGDOM

▼ *Dan: Podium (partially reconstructed) where the king sat on a throne at the entrance to the city gate*

Although Israel and Judah had been joined together for some seventy-three years under the leadership of David and Solomon, the continuation of the union was by no means assured. After the death of Solomon (930 BC), his son and successor, Rehoboam (930–913 BC), traveled north to the tribal center at Shechem in order to secure the continued allegiance of the northern tribes (1 Kings 12:1–19; 2 Chron 10:1–19). However, when the latter demanded relief from the oppressive tax burden Solomon had placed on them, Rehoboam's response was that he would increase this burden. At this point the northern tribes rejected the supremacy of the Davidic dynasty and appointed Jeroboam (930–910 BC) as their first king. Rehoboam fled for his life back to Jerusalem, and thus began the period of the divided kingdom (930–722 BC).

In the north, Jeroboam is most famous for his religious maneuver of establishing worship centers at Dan and Bethel (1 Kings 12:26–33).

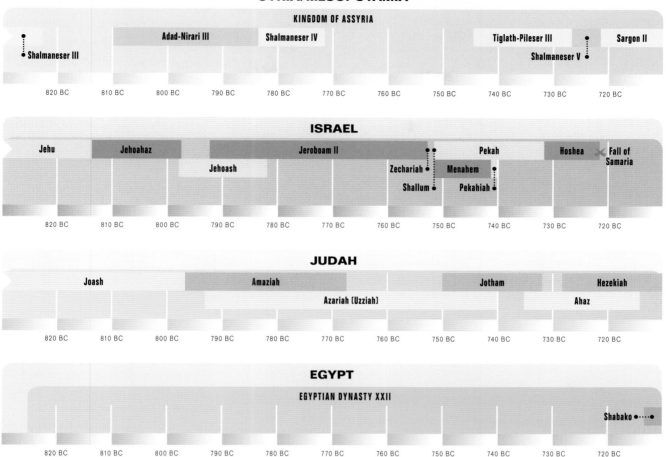

SYRIA/MESOPOTAMIA

KINGDOM OF ASSYRIA

Shalmaneser III	Adad-Nirari III		Shalmaneser IV					Tiglath-Pileser III		Sargon II
								Shalmaneser V		

820 BC 810 BC 800 BC 790 BC 780 BC 770 BC 760 BC 750 BC 740 BC 730 BC 720 BC

ISRAEL

Jehu	Jehoahaz		Jeroboam II					Pekah	Hoshea	Fall of Samaria
		Jehoash				Zechariah	Menahem			
						Shallum	Pekahiah			

820 BC 810 BC 800 BC 790 BC 780 BC 770 BC 760 BC 750 BC 740 BC 730 BC 720 BC

JUDAH

| Joash | | Amaziah | | | | | Jotham | | Hezekiah |
| | | Azariah (Uzziah) | | | | | | Ahaz | |

820 BC 810 BC 800 BC 790 BC 780 BC 770 BC 760 BC 750 BC 740 BC 730 BC 720 BC

EGYPT

EGYPTIAN DYNASTY XXII

Shabako

820 BC 810 BC 800 BC 790 BC 780 BC 770 BC 760 BC 750 BC 740 BC 730 BC 720 BC

By building sanctuaries at these two cities, which were located at the extremities of his country, he effectively defined the limits of the territory under his control. In addition, by placing a sanctuary at Bethel, which was located on the Ridge Route only eleven miles north of Jerusalem, he in effect attempted to "detour" any worshipers who might be heading to the temple in Jerusalem. Since Jeroboam appointed non-Levitical priests to serve at his shrines and since the loyalty of most of the Levites was to the sanctuary in Jerusalem and to the Davidic dynasty, many of the Levites left their allotted cities in northern Israel (see map p. 120) and moved south, to the more hospitable territory of Judah (2 Chron 11:13 – 14).

The Egyptian pharaoh Shishak (945 – 924 BC), seeing the weakness of the divided kingdom, made plans for the invasion of both Judah and Israel. In response, Rehoboam constructed fifteen fortresses in Judah (2 Chron 11:5 – 12) to ward off the attack. It is important to note that these forts were located in the Shephelah, the southern hill country, and along the edge of the Judean Desert; thus it is clear that by 925 BC, within five years of the death of the powerful Solomon, the "empire" of his son was basically confined to the Hill Country of Judah!

Shishak's invasion of Judah is briefly described in the Bible (1 Kings 14:25 – 31; 2 Chron 12:1 – 11), but his own inscription documents it much more fully. The invasion took place in Rehoboam's fifth year (925 BC) and proceeded from the coastal plain into the Hill

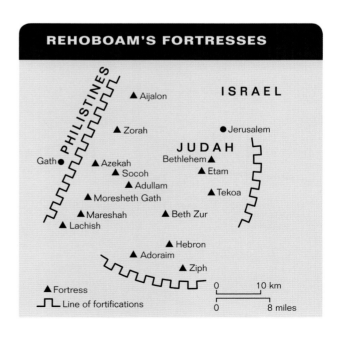

REHOBOAM'S FORTRESSES

Country of Benjamin via the Beth Horon ridge as well as along one of the ridges to the south. Although Jerusalem was threatened, it avoided capture because Rehoboam paid large sums of tribute to Shishak (1 Kings 14:25 – 28; 2 Chron 12:9 – 11). From Shishak's account of his campaign it is difficult to determine his exact line of march. Apparently he moved northward toward Shechem, pursuing Jeroboam (who previously had found refuge with Shishak in Egypt [1 Kings 11:40]) to Tirzah, down the Wadi Farah into the Jordan Valley, and farther into the hills of Gilead. Indeed, Jeroboam may have temporarily moved his capital to Penuel (12:25; Heb., but NIV "Peniel"), a city discreetly located in Transjordan away from the major thrust of Shishak's attack. In addition, the northern portion of Shishak's campaign included attacks on cities in the Jezreel Valley and the Sharon Plain.

The second section of Shishak's inscription describes his conquest of some eighty-five settlements in the southern part of Judah in the Negev. This text and recent archaeological investigations suggest that the fortresses that Solomon had built were destroyed by the Egyptian king, implying that the conveyers of

Mark Connally

▲ *Karnak, Egypt: Shishak conquering his foes lower right pleading for mercy. In the lower left are cartouches of captured enemies/cities; (1 Kings 14:25 – 28).*

▼ *Dan: High place for worship. Jeroboam originally put "golden calves" at both Bethel and Dan (1 Kings 12:28 – 30).*

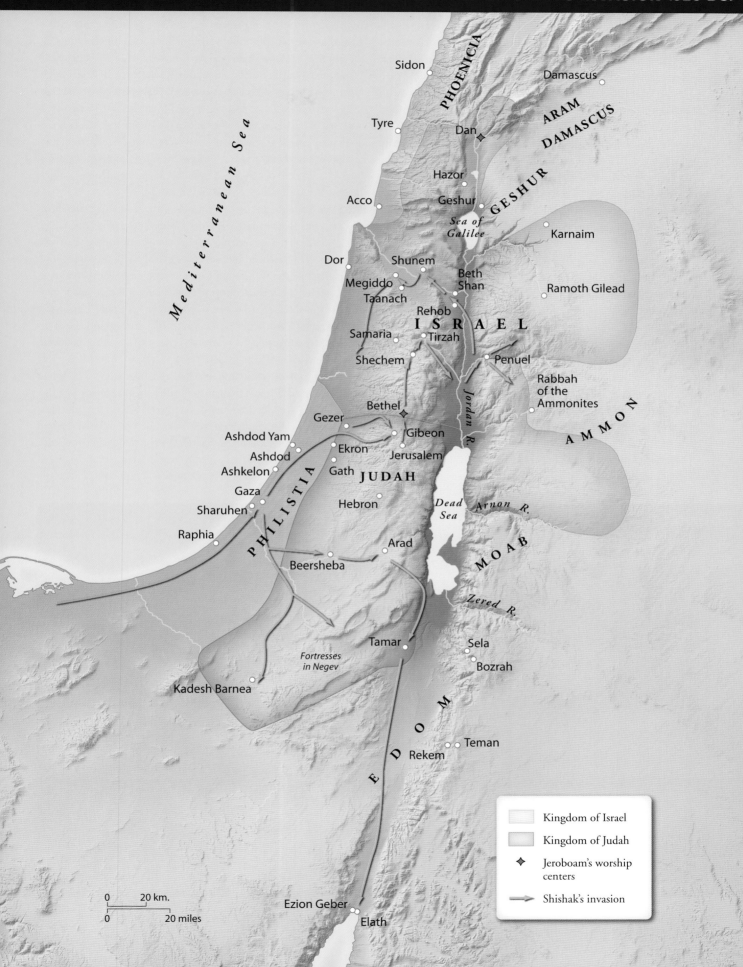

Mediterranean Sea

Sidon

Damascus

PHOENICIA

ARAM
DAMASCUS

Tyre

Dan

Hazor

GESHUR

Acco

Geshur

Sea of
Galilee

Karnaim

Dor

Shunem

Beth
Shan

Ramoth Gilead

Megiddo

Taanach

Rehob

I S R A E L

Samaria

Tirzah

Shechem

Penuel

Rabbah
of the
Ammonites

AMMON

Bethel

Gezer

Gibeon

Ashdod Yam

Ekron

Jerusalem

Ashdod

Gath

JUDAH

Ashkelon

PHILISTIA

Jordan R.

Gaza

Hebron

Dead
Sea

Arnon R.

Sharuhen

MOAB

Raphia

Arad

Beersheba

Zered R.

Fortresses
in Negev

Tamar

Sela

Bozrah

E
D
O
M

Kadesh Barnea

Teman

Rekem

Kingdom of Israel

Kingdom of Judah

◆ Jeroboam's worship
centers

→ Shishak's invasion

0   20 km.

0   20 miles

Ezion Geber

Elath

luxury goods from southern Arabia now had to follow alternate routes (through Shishak's Egypt?) to Mediterranean ports. Thus effective control of the major trade routes through Israel slipped from Israelite and Judean hands, to the benefit of the Egyptians.

Israel and Judah, although both humbled by their experiences with Shishak, still took about twenty years to resolve the dispute over their common border. Rehoboam's son Abijah (913–910 BC) made the first of several incursions, invading Israel and capturing Bethel, Jeshanah, and Ephron, as well as the surrounding villages (2 Chron 13:2–20). This northern movement of the Judean border, which brought Jeroboam's cult center at Bethel under Judean control, lasted for approximately twenty years. It was the northern king Baasha (908–886 BC) who responded by pushing the border south to Ramah, only 6 miles north of Jerusalem.

The Judean king Asa (910–869 BC) felt he needed assistance in meeting this Israelite threat and appealed to Ben-Hadad, king of Aram Damascus. Ben-Hadad, in turn, opened a northern front against Israel, invading and capturing Ijon, Dan, Abel Beth Maacah, Kinnereth, and other cities, all of which were located in the Rift Valley north of the Sea of Galilee (1 Kings 15:16–22; 2 Chron 16:1–6). Because of Ben-Hadad's attack, Baasha had to abandon his southward expansion plans, and Asa was able to push their common border northward. The resulting boundary, which left Bethel in Israel to the north and Mizpah and Geba in Judah to the south, formed the traditional boundary between the northern and southern kingdoms until the fall of the north in 722 BC.

Throughout its 209-year history (930–722 BC), the Northern Kingdom was characterized by instability. Its nineteen kings came from nine different families, and eight of its kings were either assassinated or committed suicide. This instability was also illustrated by the fact that Israel had four capitals in succession: Shechem, Penuel, Tirzah, and Samaria. Shechem was initially selected because of its long history as a tribal and religious center. Probably Penuel served as a temporary refuge for Jeroboam as he fled from Shishak. Tirzah, situated east of the watershed at the head of the Wadi Farah, next served as a secure capital on the remote (east) side of the central mountains for approximately forty years. It was at the midpoint of Omri's reign (885–874 BC) that he purchased the hill of Shemer, built a new capital there, and named it Samaria (1 Kings 16:23–24).

Since Samaria, on the western side of the mountain watershed, was situated close to an easy approach from the Sharon Plain via the Nahal Shechem, it was much more open to foreign influences than was Tirzah or even Shechem. Indeed, Israel's treaty with the king of the Sidonians indicates its outward-looking orientation. The marriage of Ahab and Jezebel, the daughter of Ethbaal, king of the Sidonians, not only cemented this alliance, but Ahab's building of a temple and altar in Samaria for Jezebel's god Baal indicates that external influences on the Northern Kingdom were not merely economic, social, and political, but religious as well (1 Kings 16:31–33).

Although Israel was able to establish friendly relations with Judah to the south and Tyre and Sidon to the northwest, its relations with the Arameans of Damascus deteriorated. There were probably several causes for this conflict. For example, a strong Israel, occupying cities such as Dan, Ijon, Abel Beth Maacah, and Hazor, in effect controlled Damascus' east–west caravan route to the Mediterranean coast (see p. 37). Also, a powerful Israel meant that it, not Damascus, controlled the lucrative

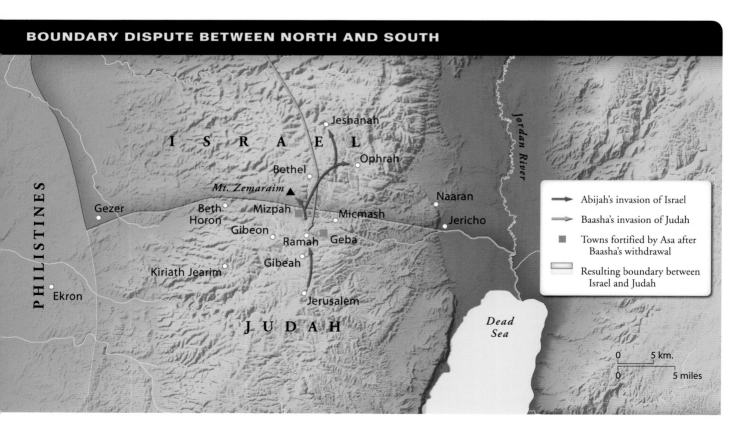

BOUNDARY DISPUTE BETWEEN NORTH AND SOUTH

Abijah's invasion of Israel

Baasha's invasion of Judah

Towns fortified by Asa after Baasha's withdrawal

Resulting boundary between Israel and Judah

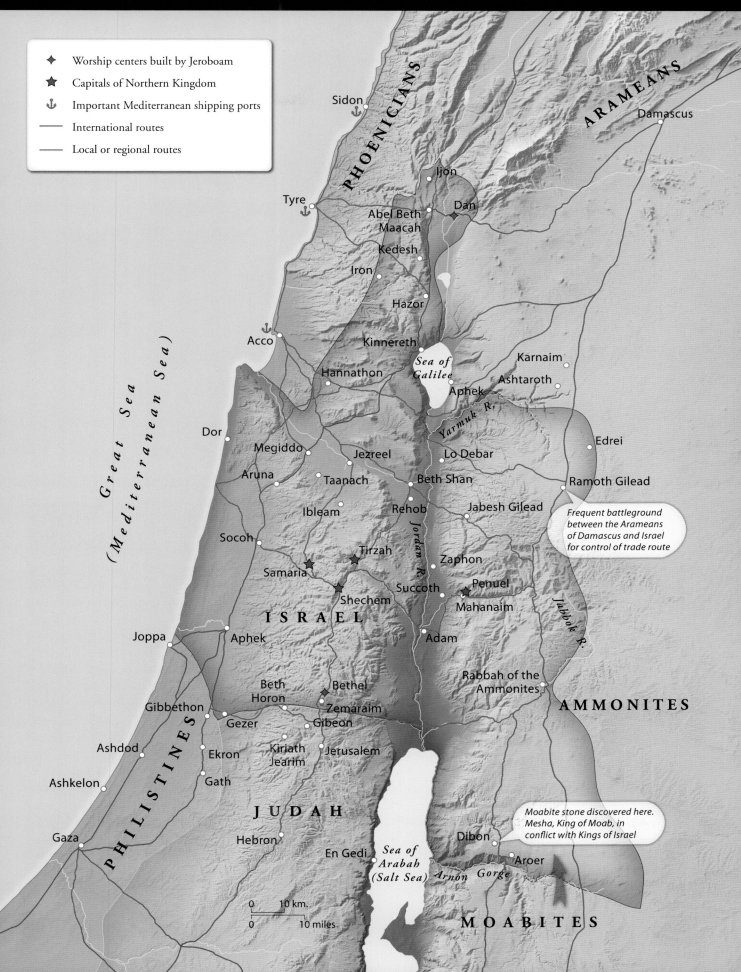

Worship centers built by Jeroboam
Capitals of Northern Kingdom
Important Mediterranean shipping ports
International routes
Local or regional routes

PHOENICIANS

ARAMEANS

Damascus

Sidon

Ijon

Tyre

Dan

Abel Beth Maacah

Kedesh

Iron

Hazor

Acco

Kinnereth

Sea of Galilee

Karnaim

Ashtaroth

Aphek

Yarmuk R.

Dor

Edrei

Megiddo

Jezreel

Lo Debar

Aruna

Taanach

Beth Shan

Ramoth Gilead

Ibleam

Rehob

Jabesh Gilead

Socoh

Jordan R.

Frequent battleground between the Arameans of Damascus and Israel for control of trade route

Tirzah

Zaphon

Jabbok R.

Samaria

Succoth

Penuel

Shechem

Mahanaim

ISRAEL

Great Sea (Mediterranean Sea)

Joppa

Aphek

Adam

Rabbah of the Ammonites

AMMONITES

Gibbethon

Beth Horon

Bethel

Gezer

Zemaraim

Gibeon

Ashdod

Ekron

Kiriath Jearim

Jerusalem

Ashkelon

Gath

JUDAH

PHILISTINES

Gaza

Hebron

En Gedi

Sea of Arabah (Salt Sea)

Dibon

Moabite stone discovered here. Mesha, King of Moab, in conflict with Kings of Israel

Aroer

Arnon Gorge

0   10 km.
0   10 miles

MOABITES

▲ *Arad, Negev: Judean worship center. Sacrificial altar and courtyard on right, broad room to the left, and most sacred room with standing stones and incense altars upper left.*

north – south incense route along the King's Highway in Transjordan. If Israel could control this route up to Ramoth Gilead, then spice and incense caravans could pass westward through Israelite territory to ports such as Acco, Tyre, and Sidon. From there, luxury goods could be loaded onto the ships of Phoenicia, Israel's new ally, and transported anywhere in the Mediterranean world. Thus, caravan transit revenues would benefit Israel rather than Damascus. It is no wonder that at least thirteen battles were fought between Israel and the Arameans; several of them occurred at or near the strategic city of Ramoth Gilead.

On a number of occasions the Arameans pressed their advantage to the gates of Samaria — for example, during the days of Ahab (874 – 853 BC; 1 Kings 20:1 – 21) and J(eh)oram (852 – 841 BC; 2 Kings 6:24 – 7:8) — but in each instance they were driven off to lands east of the Jordan River. Despite the numerous conflicts between Aram and Israel, it was necessary for these two rivals to join together along with other nations of the Levant in order to meet the growing Assyrian threat when the Assyrian king Shalmaneser III (858 – 824 BC) began pushing westward and southward. A coalition of eleven or twelve Levantine kings met him in battle at Qarqar on the Orontes in 853 BC. Shalmaneser portrays himself as victorious, but it is evident that the coalition temporarily blunted the Assyrian threat.

Significant is the fact that Hadadezer of Damascus (= biblical Ben-Hadad II) supplied the most infantry (20,000 troops), while Ahab the Israelite provided the most chariots (2,000). Although the record of the number of chariots involved seems to have been inflated by Shalmaneser's scribes, it is interesting to note that both Hadadezer and Ahab were considered to be the strongest of all the allied kings and that these two countries, which recently had been fighting each other and would do so again soon after the battle, appear as allies at Qarqar! In addition, it is well to note the appearance of Gindibu the Arab, who supplied 1,000 camels to the war effort — a fact that underlines the growing importance of the camel as a beast of burden, as well as Arabia's interest in affairs much to its north (i.e., its interest in protecting its northern spice and incense markets and terminals).

Soon after the battle of Qarqar, still in 853 BC, Ben-Hadad and Ahab were again fighting each other at Ramoth Gilead. Ahab was killed in the battle (1 Kings 22:29 – 37; 2 Chron 18:28 – 34). With the death of the powerful Ahab, Moab, Israel's vassal, revolted (2 Kings 3:4 – 27 and Mesha Stela in *ANET*, 320 – 21), gaining its independence and capturing the tableland north of the Arnon Gorge as far as Medeba, Nebo, and Bezer, at the expense of the Israelite tribes of Gad and Reuben.

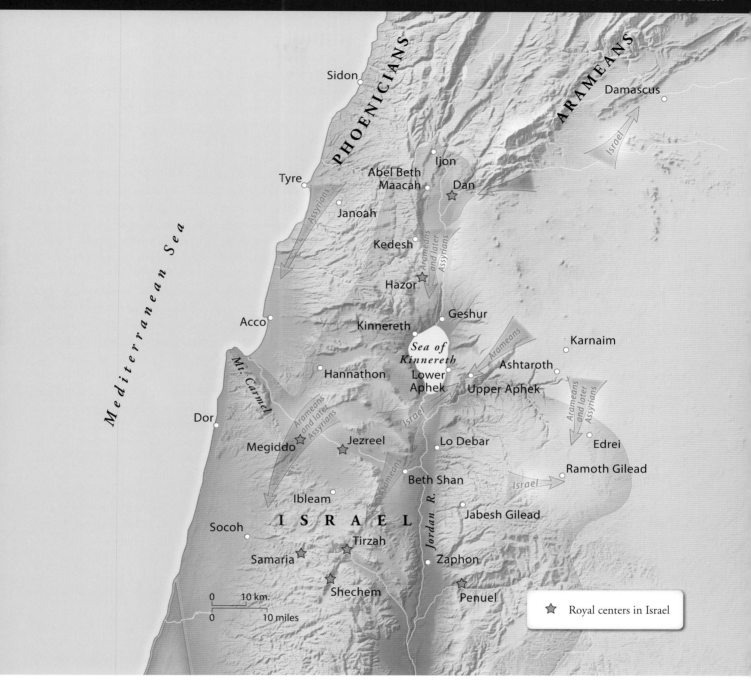

In the north, Ahab's son Joram (852 – 841 BC) bore the brunt of Aram's continuing attacks (2 Kings 6:24 – 7:8). Late in Joram's reign he was wounded while fighting the Arameans at Ramoth Gilead. While Joram was recuperating from his wounds at his winter palace in Jezreel, Jehu, the commander of Joram's army at Ramoth Gilead, staged a coup, executing not only the king but also Jezebel and eventually all the remaining descendants of the line of Omri/Ahab (841 BC; 2 Kings 9 and 10).

In contrast to the Northern Kingdom, Judah exhibited much more stability during its 345-year history — from the time of Solomon's death (930 BC) until the sack of Jerusalem by the Babylonians (586 BC). Judah had nineteen kings, all of whom were from the Davidic dynasty. In addition, at the time of Solomon's death Jerusalem had been firmly established as the religious and political capital of the Southern Kingdom. Indeed, the Levites who lived in the north moved to Judah when Israel revolted, in order to maintain their connection with the temple worship.

As noted previously, the location of Rehoboam's fortresses and the route of Shishak's invasion (925 BC) clearly indicate that the Southern Kingdom was basically confined to the Hill Country of Judah and the Shephelah. During Rehoboam's rule (930 – 913 BC) Judah was no longer able to control the trade routes that ran through the coastal plain, the Negev, and Transjordan.

The Chronicler's account of the reigns of successive Judean kings illustrates the theological principle that their expressions of fidelity and trust toward God led to blessing, prosperity, and strength, while their disobedience usually led to disaster, destruction, defeat, and eventually deportation. For example, during the early part of Asa's reign (910 – 869 BC), expressions of his trust in God were followed by victory over the invading hordes of Zerah the Ethiopian at Zephathah near Mareshah (2 Chron 14). But later, Asa's lack of trust in the power of God to deal with Baasha's aggressive move southward (p. 154) and his reliance on an alliance with the pagan Ben-Hadad of Damascus eventually led to his ignominious death (16:7 – 14).

Jehoshaphat (872 – 848 BC), Asa's successor, instituted a series of religious and legal reforms. In addition, garrison cities, forts, and store cities were constructed and manned (2 Chron 17), and internal administrative districts and governorships were established (17:2). Jehoshaphat's power was such that the Philistines to the west and the Arabians to the south and east brought him tribute (vv. 10 – 11), possibly because he was again exercising Judean control over portions of the international trade routes.

Later in the reign of Jehoshaphat, the Moabites, Ammonites, and Meunites (inhabitants of Mount Seir [v. 10]) invaded Judah from the east (2 Chron 20). They evidently crossed the Salt Sea via a ford that led from the Lisan area westward, setting up camp at Hazazon Tamar (= En Gedi). From there they proceeded up the Pass of Ziz, evidently attempting to move up the road on the ridge just north of the Nahal Arugot. Jehoshaphat, after inquiring of the Lord, went out to meet them in battle. But before his arrival, the invading forces fought among themselves and slaughtered each other, so that the Judean army merely had to collect the booty left behind. Indeed, God had gone before the faithful Judeans.

In spite of his faithfulness to God, Jehoshaphat was also noted for establishing close relations with Israel to the north. In fact, his son Jehoram married Athaliah, a daughter of Ahab and Jezebel (2 Kings 8:18; 2 Chron 21:6), and Jehoshaphat even fought alongside Ahab in the battle of Ramoth Gilead (853 BC; see p. 156). When Ahab's son Joram experienced defeat in his encounter with Mesha, the king of Moab, in the northern Mishor area, Jehoshaphat permitted Joram to pass through Judah so that he could attack Moab from the southwest. Jehoshaphat and the Edomites even joined him in this unsuccessful endeavor (2 Kings 3:6 – 27).

Further evidence of Jehoshaphat's alliance with the north was his attempt to revive Israelite/Judean shipping on the Red Sea. Unfortunately, the Tarshish ships that Jehoshaphat and Ahaziah (853 – 852 BC) had constructed to go to Ophir barely left port before being wrecked at Ezion Geber (1 Kings 22:47 – 49; 2 Chron 20:35 – 37). Indeed, every recorded joint venture between Jehoshaphat and Israel was unsuccessful (the battle at Ramoth Gilead, the invasion of Moab, shipping on the Red Sea); yet when Jehoshaphat trusted in God, he was successful (the defeat of Moabites, Ammonites, and inhabitants of Mount Seir).

Jehoram, Jehoshaphat's son, who had married Ahab's daughter Athaliah, was wicked. During his reign (853 – 841 BC) Judah experienced a number of serious reversals. To the southeast, Edom revolted, as did Libnah in the Shephelah on the west (2 Kings 8:20 – 22; 2 Chron 21:8 – 10). The Philistines and Arabians, who had formerly brought tribute to Jehoshaphat, invaded Judah and even carried off members of the royal family (2 Chron 21:16 – 17).

At Jehoram's death, his son Ahaziah continued the alliance with Israel, fighting alongside Joram at Ramoth Gilead (2 Kings 8:28; 2 Chron 22:5). However, this close alliance led to Ahaziah's death: when he was visiting the wounded Joram in Jezreel, Jehu staged a coup and killed him along with Joram, Jezebel, and the last of Ahab's descendants (2 Kings 9:27 – 28; 2 Chron 22:9).

Soon afterward, the royal Davidic line was almost annihilated by the wicked queen Athaliah (841 – 835 BC; 2 Kings 11). However, a priestly family rescued one child from the Davidic line (Joash) and secretly raised him for a number of years. At the age of seven Joash was crowned king, and the wicked Athaliah was executed (2 Kings 11:1 – 16; 2 Chron 23:1 – 15). During the early years of Joash's rule (835 – 796 BC), he and his mentor, Jehoiada the high priest, initiated religious reforms (2 Kings 12:1 – 12; 2 Chron 24:1 – 16). However, late in Joash's rule, after Jehoiada died at age 130, Joash forsook God for the worship of the Asherim and idols, and he even murdered the son of his godly advisor (2 Chron 24:17 – 22). Typically, disobedience led to disaster, and soon Joash felt the pressure of the Aramean king Hazael, who first attacked Gath in the Philistine Plain and then moved eastward into the hill country toward Jerusalem (2 Kings 12:17 – 18; 2 Chron 24:23 – 25). However, when Joash paid tribute to Hazael, the latter abandoned the attack on Jerusalem.

Disgruntled elements assassinated Joash (2 Kings 12:19 – 21; 2 Chron 24:25 – 27), and his son Amaziah ruled in his place (796 – 767 BC). Amaziah began his reign by subduing Edom and capturing Sela, renaming it Joktheel (2 Kings 14:7; 2 Chron 25:1 – 15). Soon Amaziah was worshiping captured Edomite gods, however, and his challenge to meet the Israelites in battle was answered by the Israelite king Jehoash — who defeated him in the battle of Beth Shemesh (2 Kings 14:8 – 14; 2 Chron 25:17 – 24). The Israelites continued on to Jerusalem, plundered the city, and destroyed portions of its walls. Amaziah's political support was at a low point, and as he fled Jerusalem he was assassinated by conspirators at the Judean city of Lachish (2 Kings 14:18 – 20; 2 Chron 25:26 – 28).

Amaziah was succeeded by his son Azariah (also called Uzziah; 792 – 740 BC). Azariah was judged by the ancient historians to have been a "good" king, and during his reign Judean influence spread considerably. On the west he captured Gath, Jabneh, and Ashdod (2 Chron 26:6 – 8). To the south and southwest the Arabs in Gur Baal and the Meunim paid tribute to him, as did the Ammonites on the east. In the south he constructed forts in the wilderness, subdued the Edomites, and rebuilt Elath on the Red Sea (2 Kings 14:22; 2 Chron 26:2). He thus reinstituted Judean control over the north – south trade routes that ran through the coastal plain and through Transjordan, as well as over the important east – west route that led from Edom/Arabia to the Mediterranean ports at Gaza and Ashdod. Because of this, it can be assumed that considerable revenues accrued to the Judeans, just as they had when Solomon had controlled these routes. In addition, Azariah fortified Jerusalem and built the Corner and Valley gates (2 Chron 26:9 – 15).

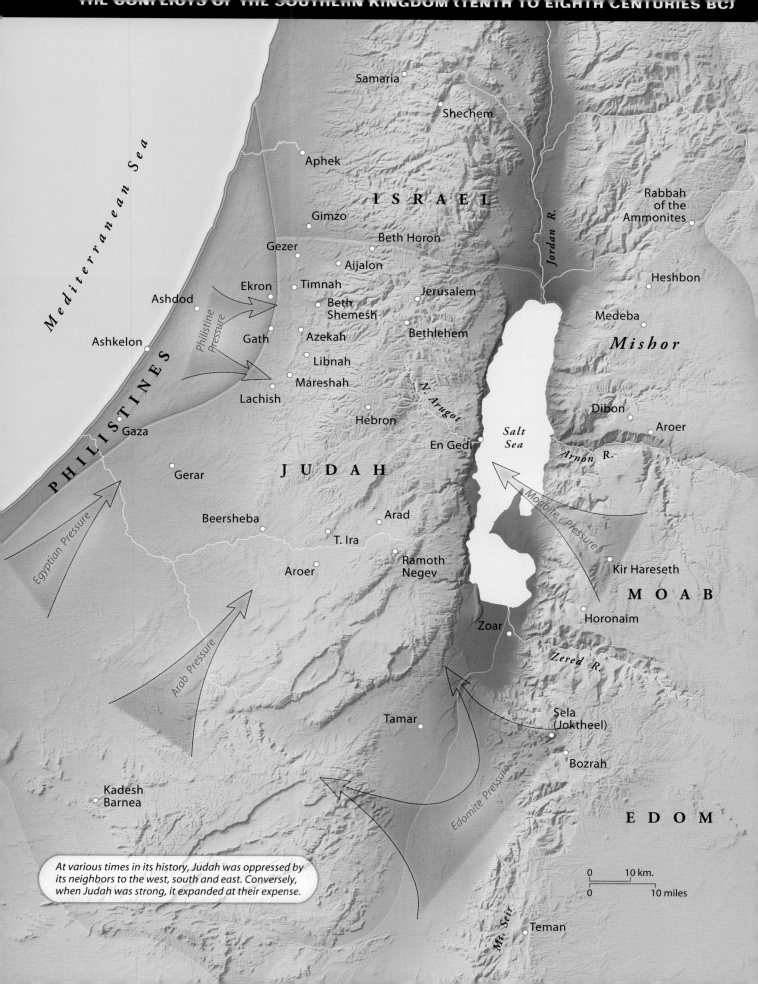

Mediterranean Sea

Samaria

Shechem

Aphek

I S R A E L

Gimzo

Beth Horon

Gezer

Aijalon

Ekron

Timnah

Philistine Pressure

Ashdod

Beth Shemesh

Jerusalem

Ashkelon

Gath

Azekah

Bethlehem

Libnah

Mareshah

N. Arugot

Lachish

Hebron

Gaza

P H I L I S T I N E S

En Gedi

Salt Sea

Gerar

J U D A H

Beersheba

Arad

Egyptian Pressure

T. Ira

Ramoth Negev

Aroer

Arab Pressure

Tamar

Kadesh Barnea

Edomite Pressure

Zoar

Rabbah of the Ammonites

Heshbon

Medeba

Mishor

Dibon

Aroer

Arnon R.

Jordan R.

Moabite Pressure

Kir Hareseth

M O A B

Horonaim

Zered R.

Sela (Joktheel)

Bozrah

E D O M

Mt. Seir

Teman

At various times in its history, Judah was oppressed by its neighbors to the west, south and east. Conversely, when Judah was strong, it expanded at their expense.

| 0 | 10 km. |

| 0 | 10 miles |

The kingship of Jotham, Azariah's son, is briefly described (2 Kings 15:32 – 38; 2 Chron 27:1 – 9). He was known for continuing in the ways of his father, and the historian notes his building projects in Jerusalem as well as his construction of cities, forts, and towers in the Judean Hill Country. However, toward the end of his reign, when Ahaz was coregent with him (apparently 735 – 732 BC), Pekah of Israel, Rezin of Aram Damascus, and probably the son of Tabeel of Transjordan attacked Judah. They were evidently attempting to replace the Davidic king with one of their own choosing, namely, the son of Tabeel, who would join them in their anti-Assyrian alliance (2 Kings 16:5 – 6; 2 Chron 28:5 – 8; Isa 7:1 – 17). The Judeans, however, appealed to the Assyrian monarch Tiglath-Pileser III (744 – 727 BC) for assistance, and he willingly responded by attacking Israel in the north (2 Kings 16:7 – 10; 15:29; 2 Chron 28:16, 20 – 21. Although the Judeans were relieved of immedi-ate Israelite/Aramean pressures, they would later suffer at the hands of their Assyrian ally.

During Ahaz's reign (735 – 715 BC), religious and political conditions in Judah deteriorated. Children were made to pass through the fire, Baals were made and worshiped, and pagan rites at the high places were revived (2 Kings 16:2 – 4; 2 Chron 28:1 – 4). These religious setbacks were paralleled by political and geographical reversals: 120,000 Judeans were killed during the Israelite/Aramean invasion (2 Chron 28:6); Edom revolted and even invaded Judah (2 Kings 16:5 – 6); the Philistines invaded the Shephelah and the Negev, capturing Beth Shemesh, Aijalon, Gederoth, Soco, Timnah, Gimzo, and nearby villages (2 Chron 28:17 – 19).

Meanwhile, in the Northern Kingdom, Jehu, after assassinating Jehoram (841 BC), continued his purge by wiping out the living male

ASSYRIAN EXPANSION

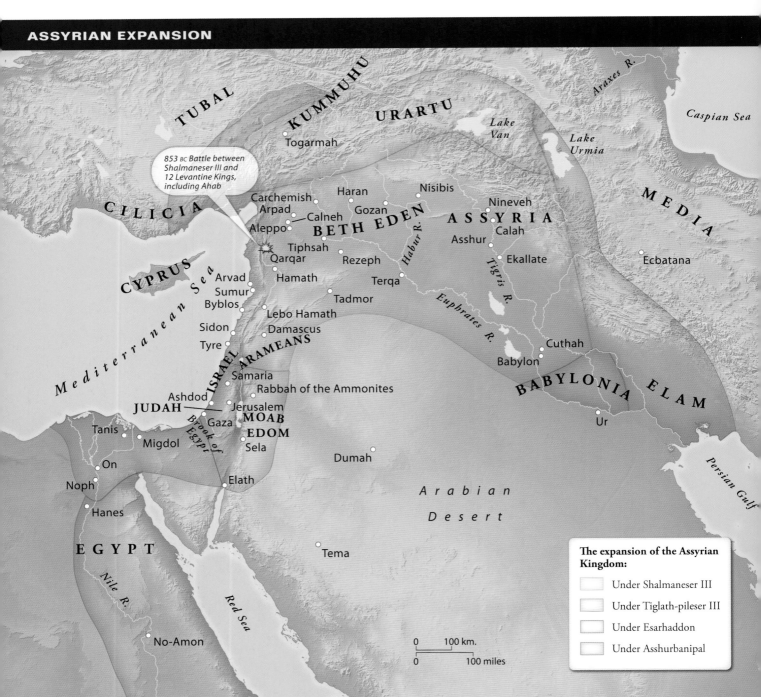

853 BC Battle between Shalmaneser III and 12 Levantine Kings, including Ahab

The expansion of the Assyrian Kingdom:

- Under Shalmaneser III
- Under Tiglath-pileser III
- Under Esarhaddon
- Under Asshurbanipal

0   100 km.
0   100 miles

Sidon

Damascus

Zarephath

Tyre

Dan

Mediterranean Sea

Hazor

Sea of Galilee

Acco

Jonah
Gath
Hepher

Gilead

Kishon R.

Mt. Carmel

Shunem

Jezreel

Ramoth
Gilead

Dothan

Kerith Ravine

Abel
Meholah
Elisha

Tishbe
Elijah

Samaria

Jordan R.

I S R A E L

Shiloh

A M M O N

Bethel

Gilgal

Ekron

Jeremiah
Anathoth

Jericho

Jerusalem

Moresheth
Gath
Micah

Ezekiel
Isaiah

J U D A H

PHILISTINES

Tekoa
Amos

Hebron

Dead Sea

M O A B

Arad

Way of Edom

Beersheba

Wilderness of Beersheba

E D O M

Legend

Tekoa
Amos   Cities of the Prophets

△   Places of importance in Elijah's prophetic activity

△   Places of importance in Elisha's prophetic activity

0   10 km.

0   10 miles

▲ *Black Obelisk (6 1/2 ft. high) panel portraying Jehu, the Israelite king, bowing down in submission to Shalmaneser III (from Calah/Nimrud in Iraq)*

descendants from the family of Omri and by destroying the priests and worshipers of Baal as well as the temple of Baal at Samaria (2 Kings 10). Although not mentioned in Scripture, it is known from Assyrian records (*ANET*, 280 – 81) that Jehu paid tribute to Shalmaneser III (858 – 824 BC). The latter had invaded Aramean territory, captured cities in Gilead, and received tribute from Jehu and the kings of Tyre and Sidon at Baal-Rosh (= Mount Carmel) in 841 BC.

Although Israel survived under the leadership of Jehu and his successors, it did experience some defeats. For example, near the end of Jehu's reign, probably in the 820s, Hazael the Aramean invaded Israel, capturing Transjordanian lands belonging to the tribes of Manasseh, Reuben, and Gad, all the way south to the Arnon Gorge (2 Kings 10:32 – 33). To the southwest Hazael attacked Gath in the Philistine Plain as well as Judah and Jerusalem (12:17 – 18; 2 Chron 24:23 – 24), indicating that he must have been able to move freely through prime Israelite territory, including the Jezreel Valley and the Sharon Plain, in order to reach these targets.

Jehu's successor, Jehoahaz (814 – 798 BC), was also oppressed by the Arameans. He also paid tribute to the Assyrians, namely to Adad-Nirari (810 – 783 BC; *ANET*, 281). Militarily, Israel was reduced to 50 horsemen, 10 chariots, and 10,000 foot soldiers (2 Kings 13:7) — a modest force compared with the reputed 2,000 chariots Ahab had mustered for the battle of Qarqar in 853 BC.

From roughly 800 BC until 740 BC the Assyrians were preoccupied elsewhere in the Near East, and the Israelite kingdom was able to flourish. The final Israelite renaissance began during the days of Jehoash (798 – 782 BC), who inflicted several defeats on the Aramean Ben-Hadad III and even overpowered arrogant Amaziah, the Judean king (see above, p. 158). The peak of Israelite expansion and prosperity was reached during the long rule of Jeroboam II (793 – 753 BC). The area of his influence stretched from Lebo Hamath in the north to Judah and the Sea of the Arabah (= Salt Sea) in the south and included control over Damascus (2 Kings 14:25 – 29). Indeed, when the territories controlled by the Judean king Azariah (see above, p. 158) and those controlled by Jeroboam II are considered together, their combined area almost reached Solomonic proportions.

Although the Bible is silent about the internal administrative structure of Israel during this era of great prosperity, sixty-three ostraca (broken pieces of pottery with writing on them) found in the excavation of the citadel at Samaria shed welcome light on this period. These ostraca record the receipt of wine and oil by persons living in Samaria from estates located in the surrounding villages. Many of the villages mentioned can be identified, and from the ostraca it is possible even to note on a map where various clans from the tribe of Manasseh were located. For the historical geographer this is a great bonanza, since the tribal description of the territory of Manasseh (Josh 16 – 17) is brief and

△ Cities captured by Tiglath-pileser III

▲ Cities captured by Shalmaneser V and Sargon II

Assyrian provinces in the days of Tiglath-pileser III

Assyrian provinces added in the days of Shalmaneser V and Sargon II

Mediterranean Sea

MASUATE

SUBITE

Sidon

Damascus

DAMASCUS

Ijon

Tyre

Abel Beth
Maacah

Janoah

Kedesh

Iron

Merom

KARNAIM

Janoah?

Hazor

Acco

Jotbathah

Kanah

*Sea of
Galilee*

Karnaim

Hannathon

Arumah

Ashtaroth

MEGIDDO

Dor

Megiddo

Beth Shan

GILEAD

Ramoth
Gilead

HAURAN

DOR

Samaria

SAMARIA

Jordan R.

Succoth

Aphek

AMMON

Joppa

Bethel

Rabbah of the
Ammonites

Gibbethon

Jericho

Bezer

Gezer

Jerusalem

Heshbon

Ashdod Yam

Ekron

Medeba

Ashdod

Gath

Azekah

Bethlehem

Beth
Diblathaim

Ashkelon

Lachish

Dibon

MOAB

Gaza

Hebron

En Gedi

Aroer

ASHDOD

JUDAH

*Dead
Sea*

Gerar

Arad

Kir Moab

Raphia

Beersheba

Aroer

Zoar

EDOM

0   10 km.

0   10 miles

Bozrah

lacks a list of cities. Thus our knowledge of the heartland of Manasseh is greatly augmented by the Samaria ostraca.

Jeroboam's two successors ruled for a combined total of seven months, and both were assassinated. It was during the reign of Menahem (752 – 742 BC) that the presence of the Assyrian menace was again felt. Menahem assessed a 50-shekel levy against every wealthy man in order to pay 1,000 talents (ca. 37 tons) of silver to Tiglath-Pileser III (2 Kings 15:19 – 20, *ANET*, 283).

It was Menahem's successor, Pekah (752 – 732 BC), who bore the brunt of the initial Assyrian onslaught. In order to meet the Assyrian threat, Pekah made an alliance with Rezin of Aram Damascus and with the influential son of Tabeel of Transjordan (Isa 7:5 – 6). These allies attempted to force the Judean king Ahaz into joining their coalition but, though able to devastate Judah and threaten Jerusalem, they were not successful (see above, p. 160). Biblical and especially Assyrian records describe Tiglath-Pileser III's invasions (2 Kings 15:29 – 30; *ANET*, 282 – 84). In 734 BC he marched down the Mediterranean coast, capturing cities all the way to Gaza and the Brook of Egypt. In 733 BC his army again returned, capturing northern Israelite cities such as Ijon, Abel Beth Maacah, Janoah, Kedesh, and Hazor (15:29), as well as cities in Naphtali (Galilee) and in Gilead. Finally in 732 BC, after having cut off the routes for possible Egyptian and Israelite support, he attacked and captured Damascus. Rezin was deposed and Pekah assassinated, and in the latter's stead, Tiglath-Pileser III put Hoshea (732 – 723 BC) on the Israelite throne as a puppet king.

Shortly after the death of Tiglath-Pileser III (727 BC) Hoshea, the last of the Israelite kings, revolted against the Assyrians, who responded by laying siege to Samaria under the leadership of Shalmaneser V (727 – 722 BC). After a siege of three years Samaria fell to the Assyrians (2 Kings 17:4 – 6; 18:9 – 11; *ANET*, 284 – 85). The biblical writer, commenting on the disaster that befell the Northern Kingdom in 722 BC, states that it was due to the fact that as a nation, the north had not heeded the warnings of their prophets — Elijah, Elisha, Jonah, Amos, Hosea — but had actively been involved in gross sins, including idolatry, sorcery, divination, and child sacrifice (17:7 – 23).

Although the Bible does not go into detail, the fall of Samaria was a very traumatic experience for the Israelites because the Assyrians were known for their savage treatment of rebels: not only looting, rape, and bondage, but also decapitation, impalement, and skinning people alive. However, not everyone was treated in this way, for Sargon II (721 – 705 BC) began a series of deportations whereby Israelites were consigned to Halah, to Gozan on the Habor River, and to towns in Media (2 Kings 17:6; 18:10 – 11). In their place, Sargon II and later Assyrian rulers settled foreign peoples from Babylon, Cuthah, Avva, Hamath, and Sepharvaim (17:24). These newcomers brought with them their worship of pagan deities, yet they also attempted to worship the "god of the land" (i.e., Yahweh; vv. 25 – 41). Evidently these newcomers, with their syncretistic blend of religions, were the forerunners of the religious/ethnic group later known as the Samaritans.

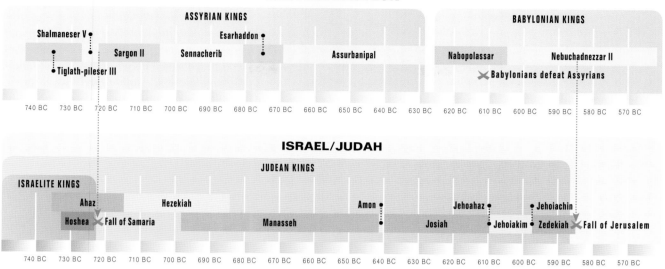

ASSYRIA/BABYLON

ASSYRIAN KINGS

Shalmaneser V •
Sargon II
Sennacherib
Esarhaddon •
Assurbanipal
• Tiglath-pileser III

BABYLONIAN KINGS

Nabopolassar
Nebuchadnezzar II
✕ Babylonians defeat Assyrians

| 740 BC | 730 BC | 720 BC | 710 BC | 700 BC | 690 BC | 680 BC | 670 BC | 660 BC | 650 BC | 640 BC | 630 BC | 620 BC | 610 BC | 600 BC | 590 BC | 580 BC | 570 BC |

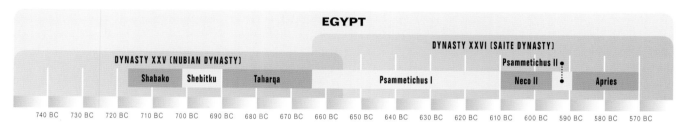

ISRAEL/JUDAH

JUDEAN KINGS

ISRAELITE KINGS

Ahaz
Hezekiah
Amon •
Jehoahaz •
• Jehoiachin
Hoshea ✕ Fall of Samaria
Manasseh
Josiah
Jehoiakim •
Zedekiah ✕ Fall of Jerusalem

| 740 BC | 730 BC | 720 BC | 710 BC | 700 BC | 690 BC | 680 BC | 670 BC | 660 BC | 650 BC | 640 BC | 630 BC | 620 BC | 610 BC | 600 BC | 590 BC | 580 BC | 570 BC |

EGYPT

DYNASTY XXVI (SAITE DYNASTY)

DYNASTY XXV (NUBIAN DYNASTY)

Shabako
Shebitku
Taharqa
Psammetichus I
Psammetichus II •
Neco II
Apries

| 740 BC | 730 BC | 720 BC | 710 BC | 700 BC | 690 BC | 680 BC | 670 BC | 660 BC | 650 BC | 640 BC | 630 BC | 620 BC | 610 BC | 600 BC | 590 BC | 580 BC | 570 BC |

JUDAH ALONE

The conquest of Damascus and the Northern Kingdom of Israel by the Assyrians brought the Judean kingdom temporary relief from military pressures between 735 BC and 722 BC, but Ahaz's importation and promotion of pagan religious practices almost ensured that the south would also fall under God's judgment. During the days of Ahaz (735–715 BC) the Edomites successfully invaded Judah and took captives (2 Chron 28:17), while on the west the Philistines captured the cities of Beth Shemesh, Aijalon, Gederoth, Soco, Timnah, and Gimzo, all located in or near the Shephelah (vv. 18–19). Judeans living in these border areas probably fled into the Hill Country of Judah — at least temporarily — in order to save their lives. It must also be assumed that many Israelites from the Northern Kingdom moved south to Judah and Jerusalem to avoid the Assyrian onslaught of Tiglath-Pileser III, Shalmaneser V, and Sargon II. The growth of Jerusalem from a city of 37 acres to 150 acres and the increase in the number of settlements in the Hill Country of Judah during the reigns of Ahaz and Hezekiah (735–686 BC) are clear indicators of this influx of population.

With the death of wicked Ahaz in 715 BC, Hezekiah's sole rule began. During the early portion of his reign, Hezekiah initiated a series of religious reforms both inside and outside Jerusalem. High places were torn down, sacred stones smashed, Asherah poles destroyed, and even the bronze snake (Nehushtan), which Moses had set up in the wilderness (Num 21:5–9), was broken into pieces (2 Kings 18:3–7; 2 Chron 29:2–19). In addition, the temple was reconsecrated, and a great Passover was celebrated in Jerusalem (2 Chron 29:20–30:27).

▼ *Jerusalem: Eighth-century wall built to defend the western hill of Jerusalem (Isa 22:10; 2 Kings 22:14). Twenty feet thick! Note the measuring stick left of center indicating the original height of the wall.*

Not only were the citizens of the Southern Kingdom included in the feast, but messengers were even sent to northern Israel, to what were now Assyrian provinces, with the result that some men of Manasseh, Zebulun, and Asher came to Jerusalem to participate.

The death of the Assyrian monarch Sargon II in 705 BC was the signal for many countries in the Near East to attempt to assert their independence. Hezekiah (728/715 – 686 BC), for example, was able to defeat the Philistines and regain territory as far as Gaza (2 Kings 18:8). In the process he evidently deposed petty kings who were still loyal to Assyria and replaced them with rulers more to his liking (e.g., Padi, the pro-Assyrian king of Ekron, was deposed and taken prisoner to Jerusalem).

Merodach-Baladan II (721 – 710 and 703 BC), the Babylonian king, was interested in encouraging anti-Assyrian activities on Assyria's southwestern flank and sent a delegation to Jerusalem, probably in 703 BC, to encourage Hezekiah in his activities. Hezekiah's proud display of his wealth to this group led to Isaiah's frightening prophecy that in the future the Babylonians would carry off his riches and his descendants (2 Kings 20:14 – 18; Isa 39:3 – 7).

Hezekiah must have anticipated that sooner or later the new Assyrian monarch, Sennacherib (704 – 681 BC), would respond to his rebellious activities. Hezekiah stationed garrisons in strategic cities throughout the Hill Country of Judah and in the Shephelah. These garrisons were provided with officers, weapons, shields, and food supplies. In the excavation

London, British Museum

▲ *Assyrians assaulting the Judean city of Lachish during the days of Hezekiah. Note the tower of the city, the battering ram, archers, torches, and lower right, Judeans being impaled on poles. Relief from Sennacherib's palace at Nineveh.*

▼ *Jerusalem: Several "false starts" near the center of Hezekiah's Tunnel where two work gangs met. The tunnel is over 1750 ft. long.*

of numerous Judean cities from this period a large number of stamped jar handles have been found. These bear impressions of four-winged scarabs or two-winged sun disks and are also inscribed in Hebrew with the word "belonging to the king," followed by the name of one of four cities: Hebron, Soco, Ziph, and an unknown city, *mmšt*. These towns were evidently the centers of the royal vineyards from which produce was shipped to garrisons stationed in various cities.

Hezekiah's preparations in and around Jerusalem were especially noteworthy. Springs in the vicinity of the city were stopped up, and a 1,750-foot-long water tunnel was dug through the solid rock in order to direct water from the Gihon Spring, located outside Jerusalem on the east, to a pool located within the city walls. This pool was also closer to the inhabitants of the newly settled western hill (see below, p. 247). In addition, Hezekiah repaired sections of Jerusalem's wall, constructed defensive towers, and reinforced the terraces on which some of the city was built (2 Chron 32:2 – 8). It was at this time that a new wall was constructed in order to include the whole of the western hill within the fortified portion of Jerusalem. Indeed, Isaiah's report that they "counted the buildings in Jerusalem and tore down houses to strengthen the wall" (Isa 22:10) has received archaeological confirmation by N. Avigad's discovery of a massive wall, 23 feet thick, located in the present-day Jewish quarter on the western hill.

Hezekiah's internal preparations were augmented by the fact that the kings of Sidon and of Ashkelon participated in the revolt. With the death of the pro-Assyrian Egyptian king Shabako in 702 BC and the accession of Shebitku (702 – 690 BC), there was probably a shift in Egyptian foreign policy to an anti-Assyrian — or at least a neutral — stance. With much of the southwestern portion of the Fertile Crescent in revolt, it was mandatory for Sennacherib to reassert his sovereignty there.

Sennacherib's response in 701 BC is one of the best-documented events in the ancient world. Scripture describes his invasion from the Judean standpoint (2 Kings 18 – 20; 2 Chron 32:1 – 23; Isa 36 – 39; probably Mic 1:8 – 16; possibly Isa 10:28 – 32). From the Assyrian standpoint the Prism of Sennacherib (*ANET*, 287 – 88) describes his invasion in great detail. In addition, stone reliefs that lined the throne room of his palace in Nineveh depict various facets of his Judean campaign, including the graphic representation of his siege of the Judean city of Lachish.

▲ *Sennacherib on his throne receiving booty and prisoners from his Judean campaign.*

After marching westward from Assyria to the Mediterranean Sea, Sennacherib proceeded south, capturing cities along the Phoenician coast as far as Acco. Continuing south, Sennacherib captured Ashkelon and cities in northern Philistia — including Beth Dagon, Joppa, Bene Berak, and Azuru. The Egyptians and Ethiopians, who had responded to Hezekiah's call for help, were defeated in the Plain of Eltekeh. It appears that Sennacherib then moved eastward, up the Sorek Valley. He captured the city of Timnah, thus cutting off the supply line between Ekron and her Judean ally, Hezekiah. This accomplished, the city of Ekron was conquered and its former king, Padi, who had been seized by Hezekiah, was reinstated. Moving south to the Elah Valley, Sennacherib captured first Azekah and then Gath.

Sennacherib proceeded farther south and assaulted the Judean stronghold of Lachish. The siege and conquest of Lachish must have been one of the high points of Sennacherib's reign, for the throne room of his palace at Nineveh was decorated with reliefs depicting this conquest. The recent archaeological excavations at Lachish have found 6 to 10 feet of ash and debris from this destruction in some areas.

From his camp at Lachish, and later from Libnah, Sennacherib sent his representatives with an army to demand the submission of Jerusalem (2 Kings 18:17). Although he was able to shut Hezekiah up as "a prisoner in Jerusalem, his royal residence, like a bird in a cage," Jerusalem never yielded to the Assyrian army. Throughout all of those difficult days, Hezekiah remained faithful in his trust in Yahweh.

In the end it was not Tirhakah, who later became the Cushite king of Egypt (2 Kings 19:9), who rescued the Judeans. Instead, their deliverance came when God sent his angel to destroy 185,000 Assyrians (19:35 – 36; 2 Chron 32:21 – 22; Isa 37:36). Obviously, Sennacherib did not mention this disastrous loss of troops in his inscriptions; instead, he emphasized how he besieged and conquered forty-six strong cities and countless villages, how he confined Hezekiah to Jerusalem, and how many of the kings of the area, Hezekiah included, sent him tribute.

Not much is known about the later portion of Hezekiah's reign (715 – 686 BC) in the aftermath of Sennacherib's invasion. In spite of Sennacherib's losses on his Judean campaign, Assyria remained powerful. During the first part of Manasseh's long reign of fifty-five years, the Assyrians were preoccupied with subduing revolts in the Babylonian and Elamite areas. In 689 BC Babylon was destroyed by Sennacherib, and it was not rebuilt until the days of his successor, Esarhaddon.

In the 670s the Assyrians were again active along the Mediterranean coastline fighting against and attempting to invade Egypt. The books of Kings and Chronicles are silent about these activities, and it must be assumed that for the most part Judah remained submissive to the Assyrians. In the biblical text Manasseh is primarily noted for the pagan religious rites that he reinstituted in Judah — including high places, altars for Baals, making an Asherah pole, worshiping the hosts of heaven, making children pass through the fire, and placing idols in the temple (2 Kings 21:2 – 9, 16; 2 Chron 33:1 – 20).

At least at one point, however, Manasseh revolted and was taken captive by the Assyrians to Babylon for a short period of time (2 Chron 33:11 – 13). Possibly this occurred during the reign of Ashurbanipal, when in 663 BC Tyre (and Judah?) revolted. The Bible briefly describes how, when Manasseh repented of his sins, he was allowed to return to Jerusalem and rebuild the wall of the city, as well as to station "military commanders in all the fortified cities of Judah" (vv. 12 – 14). These projects may reflect the fact that the Assyrians considered him a loyal vassal, who, being situated close to Egypt, would aid them in their campaigns against the pharaohs.

Amon, the wicked son of Manasseh, ruled for only two years (642 – 640 BC). He was followed by Josiah, who ruled for thirty-one years (640 – 609 BC). This young ruler certainly came under priestly and prophetic influence, for his reign was noted for a religious renaissance. In the eighth year of his reign (632 BC) Josiah sought God, and in his twelfth year (628 BC) he began to purge Judah and Jerusalem of their high places, Asherahs, and images. This purging also extended into what had been the Northern Kingdom, for towns in "Manasseh, Ephraim, and Simeon as far as Naphtali ... throughout Israel" are mentioned (2 Chron 34:3 – 7). Although the extent and nature of Josiah's influence/control in the north is not clear, the fact is that, at the time of the repair of the temple in his eighteenth year (622 BC), contributions were received not only from Judah, Benjamin, and Jerusalem, but also from Manasseh, Ephraim, and the remnant of Israel (vv. 8 – 13). All of this seems to indicate that he exerted some influence over some of the Israelites who remained in what used to be the Northern Kingdom.

Nevertheless, the heartland of the major area under his control was probably confined to the Hill Country of Benjamin and Judah, stretching from Geba in the north to Beersheba in the south (2 Kings 23:8). Although a Hebrew ostracon has been found at the coastal site of Ashdod-yam (see *ANET*, 568, for the text), indicating the presence of Hebrew-speaking residents in the area, much of the Philistine Plain had probably been under Egyptian control ever since the campaign of the powerful Psammetichus I (664 – 610 BC) to Ashdod in 635 BC passed through Philistia in 627 BC.

As mentioned above, during the final years of Manasseh's reign (696 – 642 BC) the Assyrians were increasingly preoccupied in southern Mesopotamia. In 626 BC the rebel Chaldean Nabopolassar (626 – 605 BC) captured Babylon from the Assyrians and initiated the process that would lead to the downfall of Assyria. By 620 BC Nabopolassar had established his control of southern Mesopotamia and was ready to move against Assyria itself.

However, to the northeast the powerful Medes, who had defeated the Scythians in 624 BC, were also pressuring the Assyrians. A combined Egyptian-Assyrian army was able to deter Nabopolassar's expansion plans in 616 BC, but in 614 BC the Medes actually captured and destroyed the Assyrian city of Asshur. The Medes and Babylonians then combined forces and in 612 BC conquered Nineveh. The last Assyrian ruler, Ashur-uballit II (612 – 609 BC), retreated westward and established a stronghold at Haran. There he and his Egyptian ally, Neco II (610 – 595 BC), attempted an unsuccessful defense of the city against the combined Median and Babylonian forces (610 BC). Retreating even further west, Ashur-uballit II attempted a counterattack in 609 BC. As the Egyptian pharaoh Neco was heading north to assist his Assyrian ally, the Judean king, Josiah, evidently siding with the Babylonians, attempted to impede Neco's northward march at Megiddo, just north of the Mount Carmel pass. There the godly Judean was killed in battle in 609 BC (2 Kings 23:29 – 30; 2 Chron 35:20 – 27).

The Egyptian-Assyrian coalition was defeated by a Medo-Babylonian one at Carchemish in 609 BC. This defeat was the blow that was fatal for the Assyrians. The people of Judah placed Josiah's son Jehoahaz on the throne in 609 BC, but after a reign of only three months he was deposed by Neco and taken captive to Egypt, where he died (2 Kings 23:30 – 34; 2 Chron 36:1 – 4). Pharaoh Neco placed Eliakim on the throne in Jerusalem, changing his name to Jehoiakim (2 Kings 23:34 – 36; 2 Chron 36:4 – 5; Jer 22:11 – 12).

Jehoiakim's rule of eleven years (609 – 598 BC) was anything but godly. Although specific religious sins are not mentioned, he was a constant foe of Jeremiah the prophet (see Jeremiah, passim). He also imposed heavy taxes on the people in order to send tribute to Neco (2 Kings 23:35), and he evidently squandered money to build himself a new palace. Between 608 BC and 605 BC a series of battles between Egyptian and Babylonian forces took place near the Euphrates River. In 605 BC, at Carchemish, Neco's forces were defeated by soon-to-be-king Nebuchadnezzar (605 – 562 BC), who pursued the Egyptians south to the Brook of Egypt.

Judah was forced to change allegiance and become a Babylonian rather than an Egyptian vassal. At this time (605 BC) Nebuchadnezzar deported some of the talented upper-class Judeans, taking them captive to Babylon — recorded in Scripture as the first of four deportations of Judeans (Dan 1:1; Jer 46:2; compare 52:28 – 30). With the defeat of the Egyptians at Carchemish in 605 BC, the Babylonians became heirs to

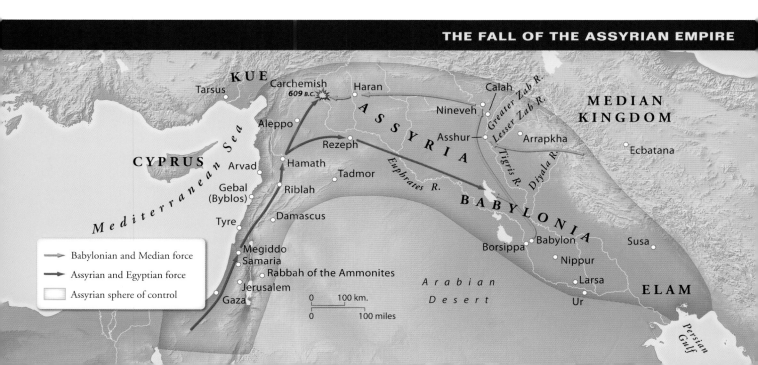

THE FALL OF THE ASSYRIAN EMPIRE

Babylonian and Median force
Assyrian and Egyptian force
Assyrian sphere of control

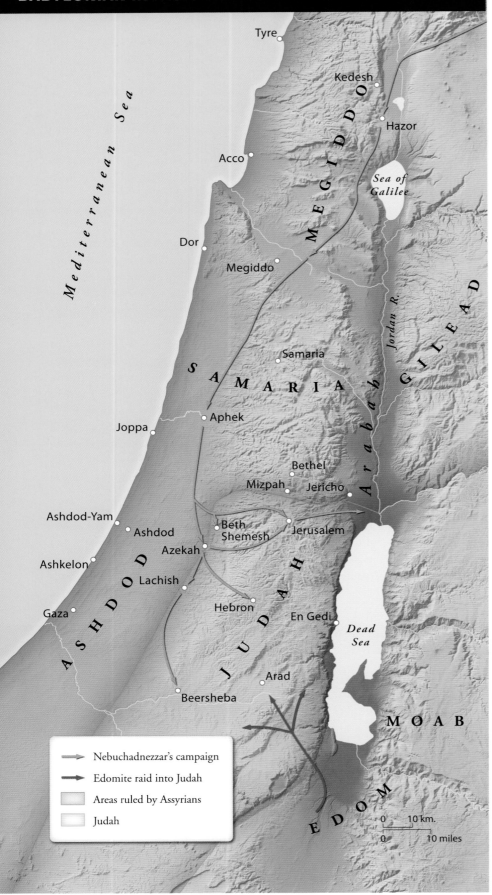

Map legend:

- → Nebuchadnezzar's campaign
- → Edomite raid into Judah
- Areas ruled by Assyrians
- Judah

territory that would eventually surpass the Assyrian Empire in extent. This new "world empire" is called the Neo-Babylonian Empire (605–539 BC).

In addition to the Babylonian oppression, the Judeans also suffered at the hands of Arameans, Moabites, and Ammonites during the reign of Jehoiakim (609–598 BC; 2 Kings 24:2–4). Toward the end of his reign, possibly after 601 BC (when Nebuchadnezzar had been repulsed at the border of Egypt by the Egyptians), Jehoiakim revolted against Babylonian overlordship.

Nebuchadnezzar was quick to respond and invaded Judah in 597 BC. Just before Nebuchadnezzar's capture of the "city of Judah" (= Jerusalem) on March 16, 597 BC, Jehoiakim died. The next king, Jehoiachin, Jehoiakim's son, ruled for only three months (598–597 BC) and presided over the fall of the city. In this second deportation (2 Kings 24:13–16), he and more than 10,000 Judeans — including officers, fighting men, craftsmen, and artisans — along with treasure from the temple and palace, were taken captive to Babylon. Jehoiachin remained a captive in Babylon until his release on March 22, 561 BC (25:27–30).

Nebuchadnezzar placed Mattaniah, Jehoiachin's uncle, on the throne and changed his name to Zedekiah. Zedekiah ruled for eleven years (597–586 BC), at first as a loyal Babylonian vassal. Then, possibly around 588 BC, when the Egyptian king Hophra/ Apries (589–570 BC) led a military expedition to Tyre and Sidon, Zedekiah joined Edom, Moab, Ammon, Tyre, and Sidon (Jer 27:1–11; and evidently Egypt) in a revolt against Babylon. Nebuchadnezzar again responded quickly, and although the details of Nebuchadnezzar's line of march into Judah are not known, we do know from both the Bible (34:6–7) and from one of the Hebrew ostraca discovered at Lachish (number IV; *ANET*, 322) that he attacked the strategically important Shephelah cities of Lachish and Azekah, just as Sennacherib, the Assyrian, had done a century earlier.

On January 15, 588 BC, Nebuchadnezzar began the siege of Jerusalem. The siege was lifted momentarily when rumors of Egyptian intervention circulated, but it was quickly reinstated. In July 586 BC Jerusalem was captured, and between August 14 and 17 the city was razed and the temple burned by Nebuzaradan, commander of the imperial guard. Zedekiah, who had fled toward the Arabah, was captured in the plain of Jericho and taken to Nebuchadnezzar's headquarters at Riblah. There his sons were executed, his eyes were put out, and he was deported to Babylon (2 Kings 25:1–7).

With the fall of Jerusalem in 586 BC the independent Israelite/Judean state, which had existed for more than four hundred years, had come to an end. The effect on God's people must have been devastating: the holy city, where God had chosen to place his name, was in ruins, the temple of the true and living God had been destroyed, sacrifices had ceased, and the pathetic heirs of the once-great Davidic dynasty were now prisoners in exile. Had God abandoned his people? What had happened to the glorious promises that had been made to Israel's ancestors? The answers to these and many other questions would slowly come to the people of God during their period of exile. It would take a "second exodus," this time from Babylon rather than from Egypt, for God to deliver his people from the catastrophe that had befallen them.

▲ *Iron Age tomb on the grounds of the École Biblique in Jerusalem. Note the burial bench and the people looking into the repository where the bones were collected.*

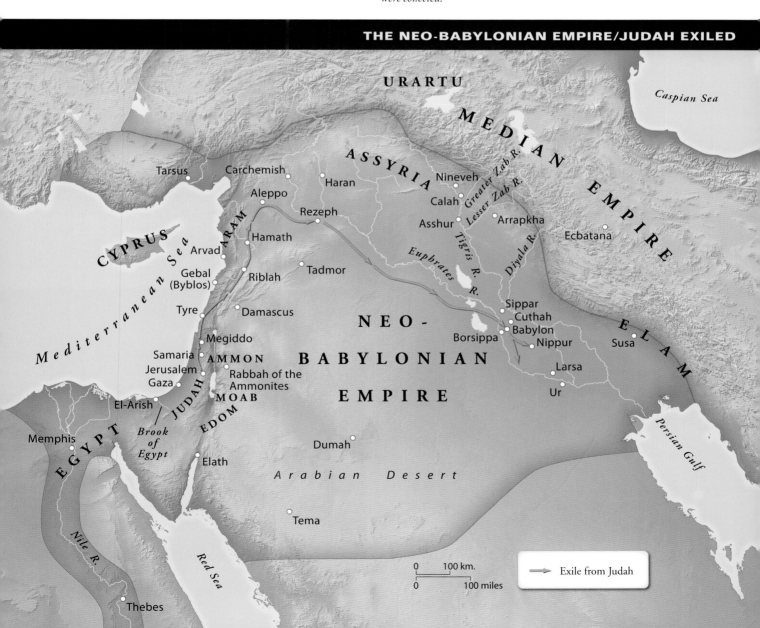

THE NEO-BABYLONIAN EMPIRE/JUDAH EXILED

Exile from Judah

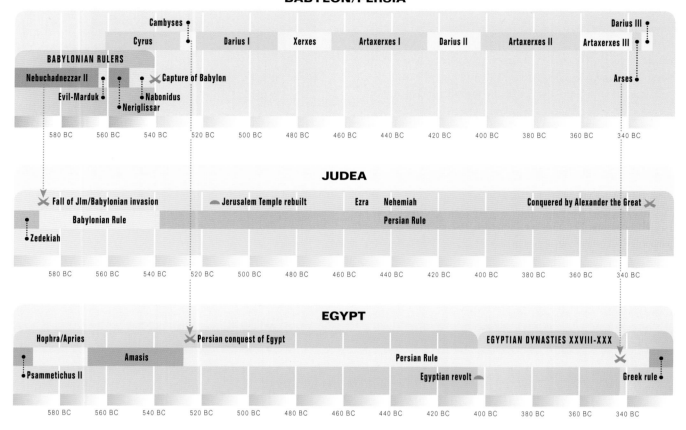

BABYLON/PERSIA

Cambyses •

Cyrus • | Darius I | Xerxes | Artaxerxes I | Darius II | Artaxerxes II | Darius III •

Artaxerxes III •

BABYLONIAN RULERS

Nebuchadnezzar II • • • ✕ Capture of Babylon

Evil-Marduk • • Nabonidus

• Neriglissar

Arses •

580 BC   560 BC   540 BC   520 BC   500 BC   480 BC   460 BC   440 BC   420 BC   400 BC   380 BC   360 BC   340 BC

JUDEA

✕ Fall of Jlm/Babylonian invasion   ▬ Jerusalem Temple rebuilt   Ezra   Nehemiah   Conquered by Alexander the Great ✕

Babylonian Rule   Persian Rule

• Zedekiah

580 BC   560 BC   540 BC   520 BC   500 BC   480 BC   460 BC   440 BC   420 BC   400 BC   380 BC   360 BC   340 BC

EGYPT

Hophra/Apries   ✕ Persian conquest of Egypt   EGYPTIAN DYNASTIES XXVIII-XXX

Amasis   Persian Rule

• Psammetichus II   Egyptian revolt ▬   Greek rule •

580 BC   560 BC   540 BC   520 BC   500 BC   480 BC   460 BC   440 BC   420 BC   400 BC   380 BC   360 BC   340 BC

EXILE AND RETURN

to remain and settle in the land, portions of the Judean community fled to Egypt, taking Jeremiah with them (42:1 – 44:30). Upon their arrival at Tahpanhes in the eastern Nile delta, Jeremiah prophesied that Nebuchadnezzar would set up his throne there and that the fleeing

▼ *Babylon: Tile figure of a mythological animal from the Ishtar Gate of Babylon, from the days of Nebuchadnezzar and Daniel*

Withe capture of Jerusalem in 586 BC by the Babylonians some Judeans were taken captive to southern Mesopotamia, some were allowed to remain in the land of Judah, while others fled to neighboring countries — including Ammon, Moab, and Egypt. As the main Babylonian force withdrew from Judah, they appointed Gedaliah to be governor of those Judeans who remained in the land (Jer 40:1 – 41:15). Gedaliah established headquarters at Mizpah, 7.5 miles north of Jerusalem. Gedaliah's authority was contested when Judeans returned from Ammon and Moab. Ishmael, who was employed by Baalis (king of the Ammonites, Jer 40:14), murdered Gedaliah, and also killed men of Shechem, Shiloh, and Samaria who had stopped at Mizpah while on their way to worship at the temple ruins in Jerusalem. Although some of the conspirators were caught and executed, Ishmael escaped to the Ammonite territory east of the Jordan River.

Since the Babylonians were still active in the area (note the thirteen-year siege of Tyre that began in 585 BC), the Judeans were fearful of Babylonian reprisals (Jer 41:16 – 17). In spite of Jeremiah's instructions

Istanbul, Museum of the Ancient Orient

Judeans would be either killed or sent into exile (43:8–13). Jeremiah is not heard from again, and he evidently died in exile in Egypt. In 582 BC the Babylonians were again in the area of Judah, and at that time several thousand additional Judeans were taken into exile (52:30). It is not certain whether this fourth and final deportation was in response to the murder of Gedaliah, but Babylonian records tell of the subduing of the peoples of Ammon — to which the murderer Ishmael had fled — and Moab in that same year.

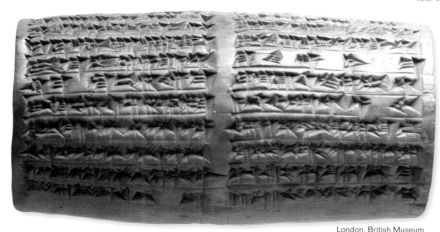

London, British Museum

▲ *Clay foundation cylinder (4 in. long) discovered at the ziggurat at Ur in southern Iraq mentioning both Nabonidus and Belshazzar his son (Daniel 5).*

Specific details regarding life in Judea between the fall of Jerusalem in 586 BC and the first return from Babylon in 538 BC are lacking. Since the Babylonians did not import foreigners to settle areas recently vacated by exiled Judeans, there probably was a decrease in the density of population in some areas. It is commonly thought that the Edomites, who during the Old Testament period lived south and east of the Dead Sea, moved, under pressure from Arabs, into the southern portion of the Hill Country of Judah, to the region between Beth Zur and Beersheba. This population group later came to be known as the Idumeans (see below, p. 186).

The Babylonian monarch Nebuchadnezzar was succeeded by Evil-Merodach (562–560 BC). It was during his reign that the Judean king Jehoiachin, who had been taken into exile in 597 BC, was released in March of 561 BC (2 Kings 25:27–30; Jer 52:31–34).

But life in Babylon for the exiled Jewish populace must have been somewhat depressing, for their religious beliefs (which were based on the promises of God) were intimately tied up with the land of Israel, from which they were exiled, and with Jerusalem, which was in ruins. In addition, the Davidic dynasty no longer ruled and its continued existence was in question, and the temple of Yahweh was in ruins (see Ps 137). The question that loomed large in the minds of the exiles was: Why?

Yet in spite of this dismal state of affairs, at least portions of the Jewish community seem to have flourished. For example, Daniel became an advisor to kings and royalty (Nebuchadnezzar, Belshazzar, and Darius the Mede); Ezekiel was free to move about and minister to the needs of his people; the exiled Judean king Jehoiachin and his entourage received rations from the Babylonians (*ANET*, 308); and it seems that, economically, portions of the community prospered (fifth-century BC cuneiform tablets

from the family of Murashu, found at Nippur, relate some of the activities of the Jews in Babylonia).

The collapse of the Babylonian empire, like that of the Assyrian one before it, was rather swift. After defeating the Medes in 550 BC at Ecbatana, Cyrus the Persian (559–530 BC) defeated a combined Babylonian, Egyptian, and Lydian army. Moving northwest into Asia Minor, Cyrus defeated Croesus, the powerful leader of the kingdom of Lydia, at Sardis (by 546 BC), with the result that Cyrus's territory stretched from Persia westward through Asia Minor to the coast of the Aegean Sea.

After defeating the Babylonians at Opis on the Tigris River, the capture of Babylon itself was relatively easy. Cyrus entered the city in

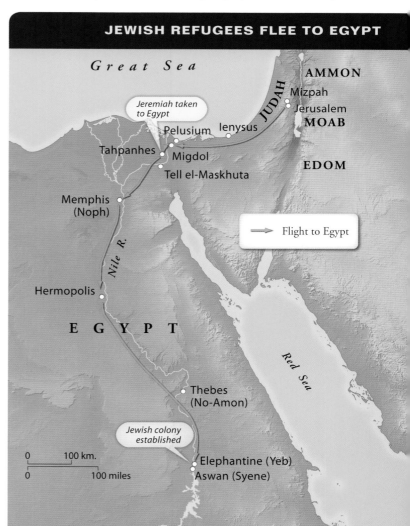

JEWISH REFUGEES FLEE TO EGYPT

Great Sea

AMMON

Mizpah

Jerusalem

JUDAH

MOAB

Jeremiah taken to Egypt

Pelusium   Ienysus

Tahpanhes

Migdol

EDOM

Tell el-Maskhuta

Memphis (Noph)

→ Flight to Egypt

Nile R.

Hermopolis

E G Y P T

Red Sea

Thebes (No-Amon)

Jewish colony established

0   100 km.

0   100 miles

Elephantine (Yeb)

Aswan (Syene)

539 BC. He was hailed as a liberator and treated the populace with leniency. His popularity was further enhanced by the restoration of the god Marduk to his preeminence after the neglect he had suffered under the last Babylonian king, Nabonidus. The Persians, like the Babylonians, treated most of their subjects with informed enlightenment. The barrel inscription of Cyrus (*ANET*, 315 – 16) recounts how, under Persian rule, some peoples who had been exiled by the Babylonians were allowed to return to their homelands and how destroyed sanctuaries were rebuilt and divine images returned to their proper places.

It is within this general framework that Cyrus's decree regarding the Jews is to be placed (2 Chron 36:22 – 23; Ezra 1:1 – 4; 6:3 – 5). This decree permitted any Jew to return to Judea. It also allowed for the temple to be rebuilt — with some royal financial support — and it commanded that the temple vessels were to be returned to Jerusalem. A total of 49,697 individuals returned to Judea with Sheshbazzar (Ezra 2:64 – 65). This was the largest of the three returns, but evidently a large number of Jews elected to remain in southern Mesopotamia, where they formed the nucleus of what would become a rather large and powerful Jewish community.

Many of the returnees listed in Ezra 2 are identified by family clans and occupations; however, within that chapter some are listed accord-

ing to their towns of origin (vv. 21 – 35). Although it is not explicitly stated, most commentators assume that these people returned to their places of origin. If this is so, then a fair portion of returnees settled in twenty-one villages located primarily in the old tribal territories of Benjamin and Dan to the northeast, north, and northwest of Jerusalem. During this first phase of the first return, around 537 BC, the sacrificial altar was reconstructed and sacrifices begun, the foundations for the temple were laid, and the Feast of Tabernacles was celebrated (Ezra 1 – 3). But because of the hostility of their enemies (4:1 – 4), work on the temple ceased until the second year of the rule of Darius (ca. 520 BC).

In the interim, Cyrus expanded his empire to the east all the way to the Indus and Oxus Rivers. Upon the death of Cyrus, Cambyses (530 – 522 BC) his son assumed the throne. After securing his position as king, Cambyses began making preparations for the invasion of Egypt. Mustering his army, possibly at Acco, Cambyses marched south along the Palestinian coast toward Egypt. An Arabian king (of Kedar?) supplied him with water and camels and in return was granted tax-free status for a territory in northern Sinai that stretched from Gaza to Ienysus (Herodotus 3.4 – 5), although the Arabian king did present a yearly "gift" of 1,000 talents of frankincense to the Persian ruler (Herodotus 3.97). Cambyses was eventually successful in capturing the whole of

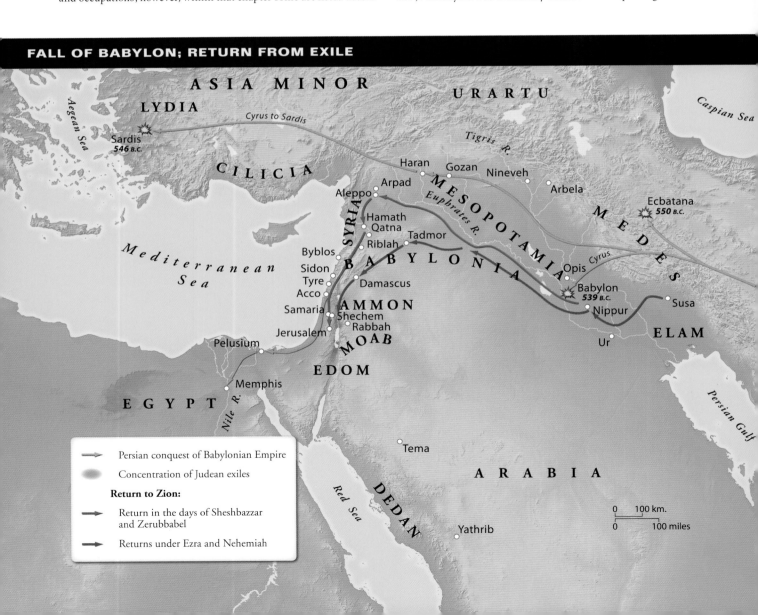

FALL OF BABYLON; RETURN FROM EXILE

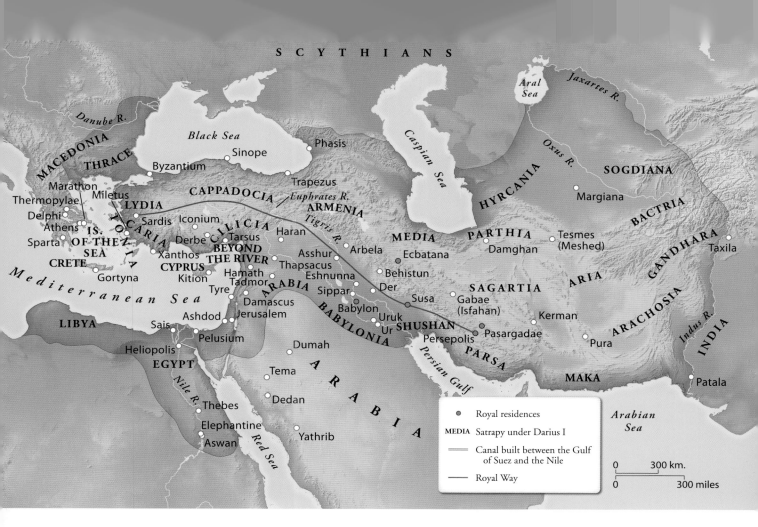

Egypt, which in turn became a Persian satrapy. While returning home, Cambyses, for some unknown reason, committed suicide near Mount Carmel in 522 BC.

Within two years Darius I (522–486 BC) was able to secure the throne by executing his rival Gautama. He also put down revolts that had sprung up all over the empire: in Media, Parthia, Hyrcania, Armenia, and Babylonia. In 519–518 BC he reconquered Egypt, and the Persian Empire entered a relatively peaceful and prosperous era.

The second phase of the first Jewish return received impetus from the urgings of the prophets Haggai and Zechariah. Under the leadership of Zerubbabel, work on the temple resumed in 520 BC and was completed by 516 BC (Ezra 4:24–6:22), at which time a joyous Passover was celebrated in Jerusalem. Certainly the providence of God was involved in the timing of the rebuilding, but it is also interesting to note that permission to rebuild was granted in 520 BC, just prior to Darius's Egyptian campaign mentioned above. Possibly Darius was trying to secure the loyalty of the Judeans by granting them their request.

Although the internal division of the Persian Empire changed from time to time, it was traditionally divided into large administrative units called satrapies; according to the Greek historian Herodotus (ca. 450 BC; 3.89–95) there were twenty such satrapies. At the time of the first return from exile Judah was evidently in a large satrapy called "Babylon and Beyond the River [Euphrates]." It seems that by 520 BC the satrapy

had been divided into at least two subunits, one of which was called "Beyond the River" (NIV "Trans-Euphrates"; mentioned seventeen times in Ezra and Nehemiah). This administrative unit, which later became a separate satrapy (number five according to Herodotus 3.90), stretched from the Upper Euphrates–Orontes River region, southward through Lebanon, Syria, and Palestine into northern Sinai; not included were desert territory to the east and southeast and the Sinai coast from Gaza to Ienysus, which were under the control of the Arabs (Herodotus 3.5, 91). From both the biblical record (Ezra 5:3, 6; 6:6, 13) and cuneiform documents, it is known that Tattenai, a subordinate of Ushtannu, was the governor of Trans-Euphrates, and it was with him that Zerubbabel and Joshua had to deal in attempting to secure permission to rebuild the temple (Ezra 5–6).

During the reign of Darius I (522–486 BC), the Persian Empire reached its greatest extent, stretching from the Indus River in the east all the way to Thrace and Macedonia in the west, and included Egypt and Libya in north Africa (satrapy six). During Darius's reign the minting of coins (the gold daric and the silver shekel) became common, the legal system of courts and judges that helped settle grievances was established, a postal system and roads (including the famous "Royal Road" that led from Sardis in Lydia to Babylon, Susa, and Persepolis) were in full operation, and even a major canal, connecting the Gulf of Suez and the Nile, was completed.

▲ *Athens: The Parthenon on the Acropolis, among other buildings, was under construction during the days of Ezra and Nehemiah (mid-fifth century BC).*

Toward the midpoint of Darius's long reign, ca. 500 BC, trouble began brewing in the Aegean area with the Ionian revolt. The revolt was suppressed by the Persians, and the city of Miletus was captured and destroyed. But in 490 BC an invading Persian army suffered a serious defeat on the Greek mainland at Marathon. Persian retaliation was not immediate, for Darius died soon after the defeat, and the first few years of his son Xerxes' reign (486 – 464 BC) were spent quelling rebellions closer to home. However, in 480 BC Xerxes tried to subdue the Greek rebels. At Thermopylae he was successful, and he went on to capture and burn Athens; but his navy was routed at Salamis, and Xerxes was forced to retreat to Asia Minor. It was soon after these events that Esther was appointed as his queen (Est 2:16 – 17; December 479 BC or January 478 BC), and the major events recorded in the book of Esther have their setting in 473 BC.

After Xerxes was assassinated, Artaxerxes I (464 – 424 BC) took the throne. Early in his reign, Egypt, with the help of Greece, revolted. This revolt (460 – 455 BC) was put down by the powerful general Megabyzus, who became the governor of the satrapy "Beyond the River." It was probably in the context of Persia's desire to secure the entrance into Egypt that additional concessions were granted to the Jews in 458 BC (Ezra 7:7 – 9). According to the royal decree (vv. 12 – 26), Ezra and an unlimited number of Jews were allowed to return to Judea. Concessions granted to the Jews included financial support from the royal treasury, the right of the Judeans to govern their own affairs on the basis of the laws of their God, the appointment of judges and civil magistrates, and the granting of tax exemptions for temple personnel.

Although the total number of Jews who returned with Ezra is not known, there were probably fewer than the 50,000 mentioned in connection with the first return, for only 1,758 adult males are mentioned in connection with the second one (8:1 – 20). The major accomplishment of the second return seems to have been a spiritual rebuilding of the Judeans — including the dissolution of intermarriages (ch. 9). During the period between the arrival of Ezra in 458 BC and that of Nehe-

miah in 445 BC, an attempt was evidently made to rebuild the wall of Jerusalem (see the letter from this time in Ezra 4:7 – 23), but because of local opposition the rebuilding was not completed.

In 446 BC the dilapidated state of Jerusalem's defenses was brought to the attention of a pious Jew named Nehemiah, who was serving in the Persian court as cupbearer to King Artaxerxes. Artaxerxes responded favorably to Nehemiah's request to supervise the rebuilding of the wall of Jerusalem, possibly because he felt that a loyal Judean populace would help counteract those forces that had supported a revolt against the Persians led by the satrap Megabyzus. Nehemiah left for Judea in 445 BC armed with a letter from the king granting him permission to rebuild the wall, commanding other "governors of Trans-Euphrates" (Neh 2:7 – 9) to grant him safe passage, and ordering Asaph, the keeper of the king's forest, to provide him with timber. Upon arriving in Jerusalem, Nehemiah surveyed the condition of the defenses of the city (vv. 11 – 16). He then rallied the Judeans behind him and, despite opposition, rebuilt the walls in fifty-two days (2:17 – 7:3).

With the walls completed, Ezra and Nehemiah led the people in a time of spiritual renewal that ultimately resulted in the signing of a covenant document whereby the people committed themselves to govern their lives in accordance with the law of Moses. After serving for twelve years as governor (Neh 5:14; 13:6), Nehemiah returned to the Persian court in 433/432 BC. He spent an undetermined period of time there and returned to Judea to serve again as governor, although his activities during his second term are not recorded.

From the biblical and extrabiblical data it is possible to gain a fair understanding of Judah's position in the Persian Empire. However, it is probable that during, or soon after, the days of Darius the large satrapy of "Babylon and Beyond the River" was divided into at least two portions, so that "Beyond the River" became a satrapy in its own right, probably with Damascus as its capital. By the time of Esther (ca. 473 BC) the Persian Empire was further divided into 127 provinces (Est 1:1; 8:9), which, in turn, were combined in various ways to make up the satrapies. It is generally assumed that the provinces in the satrapy "Beyond the River" (e.g., Hamath, Subite, Damascus, Karnaim, Hauran, Gilead, Megiddo, Samaria, some of which were mentioned by the prophet Ezekiel [47:15 – 20]) which had been established by the Assyrians and maintained by the Babylonians, were also preserved by the Persians. Judah, too, was a province (Yehud) in the satrapy "Beyond the River."

To the north of Judah was the province of Samaria, whose governor, Sanballat the Horonite, came into direct conflict with Nehemiah (Neh 2, 4, and 6, passim). The extent of the Samaritan province is not known for certain, but it probably stretched from the Jezreel Valley in the north to Judah in the south and from the Jordan River in the east to the Sharon Plain in the west. It is possible that the "plain of Ono," to which Sanballat tried to lure Nehemiah (6:2), was neutral territory, outside of both Judah and Samaria.

To the east of Judah and east of the Jordan River was the province of Transjordan, or Gilead. The governor there during the days of Nehemiah was Tobiah the Ammonite. This Tobiah was one of a series of rulers who came from a prominent Transjordanian family headquartered at Tyre of Transjordan (Iraq el-Amir) and who had close ties with Eliashib the priest in Jerusalem (Neh 13:4 – 7). On the west, some of the coastal cities may have been granted to the cities of Tyre and Sidon, but this was a very fluid situation. This territory was possibly called "Ashdod," for the "men of Ashdod" are mentioned among Nehemiah's opponents (4:7).

Besides Sanballat and Tobiah, "Geshem the Arab" (Neh 6:1) joined in the hostilities against Nehemiah. He seems to have been the same person as the Geshem whose name was inscribed on a silver bowl found at Tell el-Maskhuta, located 9 miles west of modern Ismailia in Egypt. In this inscription Geshem is called the king of Qedar (biblical Kedar; *ANET*, 657). The extent of this kingdom is not known, but it probably included a portion of the northern desert of Sinai, which stretched along the Mediterranean coast from Gaza to Ienysus, and it probably extended south and eastward toward the Red Sea and Arabia. Geshem was likely involved in controlling the overland transport of luxury goods (e.g., gold, frankincense, myrrh, pearls, and spices) that passed through his territory on their way from Arabia to urban centers in the north and west.

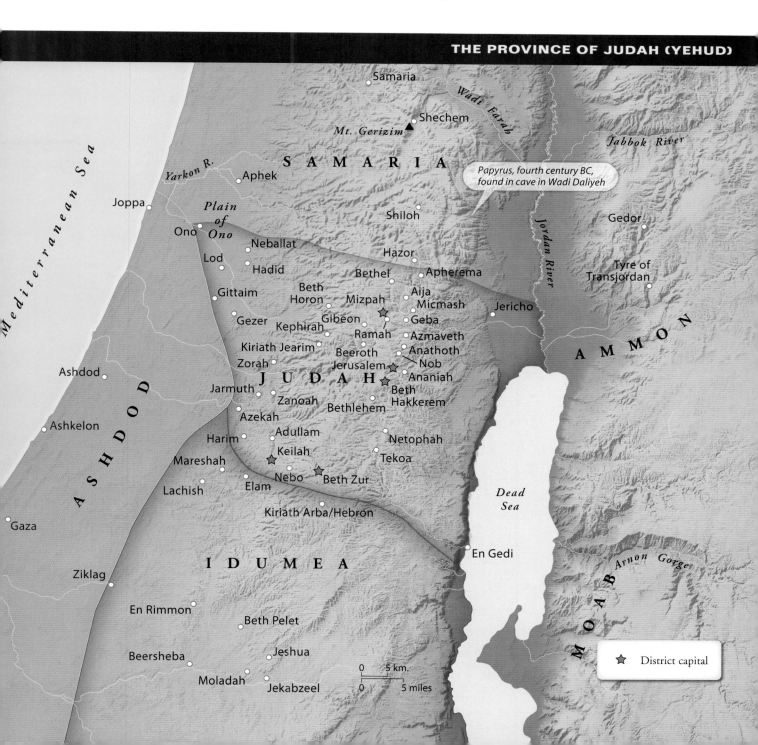

THE PROVINCE OF JUDAH (YEHUD)

Papyrus, fourth century BC, found in cave in Wadi Daliyeh

★ District capital

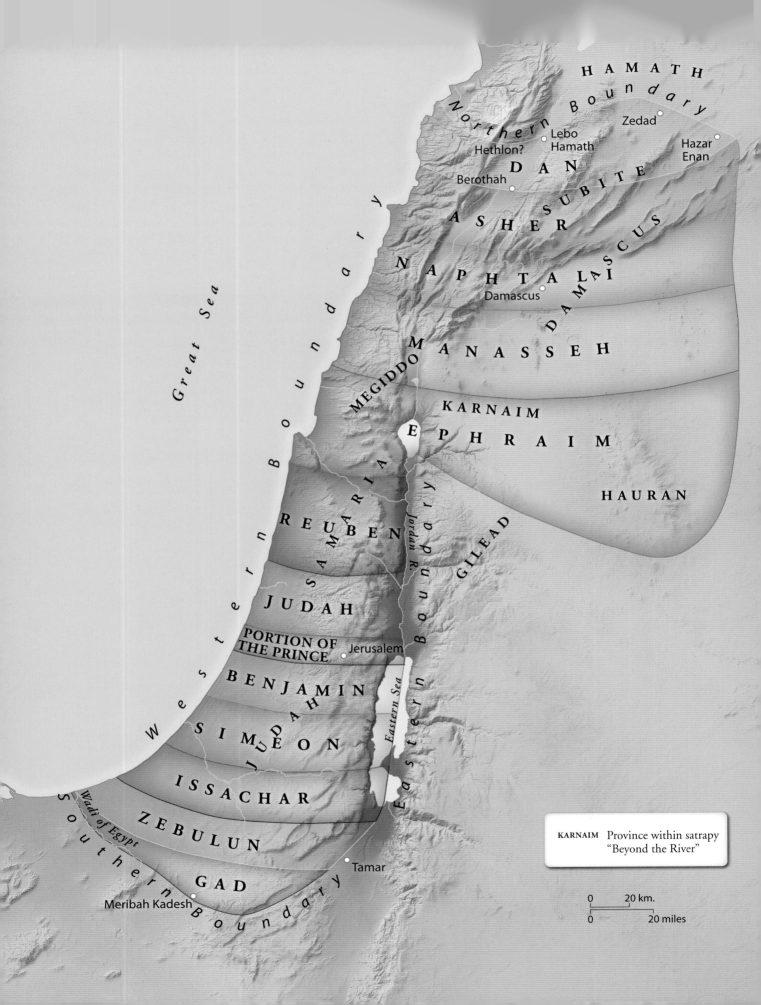

HAMATH

Northern Boundary

Zedad

Lebo
Hethlon? Hamath

Hazar
Enan

D A N

Berothah

SUBITE

A S H E R

N A P H T A L I

DAMASCUS

Damascus

DAMASCUS

Great Sea

M A N A S S E H

MEGIDDO

E

KARNAIM

Western Boundary

E P H R A I M

HAURAN

S A M A R I A

R E U B E N

Jordan R.

Eastern Boundary

G I L E A D

J U D A H

PORTION OF
THE PRINCE

Jerusalem

JUDAH

B E N J A M I N

Eastern Sea

S I M E O N

I S S A C H A R

Eastern Boundary

Z E B U L U N

Wadi of Egypt

Southern Boundary

G A D

Tamar

Meribah Kadesh

| KARNAIM | Province within satrapy "Beyond the River" |

0   20 km.

0   20 miles

As for Judah itself, it is evident that Jewish settlements were established in the old Benjamite territory north, east, and west of Jerusalem (Ezra 2:21–35; Neh 7:26–38; 11:31–36). On the west, in the Shephelah, Jews settled in En Rimmon, Zorah, Jarmuth, Zanoah, Adullam, Lachish, and Azekah (Neh 11:29–30). On the east, it is likely that Judean territory stretched to the bank of the Jordan River and to the shore of the Dead Sea. In the southern Hill Country of Judah the text mentions Jews residing in Kiriath Arba (= Hebron) and Dibon (identification unknown), and farther south, in the region where the Hill Country of Judah meets the biblical Negev. The cities of Jekabzeel, Jeshua, Moladah, Beth Pelet, Hazar Shual, and Beersheba are mentioned, as are Ziklag and Meconah, both probably located in the western Negev (11:25–28).

In addition to the above data, hints regarding the internal structure of Judah can be gleaned from Nehemiah 3. This chapter describes how certain groups of men were engaged in rebuilding the walls of Jerusalem. The villages that they came from include Keilah, Zanoah, Beth Zur, Tekoa, Beth Hakkerem, Jerusalem, Gibeon, Mizpah, and Jericho. Some terms (such as "district" and "half-district") that indicate the existence of administrative subdivisions within Judah are used in connection with Keilah, Beth Zur, Beth Hakkerem, Jerusalem, and Mizpah. Although alternative schemes are possible, the division of Judea into five districts — one located in the Shephelah (Keilah–Zanoah), three in the Hill Country of Judah (Beth Zur–Tekoa; Beth Hakkerem; and Jerusalem–Gibeah), and one in Benjamin (Mizpah–Jericho) — seems reasonable.

At the time of the returns from Babylon Judah was an independent province as a straightforward interpretation of the biblical (Ezra 5:14; Hag 1:1, 14; Neh 5:14–15; 12:26) and archaeological data (bullae, seals, jar-handle impressions, Elephantine papyri, and coins) indicates. It is even possible to reconstruct a tentative list of Judean governors, which includes Sheshbazzar, Zerubbabel, Elinathan, Yehoezer, Ahzai, Nehemiah, Bagohi, and Yehezqiyah — of which the first, second, and sixth are mentioned in the Bible.

At the end of the fifth century BC the biblical record becomes silent. Upon the death of Artaxerxes I, Darius II (423–404 BC) became ruler of the Persian Empire. Glimpses into the life of the Jewish community in Judea are furnished by papyrus documents that were found at Elephantine, a site in southern Egypt (see *ANET*, 491–92). There a Jewish colony had established itself serving as a military garrison for the Egyptians. The papyrus documents show that they had built a temple for worshiping Yahweh at Elephantine but that it had been destroyed around 410 BC. It is also evident that the Jews of Elephantine

Mark Connally

▲ *Tobiad family tomb at Tyre of Transjordan (Neh 2, 4, 6, 13) plus detail of Tobiad inscription right of door*

on occasion looked to Jerusalem for guidance, for they corresponded with Bogoas/Bagohi, the governor of Judea.

Late in the fourth century Darius III Codomannus (336–330 BC) presided over the collapse of the Persian Empire as Alexander the Macedonian moved steadily eastward, conquering province after province. It was Alexander who captured the city of Persepolis, burning the magnificent palace that had been built by Xerxes, and in 330 BC the Persian Empire came to an end when Bessus, the satrap of Bactria, killed Darius III. In the centuries following the arrival of Alexander (ca. 332 BC), the political, religious, linguistic, and cultural milieu of the Levant changed dramatically as Greco-Roman culture spread rapidly.

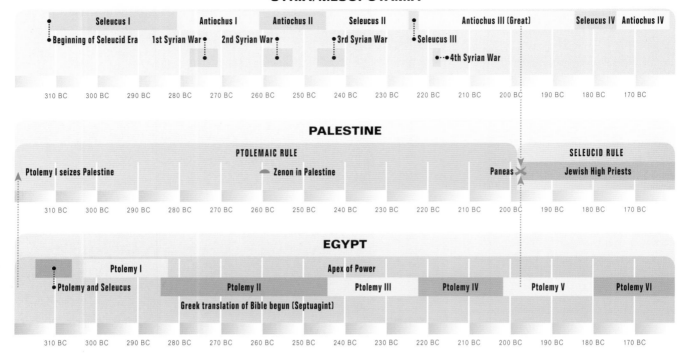

SYRIA/MESOPOTAMIA

Seleucus I | Antiochus I | Antiochus II | Seleucus II | Antiochus III (Great) | Seleucus IV | Antiochus IV

Beginning of Seleucid Era | 1st Syrian War | 2nd Syrian War | 3rd Syrian War | Seleucus III | 4th Syrian War

310 BC 300 BC 290 BC 280 BC 270 BC 260 BC 250 BC 240 BC 230 BC 220 BC 210 BC 200 BC 190 BC 180 BC 170 BC

PALESTINE

PTOLEMAIC RULE | SELEUCID RULE

Ptolemy I seizes Palestine | Zenon in Palestine | Paneas | Jewish High Priests

310 BC 300 BC 290 BC 280 BC 270 BC 260 BC 250 BC 240 BC 230 BC 220 BC 210 BC 200 BC 190 BC 180 BC 170 BC

EGYPT

Ptolemy I | Apex of Power | Ptolemy II | Ptolemy III | Ptolemy IV | Ptolemy V | Ptolemy VI

Ptolemy and Seleucus

Greek translation of Bible begun (Septuagint)

310 BC 300 BC 290 BC 280 BC 270 BC 260 BC 250 BC 240 BC 230 BC 220 BC 210 BC 200 BC 190 BC 180 BC 170 BC

THE ARRIVAL OF THE GREEKS

During the sixth, fifth, and fourth centuries BC, the power of the various Greek city-states was growing to such an extent that they were beginning to challenge Persian supremacy in Asia Minor and elsewhere. The Persians, who had invaded Greece, were kept at bay by the Greek forces, which defeated them, first at Marathon in 490 BC and then, ten years later, at Salamis. Through the remainder of the fifth century and during the first two-thirds of the fourth, the Greeks encouraged anti-Persian activities in Egypt, the Levant, and Asia Minor, but the extent of the territory under Persian rule in these areas remained virtually unchanged.

By the middle of the fourth century BC Philip II solidified his position as ruler of Macedonia in northern Greece. During his life he worked toward two great goals: the unification of the Greek city-states under his rule and the overthrow of the Persians. In the process of unifying Greece he built his Macedonian army into a small but formidable fighting force. Unfortunately, Philip was assassinated in 336 BC, after having completed plans for the invasion of Asia Minor.

At the time of his death his son Alexander ("the Great") was well positioned to carry out his father's dreams. Although only twenty years old at the time, Alexander had been prepared for his new position as leader of Macedon; he had been educated by the Greek philosopher Aristotle and had led campaigns on behalf of his father during his teenage years. Alexander began his invasion of Asia Minor by crossing the Dardanelles in 334 BC. Alexander defeated the Persian satraps of Asia Minor at the Granicus River and then marched south through the provinces of Mysia, Lydia, Caria, Lycia, and Pamphylia in a counter-

▼ *Cilician Gates, though which Darius III and Alexander the Great passed as they entered/exited Asia Minor.*

clockwise direction. At Gordium, the ancient capital of Phrygia, he met Parmenio, his general, who had proceeded directly from Sardis to Gordium. The Greeks crushed resistance along both routes, captured cities, secured treasuries, and replaced Persian overlords.

Alexander proceeded from Gordium southeast to Tarsus, the capital of Cilicia, via Ancyra and the Cilician Gates. Continuing around the northeastern corner of the Mediterranean Sea, he passed through the Syrian Gates in the Amanus Mountains before realizing that Darius, the Persian monarch, had assembled a large army at Issus behind him. Retracing his steps, Alexander met and defeated Darius III at the battle of Issus in 333 BC. In spite of considerable losses, Darius was able to flee eastward and began to regroup his forces for a later battle with Alexander.

▲ *Coin with a representation of Alexander the Great*

After his victory at Issus, Alexander marched south along the coast of the Levant, securing or seizing Aradus, Byblos, Sidon, Tyre (in 332 BC, after the construction of a causeway to the island and a seven-month siege), Gaza (after a two-month siege), and other cities. After Alexander had crossed northern Sinai, Egypt submitted to his rule, and he spent the winter of 332–331 BC there. At that time the city of Alexandria was founded on the western branch of the Nile near the Mediterranean Sea. This city became not only the capital of Egypt under the subsequent Egyptian Ptolemaic dynasty but also one of the leading commercial and intellectual centers of the Greco-Roman world.

After Alexander was declared the son of the Egyptian god Amon, which meant that he had been designated as the legitimate ruler of Egypt, he left Ptolemy in charge of administering the country and began to retrace his steps northward through the Levant. There is a tradition that Alexander visited Jerusalem but this is not certain. His treatment of the Judeans is not known but it is probably to be assumed that the Jews did not interfere with his advance along the coast and that, in turn, they were not molested. The Samaritans to the north of Judea, however, murdered their governor (Andromachus), who had been appointed by Alexander. In retaliation, Alexander destroyed the city of Samaria and resettled it with Macedonian veterans. Thus a large foreign (Greek) element was added to the basically Samaritan/Semitic populace of the region.

Heading north and then east, Alexander passed through northern Mesopotamia. To the east of the Tigris, on the plain between Gaugamela and Arbela, Darius and the Persians were soundly defeated. Darius fled eastward, but just prior to being overtaken by Alexander, he was murdered by Bessus, the satrap of Bactria, a former ally. Thus nearly 200 years of Persian rule in the Near East were brought to a close.

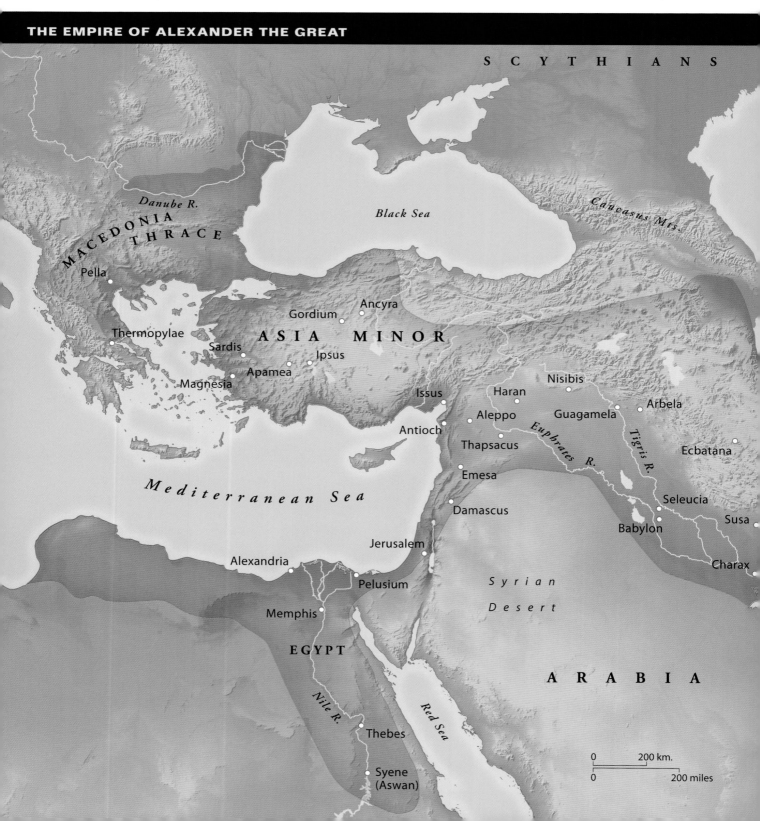

THE EMPIRE OF ALEXANDER THE GREAT

SCYTHIANS

Danube R.

Black Sea

Caucasus Mts.

MACEDONIA

THRACE

Pella

Thermopylae

Gordium

Ancyra

ASIA MINOR

Sardis

Ipsus

Apamea

Magnesia

Nisibis

Haran

Issus

Aleppo

Guagamela

Arbela

Antioch

Thapsacus

Euphrates R.

Tigris R.

Ecbatana

Emesa

Mediterranean Sea

Damascus

Seleucia

Babylon

Susa

Jerusalem

Syrian

Charax

Alexandria

Desert

Pelusium

Memphis

EGYPT

ARABIA

Nile R.

Red Sea

Thebes

Syene
(Aswan)

0   200 km.

0   200 miles

Alexander continued his march eastward into the northeastern portions of the Persian Empire, to the area of modern Afghanistan and Kashmir, and then headed south into the Indus Valley (in modern Pakistan). When his troops refused to continue east into India he embarked on a difficult march west, through the deserts and mountains of southern Persia (modern Iran), to Babylon.

Alexander may have intended to make Babylon the capital of his empire, although this is not certain; but before he was able to enjoy the fruits of a mature empire, he died in Babylon, in 323 BC, at the age of thirty-two. It was years before the full impact of his conquests took effect, but it is worth noting here that the Near East was radically changed with the arrival of Greek language and culture.

Because Alexander died without leaving a designated successor, a number of his generals and regents quarreled over whether the kingdom should remain a united whole or whether it should be divided among the competing powers. In 320 BC Ptolemy I annexed Palestine

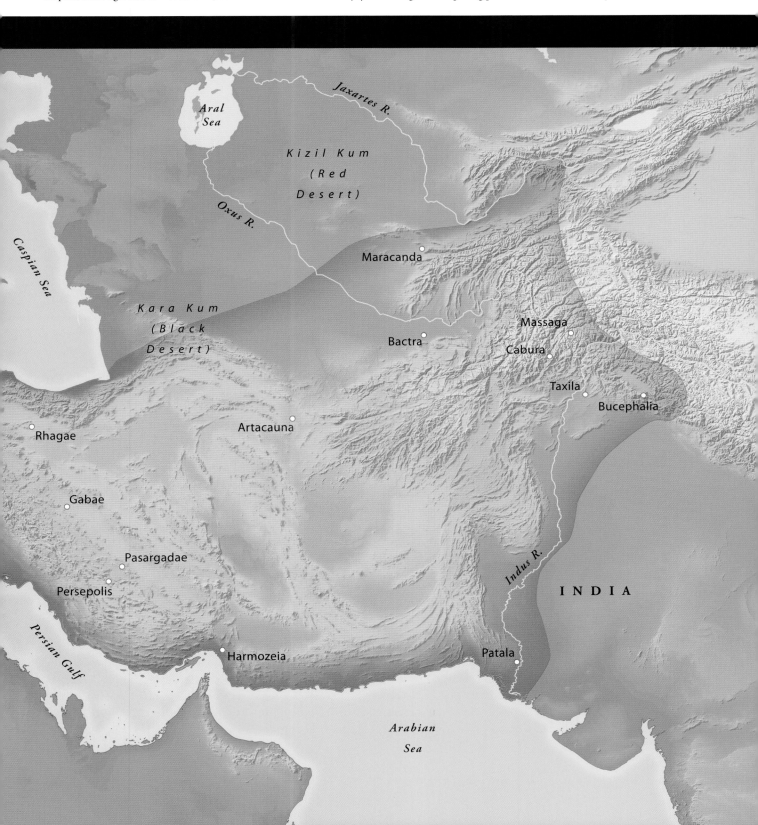

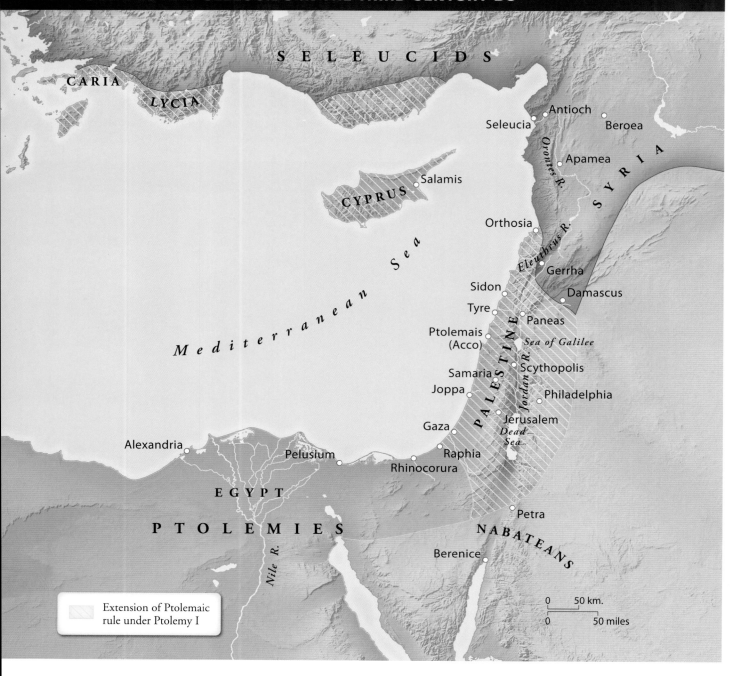

Extension of Ptolemaic rule under Ptolemy I

0   50 km.

0   50 miles

to Egypt. Following the battle of Ipsus in 301 BC, in which Antigonus and his allies were defeated, there was a period of relative peace, during which time Antipater and Cassander were established in Macedonia and Greece; Lysimachus in Thrace and Asia Minor; Seleucus I in Syria, Mesopotamia, and Persia (all the way to the Indus River); and Ptolemy I in Egypt and Palestine.

Ptolemy was able to establish himself as the king of Egypt, and thus began the rule of his Greek dynasty, which lasted for almost three centuries, from 304 BC until 30 BC. This dynasty consisted of at least sixteen different kings and queens who were called Ptolemy or Cleopatra. Not only was the king of Greek origin, but the ethnic background of much of the ruling administration was also Greek, and

many of the native, non-Greek officials soon learned the Greek language and adopted Greek customs. The capital of Egypt was moved to the recently established city of Alexandria. There Ptolemy I buried Alexander the Great, and he was able to establish the city as one of the leading commercial and intellectual centers of the world.

During the reign of Ptolemy I the famous library of Alexandria was established, as was the scholarly center called the Museum. Since the early Greek rulers could neither speak Egyptian nor read hieroglyphics, it was necessary for them to rely on translators to communicate with the local populace and to understand the cultural history of Egypt. It was in this context, during the reign of Ptolemy I (304 – 282 BC), that Manetho, the Egyptian, wrote his famous history of Egypt in Greek, and it is from his lead that Egyptian history is normally divided into periods of thirty or thirty-one dynasties.

The political and military fortunes of Ptolemy I were varied, but at one time he was able to extend Ptolemaic rule not only to the island of Cyprus but also into southern Turkey, to Caria and Lycia, and even toward the Greek mainland, to Corinth — for a brief period of time. These territories would pass in and out of Ptolemaic hands during the third century BC.

To the north of Palestine, Seleucus I (312 – 280 BC) was establishing a mighty kingdom. He even gained control of Asia Minor, so that his kingdom stretched from western Turkey, eastward through Syria, Mesopotamia, and Persia, to the Indus River. He established his capital at the city of Antioch, which was located on the Orontes River, near, but not on, the northeastern shore of the Mediterranean Sea. The influence of the Seleucid state was so pervasive that the calendrical system used in the Near East for hundreds of years was reckoned from the beginning of Seleucus I's reign in what is now known as 312 BC. During the days of Seleucus I there was a period of relative peace between the Seleucid state in the north and Egypt in the south.

Although the Ptolemies are usually considered to have been the rulers of Palestine during the third century BC, the situation in fact was much more complex. During the third century there was a series of four major wars pitting the Seleucids against the Ptolemies. These wars are called the "Syrian Wars," and the Ptolemies were, in the main, successful in defending and controlling their territory.

During the third century BC Ptolemaic interest in Palestine was primarily twofold. First, Palestine was important as a buffer state, serving as Egypt's first line of defense against Seleucid aggression. In addition, Palestine served as an agricultural hinterland, from which Egypt was able to secure quality olive oil, wines, wood products, and, at times, slaves.

The extent, depth, and precise nature of Ptolemaic rule in Palestine is difficult to determine, though it is possible to describe its basic outline. In general, Ptolemaic holdings stretched north along the Mediterranean Sea to the region of the Eleutherus River. However, inland, in Transjordan, they did not reach as far north as Damascus, which was usually under Seleucid control. The center of Ptolemaic administration was apparently located at Alexandria in Egypt, from which a governor and a director of revenues of Syria and Phoenicia controlled matters. Palestine was divided into a number of administrative units called *hyparchies* in Greek. These were basically equivalent to Persian "provinces" (*medinah* in Aramaic). In general, each hyparchy had three major officials, all of whom were Greek: a governor, a financial officer, and a person in command of the army/police.

Not much is known about the hyparchy of Judea during the third century BC. But since it had submitted to Alexander the Great and since it was not deeply involved in the Syrian Wars, it is generally assumed that few changes occurred in its size or internal administration, whose chief Jewish official was the high priest.

To the north of Judea, the hyparchy of Samaria underwent some change. The inhabitants of the city of Samaria had revolted against the

▼ *Tomb from the Hellenistic Period of a Sidonian family living at Marisa in Idumea. Paintings in the tomb were reconstructed.*

governor whom Alexander had appointed — they had burnt Andromachus alive. In retaliation, the city of Samaria was captured, its population executed or banished, and the city resettled with Macedonian veterans, thus establishing a viable Greek colony in and around the city. The remaining portion of the hyparchy of Samaria was probably dominated by the Samaritans, who maintained their religious and political institutions on and near Mount Gerizim.

To the north of Samaria were royal estates in the Jezreel Valley, and to the north of these was the hyparchy of Galilee, which is, in fact, mentioned in the Zenon papyri. The administrative center of Galilee could have been at the fortress of Itabyrium, located on the top of Mount Tabor or, more probably, at Scythopolis (the "city of the Scythians" = OT Beth Shan).

To the north and west of Galilee, along the Phoenician coast, maritime cities such as Tyre and Sidon had a high degree of independence; but they remained submissive to Ptolemaic rule, supplying ships and sailors for the formidable Ptolemaic navy as well as a merchant fleet. Indeed, during the third century BC the Ptolemies were the premier naval power in the eastern Mediterranean Sea. Farther south along the Mediterranean coast the importance of Acco, the port of Galilee, was emphasized by the fact that it was one of the few cities to receive the dynastic name, Ptolemais. Farther to the south, along the Sharon and Philistine plains, coastal cities — Dor, the Tower of Strato (later Caesarea), Joppa, Jamnia, Azotus (= OT Ashdod), Ascalon, and Gaza, among others — were influenced by Hellenistic civilization. Nevertheless, they were able to retain varying degrees of independence within the context of maintaining at least nominal loyalty to the Ptolemaic dynasty.

To the east of the Philistine Plain was the hyparchy of Idumea. Its westward orientation, toward Greek influence, is emphasized by the fact that although its territory stretched east to the Dead Sea, its capital, Marisa, was located in its western region, in the Shephelah overlooking the coastal plain.

In Transjordan, the large Assyrian/Babylonian/Persian districts were generally divided into smaller administrative units. For example, the old Hauran district was divided into Trachonitis and Auranitis. Besides the new district names, major cities were given Greek names, and a Hellenistic (or Hellenized) populace began to flourish in them: Rabbah of the Ammonites (NIV) was renamed Philadelphia, Pella/Pehel became Bernice, and in Cisjordan, as already noted, Acco became Ptolemais, Beth Shan became Scythopolis, etc. In these cities, it was the upper classes who were motivated to adopt Greek language and customs for economic as well as other reasons. However, it must be assumed that a rather large portion of the local population maintained their traditional Semitic languages (e.g., Hebrew, Aramaic) and customs, for later, after the veneer of Greco-Roman influence was removed, the Semitic influences still prevailed. For example, the name of the city of Scythopolis (Greek) eventually reverted to a form of the old name "Beth Shan" (Semitic), indicating that the local populace had continued using that name through centuries of Greco-Roman rule.

With the accession of Ptolemy IV (221 – 203 BC) to the Egyptian throne and of Antiochus III (223 – 187 BC) to the Seleucid throne, the balance of power in the Levant began to shift in favor of the Seleucids. In the Fourth Syrian War (221 – 217 BC) Antiochus was able to recapture Seleucia, the port of the capital Antioch, to push south into Galilee,

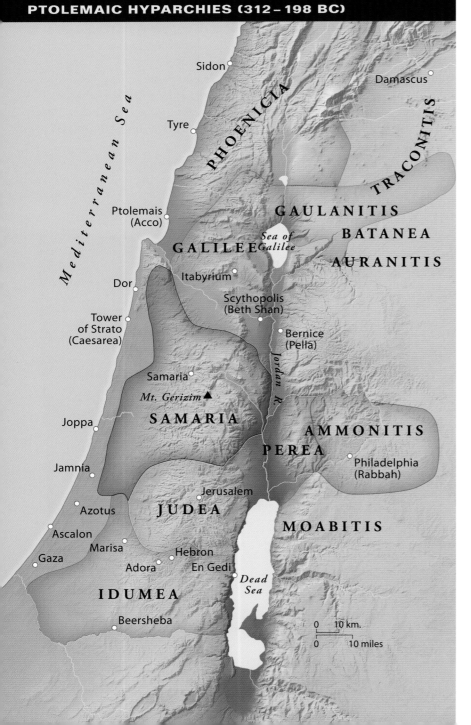

PTOLEMAIC HYPARCHIES (312 – 198 BC)

the Jezreel Valley, and Transjordan, and to advance all the way south to Gaza and Raphia. However, the final Seleucid conquest of Palestine was thwarted for two more decades, as Antiochus was defeated at Raphia and the forces of Ptolemy IV were able to recapture Palestine and Phoenicia.

For the next decade Antiochus III was preoccupied with containing revolts in Asia Minor and the eastern regions of his kingdom. In 201 BC Antiochus III led a thrust into Palestine that would eventually culminate in the transfer of Palestine from Ptolemaic to Seleucid control. However, when Antiochus III turned his attention to the conquest of Pergamum in western Asia Minor, the Egyptian general Scopas reentered Palestine, reclaiming it for Egypt. Returning to confront this new threat, Antiochus III met Scopas in the decisive battle at Paneas in 198 BC; the Egyptian army was routed, and Scopas was forced to flee to Sidon. In the aftermath of the battle Antiochus marched south through portions of Batanea in Transjordan and into Samaria and Judea. He evidently was well received by the Jewish population of Jerusalem, who aided him in the expulsion of the Ptolemaic garrison that had been stationed there. Thus began, as a result of the Battle of Paneas, close to half a century of Seleucid control of Judea (198 – 142 BC).

During the rule of Antiochus III (223 – 187 BC) and the first portion of the reign of Seleucus IV (187 – 175 BC), the Judeans seem to have prospered under the leadership of the high priest Onias III (198 – 174 BC). Because the Jewish population of Jerusalem had so readily received Antiochus III, they were granted special privileges, including the restoration of Jerusalem, limited tax exemptions, subsidies for the temple, and permission to live according to their ancestral laws (Jos. *Antiq* 12.3.3, 4 [138 – 46]).

However, the takeover of Palestine came at a time when Seleucid power was beginning to diminish. During the 190s, the Seleucids progressively lost control of their holdings in Asia Minor to an ascendant Rome (see below). Seleucid control of the eastern portion of their empire was also being challenged, and, as will be described shortly, Palestine itself entered a period of turmoil and intrigue with the revolt of the Maccabees and the rise of the Hasmonean dynasty.

Given this state of flux within the Seleucid domains, and with only a limited number of historical sources, it is possible to describe the geographical scene in Palestine in only the most general way, for it must be kept in mind that borders, spheres of influence, and whole populations of cities were constantly changing. In general, it can be asserted that the Seleucids were much more interested in promoting the advancement of Greek language, culture, and customs than were the Ptolemies before them. They accomplished this not only by introducing all of these Greek elements in existing cities but also by establishing new cities on the model of the Greek *polis*, where the adult male citizens met together to govern the affairs of the city.

The Seleucids also changed the Semitic names of cities to Greek ones, which indicates their pervasive influence in the countries that they controlled. For example, Jerusalem became Antiochia, Gerasa became Antiochia on Chrysorrhoas, Gadara became Antiochia Seleucia, Abila became Seleucia, and Ptolemais became Antiochenes. But these Seleucid names lasted for only a brief period of time, and then the names of cities often reverted back to their earlier designations.

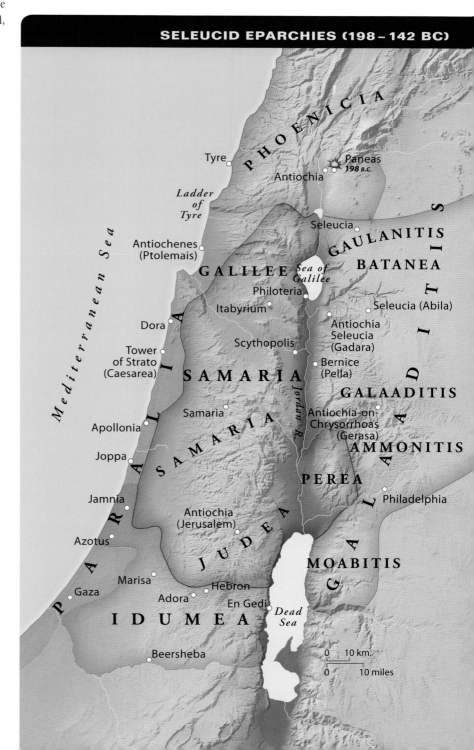

Mark Connally

▲ *Pleasure palace at Iraq el-Amir in Jordan from the Hellenistic Period.*

In addition, new cities, such as Antiochia (near Paneas and Dan) and Seleucia in Gaulanitis, were also established.

It seems that the Seleucids preferred larger administrative units than those established under Ptolemaic rule, for they combined several Ptolemaic hyparchies to form a larger unit called an "eparchy." There were four such eparchies in Palestine. One of the largest of these was the eparchy of Samaria, whose governor resided in the city of Samaria. Included in this eparchy were the districts of Judea, Samaria, Galilee, and Perea. The first three districts have been described above (pp. 185 – 186); Perea was located east of the Jordan River, just north of the Dead Sea. This small territory was bordered by Ammonitis on the east.

To the south of the eparchy of Samaria was that of Idumea. Along the coast of the Mediterranean Sea was Paralia, an elongated eparchy that stretched from the border of Egypt in the south to the Ladder of Tyre in the north, excluding the Idumean corridor to the sea in the Azotus-Jamnia region. On the east of Samaria, east of the Jordan River, was the large eparchy of Galaaditis. There in Transjordan, along the trade routes, Greek cities seem to have flourished. Eventually some of these cities united to form a league known as the Decapolis (see below, pp. 210 – 211).

During the latter part of his rule, Antiochus III experienced a series of defeats at the hands of the Romans: first at Thermopylae in 191 BC and then at Magnesia in 190 BC. In 188 BC the Treaty of Apamea was signed, in which Antiochus relinquished control of much of Asia Minor and pledged to pay a heavy tribute to Rome.

Antiochus III (the Great) was succeeded by his son Seleucus IV Philopater, who ruled from 187 to 175 BC. By adhering to the terms of the Peace of Apamea he was able to maintain good relations with Rome. During the early part of Seleucus IV's reign he maintained good relations with the Judeans, even presenting gifts to the temple in Jerusalem (2 Macc 3:3). However, later in his reign a Jew named Simon, wanting to oust the pious Onias III from the office of high priest, tried to ingratiate himself with the Seleucid authorities by encouraging Apollonius, the governor of Coele-Syria and Phoenicia, to expropriate the Jerusalem temple treasury for the benefit of the king, Seleucus IV. Heliodorus, a high official of the king, was sent to Jerusalem to confiscate the treasury, but it is reported that due to divine intervention he was not able to carry out this plot (2 Macc 3). At the death of Seleucus IV in 175 BC, Antiochus IV was able to seize the throne. With the rise of Antiochus IV, Judea entered a critical phase in its history, which will be taken up in the next chapter.

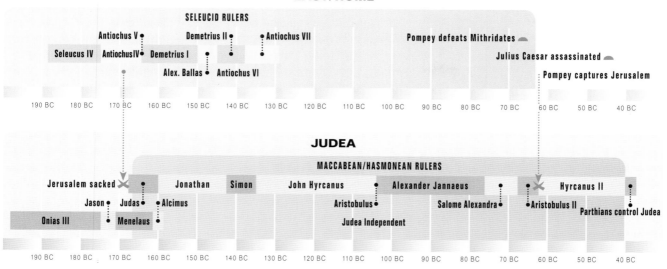

EAST/ROME

SELEUCID RULERS

Antiochus V • | Demetrius II • | • Antiochus VII | Pompey defeats Mithridates

Seleucus IV | Antiochus IV • | Demetrius I | | Julius Caesar assassinated

Alex. Ballas • | • Antiochus VI | Pompey captures Jerusalem

190 BC 180 BC 170 BC 160 BC 150 BC 140 BC 130 BC 120 BC 110 BC 100 BC 90 BC 80 BC 70 BC 60 BC 50 BC 40 BC

JUDEA

MACCABEAN/HASMONEAN RULERS

Jerusalem sacked | Jonathan | Simon | John Hyrcanus | Alexander Jannaeus | Hyrcanus II

Jason • | Judas • | • Alcimus | Aristobulus • | Salome Alexandra • | • Aristobulus II

Onias III | Menelaus | | Judea Independent | | Parthians control Judea

190 BC 180 BC 170 BC 160 BC 150 BC 140 BC 130 BC 120 BC 110 BC 100 BC 90 BC 80 BC 70 BC 60 BC 50 BC 40 BC

EGYPT

Ptolemy VI

Jewish Temple built at Leontopolis in Delta

190 BC 180 BC 170 BC 160 BC 150 BC 140 BC 130 BC 120 BC 110 BC 100 BC 90 BC 80 BC 70 BC 60 BC 50 BC 40 BC

THE MACCABEAN REVOLT AND THE HASMONEAN DYNASTY

adopted the new Hellenistic lifestyle (2 Macc 4), which meant breaking with their religious, cultural, and linguistic heritage. Soon afterward, around 171 BC, Menelaus, another Hellenizer, had Jason removed and himself appointed to the high priesthood by offering Antiochus IV a larger bribe. He even had Onias III murdered. Thus official Judaism

▼ *Temple of Zeus at Gerasa (Jordan) — one of the Greco-Roman cities of the Decapolis. Originally built in the Hellenistic/Early Roman period.*

With the rise of Antiochus IV in 175 BC a chain of events began that culminated in the establishment of an independent Jewish state in 142 BC, a state that lasted until the capture of Jerusalem by the Romans in 63 BC. These events are of more than passing interest, for they had a direct influence on Jewish life and practice for the next two centuries and have had a profound effect on Judaism to the present day.

Antiochus IV Epiphanes (175–163 BC) did not encounter much difficulty in seizing the Seleucid throne after the assassination of Seleucus IV. To counter the rising threat of a powerful Rome, Antiochus IV attempted to solidify his kingdom by uniting its diverse elements under the banner of Hellenism. In Judea certain elements of the population supported the Hellenizing process, and these gained more influence after Jason deposed the pious Onias III (2 Macc 4:7) and had himself appointed high priest by bribing Antiochus IV. Many Jews quickly reached a low point as the office of high priest was sold to the highest bidder.

Antiochus IV successfully invaded Egypt in 169 BC but withdrew the same year, leaving a teenager (Ptolemy VI Philometer) as king of Egypt. In Judea it was rumored that Antiochus IV had died while in Egypt, and Jason, the former high priest, who thought that this would be an opportune moment to regain his lost position, attacked Jerusalem in an attempt to depose Menelaus. Antiochus responded by beating back the attack and recapturing Jerusalem. In the process, thousands of Jews were killed or sold into slavery, the temple treasury was plundered, and Antiochus IV, a Gentile, even entered the most sacred room of the temple, the Most Holy Place; access to this room was limited to the high priest, and then on only one day of the year, the Day of Atonement.

The following year (168 BC) Antiochus IV mounted a second attack on Egypt. As he advanced on the city of Alexandria, he was met by the Roman legate C. Popilius Laenas, who presented him with

THE MACCABEAN REVOLT (167–152 BC)

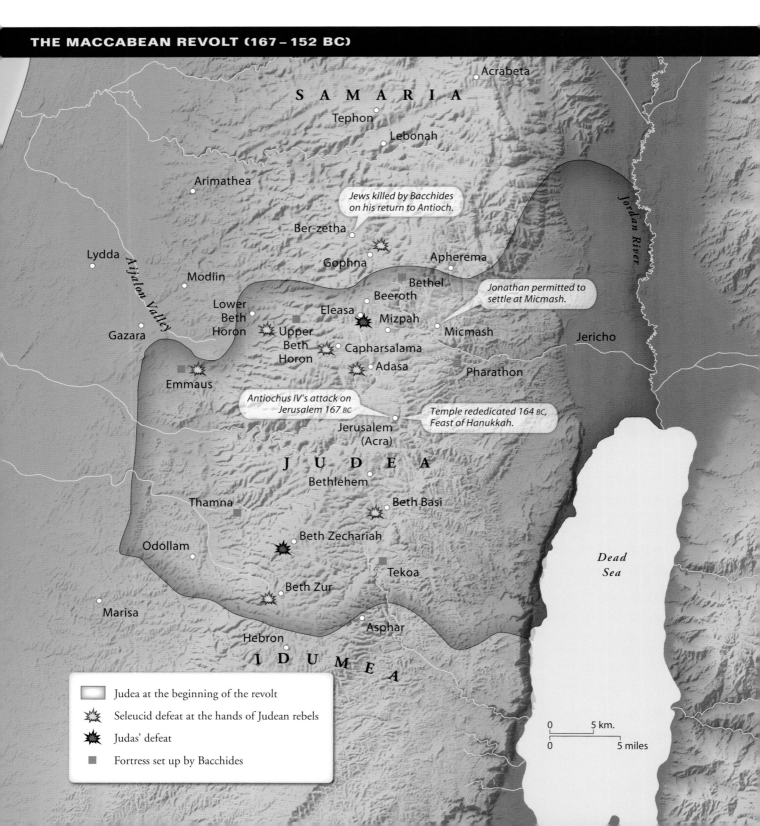

SAMARIA

Acrabeta

Tephon

Lebonah

Arimathea

Jews killed by Bacchides on his return to Antioch.

Ber-zetha

Gophna

Apherema

Lydda

Aijalon Valley

Modiin

Bethel

Beeroth

Eleasa

Mizpah

Jonathan permitted to settle at Micmash.

Lower Beth Horon

Upper Beth Horon

Capharsalama

Micmash

Jericho

Gazara

Adasa

Pharathon

Emmaus

Antiochus IV's attack on Jerusalem 167 BC

Temple rededicated 164 BC, Feast of Hanukkah.

Jerusalem (Acra)

J U D E A

Bethlehem

Thamna

Beth Basi

Beth Zechariah

Odollam

Beth Zur

Tekoa

Dead Sea

Marisa

Asphar

Hebron

I D U M E A

Jordan River

Judea at the beginning of the revolt

Seleucid defeat at the hands of Judean rebels

Judas' defeat

Fortress set up by Bacchides

0   5 km.

0   5 miles

▲ *Tombs of the "sons of Hezir" (left) and of "Zechariah" (center) in the Kidron Valley, Jerusalem. Note the Doric and Ionic columns, and the pyramid-shaped roof, indicating Greek and Egyptian influence on the country.*

Rome's ultimatum that he leave Egypt immediately. Antiochus IV tried to stall for time, but Laenas drew a circle around him in the sand and ordered him to make his decision before leaving the circle. In the face of Rome's threat, a humiliated Antiochus IV and his forces left immediately.

When Egypt denied him, Antiochus IV evidently decided to strengthen the southern approach to his kingdom by solidifying his position in Palestine. In 167 BC he dispatched troops to Jerusalem, where part of the population was slaughtered. In an attempt to Hellenize the population further, Jews were commanded to worship Zeus and other pagan deities, to burn their copies of the Torah, and to forsake the laws of their God (1 Macc 1:41 – 64). They were forbidden to observe the Sabbath, to celebrate their feasts, to sacrifice to God, and to circumcise their children. In addition, portions of the walls of Jerusalem were torn down and a pagan citadel, called the Acra, was constructed at a strategic location within the city. The temple in Jerusalem was turned into a temple of Olympian Zeus, and on December 16, 167 BC, an unclean sacrifice was offered to him (2 Macc 6:1 – 11).

During the latter portion of his reign (167 – 164 BC), Antiochus IV undertook a series of campaigns to the east. He eventually died in Persia in 164 BC. While Antiochus was away on these campaigns, Lysias was left in charge of the territory west of the Euphrates River; thus it was he who had to deal with a serious rebellion in Judea. In 167 BC a delegate of Antiochus IV attempted to force Mattathias, a priest who lived in Modiin, to sacrifice to a pagan deity. Mattathias refused, but another Jew volunteered to perform the rite. Outraged, Mattathias killed both the Seleucid delegate and the errant Jew, and thus the Maccabean revolution began (1 Macc 2:1 – 48). The aged Mattathias soon died a natural death, leaving his five sons to carry the revolutionary torch (1 Macc 2:49 – 70).

The first and foremost leader of the revolution was Judas, the middle son, who was also called Maccabeus (probably "the hammerer"). Because of the prominence of Judas and the uniqueness of his nickname, this revolt is often called the Maccabean Revolt; however, the period of the revolt includes not only the days of his leadership but also those of his brother Jonathan (160 – 142 BC) and to some extent those of his brother Simon (142 – 135 BC) as well. Gaining the support of the Hasidim, the "pious ones," who were true to ancient Jewish beliefs and practices, Judas and his followers went throughout the countryside, tearing down pagan altars and circumcising Jewish children.

The first attempt at suppressing the revolt was undertaken by Apollonius, the governor of the region and commander of the garrison at Samaria. As he marched south toward Judea, he was met by Judas and his troops and was defeated and killed (1 Macc 3:1 – 12). Soon afterward Seron, a Seleucid general, was sent to suppress the uprising. While attempting to enter the hill country from the coastal plain via the traditional Beth Horon road, Seron and his army were routed by Judas and his forces at the ascent of Beth Horon (1 Macc 3:13 – 26).

In 165 BC Lysias, the superior of Seron, assembled a large army that moved south along the coastal plain and set up camp at Emmaus, where the road to Jerusalem begins its ascent into the hill country. Judas mustered his warriors at the old tribal center of Mizpah in the Hill Country of Benjamin. While Gorgias, a Seleucid commander, and a portion of the Syrian army marched into the hills on a search-and-destroy mission, Judas and his men descended on the portion of the Seleucid army that had remained at Emmaus in the Valley of Aijalon. The surprised Seleucids were defeated and retreated to the coast, as did Gorgias and his troops after returning to their sacked camp at Emmaus (1 Macc 3:27 – 4:25).

After these three initial defeats, Lysias himself became personally involved in leading the Seleucid troops in a fourth campaign against the Maccabees (164 BC). Moving again south along the coastal plain, Lysias did not repeat the mistake of the two previous attacks — attempting to enter the hill country via the Aijalon Valley and Beth Horon road. Instead he continued south, to the friendly Hellenistic city of Marisa, and from there entered the Hill Country of Judah in the Adora/Hebron area, an Idumean region also friendly to the Seleucids.

As Lysias marched north, Judas marched south along the watershed. The two armies met in battle at Beth Zur, a village located on the southern boundary of Judea. There, to the Judeans' great delight, the army of Lysias was routed (1 Macc 4:26 – 35). Flushed with victory, Judas and his warriors marched to Jerusalem and recaptured the whole city, save the Acra, which remained in the hands of the Hellenizers. In the temple area, Judas commanded that the pagan altar be torn down and Yahweh's altar rebuilt. The temple was cleansed, and on Chislev 25 (December 14), 164 BC, the temple was rededicated and proper Jewish sacrifices were resumed. Jews have commemorated this event through the ages as the Feast of Hanukkah — the Feast of Dedication (vv. 36 – 61).

After the death of Antiochus IV in 164 BC the Seleucid leaders were preoccupied with who would succeed him, and Judas and his brothers were able to expand their influence in and around Judea. Jews who were being oppressed in Acrabeta, in the district of Samaria north of Jerusalem, and by the Beonites in Transjordan were delivered from their plight by Judas. To the northeast, Jews in the Gilead and Golan regions, in the villages of Bostra, Dathema, Alema, Caspein, Maked, Bosor, Charax, Raphon, and Carnaim, as well as those in Ephron and Scythopolis, also received relief from their oppressors because of Judas's successful military campaign in the area. While Judas was fighting in Transjordan, his brother Simon led an expedition to the Sharon Plain (Arbata), the Jezreel Valley, and western Galilee, relieving pressure on the Jews who lived in those areas. Finally, Judas himself led his men on raids to the south and southwest of Judea, harassing and/or capturing such cities as Hebron, Marisa, Azotus, and Joppa; yet in the end he was unable to hold these gains and was eventually forced to retreat to Odollam (= OT Adullam) in the Shephelah (1 Macc 5:1 – 68; 2 Macc 12:38).

Despite these victories of Judas there was still a large influential portion of the populace who desired closer ties with the Seleucids and preferred the Hellenistic lifestyle. These Hellenizers appealed to Antiochus V for help, and Lysias was again sent to Judea with a large army. His line of march was similar to the one he had used previously: coastal

plain to Marisa to Hebron to Beth Zur. Then, continuing north along the watershed of the Hill Country of Judah, he met Judas at the village of Beth Zechariah. In the battle Lysias made use of war elephants, to the detriment of Judas's forces. In fact, one of Judas's brothers, Eleazar, was killed when one of the elephants fell on him.

Lysias was victorious at Beth Zechariah and proceeded to Jerusalem while Judas and his troops fled to the remote regions of the country (1 Macc 6:18 – 54). However, as Lysias was laying siege to Jerusalem he received word that his rival, Philip, was marching from Persia to Syria to claim the Seleucid throne. Needing to return to Antioch, Lysias made peace with Judas, guaranteeing the Jews religious freedom (vv. 55 – 63) but demanding that the walls of Jerusalem be torn down. Thus, in spite of Judas's initial defeat, at least the religious gains of the revolt were preserved. However, Lysias accused Menelaus, the high priest, of being an instigator of the rebellion, and he was executed in Beroea and replaced by Alcimus, another Hellenizing high priest (2 Macc 13:3 – 8).

The situation in Antioch was chaotic, and in the end Demetrius I, the son of Seleucus IV, who had been held hostage in Rome, escaped and returned to Antioch and deposed and executed all his rivals (1 Macc 7:1 – 4). Demetrius I confirmed Alcimus as high priest, but Alcimus still had to be established in Jerusalem. This was done by sending a military force to Jerusalem under the leadership of Bacchides, a powerful general in charge of all the territory west of the Euphrates River. The Hellenists of Jerusalem were, of course, willing to accept Demetrius I's rule; but surprisingly, so were the Hasidim (the "pious ones"), because Alcimus was of legitimate, Aaronic, priestly descent. But Alcimus, after he had been installed as high priest, executed sixty Hasidim, and Bacchides killed even more Jews at Beth Zaith as he was returning to Antioch (1 Macc 7:5 – 25). With this turn of events the Hasidim returned to Judas, and the Maccabees began to pressure Alcimus and the Hellenizers.

Nicanor, the Syrian commander whom Bacchides had left in charge of the garrison at Jerusalem, tried to relieve the Maccabean pressure by venturing out of Jerusalem in an attempt to open the road that led from Jerusalem to the coastal plain. He was met and defeated by Judas at Capharsalama, 6 miles northwest of Jerusalem. Later, Nicanor again attempted to open the road, but this time he himself was killed in battle at Adasa (1 Macc 7:26 – 50).

Judas knew that the powerful Bacchides would respond, and so he appealed to Rome for help, but this assistance would not arrive in time. In 160 BC Bacchides began his advance to Jerusalem. Passing south along the western shore of the Sea of Galilee he killed the Jews living in Arbela and proceeded through Samaria to Jerusalem. After arriving in Jerusalem, Bacchides marched northwest out of the city to secure the Beth Horon road to the coast. Judas and his troops fought him near the village of Eleasa, and there the Maccabean forces were decimated. Judas was killed, and the surviving members of his force fled to the Judean Wilderness, to the well of Asphar in the Wilderness of Tekoa. There, Judas's brother Jonathan attempted to rebuild the Maccabean fighting force.

Bacchides, for his part, established a chain of fortresses in Judea at places such as Bethel, Jericho, Pharathon, Upper Beth Horon, Emmaus,

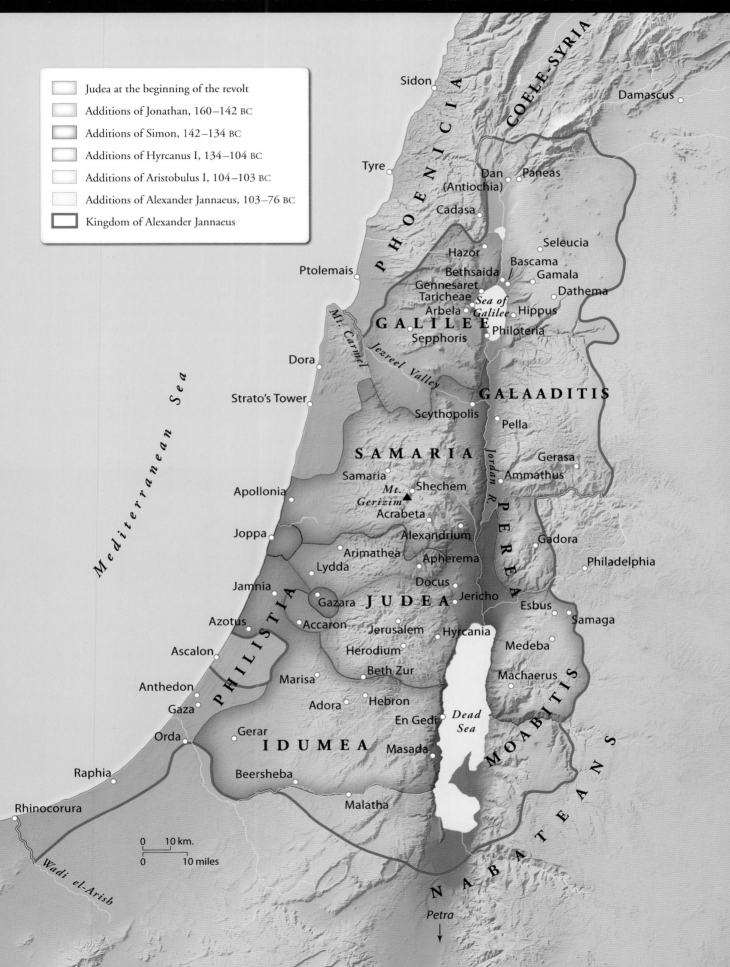

Judea at the beginning of the revolt
Additions of Jonathan, 160–142 BC
Additions of Simon, 142–134 BC
Additions of Hyrcanus I, 134–104 BC
Additions of Aristobulus I, 104–103 BC
Additions of Alexander Jannaeus, 103–76 BC
Kingdom of Alexander Jannaeus

Mediterranean Sea

Sidon
Damascus

COELE-SYRIA

PHOENICIA

Tyre
Dan
(Antiochia)
Paneas

Cadasa
Seleucia

Hazor
Bascama
Gamala
Bethsaida
Gennesaret
Dathema
Taricheae
Sea of
Galilee
Arbela
Hippus
GALILEE
Philoteria
Sepphoris

Ptolemais

Dora
Mt. Carmel
Jezreel Valley
GALAADITIS

Strato's Tower
Scythopolis
Pella

SAMARIA
Gerasa
Samaria
Ammathus
Mt.
Shechem
Gerizim
Acrabeta
Jordan
PEREA

Apollonia
Alexandrium
Gadora

Joppa
Arimathea
Apherema
Philadelphia
Lydda
Docus

Jamnia
Jericho
Gazara
JUDEA
PHILISTIA
Esbus
Accaron
Samaga
Azotus
Jerusalem
Hyrcania
Herodium
Medeba
Ascalon
Beth Zur
Machaerus
Anthedon
Marisa
Dead
MOABITIS
Gaza
Adora
Hebron
Sea
Orda
En Gedi
Gerar
IDUMEA
Masada
Raphia
Beersheba
Rhinocorura

NABATEANS

Malatha

0   10 km.
0   10 miles

Wadi el-Arish

Petra

Thamna, and Tekoa. Thus the revolt that had begun with Mattathias's deed at Modiin in 167 BC and had accomplished a number of significant goals (the temple area was regained, sacrifices had been reinstituted, military victories over superior forces had been won, and for a brief period of time minor territorial advances had been made) was in danger of collapse by 160 BC. The charismatic leader Judas was dead, his forces dispersed, and the enemy securely in control of the countryside.

In 159 BC Alcimus, the high priest, died, and the Seleucids did not appoint a successor. Officially, Judea was under direct Syrian control; Bacchides remained in Jerusalem and from there governed a subdued populace. But in 157 BC Bacchides returned to Antioch, and Judas's brother Jonathan took advantage of his absence to resume hostilities. Jonathan and his troops seized the abandoned fortress of Beth Basi, 1.3 miles southeast of Bethlehem. Bacchides returned to Judea in 156 BC to deal with this new phase of the uprising. After a series of battles, Bacchides withdrew his forces from Judea and Jonathan was able to regain at least limited control of the countryside, although both Jerusalem and Beth Zur remained in the hands of the Hellenists.

This arrangement lasted for several years until Jewish-Seleucid relations entered a new phase in 152 BC. At that time Alexander Balas, who claimed to be the son of Antiochus IV, challenged the kingship of Demetrius I. In their rivalry both men attempted to win the loyalty of Jonathan by granting him all sorts of concessions: money, tax exemptions, territory, permission to rebuild the walls of Jerusalem, symbols of kingship, and the office of high priest. Jonathan eventually sided with Alexander Balas. At the Feast of Tabernacles in 152 BC Jonathan wore the high priestly garments; thus the office was filled once again, this time by a Hasmonean. (The name Hasmonean was taken from a certain Hasman, who was a priest from the family of Joarib [1 Macc 2:1; 1 Chron 9:10; 24:7] and who was the great-grandfather of Mattathias, the father of Jonathan and Simon and their brothers.)

Although Alexander Balas was able to defeat and kill Demetrius I, his struggle for the throne was not over, for Demetrius II returned from his refuge in Cnidus (in southwestern Turkey) to challenge Alexander Balas.

In 145 BC the internal Seleucid strife came to a head when the troops of Alexander Balas met Demetrius II in battle near Antioch. Alexander Balas was killed and Demetrius II became the sole ruler of Syria. Although Jonathan and Demetrius II had been enemies, Jonathan sent him gifts, and Demetrius II, impressed with Jonathan's prowess and audacity, confirmed him as high priest and gave him the title "Friend of the King," and he granted him control of three districts of Samaria — this in addition to territory that he secured during this period of conflict, including Joppa, Azotus, Ascalon, and the districts of Accaron, Lydda, Arimathea, and Apherema.

The rule of Demetrius II was challenged by Tryphon, a general who backed Antiochus VI, then a child, for the kingship. Jonathan sided with Tryphon and defeated Demetrius II at Cadasa in Upper Galilee and then

▼ *Qumran Cave IV: Numerous Dead Sea Scroll fragments were found in this cave.*

later in the Plain of Hamath, far to the north of Palestine. However, Tryphon feared a powerful Jonathan, and captured him by trickery and eventually executed him.

Simon, Jonathan's brother, naturally allied himself with Tryphon's opponent, Demetrius II. The latter, in recognition of Simon's loyalty, sent him a letter confirming Judea's complete independence. Thus 142 BC marked the official independence of the Judean state — the first time that it had been officially free from foreign domination since 586 BC, when Jerusalem had fallen to the Babylonians.

Upon the death of Demetrius II his brother Antiochus VII ruled in his stead and he struggled for power with Tryphon until the latter committed suicide. In the meantime Antiochus VII had granted Simon permission to mint coins, a sign of Judea's growing independence, and the Jews conferred on Simon the position of governor and of high priest "forever, until a trustworthy prophet should arise" (1 Macc 14:25 – 43). Thus, it was during Simon's reign that the Hasmonean dynasty (142 – 63 BC), in a narrow sense, began to rule.

However, even at this high point for the Hasmoneans, pro-Antiochus VII elements were working in Judea, attempting to unseat Simon. Tragically, one of these attempts was successful. Ptolemy, the son-in-law of Simon and governor of Jericho, invited Simon and his sons to a feast at the fortress of Docus and there assassinated Simon and all his sons. Only John Hyrcanus, who was absent from the banquet, escaped the slaughter. In the aftermath of the assassinations, Antiochus VII allowed John Hyrcanus to maintain his position as governor and high priest.

When Antiochus VII was killed in battle, strong Seleucid rule effectively ceased, and the Judeans engaged in expansionist activities. In 128 BC John Hyrcanus was able to seize Medeba and Samaga in Transjordan, which gave him at least partial control of the Transjordanian Highway. In the same year he attacked the Samaritans, who had been harassing the Jews, and destroyed their temple on Mount Gerizim. John Hyrcanus also established an alliance with Rome, and Rome confirmed his independence. In 125 BC he was able to move against Idumea, to the south of Judea, capturing the whole toparchy — including Hebron, Adora, and Marisa. The Idumeans were forced to convert to Judaism; thus the territory now controlled by John Hyrcanus stretched south all the way to Beersheba. Later in his reign, John Hyrcanus attacked the city of Samaria (ca. 108 – 107 BC), and by the time of his death in 104 BC, all of Samaria up to the Jezreel Valley was under Judean control.

After his long and successful reign, John Hyrcanus was replaced by his son Aristobulus I, who promptly neutralized other contenders for the positions of governor and high priest. Aristobulus I ruled for only a year (104 – 103 BC), but during that time he made a significant move north, conquering Galilee and forcing its non-Jewish inhabitants to convert to Judaism. Upon his death, his wife Salome Alexandra released Aristobulus's three brothers from prison and appointed Alexander Jannaeus as king and high priest. She in turn married Alexander Jannaeus, in spite of the fact that the high priest was supposed to marry only a virgin.

Alexander Jannaeus's long reign (103 – 76 BC) could in some ways be considered the high point of Hasmonean power; yet it was marred by internal discord. During his rule the Hasmonean kingdom stretched, at least briefly, from Dan (Antiochia) in the north to Beersheba in the south, which meant that the traditional heartland of Israel was now under Jewish control.

To the southeast of Judea were the Nabateans, an Arab people who had replaced the Edomites in Mount Seir and had made the rock-cut city of Petra their capital. They were anxious to be the masters of the trade routes that led from southern Arabia north to Damascus and west to the Mediterranean Sea. Not only were products of Arabia (such as gold, frankincense, and myrrh) carried along these routes, but also spices from India and silk from China. The price of these goods at times doubled while in Nabatean hands. Since Alexander Jannaeus's kingdom sat astride the Transjordanian route to Damascus, he presently controlled the northward movement of trade goods; he also controlled the Mediterranean coastline all the way to Rhinocorura, south of Gaza on the Wadi el-Arish. Thus, the natural flow of goods from the southeast to the northwest, to the port of Gaza, was monitored by Alexander Jannaeus. As a result, several times during his governorship he met the Nabatean kings in battle, contesting supremacy over the trade routes.

On the negative side, during this time the internal conflict between the Sadducees and Pharisees came to a head. The governor/high priest sided with the Sadducees and, on occasion, went out of his way to offend the Pharisees. For example, during the celebration of the Feast of Tabernacles, instead of pouring the sacred water on the altar — as prescribed in Pharisaic tradition — he poured it on his feet. The worshipers at the temple responded by pelting him with lemons; Alexander, in turn, responded by massacring some 6,000 Jews. Indeed, using foreign mercenaries, Alexander Jannaeus fought with his own countrymen over a six-year period, with the result that almost 50,000 Jews were eventually killed in the conflict.

In 88 BC the Pharisees, who were the spiritual descendants of the Hasidim (the "pious ones"), were driven to appeal to the Seleucid ruler Demetrius III. The abortive rebellion failed and Alexander Jannaeus regained control of Jerusalem. He did not treat his opponents mercifully. Eight hundred Pharisees were crucified while their wives and children were put to death before their eyes and while Alexander openly engaged in an orgy with his concubines. This tragic incident illustrates the fact that the conflict between the Pharisees and Sadducees, evident even on the pages of the New Testament, was more than a theological dispute; its history was punctuated with literal life-and-death matters.

Before Alexander Jannaeus died, he realized that the Pharisees enjoyed popular support, and on his deathbed he instructed his wife Salome Alexandra to make peace with them. After his death, Salome Alexandra assumed civil rule (76 – 67 BC). Since she could not become high priest, she appointed Alexander Jannaeus's son Hyrcanus II to the position. Salome Alexandra made peace with the Pharisees as her husband had instructed her to do, and her rule as a whole was characterized by peace. However, there was constant agitation by Aristobulus II, the youngest son of Alexander Jannaeus, who wanted to have himself appointed high priest in place of his brother Hyrcanus II.

Within several months of the death of Salome Alexandra, Aristobulus II ousted Hyrcanus II from the highpriestly office and had him-

▲ *Mt. Gerizim: Monumental staircase from the second century BC that led up to the Samaritan temple on the top of Mt. Gerizim. The temple was destroyed by John Hyrcanus in 110 BC (compare John 4:20).*

self appointed to that position, as well as to the governorship (67–63 BC). During his tenure as high priest Aristobulus II and Hyrcanus II, with Idumean backing, vied for religious and political control of Judea.

During this internal conflict the Romans were making their presence felt in the area, and both Aristobulus II and Hyrcanus II appealed to the Romans for support. When Pompey himself arrived in Damascus (63 BC), the two brothers appeared before him, each hoping to be appointed high priest and ruler; this time, however, the Pharisees appeared also, requesting the abolition of Hasmonean rule.

Pompey seems to have delayed his decision as he undertook a campaign against the Nabateans. However, when Aristobulus II abandoned this campaign, Pompey turned on him and marched into Judea. Pompey marched south through the Jordan Valley to Jericho and from there on up to Jerusalem. The backers of Hyrcanus II opened the gates of the upper city (the western hill) to Pompey, but the followers of Aristobulus II held out on the temple mount. After a three-month siege the Romans entered the temple area, and 12,000 Jews were killed. In spite of the battle, the priests continued to offer sacrifices to the very end.

After the capture of the temple, Pompey himself entered the Most Holy Place, but soon he ordered the temple to be cleansed and sacrifices to be reinstated. Aristobulus II and most of his family were taken as prisoners to Rome, and Hyrcanus II was installed as high priest for a second time (63–40 BC; he previously served from 76 to 67 BC, see above), though with much more limited powers, for Judea and Jerusalem were now firmly under Roman control.

Thus in 63 BC the Hasmonean state, independent since 142 BC, officially ceased to exist. For the next century and a half, many Jews suffering under Roman rule nostalgically looked back on the Maccabean revolt and Hasmonean rule as the "golden days," when God had worked on behalf of his people. They remembered those years as a time when a "united" Jewish community drove off the Seleucids, a great world power. These selective memories would fuel the hearts of the Jews in AD 66 and AD 132, as the few and the weak, just as in the old days, would again attempt to throw off the yoke of their oppressors.

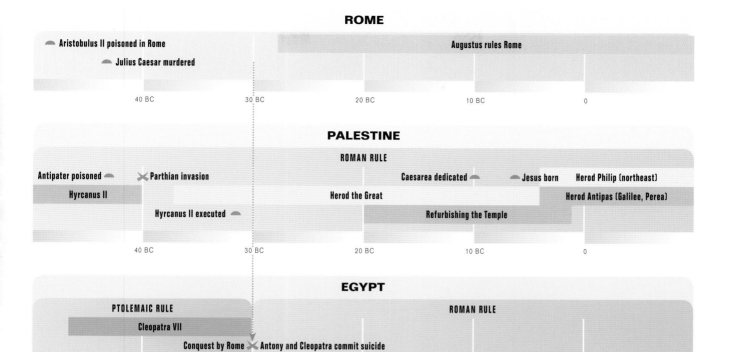

ROME

Aristobulus II poisoned in Rome
Julius Caesar murdered
Augustus rules Rome

40 BC   30 BC   20 BC   10 BC   0

PALESTINE

ROMAN RULE

Antipater poisoned   Parthian invasion   Caesarea dedicated   Jesus born   Herod Philip (northeast)

Hyrcanus II   Herod the Great   Herod Antipas (Galilee, Perea)

Hyrcanus II executed   Refurbishing the Temple

40 BC   30 BC   20 BC   10 BC   0

EGYPT

PTOLEMAIC RULE   ROMAN RULE

Cleopatra VII

Conquest by Rome   Antony and Cleopatra commit suicide

40 BC   30 BC   20 BC   10 BC   0

EARLY ROMAN RULE IN PALESTINE

When Pompey withdrew from the Near East, he left behind a proconsul to govern the province of Syria, and the sphere of Jewish control was greatly reduced. To the north of Judea, the Romans granted limited independence to the Samaritans, and the Valley of Esdraelon was detached from Galilee. Along the Mediterranean coast, cities from Raphia in the south to Dora in the north were granted autonomous status directly under the proconsul. In fact, the Jewish port of Joppa was detached from Judea.

Cities to the east of the Jordan River — including Gadara, Hippus, Abila, Dium, Pella, and Gerasa — along with Scythopolis to the west of the Jordan, were also freed from Jewish control, and their Gentile populations (which had been exiled by the Maccabees and Hasmoneans) were encouraged to return. Subsequently some of these Greek cities banded together into a league called the Decapolis ("ten cities"). The date of origin of this league of ten cities (the number varied from time to time) is disputed. The Romans may have encouraged the growth of these Greco-Roman cities in order to protect and control the Transjordanian Highway and its connections to the Mediterranean, to protect the settled areas from raiding desert tribes, and to counterbalance the strong Jewish population west of the Jordan River. To the north and east of the Sea of Galilee, the Itureans were granted the Lake Semechonitis, Paneas, and Gaulanitis areas but lost their outlets to the Mediterranean Sea. In the aftermath of these changes, Jewish territory was limited to Judea proper, to the mountains and desert south and east of Judea (= eastern Idumea), to Perea, and to a portion of Galilee.

The Romans appointed Hyrcanus II (63 – 40 BC) as high priest and left him in charge of Jewish affairs. Early in the tenure of Hyrcanus II his brother Aristobulus II made several attempts to seize power in Palestine, but each time Gabinius, the proconsul of Syria (57 – 55 BC), was successful in subduing the rebels.

As proconsul, Gabinius oppressed those under him, and at one time he attempted to divide Jewish territory into five districts that had their headquarters at Sepphoris (Galilee), Ammathus (Perea), Jericho, Jerusalem, and Adora (eastern Idumea). However, this arrangement was soon cancelled by decrees of Caesar. Gabinius's replacement, M. Licinius Crassus (54 – 53 BC), continued to oppress the Jews and confiscated 2,000 talents of gold and 8,000 talents of precious objects from the temple in Jerusalem.

During the rule of Hyrcanus II (63 – 40 BC), the Roman Empire was racked by civil strife, which began when Caesar crossed the Rubicon in 49 BC and lasted until the death of Antony in 30 BC. In 49 BC Caesar enlisted the help of the Hasmonean Aristobulus II to fight Pompey's followers in Palestine. But Aristobulus II was poisoned before he was able to leave Rome, and soon after that one of his sons, Alexander, was beheaded at Pompey's command. By 48 BC Caesar was gaining the

Tyre

Paneas

ITUREANS

L. Semechonitis

PHOENICIA

GAULANITIS

Ptolemais

GALILEE

Sea of Galilee

Sepphoris

Hippus

Philoteria

Dium

Valley of Esdraelon

Gadara

Abila

Dora

DECAPOLIS

Strato's Tower

Scythopolis

Pella

Mediterranean Sea

Samaria

Gerasa

Shechem

Ammathus

Mt. Gerizim

SAMARITANS

Jordan R.

Apollonia

Alexandrium

Philadelphia

Joppa

Thamna

PEREA

Lydda

Gophna

JUDEA

Jamnia

Emmaus

Jericho

Esbus

Azotus

Jerusalem

Medeba

Ascalon

Machaerus

Hebron

Gaza

Adora

IDUMEA

Dead Sea

Masada

Raphia

N A B A T E A N S

■ Autonomous cities directly
under the proconsul

□ Sphere of Jewish control

0   10 km.

0   10 miles

upper hand against Pompey, pursuing him to Egypt. Caesar received welcome assistance when Hyrcanus II, the high priest in Jerusalem, instructed the Jews of Egypt to support Caesar. In addition, Antipater the Idumean, the power behind Hyrcanus, aided Caesar's ally, Mithridates of Asia Minor, in the siege and capture of Pelusium in the eastern delta of the Nile. In return for their assistance, Caesar confirmed Hyrcanus II as high priest and ethnarch and appointed his colleague Antipater as procurator. Caesar also granted permission to rebuild the walls of Jerusalem, and it is possible that the port of Joppa and the Valley of Esdraelon were returned to Jewish control at about this time.

Antipater secured his hold on the country by appointing his sons Phasael and Herod as governors in Jerusalem and Galilee. Herod demonstrated his ability to rule by vigorously subduing rebellious elements in Galilee.

In 44 BC Julius Caesar was murdered, and civil war resumed in Rome. Cassius — who was one of the contenders in the civil war and who had become ruler of Syria — imposed a heavy tribute on his subjects, including the Judeans, in order to finance his war efforts. Both Antipater and Herod were useful to him in raising taxes in Palestine.

With the defeat of Cassius at Philippi in 42 BC, Antony became master of Roman holdings in Asia. On two occasions Jewish leaders from Jerusalem requested him to dismiss Phasael and Herod, but Hyrcanus II was able to have them both confirmed in their positions. The concessions gained by Hyrcanus were short-lived, however, for in 40 BC the Parthians invaded Palestine. The Parthians installed Antigonus II, a Hasmonean, as king and high priest in Jerusalem; Hyrcanus II's ears were mutilated, so that he was no longer fit to be high priest, and he was taken to Parthia as a prisoner.

Phasael, Herod's brother, committed suicide while in Parthian captivity. As for Herod, he fled south from Jerusalem toward Tekoa and Orhesa, then eastward to the rock fortress of Masada. Leaving Joseph his brother there to defend his family, Herod continued on to Egypt. Braving fall storms on the Mediterranean, Herod made his way to Rome. There he was warmly received by Octavian and Antony, who were able to persuade the senate to appoint him king of Judea and to add Samaria and western Idumea to his realm.

The first portion of Herod's rule — from 40 BC until 37 BC — was a period of conquest, during which Herod fought to gain control of the territory that the Romans had granted him. By the time he landed at Ptolemais in 39 BC the Parthians had already retreated from Palestine; however, much of the Jewish population was anti-Herodian, and Antigonus II, the Hasmonean, ruled in Jerusalem.

Herod began his conquest of Palestine by moving south along the coastal plain, capturing the port city of Joppa, which had been in Jewish hands. He then proceeded inland to rescue his family from Masada, the fortress on the western shore of the Dead Sea where they had sought refuge while he was in Rome. This accomplished, he marched against Antigonus II in Jerusalem, but his first attempt to capture the city failed. Herod then left off the siege of Jerusalem and marched through Samaria to Galilee. After capturing the city of Sepphoris in a snowstorm, he spent the winter of 39/38 BC there. During this period Galilee was subdued by Herod's forces, in spite of strong local opposition.

In 38 BC Herod visited his patron Antony in Samosata (a city in ancient Commagene, located in what is now southeastern Turkey) in order to secure support for his military activities. While he was away, the Roman commander Macherus moved eastward from the coast of Palestine and captured Emmaus, which guarded the roads that led up to Jerusalem from the west. At about the same time, Herod's younger brother, Pheroras, moved south down the Jordan Valley, and after refortifying the Alexandrium he set up his camp at Jericho to secure the eastern approaches to Jerusalem.

Upon his return to Palestine, Herod led his troops from Jericho up into the mountains north of Jerusalem. There, at Isana, Antigonus's commander Pappus was defeated. With this victory the noose around Jerusalem was being tightened: Herod's people, the Idumeans, controlled the territory to the south of Judea and Jerusalem, while the Romans held Emmaus on the west, and Herod controlled the Benjamin Plateau to the north and Jericho on the east. The actual siege of Jerusalem began in the winter of 38/37 BC, and the city fell into Herod's hands in the summer of 37 BC. In the end, Antigonus was beheaded and Herod was established as ruler in Jerusalem.

Following this period of conquest there was a period of consolidation of Herod's kingdom (37 BC until 25 BC). Internally, Herod faced considerable opposition from the Pharisees, from remnants of the Hasmonean family, and from portions of the populace and aristocracy. Throughout this period of consolidation, opponents were systematically eliminated or neutralized. For example, Herod executed forty-five members of the Sadducean aristocracy who had been supporters of Antigonus and confiscated their property to pay the demands of his overlord, Antony.

Herod, of Idumean descent, was never accepted by the Jewish population at large as a true Jew. In fact, the "Jewishness" of all Idumeans was suspect, because they had been forcibly converted to Judaism during the reign of John Hyrcanus (134–104 BC). In an attempt to legitimize his claim to the kingship, Herod married the Hasmonean Mariamne during the siege of Jerusalem in 38 BC. Although Mariamne was never directly implicated in any intrigues against her husband, her Hasmonean relatives were. Indeed, her mother, Alexandra, was able to secure a Hasmonean foothold in the religious-political structure of the government by managing to have her seventeen-year-old son Aristobulus appointed as high priest. Aristobulus was popular with the people, who saw him as a legitimate Jewish replacement for Herod, the Idumean. Enraged by this, Herod arranged for some of his friends to hold the young Aristobulus under water too long while they were swimming in one of the pools in Jericho. In spite of Herod's feigned grief over the "accidental" death of Aristobulus, it was well known that he, in fact, was the instigator of the deed.

During this period of consolidation additional members of the Hasmonean family were eliminated. Herod even had Mariamne, his beloved wife, executed in 29 BC out of intense jealousy, incited by rumors that were spread by his sister Salome and his mother, Cyprus. Within a year, Mariamne's mother, Alexandra, who had attempted to seize fortified positions in Jerusalem, was executed as well. The former high priest and friend of Herod, the aged Hyrcanus, had been executed previously (ca. 31 BC), and soon his male descendants were

Sidon

Damascus

Mediterranean Sea

Tyre

S Y R I A

P H O E N I C I A

U L A T H A

Paneas

Meroth

ITUREANS

Ptolemais

GAULANITIS

GALILEE

Sea of Galilee

B A T A N E A

TRACONITIS

Gaba

Tiberias

Hippus

Dion

A U R A N I T I S

Mt. Carmel

Sepphoris

Abila

Jezreel Valley

Gadara

Caesarea Maritima
(Strato's Tower)

Scythopolis

D E C A P O L I S

S A M A R I A

Pella

Jordan R.

Sebaste
(Samaria)

Gerasa

▲ *Mt. Gerizim*

Ammathus

Alexandrium

P E R E A

Joppa

Antipatris

Phasaelis

Philadelphia

Jamnia

J U D E A

Jericho

Esbus ✦

Emmaus

Jerusalem

Cypros

Azotus

Bethlehem

Ascalon

Hyrcania

Herodium

Betogabris

Tekoa

Callirrhoe

Anthedon

Machaerus

Gaza

Hebron

Orhesa

*Dead
Sea*

I D U M E A

Beersheba

Masada

Malatha

N A B A T E A N S

✦	Military colony founded by Herod
▪	Herodian fortress
	Herod's kingdom at the start of his reign
	Additions to Herod's kingdom

0   10 km.

0   10 miles

also eliminated, partially on the basis of rumors spread by Salome and Cyprus. Thus by the end of this period of consolidation (ca. 25 BC), most of the internal threats to his kingship had been removed.

In addition, Herod had to face a formidable external threat from Egypt in the person of Cleopatra, who had designs on reviving and expanding the Ptolemaic empire into Palestine and Arabia. Antony, her lover and the master of the east, submitted to her requests and granted her large portions of Herod's territory and of Arabia in 35 BC. In this way she gained control of the coastlands of Phoenicia and Philistia, save for the cities of Tyre and Sidon. This severed Herod's direct connections to the Mediterranean Sea and thence to Rome.

To the east of Herod's kingdom Antony gave Cleopatra the rich oasis of Jericho, where there were palm and balsam plantations and where perfumes were manufactured. Because she controlled both the Mediterranean coast and part of Arabia, Cleopatra was, in fact, able to control the lucrative spice and incense route that led from southern Arabia to the sea and from there to Rome.

Cleopatra seems to have had designs on taking total control of Herod's kingdom, and she might eventually have been successful had it not been for significant changes internationally. Civil strife within the Roman Empire led to Antony's defeat at the hand of Octavian in 31 BC at the naval battle near Actium (Greece), and in 30 BC she and Antony committed suicide rather than face the wrath of Rome. While Antony and Cleopatra were losing power, Herod skillfully changed his allegiance from Antony to Octavian so that when the latter emerged victorious, Herod was in a position to benefit. And benefit he did, for Octavian returned to him the territories that he had lost to Cleopatra, including such cities as Jericho, Gadara, and Hippus on the east, Samaria in the central portion of his realm, and Gaza, Anthedon, Joppa, and Strato's Tower along the Mediterranean coast. Thus, although at the midpoint of this period of consolidation Herod was in danger of losing his whole kingdom to Cleopatra, he was able to emerge from the crisis in a strong position, both politically and territorially.

The third period of Herod's reign (25 – 14 BC) can be characterized as a period of prosperity, construction, and additional territorial gains. It is interesting to note that Herod did not add to his kingdom primarily through military conquest but rather through territorial grants from his Roman overlords. As noted above, in 30 BC Octavian returned to him cities that had been given to Cleopatra by Mark Antony. In 23 BC Herod was entrusted with the task of subduing a band of marauders in Batanea, Auranitis, and Trachonitis; having accomplished this, these territories were added to his realm. Then in 20 BC, Gaulanitis, Ulatha, and Paneas, some of the holdings of Zenododrus, the last king of the Itureans, were transferred to Herod's rule.

At that point Herod's kingdom had reached its greatest extent. In the south, its common boundary with the Nabatean kingdom ran somewhere in the vicinity of the Nahal Beersheba near the fortress Malatha. Along the Mediterranean Sea his territory stretched from south of Gaza to slightly north of Caesarea, his holdings only interrupted by the free city of Ascalon. Although the tip of Mount Carmel and the Plain of Acco were excluded from his kingdom, large portions of Lower and Upper Galilee were included. The northernmost limit of his kingdom was at Paneas. To the northeast his kingdom included Gaulanitis, Batanea, Auranitis, and Trachonitis. Herod did not hold the cities of the Decapolis nor the kingdom of the Nabateans, which lay east of the Jordan River and Dead Sea. However, Perea, a strip of land immediately east of the Jordan, stretching from approximately Ammathus in the north to Machaerus in the south, was included in his kingdom.

The internal organization of Herod's realm can only be outlined, but it included large traditional administrative units — Idumea, Judea, Samaria, Galilee, and Perea. In addition, newly acquired territories (including Gaulanitis, Batanea, Auranitis, and Trachonitis) were administered directly by the king as military districts. In these largely uninhabited regions Herod was able to settle people who would be indebted and loyal to him. Finally, there were royal estates scattered throughout the country, the most important of which were situated in the fertile Jezreel Valley and in and around Jericho.

Because of his early fear of Cleopatra and his constant fear that his Jewish subjects might revolt, or even that his Nabatean neighbors might attack, Herod took a number of measures to secure his kingdom. He established at least two military colonies: one at Gaba, which guarded the northwest entrance to the Jezreel Valley, and one at Esbus (= OT Heshbon), which protected his eastern frontier from Nabatean aggression. In addition, Herod built or rebuilt a string of fortresses in or near the wilderness that overlooked the Rift Valley from the west. These included the Alexandrium, Cypros, Hyrcania, Herodium, Masada, and Malatha, as well as Machaerus east of the Dead Sea. These fortresses were normally furnished with royal amenities (such as palaces and baths), and troops were garrisoned in or near them. They not only could be used to control nearby territory but also could serve as places of retreat and security if Herod had to flee the country (e.g., Masada) or as prisons (e.g., Hyrcania), and one even served as a mausoleum (the Herodium).

Another technique Herod used to neutralize the potential threat from the Jewish population was to build and rebuild cities along Greco-Roman lines and settle Gentiles in them. For example, Samaria was rebuilt, renamed Sebaste (the Greek name for Augustus, the emperor), and settled with Gentiles. There, as well as in other Greco-Roman cities, Herod built temples to Augustus and to pagan deities, as well as theaters, amphitheaters, and other structures. Thus, he could retreat from the Jewish population with their restrictive scruples to places like Sebaste in order to indulge in a pagan lifestyle. It is easy to understand why Alexander and Aristobulus, the two sons of his Hasmonean wife, Mariamne, were executed at Samaria rather than in a predominantly Jewish city such as Jerusalem.

With the return of most of the Mediterranean coastline, Herod moved to reestablish a secure port for himself from which he could maintain constant contact with Rome. He was also anxious to have a port in the northern part of his country from which he could export grain crops to Rome and elsewhere from his Gaulanitis, Batanea, Auranitis, and Trachonitis territories and from his royal estates in the Jezreel Valley. There were several cities that could have served as ports for Herod, but of the two major candidates, Joppa was predominantly Jewish in character and too far south while Ptolemais in the north was

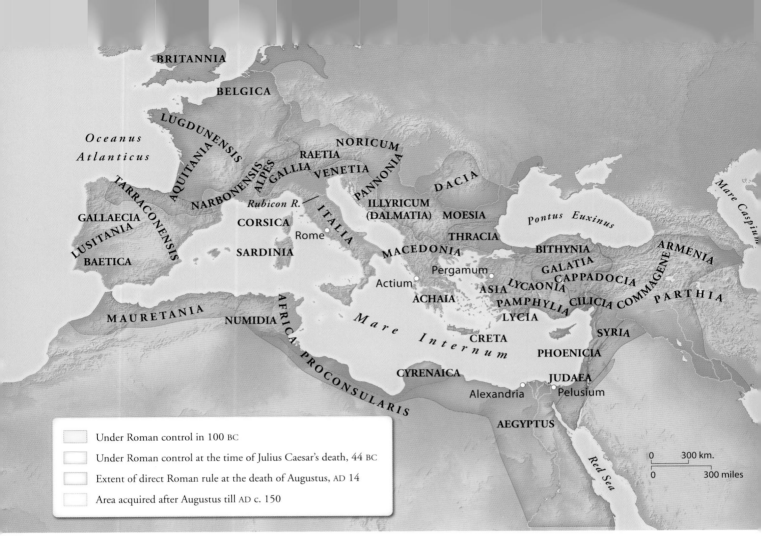

▼ *Model of the temple in Jerusalem that Herod the Great refurbished and which was standing in the days of Jesus — up until its destruction by the Romans in AD 70.*

in Phoenician and Syrian hands. Thus he selected a small landing called Strato's Tower as the place where he constructed a new port to suit his purposes.

Strato's Tower was well situated, for an easy pass through the Mount Carmel range connected it with the Jezreel Valley and the rich agricultural areas northeast of the Sea of Galilee. In addition, it had convenient connections with Sebaste and Antipatris to the east and south. Thus Herod built Caesarea Maritima at Strato's Tower, naming it after the emperor. To do this, swamp land was drained and fresh water was brought, via tunnels and aqueducts, from springs located at the foot of Mount Carmel, 12 miles to the northeast. A huge port, larger than Piraeus, the port of Athens, was built, including breakwaters, docks, and quays. Magnificent buildings were constructed in the city, such as a temple to Augustus, a theater, an amphitheater, and other public structures. It took more than twelve years to build the city, which was dedicated in 10 BC with special cultural activities, including athletic contests, theatrical performances, and gladiator fights, to mark the occasion. Over the ensuing years Caesarea continued to expand and grow in importance, so that it soon became the capital of the country, a position it would hold for almost 600 years.

▲ *Herodium: Interior courtyard of Herod the Great's pleasure palace. Note large circular eastern tower. Apartments once ringed the area.*

Herod lavished a great deal of his attention and energy on Jerusalem. There, in 20 BC, he initiated one of his great building projects — the refurbishing of the entire temple area. The temple platform was almost doubled in size, and great colonnades were built around its perimeter. Even the sacred temple itself, along with related buildings, was rebuilt and beautified. However, always wanting to remain in control of the populace, Herod also strengthened the Antonia fortress, which overlooked and controlled all of the temple precincts. For himself he built a magnificent palace on the western edge of the western hill and fortified the approach to it from the north by building three huge towers, which he named after Hippicus (a friend), Mariamne (his wife, whom he had executed), and Phasael (his brother). In addition, he either built or rebuilt the theater, amphitheater, and stadium, and he probably constructed a portion of the aqueduct that brought fresh water to Jerusalem from springs to the south.

During this prosperous period strong contacts were maintained with Rome, where two of his sons were being educated. Herod him-self was busy either visiting or hosting close personal friends of the emperor, and the period seems to have reached its climax with the visit of the emperor's close associate, Agrippa, to Herod's kingdom in 15 BC.

During the fourth period of Herod's rule (15 BC – 4 BC) major building projects, such as the remodeling of the temple, continued, while others (e.g., the construction of Caesarea) were completed. Friendly relations were maintained with Rome. The major concern of this era, however, was the question of which of Herod's sons would succeed their aging father as king.

The prime contenders for the throne were Antipater, the son of Doris, and Alexander and Aristobulus, two sons of Mariamne. During this time of intrigue, plots, slander, and duplicity, Herod drew up at least six wills, naming first one and then another of his sons as his successor. But in the end, Herod had Alexander and Aristobulus executed

▼ *Herodium: Seven miles south of Jerusalem. A combination pleasure palace, fort, and mausoleum built by Herod the Great.*

▲ *Herodium: Recently discovered foundation monument of the tomb of Herod the Great*

by strangulation at Samaria (7 BC), and ordered the death of Antipater only five days prior to his own death in 4 BC.

Toward the end of this period Herod's health deteriorated rapidly. Even the hot baths of Callirrhoe, east of the Dead Sea, could not bring relief to the agony of his incurable disease, and Herod, in much pain, died in Jericho in the spring of 4 BC. Even though the Jewish population rejoiced at his death, his family and soldiers gave him a lavish funeral, carrying his body with great pomp in a jewel-studded gold coffin from Jericho to his mausoleum, the Herodium.

At his death, Herod left behind a kingdom that was economically and materially prosperous. Many of the building projects that he had undertaken to aggrandize himself and to ingratiate himself with Rome were of such magnitude that remains of them have been preserved down through the centuries. Throughout his reign he attempted to maintain good relations with the Romans; because of this there was a measure of peace and stability in his kingdom. However, this peace was maintained at a price, for those of his subjects who agitated against his government or against Roman rule were dealt with harshly. Even among his own family he executed two of his ten wives, at least three of his sons, his brother-in-law, and one of his wives' grandfather.

Thus, when Magi from the east appeared in Jerusalem in Herod's presence asking, "Where is the one who has been born king of the Jews?" it is no wonder that "when King Herod heard this he was disturbed, and all Jerusalem with him" (Matt 2:1 – 3). His slaughter of the baby boys of Bethlehem to remove a possible threat to his throne (vv. 16 – 18) was certainly in keeping with his character. Thus, from one standpoint, he could be called "Herod the Great," but from other standpoints he might be called "Herod the Despicable." And from an eternal standpoint, the lasting achievements of the one buried in the Herodium — a magnificent stone monument — can in no way be compared with those of the One born at Bethlehem, almost within the shadow of the Herodium.

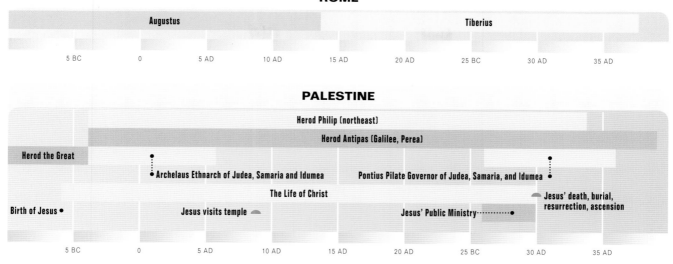

ROME

Augustus					Tiberius			
5 BC	0	5 AD	10 AD	15 AD	20 AD	25 BC	30 AD	35 AD

PALESTINE

Herod Philip (northeast)

Herod Antipas (Galilee, Perea)

Herod the Great

Archelaus Ethnarch of Judea, Samaria and Idumea

Pontius Pilate Governor of Judea, Samaria, and Idumea

The Life of Christ

Jesus visits temple

Jesus' death, burial, resurrection, ascension

Birth of Jesus •

Jesus' Public Ministry

| 5 BC | 0 | 5 AD | 10 AD | 15 AD | 20 AD | 25 BC | 30 AD | 35 AD |

THE LIFE OF CHRIST

At the time of the death of Herod the Great the question of who would succeed him had still not been resolved, although his sixth and final will designated Archelaus as king of Idumea, Judea, and Samaria; Antipas as ruler in Galilee and Perea; and Philip as governor of the lands northeast of the Sea of Galilee. But before Archelaus could assume the kingship, his appointment had to be confirmed, and to that end both he and Antipas traveled to Rome. In the meantime, anti-Herodian and anti-Roman elements of the population revolted, and it took the personal intervention of Varus, the legate of Syria, plus three legions, to suppress the rebellions.

In Rome, Archelaus, Antipas, and Philip were confirmed as rulers over their designated territories. However, Archelaus was not given the title of king but rather the lesser title of "ethnarch" (meaning "ruler of the nation"). Archelaus's ten-year rule (4 BC – AD 6) was a brutal one, to the extent that he even slaughtered some of his Jewish and Samaritan subjects. In the Jordan Valley he built a village north of Jericho and had the audacity to name it after himself — Archelais. It was during his unjust rule that Mary, Joseph, and the baby Jesus returned to Galilee from Egypt. It is no wonder that they avoided returning to Judea, for they were "afraid to go there" when they heard that Archelaus was ruling in place of his father (Matt 2:19 – 23). Instead, they proceeded to Galilee, the territory of Antipas, and settled in the village of Nazareth.

Herod Antipas (4 BC – AD 39) ruled over both Galilee and Perea. Each of these two disjointed territories had a considerable Jewish population, which he was able to keep in check. Soon after taking control of

Galilee and Perea, he rebuilt the city of Sepphoris (3 BC – AD 10), which served as his first capital.

The Galilee ruled by Antipas stretched from the Esdraelon Valley in the south to the Baca/Gischala area in the north, and from Thella in the Rift Valley on the east to Chabulon overlooking the Plain of Acco on the west (Josephus, *War* 3.3.1 – 2 [35 – 43]). The area north of Bersabe and Kefar Hananiya was higher in elevation and was called Upper Galilee by Josephus. This region of mountains and deep valleys was remote, and the outside contacts of its inhabitants may have been directed as much toward Tyre to the northwest as to Ptolemais

▼ *Nazareth: Church of the Annunciation surrounded by the hills rising above Nazareth*

Sidon

Damascus

Tyre

TYRE

ULATHA

Caesarea
Philippi

Raphana

Gischala

Thella

GAULANITIS

BATANEA

TRACONITIS

Ptolemais
(Acco)

UPPER GALILEE

Baca

Bersabe

Kefar Hananiya

Chabulon

Gabara

Taricheae

Sea of Galilee

Jotapata

LOWER
GALILEE

Hippus

Dion

Canatha

Sepphoris

Kafr Kana

Tiberias

AURANITIS

Mediterranean Sea

Dora

Esdraelon Valley

Nazareth
Japhia

Gadara

Abila

Edrei

DECAPOLIS

Bostra

Caesarea

Scythopolis

Pella

Sebaste

Jordan R.

Gerasa

Apollonia

Mt. Gerizim

Ammathus

SAMARIA

PEREA

Joppa

Antipatris

Lydda

Archelais

Gadora

Jamnia

JUDEA

Neara

Philadelphia

Jericho

Abila

Azotus

Emmaus
(Nicopolis)

Beth-ramatha

Ascalon

Jerusalem

Bethany, on the
other side of
the Jordan

IDUMEA

Gaza

Hebron

Machaerus

Dead
Sea

0   10 km.

0   10 miles

Masada

NABATEANS

Beersheba

◆ Cities of the Decapolis (Pliny)

Territory under Antipas

Territory under Philip

Territory under Procurator of Judea

Territory under the Proconsul of Syria

directly to the west. To the south, Lower Galilee was much more open to outside influence, and its broad, spacious valleys provided good land for growing grain crops. Josephus, writing of both Upper and Lower Galilee, describes them as having soil so fertile that even the laziest of their inhabitants could make a good living.

Galilee must have been intensively cultivated and densely populated in Jesus' day, although Josephus's statistic of 15,000 people living in even the smallest villages (which would imply that Galilee had 3,000,000 inhabitants) must be grossly inflated (*War* 3.3.2 [41 – 44]). There may, in fact, have been a population of approximately 300,000 living in the two hundred or so cities and villages that Josephus says were in existence. As Jesus was growing up, the largest of the cities was Sepphoris, which may have had a population of 50,000.

▲ *Remains of a first-century boat that was excavated on the northwestern shore of the Sea of Galilee*

Antipas had furnished the city with Greco-Roman institutions (e.g., a theater), and its beauty was such that Josephus called it the "ornament of all Galilee" (*Antiq* 18.2.1 [27]). This city overlooked valuable farmland and was close to an important east – west route that connected the cities of the Decapolis and the Gaulanitis regions with the port of Ptolemais. Other large cities of Galilee included Tiberias, Taricheae, Gabara, Jotapata, and Japhia.

Jesus was raised in the small village of Nazareth, only 3.5 miles southeast of the capital, Sepphoris. Although Nazareth itself was small and insignificant, its residents probably had numerous contacts with their more cosmopolitan neighbors. In all probability they came into contact with some of the caravans and Greek-speaking Gentile traders who passed through Sepphoris on the north and/or the Esdraelon Valley (= OT Jezreel Valley) on the south.

When Jesus began to minister at about the age of thirty (ca. AD 26), he spent much more time in Lower than in Upper Galilee. It is recorded in Luke 4:16 – 30 that during his public ministry, Jesus visited and taught in his hometown of Nazareth, where he was not welcome. He also ministered at Cana of Galilee, which has been identified as the antiquity site of Khirbet Qana, located 8 miles north of Nazareth, although first-century remains have been found at the more traditional site of Kafr Kana (4 mi. northeast of Nazareth). It was in Cana that he performed his first miracle — turning water into wine

at a wedding feast (John 2:1 – 11). On another occasion, also in Cana, he healed the son of a Roman official who had come to him from Capernaum (4:43 – 54), and it appears that Nathanael, one of his disciples, was originally from Cana (21:2).

It is about 12 miles from Cana to the Sea of Galilee, about a six-hour walk. It was there, along the northern shore of the Sea of Galilee, that Jesus spent much of his time during his public ministry. In those days (ca. AD 26 – 30) the largest city on the lake was the newly built city of Tiberias (Josephus, *Ant.* 18.2.3 [36 – 38]). Herod Antipas had

▼ *A full-scale model of the excavated Galilee Boat. This multipurpose boat could carry about 15 persons total.*

constructed this city between AD 18 and 22 and had transferred his capital there from Sepphoris. He had adorned the city with a luxurious palace for himself, a stadium, a synagogue, and possibly a city wall. He had hoped to settle Jews in the city, but religious Jews were reluctant to live there because it was rumored to have been built over a cemetery; thus the city must have had a mixed Jewish and Gentile population at best. To the south of the city were hot baths noted for their therapeutic properties. The Gospels never record that Jesus entered Tiberias, although the Sea of Galilee is called the "Sea of Tiberias" in John 6:1 and 21:1, and, on one occasion, boats from Tiberias are said to have arrived with passengers wanting to see Jesus (6:23).

Four miles to the northwest of Tiberias, along the western shore of the sea, is the probable site of the New Testament city of Magadan (Matt 15:39). This is the place visited by Jesus after the feeding of the 4,000 on the other side of the lake. The parallel text in Mark (8:10) has "Dalmanutha" in place of "Magadan," and it is not certain whether the two places were identical. Possibly Dalmanutha was the port of Magadan. Compounding the uncertainty regarding the name and location of the site is the fact that the KJV, using late Greek manuscripts, reads "Magdala" in Matthew, and early Christian tradition held the site to be the home of Mary Magdalene. In addition, the same site seems to be called "Taricheae" ("the place of salted fish") by Josephus, who refers to Taricheae as a large city on the shore of the sea and the location of a bloody battle between Jews and Romans during the first Jewish revolt (ca. AD 66 – 70; Josephus, *War* 3.10.1 – 10 [462 – 542]).

Proceeding from Magadan 6 miles in a clockwise direction around the north shore of the Sea of Galilee, one comes to the village of Caper-

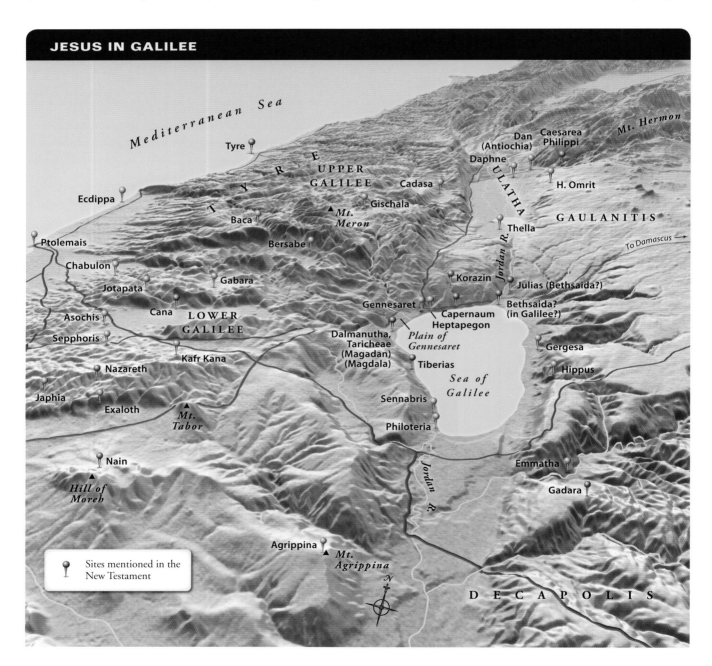

JESUS IN GALILEE

naum. Apart from Jerusalem, this is the most important of all the villages mentioned in the gospel accounts, for it was here that Jesus established his headquarters for the major portion of his public ministry. Three of his disciples were from Capernaum, and Peter and Andrew had evidently moved there from Bethsaida (Mark 1:29). According to archaeological investigations, the village spread out for a quarter of a mile along the lakeshore. It is probable that fishing was the major occupation of the inhabitants, and it is possible that basalt implements (e.g., olive presses and grain grinders) were produced there as well.

Capernaum sat astride the International Route that ran from the Mediterranean Sea to Transjordan and Damascus, and it seems that a custom station was located there because of its proximity to the Jordan River and Philip's territory (Matt 9:9). The town was large or important enough for a Roman centurion

▲ *Black basalt foundation wall of an earlier synagogue upon which the rebuilt white walls of the sixth-century synagogue at Capernaum are visible*

and his troops to be stationed there (8:5–9), as was a royal official (John 4:46). Jesus healed many there, including the servant of the centurion (Matt 8:5–13), the paralytic who was let down through the roof of a dwelling (Mark 2:1–12), and Peter's mother-in-law, who suffered from a fever (1:29–31). The Franciscans, who now own most of the site of Capernaum, have excavated a beautiful white limestone synagogue that dates from the fourth century AD; underneath it they have discovered the massive foundation walls of a black basalt synagogue that preceded it. This earlier synagogue probably dates back to the days of Jesus and was the one in which he preached while at Capernaum. Early Christian presence at the site is evidenced by the remains of several churches that were built over a house thought to have been the house of Peter.

▼ *"Sowers Cove" (Matt 13:1-2) on the northwestern shore of the Sea of Galilee.*

▲ *View of the Plain of Gennesaret and the Arbel Cliffs from the Mount of Beatitudes*

Although it is difficult to pinpoint exactly the location of many of Jesus' activities in the neighboring countryside, by the fourth century Christian tradition had localized the site of the Sermon on the Mount (Matt 5–7), the feeding of the 5,000 (14:13–21), and the appearance of the resurrected Lord to his disciples (John 21) near the place of seven springs — Heptapegon (Tabgha). This area, which is approximately 1.8 miles west of Capernaum, may indeed have been the site of these events, although the feeding of the five thousand probably occurred in the area northeast of the Sea of Galilee. In addition, between Capernaum and Tabgha there is a small bay on the seashore in the shape of a natural theater that may have been the spot where Jesus "got into a boat and sat in it, while all the people stood on the shore. Then he told them many things in parables, saying . . ." (Matt 13:2–3).

Two miles to the northwest of Capernaum are the remains of the city of Korazin. Although it is only mentioned as being cursed by Jesus because of its failure to repent, it is implied that Jesus had visited the place, for miracles had been performed there (Matt 11:20–24). The city is located in a basalt region, and all of the buildings were made out of the hard black rock. Excavations at Korazin have not yielded many finds from the time of Jesus, for many of the preserved remains were built in the second to the fourth centuries AD or later, including the black basalt synagogue.

The third important village/city that Jesus visited is Bethsaida. A possible location of this city is the mound called et-Tell, which is located east of the Jordan River about 1.5 miles before it enters the Sea of Galilee. This city was built by Philip, the son of Herod the Great, early in his reign (4 BC–AD 34), probably as a way station on the International Highway that led to the Mediterranean and as a port for himself on the Sea of Galilee (the ancient shoreline may have been farther north than it is now). He named the city Julias, after Julia, the daughter of Augustus. According to Josephus, Philip himself died there at the end of his peaceful rule (Josephus, *Ant.* 18.4.6 [108]).

Recently, additional support has been marshaled for the theory that there were in fact two Bethsaidas, one in Philip's territory (Bethsaida-Julias) and the other "in Galilee" (John 12:21). The latter has been tentatively identified with the small site of Araj, located close to the shore of the Sea of Galilee. It is proposed that at the time of Jesus the Jordan followed a more easterly course than it does now, so that Araj at that time was west of the river, i.e., "in Galilee." Only Jerusalem and Capernaum are mentioned more frequently in the Gospels than Bethsaida. Bethsaida was the early home of Peter, Andrew, and Philip (1:44; 12:21). There a blind man was healed (Mark 8:22–26), and in a deserted place nearby, probably along the northeast shore of the sea, the feeding of the 4,000 took place. It too, along with Korazin and Capernaum, was cursed by Jesus because of the unbelief of its inhabitants.

To the northeast of Bethsaida lay Philip's territory, which included Gaulanitis, Batanea, Auranitis, Traconitis, and Ulatha. During the days of Jesus, most of this territory was settled by Gentiles — except western Gaulanitis, which may have been Jewish — but it is evident that Jesus did not spend much time there. However, on at least one occasion he traveled with his disciples to the vicinity of Caesarea Philippi, about 25 miles north of Bethsaida. There, at the headwaters of the Jordan and at the foot of Mount Hermon, Herod the Great had built a white marble temple in honor of the emperor; and there his successor, Philip, built a large city that he named after the emperor — adding his own name to the title to distinguish it from Caesarea Maritima.

Philip made Caesarea Philippi the capital of his territory, and it must have been a thriving city, for it was situated along the highway that led from Damascus to Tyre and Sidon. It was in the vicinity of Caesarea Philippi that Peter made his "Great Confession," stating that he believed that Jesus was "the Christ, the Son of the living God" (Matt 16:13–20). Soon afterward Jesus was transfigured in the presence of Peter, James, and John (Matt 17:1–8; Mark 9:2–8; Luke 9:28–36). Because of the close sequence of events in the gospel narratives, it is possible that the transfiguration also occurred in this region, perhaps on Mount Hermon, or on one of the nearby peaks, such as the Hermonit. Other candidates for the place of transfiguration have been suggested, including Mount Meiron (Jebel Jarmak) and Mount Tabor, but none of them command as much textual or geographical support as the Mount Hermon vicinity.

To the south of Philip's territory was a region that came to be known as the Decapolis. This was a group of cities that were either founded or rebuilt as Greco-Roman cities during the Hellenistic and Roman periods. Their populations were mainly Gentile, although when they were controlled by the Jews for a short period of time — from the days of Alexander Jannaeus (103–76 BC) until the arrival of Pompey (63 BC) — they contained a large Jewish population. This group of cities was evidently originally ten in number — hence the name Decapolis ("ten cities") — but it often included more than ten cities in later years. It included, east of the Rift Valley from north to south: Damascus, Raphana, Canatha, Dion, Hippus, Abila, Gadara, Edrei, Bostra, Pella, Gerasa, and Philadelphia (modern Amman), as well as Scythopolis west of the Jordan.

The Decapolis was not an area frequently visited by Jesus, for he had said that his mission was primarily to the "lost sheep of Israel" (Matt 15:24), but on one occasion he healed two demon-possessed men (Matt 8:28), one of whom, after having been healed, went into the Decapolis to tell of all that Jesus had done for him (Mark 5:20). This event is recorded in all three of the Synoptic Gospels, but Mark and Luke seem to mention only the more prominent of the two men who were healed (Mark 5:2; Luke 8:27).

Aside from the problem of the number of men healed, various Greek manuscripts offer different readings as to the place where the healing(s) occurred: "the region of the Gerasenes [or Gadarenes or Gergesenes]" (Mark 5:1; Luke 8:26, cf. Matt 8:28). The identification of the site of the healing with the Decapolis city Gerasa (modern Jerash) is problematic, for Jerash is situated 35 miles south of the Sea of Galilee — too far south to be the correct site. The placement of the event near the Decapolis city of Gadara (modern Umm Qeis) is more plausible, since it is only 6 miles

southeast of the sea and would be more likely to have possessed lakeshore territory than Gerasa; however, this is not certain either. Since the fifth century, Christian tradition has placed the event at Gergesa (modern Kursi), which is on the eastern shore of the Sea of Galilee directly opposite Taricheae and Tiberias. There a monastery was founded to commemorate the healing, and it is easy to imagine a herd of swine (an indication that this was Gentile territory) hurtling down the nearby hills into the sea.

To the south and west of the Decapolis was the region called Perea. This is a shortened form of a Greek phrase that can be translated as "other side of the Jordan" or "regions across the Jordan." According to Josephus (*War* 3.3.3 [46]), it was bounded on the north by Pella and stretched to the south of Machaerus. On the west the Jordan River was its boundary, and on the east it approached, but did not include, the Decapolis city of Philadelphia. From the time of Jonathan's capture of the area (152 BC), the region was settled by Jews. Its capital was Gadora, and other prominent cities or forts included Ammathus, Abila, Bethramatha (Livias/Julias), Callirrhoe, and Machaerus. Herod Antipas was granted this territory after the death of his father and controlled both it and Galilee. Perea, Galilee, and Judea are called "the three Jewish provinces" in the Mishnah (written down around AD 200).

It is evident that Jesus ministered in Perea, since Luke 9:51 – 18:34 places quite a few events there. In addition, John was baptizing "at

▼ *Rock-cut sanctuary of Pan at Caesarea Philippi — near where Peter affirmed that Jesus was the Messiah (Matt. 16)*

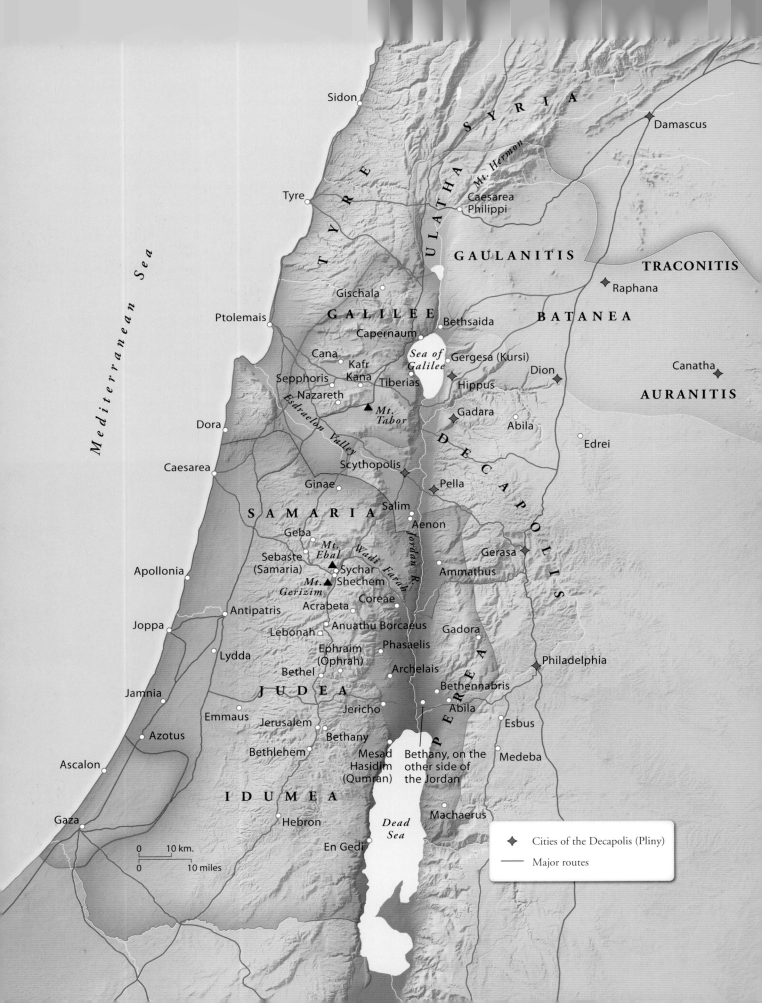

Sidon

Damascus

TYRE

Tyre

ULATHA

Mt. Hermon

SYRIA

Caesarea
Philippi

GAULANITIS

TRACONITIS

Raphana

Gischala

GALILEE

Bethsaida

BATANEA

Ptolemais

Capernaum

Cana

Kafr
Kana

Sea of
Galilee

Gergesa (Kursi)

Dion

Canatha

Sepphoris

Tiberias

Hippus

AURANITIS

Nazareth

Mt.
Tabor

Gadara

Abila

Dora

Esdraelon Valley

DECAPOLIS

Edrei

Caesarea

Scythopolis

Ginae

Pella

Salim

SAMARIA

Aenon

Geba

Gerasa

Sebaste
(Samaria)

Mt.
Ebal

Sychar

Wadi Farah

Jordan R.

Ammathus

Mt.
Gerizim

Shechem

Coreae

Gadora

Acrabeta

Antipatris

Lebonah

Anuathu Borcaeus

Phasaelis

Philadelphia

Apollonia

Joppa

Ephraim
(Ophrah)

PEREA

Lydda

Bethel

Archelais

Bethennabris

JUDEA

Jericho

Abila

Esbus

Jamnia

Emmaus

Jerusalem

Bethany

Azotus

Bethlehem

Mesad
Hasidim
(Qumran)

Bethany, on the
other side of
the Jordan

Medeba

Ascalon

IDUMEA

Machaerus

Gaza

Hebron

Dead
Sea

En Gedi

Mediterranean Sea

0   10 km.

0   10 miles

◆   Cities of the Decapolis (Pliny)

─── Major routes

Bethany on the other side of the Jordan" (John 1:28). This Bethany is difficult to locate precisely, but it may have been in the vicinity of Bethennabris or a spot closer to the Jordan, near the Wadi el-Kharrar, although a more northerly area is sometimes suggested. It was probably in the Wadi el-Kharrar, area that Jesus was baptized.

Later, the gospel writer notes that John was "baptizing at Aenon near Salim, because there was plenty of water" (3:23). This Aenon ("springs") is also difficult to identify with certainty. Some authorities suppose that John had not moved from his earlier spot and that Aenon should be located in the region of "Bethany beyond the Jordan." However, there is no Salim close by. Others, noting its literary context in John 3 – 4, locate the place in Samaria at the headwaters of the Wadi Farah, 5.5 miles north of the village of Salim, at or near the Old Testament site of Tirzah. But it seems best to follow the suggestion of the early church historian Eusebius (fourth century AD) that it was located in the Jordan Valley near Salim, 7 miles south of Scythopolis, just west of the Jordan River. This would place John's activities in the territory of the Decapolis, just outside the reach of Herod Antipas (who had been annoyed by his preaching) and of Pilate (who might have considered him a revolutionary).

In the end, Herod Antipas beheaded John because of his condemnation of Herod's divorce and remarriage. According to Josephus, the execution took place at the fortress of Machaerus, at the southern extremity of Perea (away from John's followers? Josephus, *Ant.* 18.5.2 [116 – 19]). However, Josephus may be wrong in his assessment, for a "birthday banquet" would more probably be held in Herod's palace in Tiberias.

Jews living in Perea probably had close contacts with Jerusalem, for it was an easy matter to cross the fords of the Jordan opposite Jericho, near where John had originally been baptizing, and then to pass through Jericho on the way up to Jerusalem. Since the days of Alexander the Great, Jericho had been something of a royal estate, generally owned and operated by the ruling monarch. The Hasmonean rulers had built palaces there; Cleopatra had rented it to Herod for a few years; and Herod and his son Archelaus built and rebuilt aqueducts, plantations, fortresses, palaces, and pools in the area. Herod had his wife's brother (Aristobulus) murdered in Jericho, and Herod himself died there. In Jesus' day the Romans controlled Jericho, and its plantations were spread out over a large area.

Jericho figures prominently in the Gospels. Jesus mentioned it in the parable of the Good Samaritan (Luke 10:25 – 37) and passed through it himself as he traveled from Perea to Bethany on his way to raise Lazarus from the dead (John 10:40 – 11:54). It was there that two blind men (Matt 20:29 – 34; Mark 10:46 – 52; Luke 18:35 – 43), including Bartimaeus (Mark 10:46), were healed, and it was there that

▼ *Wilderness of Judah east of Jerusalem. Jesus fasted for forty days in this area and passed through it on his way from Jericho to Jerusalem.*

▲ *Church and olive grove at the traditional site of the Garden of Gethsemane at the western foot of the Mount of Olives.*

Zacchaeus, the tax collector, was told to come down from a tree so that Jesus could dine with him in his house (Luke 19:1 – 10).

From Jericho a well-traveled road ran up to Jerusalem through the dry, chalky wilderness. Along this fifteen-mile stretch of road the parable of the Good Samaritan has its setting. After an uphill walk of six to eight hours from Jericho, one approached the back (east) side of the Mount of Olives. Slightly to the south of the old Roman road was the village of Bethany, the home of Mary, Martha, and Lazarus, today called el-Azariyeh (named after Lazarus). Jesus stayed in the house of these friends on many occasions, and events such as the teaching of Mary, the raising of Lazarus, and the anointing with precious oil took place there. It was in the Bethany/Bethphage area that Jesus mounted a colt and rode it into Jerusalem (on what is now commemorated as Palm Sunday), and later that same week he and the disciples made the same trip together for the last time, to the Upper Room in Jerusalem. During that final week of Jesus' life he spent several days teaching in Jerusalem, but it appears that he returned to Bethany every night.

The road from the east was, of course, not the only road leading into and out of Jerusalem. The Ridge Route led southwest out of Jerusalem toward Bethlehem, Hebron, and Beersheba. It passed through Judea to Idumea and then on into the Negev. From there one could proceed east to Transjordan and Arabia, southwest to Sinai, or west toward Egypt.

Other than the birth narrative, the Gospels do not record any activity of Jesus south of Jerusalem.

To the north of Jerusalem the Ridge Route headed toward Shechem. In Jesus' day the territory of Judea stretched some thirty-five miles north of Jerusalem to the area of Anuathu Borcaeus and Acrabeta. It was probably early in his public ministry in this region that "Jesus and his disciples went out into the Judean countryside [north of Jerusalem], where he spent some time with them, and baptized ... although in fact it was not Jesus who baptized, but his disciples" (John 3:22; 4:2). Late in his ministry, after raising Lazarus and after learning of a plot on his life, Jesus withdrew with his disciples to this same area, to a "village called Ephraim" (11:54). Although New Testament Ephraim has not been identified with absolute certainty, it is probably to be equated with Old Testament Ophrah (modern et-Taiyiba). This village was located east of the watershed on the edge of the wilderness and could have provided a convenient place of retreat.

To the north of Anuathu Borcaeus was the district of Samaria, which reached to the southern edge of the Esdraelon Valley, to the village of Ginae (modern Jenin). This district, like Judea and Idumea to the south, was governed in Jesus' day by the Roman official Pontius Pilate. The district was named after the Old Testament city of Samaria (then called Sebaste), and the Samaritans dominated large portions of the area. An important route that ran through the district of Samaria was used by the Jewish inhabitants of Galilee on their pilgrimages to Jerusalem (Josephua, *Ant.* 20.6.1 [118]). According to Josephus, this journey took three days at a minimum (*Life* 52 [269]), which implies at least two overnight stays along the way. After crossing the Valley of Esdraelon, heading south, Jewish pilgrims entered Samaria at Ginae. It was probably in this area, "along the border between Samaria and Galilee" (Luke 17:11), that Jesus met and healed ten lepers, one of whom was a Samaritan (vv. 12 – 19). Jesus may have been traveling "along the border" because he and his disciples had encountered Samaritan opposition as they attempted to enter Samaria; Josephus records that on at least one occasion Jews had met stiff resistance at Ginae (*Antiq* 20.6.1 [118]; *War* 2.12.3 [232]).

Typically, Jewish pilgrims would have continued south from Ginae toward Shechem. It seems reasonable to assume that they would have had to spend the night in the area of Geba, about a day's walk from Galilee. Where Jewish pilgrims or travelers would find acceptable accommodations is difficult to imagine. It does not seem reasonable to expect that they would have stayed in Samaritan or Gentile homes, so they may have had to camp out in the open. From the Geba region, the caravans would have continued south, passing Mount Ebal and Mount Gerizim on the next day's journey. Exiting the district of Samaria, they would have entered Judea before settling in for the night, possibly in the el-Lubban (= OT Lebonah) region. The third and final day of their journey would have taken them on into Jerusalem. It was probably near el-Lubban, at the overnight stop nearest Jerusalem, that Jesus' parents realized that he was missing from the caravan that was heading back to Galilee (Luke 2:41 – 50). Indeed, Jesus must have traveled this route frequently for "every year his parents went to Jerusalem for the Feast of the Passover" (v. 41).

On one memorable occasion early in his public ministry, Jesus stopped at "Jacob's well" near the town of Sychar (modern Askar) at midday (John 4:4 – 6); note that it is about a half-day's journey north from the el-Lubban overnight stop to Sychar. While his disciples went off to the village to purchase food, he met the Samaritan woman at the well. There, near the foot of the Samaritan holy mountain, Mount Gerizim, he pointed her to the real source of living water so that she, and others like her, could worship God in spirit and truth (vv. 4 – 42).

Although Jesus traveled the roads north and east of Jerusalem on a number of occasions, only one event is placed west of the city. This

was his appearance to the two disciples on the road to Emmaus (Luke 24:13 – 35). According to the best Greek manuscripts, Emmaus was 60 stadia (ca. 7 mi.) from Jerusalem. Since the Crusader period (AD 1099 – 1291), the villages of Abu-Ghosh and el-Qubeibeh have both been suggested as the biblical Emmaus, since they are each about 7 miles from Jerusalem. However, both traditions are late (Middle Ages).

A third site that is considered a candidate for the biblical Emmaus is near modern Qaloniya/Motza. This site is 3.5 miles west of Jerusalem on the Roman road that led from Jerusalem to Joppa. One ancient source that refers to it as Emmaus says that Vespasian, the Roman emperor, ordered the settling of 800 retired veterans in a colony there (Josephus, *War* 7.6.6 [216 – 17]) — hence the newer name Colonia/Qaloniya. If this is the biblical site, then the distance in Luke 24:13 (7 mi.) is the distance from Jerusalem to Emmaus and back, i.e., the distance of a round trip. However, this is a questionable interpretation of the passage in Luke.

A fourth possible site for biblical Emmaus is the city of Emmaus/Nicopolis, which was well known in Roman times. The name of the ancient city was preserved in the now-destroyed Arab village of Imwas, which overlooked the strategically important Aijalon Valley. However, this site is about 19 miles from Jerusalem. On the other hand, one important Greek manuscript reads "160 stadia" (= ca. 20 mi.) instead of "60 stadia" (= 7 mi.), which would place the biblical city at the same site as this well-known Roman city. It seems that the Imwas-Emmaus identification is the most probable, but it is not without difficulties.

It was back in the Jerusalem area, on the Mount of Olives, that Jesus departed from his disciples, concluding his earthly ministry (for Jesus in Jerusalem, see pp. 250 – 253). It is quite amazing to reflect on the worldwide significance of the message and work of this first-century itinerant Jewish prophet, especially when one considers that he only ministered for three or four years, that he left behind only a small band of loyal followers, and that his ministry was primarily confined to a rather small province of the Roman Empire, located on its distant southeastern frontier. But the New Testament writers were anxious to establish that it was not through the might of Herod the Great, nor through the power of the Roman emperors, but through Jesus that all of the nations of the earth would be blessed (Gen 12:3; Gal 3:6 – 15).

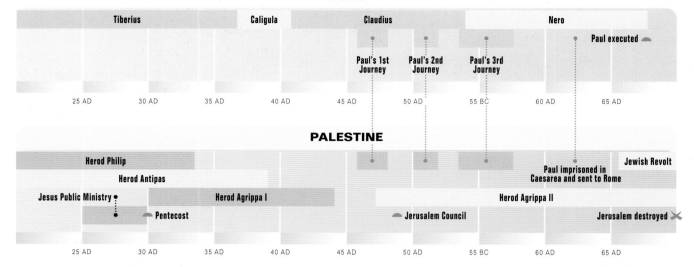

ROME

Tiberius	Caligula	Claudius	Nero

Paul executed

Paul's 1st Journey   Paul's 2nd Journey   Paul's 3rd Journey

25 AD   30 AD   35 AD   40 AD   45 AD   50 AD   55 BC   60 AD   65 AD

PALESTINE

Herod Philip

Herod Antipas

Jewish Revolt

Jesus Public Ministry   Herod Agrippa I   Herod Agrippa II

Pentecost   Jerusalem Council   Paul imprisoned in Caesarea and sent to Rome   Jerusalem destroyed

25 AD   30 AD   35 AD   40 AD   45 AD   50 AD   55 BC   60 AD   65 AD

THE EXPANSION OF THE CHURCH IN PALESTINE

At the time of Jesus' death, Pilate was the Roman governor of Idumea, Judea, and Samaria, while Herod the Great's sons Antipas and Philip still held their respective positions in the north. Philip died in AD 34, and his territory (Gaulanitis, Batanea, Auranitis, and Trachonitis) was transferred to the control of the province of Syria. But in AD 37 the districts were detached from Syria when the Roman emperor Caligula appointed Herod Agrippa I (AD 30 – 44) as the ruler over Philip's old domain. When in AD 39 Antipas foolishly requested an improvement of his position, he was banished to Gaul, and Galilee and Perea were added to the realm of Herod Agrippa I. In AD 41 the emperor Claudius, grateful for Agrippa's assistance in helping him secure the throne, added Samaria, Judea, and Idumea to his holdings. With these additions, Agrippa's kingdom was as extensive as that of his grandfather, Herod the Great. However, his large realm did not hold together for long, for at the peak of his power Agrippa was struck down with a terminal illness, and he died in Caesarea in AD 44 (Acts 12:19 – 23; Josephus, *Ant.* 19.8.2 [343 – 52]).

Scripture records a select number of events from the history of the early church between the resurrection of Jesus and the death of Agrippa I (AD 44; Acts 1 – 12). The sequence followed in the book of Acts tells of the growth of Christianity from its beginning in Jerusalem, through the expansion into Judea and Samaria, to its spread throughout the Roman world (1:8).

After Jesus' ascension the disciples gathered in Jerusalem. Acts 2 records this special manifestation of the Holy Spirit on the disciples at the feast of Pentecost, fifty days after Passover. At the time, Jerusalem was filled with Jewish pilgrims who had come from various parts of the Roman world, from the Parthian Empire on the east, and from Africa on the southwest. The pilgrims had come to celebrate the three great Jewish festivals: Passover/Unleavened Bread, the Feast of Weeks, and the Feast of Tabernacles, and were staying for a period of several months. After hearing unlearned Galileans witness to them of the living Christ in their native languages, a number of them were converted to the new faith.

The early chapters of Acts describe the growth and persecution of the church in Jerusalem (chs. 2 – 7). Opposition to the early church grew as an increasing number of Greek-speaking Jews (Hellenists) joined the ranks of the followers of Jesus, and as some of them joined

▼ *Jerusalem: Paved street with shops on the west side of the temple mount (vertical wall on right). The tumble of boulders is from the destruction of Jerusalem in AD 70 by the Romans.*

others in making explicit the implications of their newfound faith in its relationship to Judaism. This persecution culminated with the stoning of Stephen. With that, a sizeable segment of the Jerusalem church scattered throughout Judea and Samaria (8:1), and as the believers dispersed they shared their faith with others. One of these enthusiastic preachers was Philip, who "went *down* to a city in Samaria" (8:5; one always goes "up to" or "down from" Jerusalem). In this unnamed city — it probably was not Samaria/Sebaste, which was predominantly Gentile — inhabitants of the district of Samaria were converted and received the Holy Spirit. Indeed, even Peter and John (pillars in the Jerusalem church, Gal 2:9), who had come to pray for the new converts, willingly preached the gospel in Samaritan villages as they returned to Jerusalem (Acts 8:25).

Acts 8 also describes Philip's activities south and west of Jerusalem. There, Philip met the Ethiopian official who was reading from Isaiah 53 as he returned to Africa. After Philip explained the meaning of the passage, the Ethiopian believed, was baptized, and "went on his way rejoicing" (vv. 26–39). The site of this event is difficult to locate precisely, but since the Ethiopian was riding in a chariot it seems that he must have been traveling on a developed road. It may be that he was traveling on the road that led from Bethlehem to the Valley of Elah, the route that David had taken when he carried supplies to his brothers (1 Sam 17) and the one that the Romans eventually paved and marked with milestones. This road led south from the Valley of Elah through the low rolling hills of the Shephelah to Betogabris and continued from there west to Gaza. From Gaza a road continued west across northern Sinai into Africa. It seems reasonable to assume that Philip continued preaching in the Shephelah before moving out into the Philistine Plain

(Azotus; Acts 8:40) and then north, along the coast to Caesarea, where he eventually settled (21:8).

Others from Jerusalem were also active in the coastal plain area. Peter, for example, healed Aeneas at Lydda and then moved on to the Jewish port of Joppa, where he raised Tabitha from the dead (Acts 9:32–42). At Joppa Peter received his vision of the unclean animals and soon afterward he accepted an invitation to go to the house of Cornelius, a centurion living in Caesarea. As Peter shared the gospel, Cornelius and others gathered with him believed. Thus it was at Caesarea, the huge port city that Herod had built, the city of the residence of the Roman governor (Pilate may still have been procurator), that the gospel began to make inroads into the Gentile world on the shores of the Mediterranean Sea.

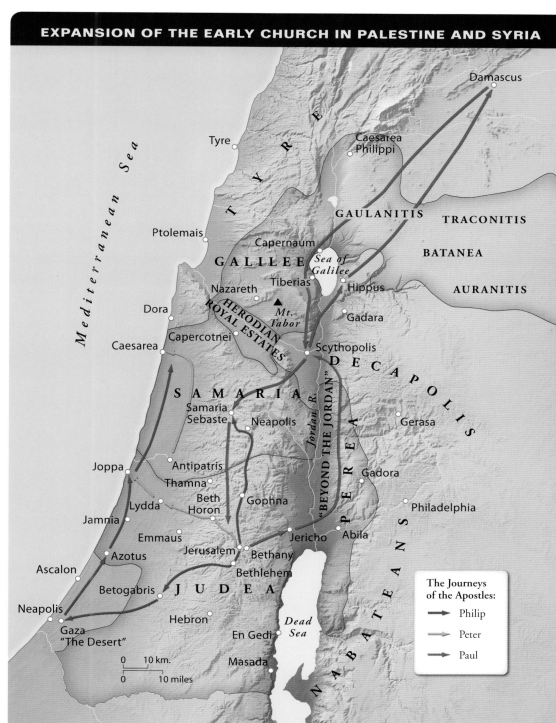

EXPANSION OF THE EARLY CHURCH IN PALESTINE AND SYRIA

The Journeys of the Apostles:
→ Philip
→ Peter
→ Paul

▲ *Gamala: Breach in the defensive wall made by the Romans when they captured the city – AD 67*

Meanwhile, the persecution of the church continued in Jerusalem and Judea. It was when Saul, a zealous Pharisee, armed with official sanction, was traveling to Damascus in order to persecute the believers there that the risen Lord appeared to him. The route he took cannot be determined, nor can the spot where the Lord appeared to him be located, although he was probably in northern Transjordan when the event occurred. After spending a short time in Damascus, Paul retired to Arabia (= the area of the Nabatean king Aretas IV), where

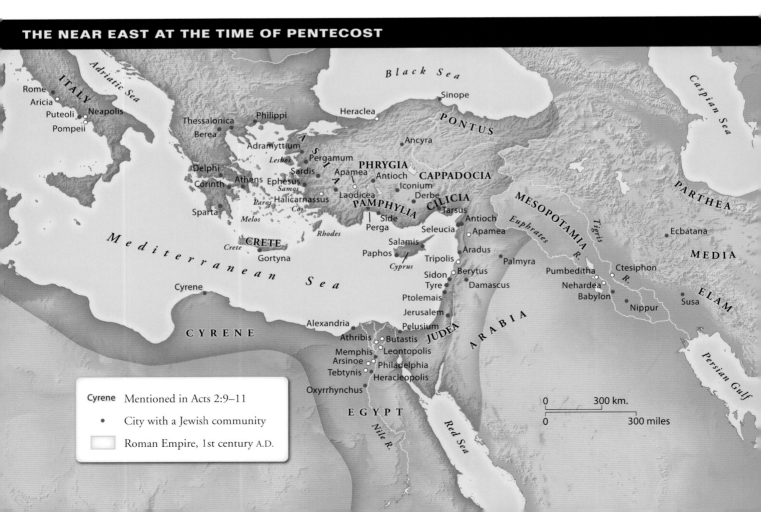

THE NEAR EAST AT THE TIME OF PENTECOST

Cyrene Mentioned in Acts 2:9–11

• City with a Jewish community

☐ Roman Empire, 1st century A.D.

he remained for three years (Gal 1:17). After briefly returning to Damascus, he traveled to Jerusalem for a two-week visit (Acts 9:26 – 29; Gal 1:18 – 19), and then he departed for Tarsus, his home city in Cilicia (southeastern Turkey).

In the book of Acts, little attention is given to the events between the conversion of Saul (ca. AD 32/33) and the death of Agrippa I (AD 44), but during that time the church continued to grow and opposition to it increased.

After the death of Agrippa I at Caesarea in AD 44, inept and offensive procurators ruled much of Palestine. Various Jewish groups attempted to revolt but none was successful. In the meantime, Herod Agrippa II was being granted more and more territory by the Romans, so that by the time of the Jewish revolt (AD 66 – 70) he was in control of Gaulanitis, Batanea, Auranitis, Trachonitis, and portions of Galilee.

During this time (i.e., AD 44 – 66), Paul was traveling on his three great missionary journeys (see the next chapter), and he returned to Judea for visits after his second and third journeys. After his third journey he was imprisoned in Jerusalem, having been accused of bringing a Gentile into the temple area. Because of a plot on Paul's life he was transferred by night to Caesarea. The military escort that accompanied him traveled by horseback to Caesarea via Antipatris (= OT Aphek). The route from Jerusalem to Antipatris can only be surmised; possibly the detachment followed the old Beth Horon road, or they may have taken the road that ran further north through Gophna and Thamna. In any event, Paul spent over two years imprisoned at Caesarea. During that time he appeared before two different procurators, Felix and Festus, as well as before the ascending Jewish king, Agrippa II. In the end, Paul appealed to Caesar and was escorted to Rome under guard (see next chapter).

During this period the Roman governors of Palestine continued to offend Jewish sensitivities and did little to restrain the Gentiles of Caesarea who were harassing the Jews. It was during the tenure of Florus (AD 64 – 66) that Jews all over Palestine revolted: in Caesarea, in Jerusalem, and in Galilee. In Jerusalem the Temple Mount and the Antonia Fortress were seized, and Roman forces were shut up in the three towers just north of Herod's palace on the western hill. By the end of the summer of AD 67 all of Jerusalem was under Jewish control. In response, the Roman legate of Syria marched south with the Twelfth Legion and attempted to regain Jerusalem. But the attempt failed, and the legion was slaughtered while withdrawing down the Beth Horon road.

Although the Jewish rebel forces faced serious internal divisions throughout the revolt, a Jewish government was established, coins were struck, and military districts (commands) were set up. The most famous of those commanders was Josephus, the son of a priest, who was

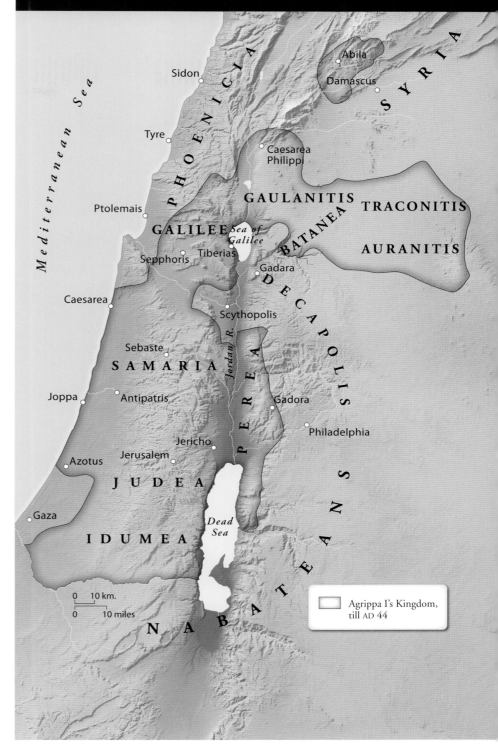

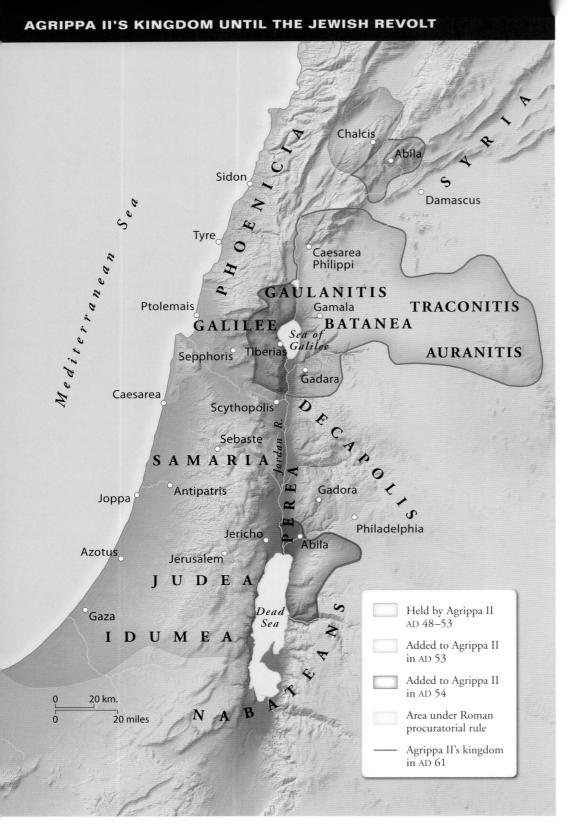

Chalcis

Abila

SYRIA

Sidon

Damascus

Tyre

PHOENICIA

Caesarea
Philippi

GAULANITIS

Ptolemais

Gamala

TRACONITIS

GALILEE

BATANEA

Sea of
Galilee

AURANITIS

Sepphoris

Tiberias

Gadara

Caesarea

Scythopolis

DECAPOLIS

Sebaste

Jordan R.

SAMARIA

PEREA

Joppa

Antipatris

Gadora

Jericho

Philadelphia

Azotus

Abila

Jerusalem

JUDEA

Dead
Sea

Gaza

IDUMEA

NABATEANS

Mediterranean Sea

0   20 km.
0   20 miles

Legend:
- Held by Agrippa II AD 48–53
- Added to Agrippa II in AD 53
- Added to Agrippa II in AD 54
- Area under Roman procuratorial rule
- Agrippa II's kingdom in AD 61

objective was to secure the northern part of the country, specifically Lower Galilee (map p. 222). After retaking Sepphoris, he laid siege to the fortress of Jotapata. Although most of the defending garrison died, their commander, Josephus, saved his life by surrendering to the Romans. Approaching the Sea of Galilee from the south, Vespasian proceeded through Tiberias to Taricheae. There he met stiff Jewish resistance but defeated the rebel forces after bloody land and sea battles. East of the Sea of Galilee, at Gamala, the Jewish forces also fought valiantly, but again the Romans prevailed, and most of the defenders were slaughtered. By the end of AD 67 all of Galilee was under Roman control.

In the meantime, Roman troops marched south along the coast, capturing the Jewish port of Joppa and destroying the small Jewish navy. Farther south, such cities as Jamnia and Azotus were captured, and to the east the Romans secured Samaritan territory in the Mount Ebal and Gerizim region.

In the spring of AD 68 fighting resumed, and it appears that Vespasian's goal was first to isolate and then to capture Jerusalem. To this end the Romans invaded Perea, the Jewish territory to the east of the Jordan River. In addition, the Fifth Legion moved south, along the coast, and captured the Shephelah. The Fifth Legion was stationed at

entrusted with the defense of Galilee. Later Josephus became famous as a historian of this and earlier periods.

The emperor, Nero (AD 54–68), sent his general Vespasian to crush the revolt. Vespasian established his headquarters in Ptolemais and set out to deploy the Tenth, the Twelfth, and the Fifteenth legions. His first

Emmaus to guard the western approaches to Jerusalem. The Romans also marched eastward through Samaria, down the Wadi Farah to Coreae, and from there south, down the west side of the Jordan Valley, to Jericho. After Jericho was captured, the Tenth Legion was stationed there to guard the eastern approach to Jerusalem. Raiding parties from

Jericho were sent out into northern Judea against such cities as Acrabeta and Gerasa.

In June of AD 68 Nero, the emperor, committed suicide, and although the leadership in Rome was in turmoil (there were three rulers within a one-year period), Vespasian was able to keep up the pressure in Judea and by mid-summer of AD 69 only Jerusalem, the Judean Desert, Masada, and Machaerus remained in Jewish hands.

In the summer of AD 69 Vespasian's troops declared him emperor. Vespasian returned to Rome to assume the throne, and in the summer of AD 70 it was his son, Titus, who captured Jerusalem. Jerusalem was under siege from the end of May until the middle of August; Josephus describes the intense suffering of the inhabitants. In August the Temple Mount was taken and on the ninth of Ab (ca. August 28) the Romans set fire to the temple. The upper (western) portion of Jerusalem held out for a few weeks longer but soon fell to the Romans. Although Jerusalem had been captured and destroyed, it was not abandoned, and the Romans stationed the Tenth Legion there in order to prevent further insurrection.

▲ *Greek inscription, found in Jerusalem, forbidding Gentiles to enter into the more sacred precincts of the temple — Paul was accused of violating this ban (Acts 21:27 – 29)*

▼ *Gamala: View to west southwest of the city where 9,000 Jews died attempting to defend the city against the Romans — AD 67*

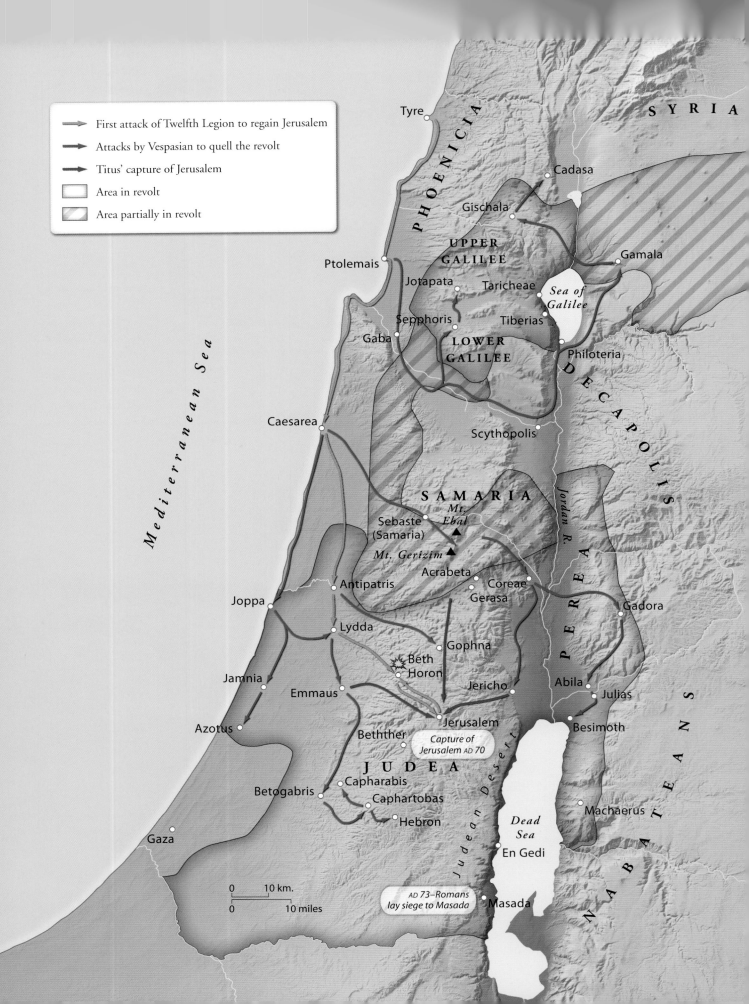

First attack of Twelfth Legion to regain Jerusalem

Attacks by Vespasian to quell the revolt

Titus' capture of Jerusalem

Area in revolt

Area partially in revolt

Mediterranean Sea

SYRIA

Tyre

PHOENICIA

Cadasa

Gischala

UPPER GALILEE

Ptolemais

Jotapata   Taricheae   *Sea of Galilee*   Gamala

Sepphoris   Tiberias

Gaba   **LOWER GALILEE**   Philoteria

DECAPOLIS

Caesarea   Scythopolis

Jordan R.

SAMARIA   *Mt. Ebal* ▲

Sebaste (Samaria)

Mt. Gerizim ▲

Acrabeta

P E R E A

Antipatris   Coreae   Gadora

Gerasa

Joppa

Lydda   Gophna

Beth Horon

Jericho   Abila

Jamnia   Julias

Emmaus   Besimoth

Azotus   Jerusalem

Bethther   *Capture of Jerusalem AD 70*

J U D E A   Capharabis   *Judean Desert*

Betogabris   Caphartobas   *Dead Sea*   Machaerus

Hebron   En Gedi

Gaza

N A B A T E A N S

0   10 km.

0   10 miles

AD 73–Romans lay siege to Masada

Masada

The capture of Jerusalem in AD 70 marks the end of the Jewish revolt, although the Romans had to engage in mopping-up operations, particularly in the Judean Desert. In AD 73 they had to lay siege to the massive rock fortress of Masada, located near the western shore of the Dead Sea. There, 960 of the 967 Jewish defenders decided to commit suicide and to die as free individuals rather than to be captured by the Romans. Again, it is Josephus who describes that tragic moment in Jewish history.

The destruction of the temple and the cessation of sacrificial worship necessitated a dramatic change in Jewish life. The Sanhedrin, now dominated by the Pharisees, reconstituted itself at Jamnia and gave new direction to Jewish life and practice. Jewish life was now focused in the villages and cities outside of Jerusalem throughout Palestine.

In the early second century (AD 132 – 35) the Jewish people again

London, British Museum

▲ *Head of Titus: As general, he destroyed the temple and Jerusalem and later (AD 79 – 81) ruled as emperor of Rome.*

attempted to throw off the Roman yoke. This second revolt, in contrast to the first, appears to have been carefully planned in advance. While the Romans were preoccupied in other parts of the empire, Simeon Bar Kokhba was declared the leader of the Jewish people by the highly respected Rabbi Akiba. The rebellion had its center in and around Jerusalem and Judea. An effective Jewish administration was set up, and even coins were struck.

Initially the Jews enjoyed a number of military successes, evidently even wiping out a Roman legion. However, the Romans again responded in strength and sent several legions into the region to suppress the revolt. Bar Kokhba had to abandon Jerusalem and retreat to Bethther, 7 miles southwest of Jerusalem. The Romans laid siege to his fortress, and he and his garrison were annihilated. With the end of the second Jewish revolt, the emperor Hadrian ordered Jerusalem to be destroyed and rebuilt as a Roman colony named Aelia Capitolina, with Jews forbidden entrance into the city. In order to further eradicate the "Jewishness" of the land, its name was changed from Judea to Palestine.

▼ *Rome: Arch of Titus at the entrance to the Forum depicting the menorah being displayed in Rome after the Jerusalem temple had been destroyed in AD 70.*

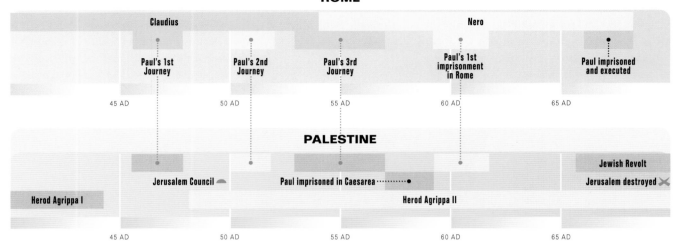

Claudius

Nero

Paul's 1st
Journey

Paul's 2nd
Journey

Paul's 3rd
Journey

Paul's 1st
imprisonment
in Rome

Paul imprisoned
and executed

45 AD   50 AD   55 AD   60 AD   65 AD

PALESTINE

Jewish Revolt

Jerusalem Council

Paul imprisoned in Caesarea

Jerusalem destroyed

Herod Agrippa I

Herod Agrippa II

45 AD   50 AD   55 AD   60 AD   65 AD

THE JOURNEYS OF PAUL

The Early Life of Saul

During the period between the ascension of Jesus and the fall of Jerusalem to the Romans (AD 30 – 70), the gospel message was preached not only in Jerusalem, Judea, and Samaria (Acts 1 – 12), but also throughout the Roman world (Acts 13 – 28). The major figure involved in this enterprise was Paul. Born in the Greco-Roman city of Tarsus, on the southeastern coast of Asia Minor, Saul (later called Paul) was sent to Jerusalem to study under the famous teacher, Rabbi Gamaliel (22:3; cf. 5:34). While in Jerusalem, Saul witnessed the stoning of Stephen (ca. AD 35), but soon afterward he was converted to Christianity on the road to Damascus (Acts 9). After spending three years in Arabia and making short visits to Damascus and Jerusalem, Saul returned to Tarsus, from which he apparently ministered in both Cilicia and northern Syria.

Antioch was the major city of Syria. It was located to the southeast of Tarsus and was one of three chief cities in the eastern Mediterranean. Located in northern Syria on the Orontes River, about 16 miles from the sea, Antioch was founded by Seleucus I (Nicator; 312 – 280 BC). Over the years it had developed into a leading commercial center, since it was located at the western end of land routes leading from Mesopotamia to the Mediterranean and because it also served as the northwestern terminus for the spice and incense route from southern Arabia. For many years it had been the capital of the Seleucid kingdom, and it became the capital of the Roman province of Syria, serving as Rome's last major outpost vis-à-vis the kingdom of Parthia to the east. It was mainly Gentile in character, but a considerable number of Jews lived there, for they

had received property rights and citizenship because of the assistance they had rendered to various Seleucid causes.

It is probable that some Christians who fled Judea because of persecution sought refuge in Antioch and began to share their faith with its inhabitants. Because of the success of this new faith, the Jerusalem church sent Barnabas to investigate the situation (Acts 11:22). After spending some time in Antioch, Barnabas, who had assisted Saul previously, went to Tarsus to seek Saul's help in the ministry at Antioch. For a few years (ca. AD 43 – 45), Barnabas and Saul ministered together at Antioch, the place where believers were first called "Christians" (v. 26).

▼ *Tarsus: One of the main streets from the days of Paul. Note the basalt paving stones, white limestone curbing, and the remains of buildings that line the right side of the street.*

Paul's First Missionary Journey (Acts 13:4 – 14:28)

After spending several years in Antioch, during which time Barnabas and Saul had made a trip to Jerusalem to deliver a gift for famine relief (Acts 11:27 – 30), they were set aside by the church in Antioch to be ministers to other parts of the Roman world. It was from Antioch's important harbor, Seleucia, that Saul set out on the first of his three missionary journeys (ca. AD 46 – 48). Barnabas and Saul — accompanied by John Mark, Barnabas's cousin — set sail for the island of Cyprus some 80 miles to the west, landing at Salamis, which was located on the eastern edge of the island. There at Salamis, the largest city on the island, they preached in the synagogue before moving over-land to Cyprus's administrative capital, Paphos, on the southwestern edge of the island. At Paphos, the proconsul of Cyprus, Sergius Paulus, was converted after having heard the gospel and having seen the judgment that fell on the sorcerer Elymas.

From Paphos, Paul and Barnabas sailed 170 miles northwest to ancient Pamphylia, on the southern coast of modern Turkey. Bypassing the port of Attalia, they proceeded up the Kestros River to Perga, one of the largest cities of the province. At Perga, for unknown reasons — jealousy? sickness? homesickness? fear? — John Mark left them and returned to Jerusalem (Acts 13:13). Paul and Barnabas evidently did not remain long in Perga but headed north, up into the rugged Taurus Mountains. After climbing the steep, forested southern slopes, they entered the mountainous area called Pisidia. From there they

▲ *Perga: Hellenistic towers at Perga — Paul, Barnabas, and John Mark visited the city on Paul's first journey (Acts 13:13; 14:25)*

continued north into Phrygia, an area that had been settled by peoples from Macedonia or Thrace.

The districts of Pisidia and Phrygia were closely related, so that the city of Antioch, technically in Phrygia, was also associated with Pisidia and thus came to be known as "Pisidian Antioch" (Acts 13:14). This Antioch, situated at the base of a mountain range, sat astride the great road that led from Syria to the Ionian coast. Antioch also served as the administrative center of southern Galatia, the large Roman province in which Pisidia and Phrygia were located. Paul and Barnabas preached in the synagogue for several Sabbaths. Although the Jews were not overly receptive to their message, the Gentiles were, and the gospel message spread "through the whole region" (v. 49). However, opposition soon developed, and Paul and Barnabas were expelled from the city.

Moving 90 miles to the southeast, they arrived at Iconium, another important commercial center on the road from Syria to Asia. There, Paul and Barnabas again preached in the synagogue, and a large number of Jews and Gentiles believed. Once again, however, opposition and threats on their lives forced them to flee, this time to the Lycaonian cities of Lystra and Derbe.

▲ *Sergius Paulus inscription from Pisidian Antioch. Possibly Sergius Paulus whom Paul converted at Paphos on Cyprus (Acts 13) had landholdings here.*

Possibly, Paul and Barnabas could have thought that Lystra would serve as a place of retreat from the antagonism they had been encountering, since it was not one of the larger cities of the region. But after Paul and Barnabas healed a man who had been crippled from birth, the Lycaonians thought that the gods were visiting them, and they identified Barnabas with Zeus and Paul with Hermes. Paul and Barnabas dissuaded the townspeople from worshiping them, but when Jews from Antioch and Iconium arrived, they incited the populace to stone Paul. Then they dragged him out of the city, thinking that he was dead. It is interesting to note that there probably was not a strong Jewish presence in Lystra, since no synagogue is mentioned, although apparently Timothy, whose mother was Jewish, was from the city (Acts 16:1). In addition, the lack of order and justice, evidenced in the treatment of Paul, seems to imply that Roman presence was minimal, or at least that justice was greatly influenced by local prejudices; this was true in spite of the fact that Lystra had been made a Roman colony.

From Lystra, Paul and Barnabas traveled east 60 miles to Derbe at the eastern edge of the province of Galatia. After preaching in the city, Paul and Barnabas did not take the short route via the Cilician

Gates and Tarsus back to Syrian Antioch; instead, they revisited and strengthened the churches that they had just established in Lystra, Iconium, and Pisidian Antioch. Retracing their steps south through the Taurus Mountains, they crossed the plain of Pamphylia to the port of Attalia and from there sailed back to Antioch. Thus on Paul's first journey, covering about 1,200 miles, important churches were established on Cyprus and in southern Galatia, and from strategically placed cities such as Pisidian Antioch and Iconium, the new converts could influence many who passed along the thoroughfares of Asia. On Paul's first trip, some Jews, but mainly Gentiles, were brought into the church.

▲ *Corinth: The temple of Apollo was over five hundred years old by the time that Paul visited Corinth on his second journey.*

Although Jewish proselytes had been joining the church since its earliest days, it was through the ministry of Paul and Barnabas that large numbers of Gentiles first began entering the church directly. This raised the question of the Gentile converts' relationship to the Mosaic law. To help answer this, Paul and Barnabas traveled to Jerusalem to participate in an apostolic conference that had been called to discuss the issue (Acts 15; ca. AD 49/50). After major presentations by believing Pharisees, Peter, and Barnabas and Paul, James pronounced the decision that only minimal ritual demands would be requested of Gentile converts so as to promote unity between believing Jews and Gentiles. Armed with this verdict, Paul and Barnabas returned to Antioch to minister there.

Paul's Second Missionary Journey (Acts 15:36 – 18:22)

Both Paul and Barnabas were anxious to revisit the churches they had established on their first journey, but they disagreed on the question of the suitability of John Mark as a traveling companion. Because of this disagreement, they decided to separate: Barnabas took John Mark and traveled to Cyprus, while Paul took Silas and headed toward Asia Minor. On this second journey (ca. AD 50 – 52; map p. 228), Paul and Silas went north from Antioch and traveled through the Amanus Mountains to Cilicia. In all probability they stopped at Paul's home city of Tarsus and then headed northwest through the Cilician Gates, a pass through the Taurus Mountains. After that they continued west, sharing the decision of the Jerusalem council regarding the status of Gentile converts with the churches at Derbe, Lystra, Iconium, and Pisidian Antioch. Along the way, at Lystra, Timothy joined Paul and Silas on their westward journey.

Soon after leaving Pisidian Antioch they had to choose whether to continue on the road that led west to the Ionian coast (through districts such as Caria and Lycia of the province of Asia) or to head north toward Dorylaeum. Choosing the latter route, they traveled north through Phrygia, preaching the gospel as they went. Passing southwest of Bithynia, they turned northwest to Nicaea or Prusa and from there continued west along the northern border of Mysia, another district of the province of Asia, to the port city of Troas.

Troas enjoyed great prosperity as a Roman colony, situated near the Hellespont (the modern Straits of Dardanelles, which connects the Black Sea and the Sea of Marmara with the Aegean Sea), at the end of a caravan route from the east and close to Europe. However, Troas was not to be the terminus of the Pauline journey, for here Paul evidently met Luke, a physician and the author of Luke-Acts, and, in response to a vision of a man from Macedonia, Paul and his party (now including Luke; Acts 16:11) sailed from Troas to Europe.

After anchoring for the night at the island of Samothrace, the ship landed in Macedonia at Neapolis (modern Kavalla in Greece). From Neapolis the party continued 6 miles inland to the prominent city of Philippi. Founded in the Hellenistic period and named after Philip

▼ *Erastus inscription at Corinth: Erastus was a wealthy official in Corinth and this inscription tells how he laid the street at his own expense (Rom 16:23).*

▲ *Patara: Probably the oldest preserved lighthouse in the world. Paul changed ships at Patara on his return from his third journey (Acts 21:1)*

PAUL'S SECOND MISSIONARY JOURNEY

THRACE

Black Sea

MACEDONIA

Philippi
Amphipolis · Neapolis
Thessalonica
Berea · Apollonia
Samothrace

BITHYNIA AND PONTUS

Byzantium

Sea of Marmara

Nicaea

THESSALIA

Thermaic Gulf

Hellespont

Prusa

Troas

MYSIA

Dorylaeum

GALATIA

CAPPADOCIA

ACHAIA

Aegean Sea

Thyatira

ASIA

PHRYGIA

Pisidian Antioch

Lechaeum · Athens
Corinth
Cenchrea · *Saronic Gulf*

Ephesus

CARIA

Iconium

Lystra · Derbe

Cilician Gates

PELOPONNESE

PAMPHYLIA

Taurus Mts.

CILICIA

Amanus Mts.

Tarsus

LYCIA

Rhodes

Patara

Seleucia Pieria · Antioch

CRETE

CYPRUS

SYRIA

Orontes R.

Mediterranean Sea

Sidon

Tyre · Damascus

PHOENICIA

0   100 km.
0   100 miles

— Route of the Via Egnatia

Caesarea

PALESTINE

ARABIA

Jerusalem

of Macedon, the father of Alexander the Great, Philippi was situated on the Via Egnatia — an important Roman road that led from Dyrrachium, on the western coast of modern Albania on the Adriatic Sea, to Byzantium, almost 700 miles to the east. This strategic road was part of the system that connected Rome with its eastern provinces. As a Roman colony, Philippi seems to have been predominantly populated by Gentiles, for evidently the Jewish population was not large enough to warrant a synagogue.

At a place of prayer used by Jewish women by the River Gangites, west of the city, Paul met Lydia, the purple-cloth dealer from Thyatira. Paul ministered in the city while staying at her house. In Philippi, Paul healed a slave girl who was demon-possessed, and as a result her owners had Paul and Silas thrown into prison. After an earthquake in the middle of the night and the subsequent conversion of the prison guard and his family, the leaders of Philippi begged Paul and Silas to leave the city, and the two men complied. Since the "we" section (Acts 16:10 – 17) that began at Troas ends here, it may be assumed that Luke remained behind to minister at Philippi. It seems that the ministry at Philippi was successful, for in Paul's letter to the church there, he expresses warm feelings for the people and does not reprimand the members for doctrinal divergences.

From Philippi Paul and Silas traveled 68 miles west along the Via Egnatia to Thessalonica via Amphipolis and Apollonia. Thessalonica was founded in 315 BC and was named by Cassander for his wife. Located at the northern edge of the Thermaic Gulf (Thessalonica's earlier name had been Therma — "Hot Spring"), it had become not only a district capital but also the chief port for all of Macedonia. Paul and Silas preached in the synagogue for three Sabbaths, and a number of Jews, God-fearing Greeks, and prominent women believed.

But opposition developed once again. A mob stormed the house of Jason, where Paul was staying, so Paul and Silas departed that evening. It seems that even after that short, three-week stay, Christianity took root in Thessalonica, as Paul's two letters to the church there indicate. Evidently, the gospel rapidly spread from Thessalonica into the hinterland, for within the next eighteen months Paul wrote and commended the church that "the Lord's message rang out from you not only in Macedonia and Achaia — your faith in God has become known everywhere" (1 Thess 1:8).

Paul and Silas next moved 50 miles southwest to Berea (modern Verria), where they entered the synagogue and preached. The people of Berea were known for their desire to study the Scriptures, and a number of Jews, as well as Greek men and women, believed. However, Jews from Thessalonica agitated the crowds of Berea, so Paul departed for the coast, leaving Silas and Timothy behind. Paul evidently traveled by ship to Athens, where he awaited the arrival of his companions.

The Athens that Paul visited was no longer the administrative capital of southern Greece, which was now the Roman province of Achaia — Corinth held that honor — but it was still renowned as a cultural and intellectual center. Since its Golden Age in the fifth century BC, it had been filled with temples and altars to all kinds of gods. In addition, temples, altars, and especially tombs lined the road that ran from Piraeus, the port of Athens, to the city itself. The Agora (marketplace) and the Acropolis to the southeast were also filled with shrines. Paul preached in the marketplace as well as in the synagogue, and as a result was invited by a group of Epicurean and Stoic philosophers to address the philosophical assembly called the Areopagus (= "hill of Ares").

By the time of Paul's appearance in Athens, the meeting place of this group had moved from the well-known rocky hill west of the Acropolis to the Royal Colonnade in the northwestern corner of the Agora. There, Paul presented the gospel of Jesus to that learned group (Acts 17:22 – 34). Although a number of Athenians were converted, there does not seem to have been a ready acceptance of the gospel, so Paul left for Corinth, the capital of Achaia, 40 miles to the west of Athens.

Corinth was a bustling administrative and commercial center at that time. It had been sacked by the Romans in 146 BC, but was refounded as a Roman colony in 44 BC. It owed its prosperity to its geographical location: it was just south of the narrow isthmus (4 mi. wide) that connected the Greek mainland with the Peloponnese, and all traffic between them had to pass through Corinth. In addition, the ancients preferred to portage passengers and cargo across the isthmus rather than take the longer and more dangerous trip by sea around the southern tip of the Peloponnese.

▲ *Ephesus: View toward the now-silted harbor from the theater at Ephesus (Acts 19:23 – 41).*

Corinth controlled the port of Cenchrea on the Saronic Gulf to the east (to Asia) and its counterpart, Lechaeum on the Corinthian Gulf, to the northwest (to Italy), so it certainly drew revenue from the trade that flowed through the isthmus.

In addition, crowds and revenue were drawn to the city because of the biannual, Panhellenic, athletic, musical, and poetic contests that were held at Isthmia, only 6 miles to the east of Corinth. As a "tentmaker" (Acts 18:3), Paul, along with Aquila and Priscilla and many others, may have serviced the sailing ships and made tents and shelters for the visitors to the games. Note also the "athletic imagery" that Paul uses in his letters.

In this boomtown famous for its immorality (Strabo states that there were 1,000 prostitutes for the temple of Aphrodite alone), Paul ministered for over eighteen months. Here the gospel was eagerly received and the church was established. Paul probably penned both of his letters to the church at Thessalonica during this stay in Corinth. In time, Paul was accused by disgruntled Jews of violating their religious law, but Gallio, the proconsul of Achaia, would not accept their complaints.

After his extended stay, Paul, in the company of Priscilla and Aquila, with whom he had been living, left Corinth

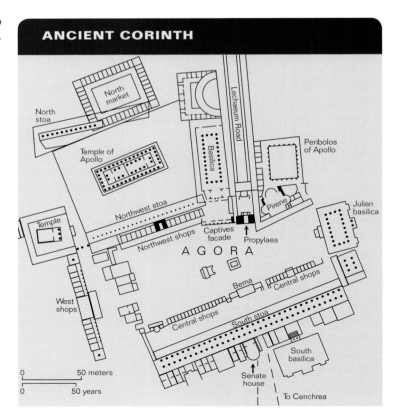

via the port of Cenchrea for the 250-mile sea journey to Ephesus. Stopping in Ephesus for a brief time, he left Priscilla and Aquila behind and set sail for Caesarea, some 600 miles distant. Upon arriving at Caesarea, he went up to Jerusalem and reported to the church the results of his journey. Then he returned north to his starting point and home base, Syrian Antioch. On this second journey, traveling about 2,800 miles, Paul took the opportunity to revisit the Galatian churches and to deal with their doctrinal problems (see Paul's letter to the Galatians) in light of the Jerusalem Council, and to preach the gospel in Europe — that is, in Macedonia and Achaia.

Paul's Third Missionary Journey (Acts 18:23 – 21:14)

After spending some time in Syrian Antioch, Paul set out on his third journey (ca. AD 53 – 57). He probably began by retracing the route of his second journey through Galatia and Phrygia to Pisidian Antioch. But this time, instead of heading north, he continued west into the province of Asia, to Ephesus. This great city was one of the three major cities of the eastern Mediterranean (the other two were Syrian Antioch and Alexandria, Egypt). It had become an important commercial center, since caravan routes from the east converged there, and it was from there that shipping lanes to the west originated. Paul spent three years ministering in Ephesus "to the Jews first but also to the Greeks," and it is probable that he or his converts carried the gospel message to additional cities in Asia. Certainly many of the churches in Asian cities addressed by John in the first three chapters of Revelation — Ephesus, Smyrna, Pergamum, Thyatira, Sardis, Philadelphia, and Laodicea — were founded and/or nurtured by Paul.

Besides being an important transportation hub, Ephesus contained the temple of the fertility goddess Artemis (called Diana by the Romans), which was so large and so magnificent that it was considered one of the wonders of the ancient world. Because of the success of the gospel in Ephesus, there was a serious decline in business associated with the worship of the goddess Artemis. Thus Demetrius, a silversmith, incited the Ephesians, who had assembled in the 24,000-seat theater, against Paul and other Christians, but the town clerk was able to dissuade the mob from carrying out any illegal acts. During his stay at Ephesus Paul may have written 1 Corinthians.

The Seven Churches of Asia (Rev. 1–3): addressed by John, some founded by Paul

Route of the Via Egnatia

▲ *Assos: The temple of Athena. Paul visited Assos on the return leg of his third journey (Acts 20:13).*

Soon afterward, Paul apparently set out for Macedonia, probably via Troas (Acts 20:1; 2 Cor 2:12 – 13). In Macedonia, he likely revisited the churches at Philippi, Thessalonica, and Berea, and it may well have been at this time that he continued westward along the Via Egnatia to Illyricum (Rom 15:19). Sometime during this period it appears that he wrote 2 Corinthians in preparation for his arrival at Corinth. After ministering in Macedonia, Paul continued south to Achaia and spent three months wintering in Corinth. It was probably at this time that he wrote his famous letter to the church at Rome, advising them of his intention to visit them as he proceeded to the western parts of the empire.

When spring came Paul planned to sail directly to Judea in order to reach Jerusalem by Pentecost, but because of a threat on his life he abandoned that plan and instead traveled overland to Philippi (Acts 20:3 – 6). After the Feast of Unleavened Bread, he sailed for Troas (on a local cargo ship?), a trip that took five days. After spending seven days at Troas (where Eutychus was raised from the dead after falling out of a window), Paul walked the 20 miles overland to Assos while the remainder of his party traveled by ship. Sailing down the Ionian coast, they stopped at Mitylene (a port on the island of Lesbos), at the islands

of Kios and Samos, and at Miletus. Miletus had formerly been one of the chief ports of Asia, but because of the silting of its harbor, it was beginning to lose its importance to Ephesus. When Paul reached Miletus, he summoned the elders of the Ephesian church, who traveled 30 miles south to Miletus to spend a few days with their beloved teacher.

After a tearful good-bye, Paul and his party set sail from Miletus. After reaching the island of Cos, they continued to Rhodes and on to the southern shore of Asia Minor, to Patara, a seaport in the province of Lycia. Boarding a cargo vessel that was heading for Phoenicia, they sailed past Cyprus and landed at Tyre. While cargo was being unloaded, Paul had a chance to spend seven days with the disciples there; then from Tyre they continued on to Caesarea, stopping briefly at Ptolemais along the way.

Paul, against the advice of some Christians, continued on to Jerusalem, where he greeted the elders of the church and completed the purification rites associated with a vow that he had made. Although he may have intended to return to Antioch as usual, Jerusalem was, in fact, the terminal point of his third journey (ca. 2,700 miles), for it was here that he was arrested (see above, p. 219). After being transported to Caesarea, he remained imprisoned there for a few years (ca. AD 57 – 59) before finally appealing to Caesar for justice.

Paul's Journey to Rome
(Acts 27:1 – 28:16)

For the trip to Rome, Paul was placed in the custody of a centurion named Julius. Along with a small party that included Aristarchus and probably Luke, Paul was placed on a ship from Adramyttium (the name of a harbor on the west coast of Asia Minor). The ship's route took it to Sidon and then on to the southern coast of Asia Minor (Turkey), to Myra, a major port of call for grain ships bound for Rome. There Paul and his party transferred onto a cargo ship that was to sail directly to Italy. Because of adverse winds, the ship was not able to reach Cnidus on the southwestern tip of Asia Minor, and rather than attempting to cross the open sea to the southern tip of the Peloponnese, it sailed south, intending to pass on the south side of the island of Crete.

Sailing along the southern coast of Crete, the ship passed up a winter anchorage at Fair Havens in an attempt to reach what the captain thought was a more desirable port at Phoenix, fifty miles further west. Since the fast (= Day of Atonement; Acts 27:9) was already over, it must have been late fall, when sailing on the Mediterranean becomes precarious.

Indeed, a "northeaster" (Acts 27:14 NIV) wind sprang up, and the ship was driven off course. As it was passing on the lee side of the island of Cauda (ca. 23 mi. south of Crete), the ship had to be lashed together, and as the storm continued there was danger that the ship would run aground and be wrecked on the sandbars of Syrtis, off North Africa. Instead the ship eventually ran aground on the island of Malta, 60 miles south of Sicily. All 276 passengers were saved, but the ship and its cargo were lost.

After wintering three months on the island, they boarded another Alexandrian ship and sailed to Italy via Syracuse and Rhegium. They landed at Puteoli (modern Puzzuoli) in spite of the fact that it was 140 miles south of Rome, as Puteoli was a favorite terminus for passengers on ships coming from the east. A good road connected it with Rome. Along this road, at the Forum of Appius (43 mi. south of Rome) and at Three Taverns (33 mi. south), Paul was greeted by Christians from the capital.

The Rome of Paul's day was a huge city with a population of more than 1,000,000. It was situated on the banks of the Tiber River, 18 miles inland from the sea. It had spread beyond its traditional seven hills and even beyond its ancient fortifications. As the capital of the Roman

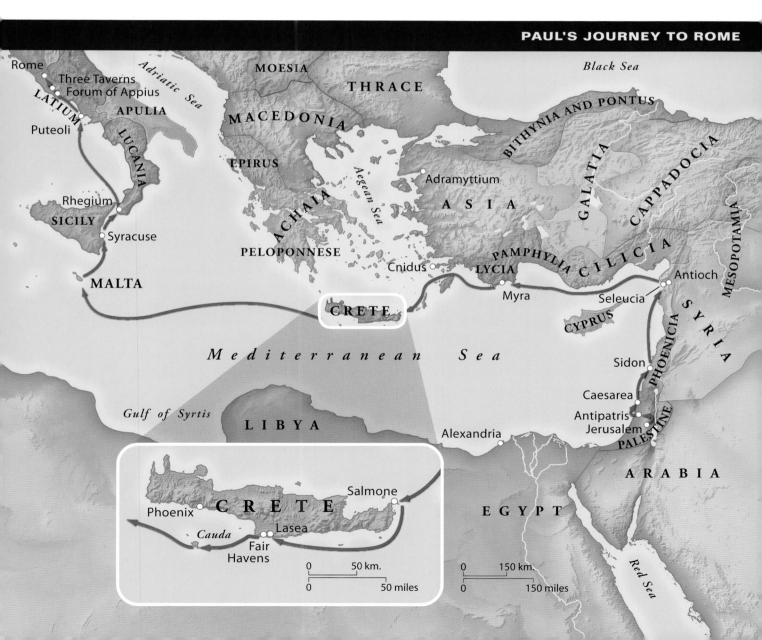

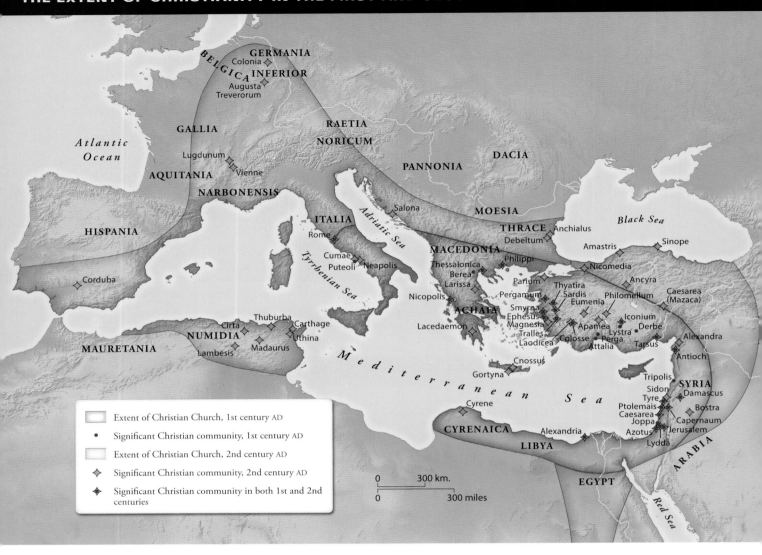

Empire, it boasted the imperial palaces on the Palatine Hill and the temples of Jupiter and Juno on the Capitoline Hill. Below these hills was the Forum, with additional temples and public buildings. Theaters, amphitheaters, hippodromes, and other monuments graced the city, but its beauty was tempered by the fact that over half its population were slaves. In addition, a large portion of the nonslaves lived in squalid conditions in high-rise apartment buildings of four and five stories and were dependent on the free distribution of food for their subsistence.

The book of Acts concludes with Paul having resided in Rome for two years (ca. AD 59–61/62), under house arrest, without ever having

had his case go to trial. From that point on we are dependent on early Christian tradition for our knowledge of the activities and fate of Paul. According to tradition, Paul was released from prison around AD 62 and traveled to various parts of the Mediterranean world — probably to Crete (Titus 1:5) and possibly to Spain. Also according to tradition, he was arrested and imprisoned again in AD 67/68, and it was during this time that he penned his final letter (2 Timothy). Finally, tradition claims that during the time of the Neronian persecution (ca. AD 68), Paul was executed outside the walls of Rome, along the road to Ostia.

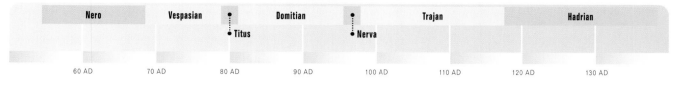

ROMAN EMPERORS

| Nero | Vespasian | Titus | Domitian | Nerva | Trajan | Hadrian |

| 60 AD | 70 AD | 80 AD | 90 AD | 100 AD | 110 AD | 120 AD | 130 AD |

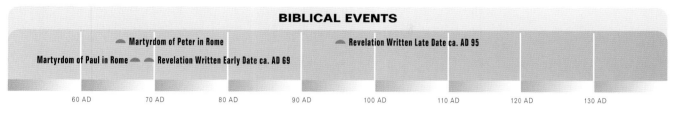

BIBLICAL EVENTS

Martyrdom of Peter in Rome

Revelation Written Late Date ca. AD 95

Martyrdom of Paul in Rome — Revelation Written Early Date ca. AD 69

| 60 AD | 70 AD | 80 AD | 90 AD | 100 AD | 110 AD | 120 AD | 130 AD |

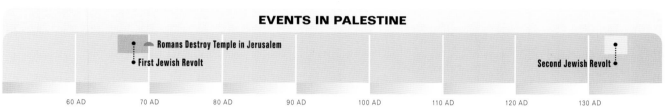

EVENTS IN PALESTINE

Romans Destroy Temple in Jerusalem

First Jewish Revolt

Second Jewish Revolt

| 60 AD | 70 AD | 80 AD | 90 AD | 100 AD | 110 AD | 120 AD | 130 AD |

THE SEVEN CHURCHES OF REVELATION

In Revelation 1 – 3, John addresses seven churches located in the Roman province of Asia. Beginning at Ephesus, the churches are addressed in a clockwise fashion — ending at Laodicea. This may be the order that the carrier of the book/scroll followed on the initial public reading of the book.

Ephesus (1:11; 2:1 – 7)

Ephesus is the first of the seven churches that John addresses. It was the closest of the churches to the desolate island of Patmos, where John may have been exiled (1:9) and where he received his "revelation" (1:10ff.). At the time of his writing the book (second half of the first century AD), Ephesus was the capital of the Roman province of Asia, and it may have been the place where John actually penned the book after being released from exile on Patmos.

Ephesus was the third or fourth largest city in the Roman Empire — behind Rome and Alexandria and maybe Syrian Antioch — boasting a population of about 250,000 people. It was an important trade center, for some of the major trade routes from the east terminated there, and from Ephesus, their cargo was shipped on the Aegean and Mediterranean seas to other parts of the empire — especially to Rome itself.

In addition, Ephesus was home to numerous temples and cults. The temple of the goddess Artemis was one of the seven wonders of the ancient world and drew pilgrims from all over the empire for her cult festival in the spring. It also housed temples dedicated to Augustus and Rome as well as a large temple called the "Sebastos," which was dedicated to the emperors from John's time, namely, to Vespasian, Titus, and Domitian. It is interesting that the book of Ephesians seems to address the priority of worshiping Jesus, rather than Caesar, as "Lord."

Some early Christian traditions associate the apostle John with Ephesus, and he may have written his gospel and his three letters while

▼ *Patmos: The barren Aegean island to which John was exiled (Rev 1:9)*

▲ *Ephesus: In the foreground, the meager remains of the Artemis temple (Rev 2:1 – 7) — once one of the Seven Wonders of the World! In the background a mosque and on the horizon a citadel and the Church of St. John*

living at Ephesus. He is said to have been buried there, and the remnants of a basilica-style church built by Justinian marks John's traditional burial site. During the Byzantine period Christianity flourished at Ephesus, and the Third Ecumenical Council was held there in AD 431.

The Austrians have excavated and restored Ephesus for over 100 years and the visible archaeological remains are extensive!

Smyrna (1:11; 2:8 – 11)

The second church addressed by John is in Smyrna (modern Izmir), located 36 miles north of Ephesus. This Roman city was situated on a deep sheltered bay of the Aegean Sea at the western end of the Hermus River Valley. It was the chief terminus of the major ancient road that came from the east — from Syria, India, and even China via the cities of Laodicea, Philadelphia, and Sardis.

After the death of Alexander the Great (323 BC), Smyrna grew in size, and it was the first city in Asia Minor to establish a temple honoring Rome (198 BC). In AD 26 Tiberius granted it permission to establish a temple for worshiping the emperor — at the time, this was only the second such city in Asia to be granted this honor (the first was Pergamum).

It is not known when Christianity was established in Smyrna, but the new faith may have taken root there during Paul's three-year stay in Ephesus. In the book of Revelation the church at Smyrna is encouraged to remain faithful in light of the coming persecution. The early church father Ignatius was deeply involved with the church at Smyrna. He wrote one letter to the church itself, one to its bishop, Polycarp, and four additional letters to other churches (Tralles, Magnesia, Ephesus, and Rome) while residing in Smyrna. Ignatius was eventually martyred in Rome in AD 107. Polycarp was martyred in the stadium of Smyrna at the age of 85 (AD 156). The "crown of life" mentioned in Revelation 2:10 may refer to the wreath given to a victorious athlete, or it may refer to the citadel that towered over the city.

Although the modern city of Izmir is built over the ancient remains of Smyrna, Turkish excavations are under way at various points in the city — especially in the ancient agora/forum area.

Pergamum (1:11; 2:12 – 17)

The third city addressed by John is Pergamum. This is the only place in the Bible where Pergamum is mentioned. Pergamum was a magnificent city that had a population of about 100,000 in John's day.

The upper city (acropolis) rises about 900 feet above the surrounding plain, and excavations have located temples and altars dedicated to Zeus, Athena, Dionysus, and Demeter, and to the emperor Trajan. The modern town of Bergama covers most of the lower city, but significant remains of the Asclepieion (an ancient healing center) and a variety of other structures are still visible.

Pergamum is located 16.5 miles inland from the Aegean Sea and thus was not a port city like its rivals — Smyrna and Ephesus. This powerful city and its territory were willed to Rome on the death of Attalus III in 133 BC. In 129 BC, when Rome created the province of Asia, Pergamum became its early capital. In 29 BC it was the first city to be granted permission to establish a temple dedicated to a Roman emperor — namely, to Augustus (ruled 27 BC to AD 14).

In Revelation 2:13 it is said that "Satan has his throne" there. It is not certain whether this refers to the fact that Pergamum was a center of emperor worship (Augustus and eventually Trajan and later Caracalla were worshiped there), to the famous Zeus altar (now reconstructed in Berlin), to the Asclepieion (Asclepius was called "savior"), to the fact that it housed temples to at least eight(!) deities, or possibly to some combination of the above.

In addition, John writes that Antipas was martyred there (2:13; by the Roman sword); but it should be noted that it is King Jesus who wields the real "double-edged sword" (2:12), not Caesar or his representatives. On the negative side, the church is warned that some in the church must give up eating "food sacrificed to idols" and immorality (2:14 – 16).

Thyatira (1:11; 2:18 – 29)

Thyatira, the fourth of the seven churches addressed by John (modern Akhisar), is located 42 miles inland from the Aegean Sea. Although it was an important city in its own right, it was situated on a low corridor that connected the Hermus Valley (Philadelphia and Sardis) and the Caicus

▲ Pergamum: The great theater and beyond it, by the tree, the altar of Zeus (the throne of Satan? Rev 2:13)

Valley (Pergamum) so that it could serve as an outpost to protect Pergamum or as a base from which to attack Pergamum. In any case, it served as a commercial center connecting Pergamum with Attalia to the southeast, and with trading centers in the Middle East.

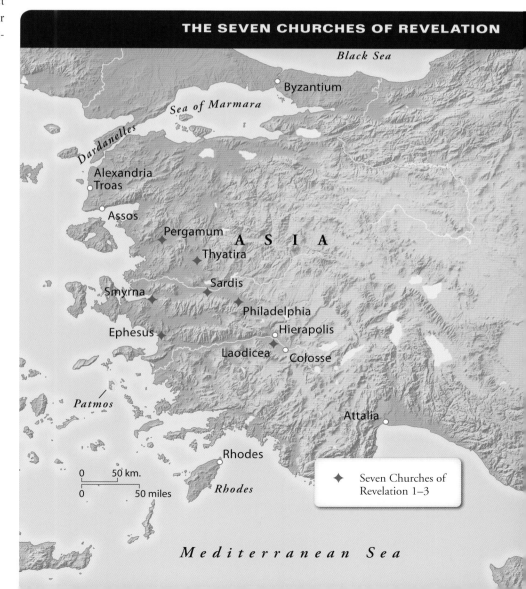

THE SEVEN CHURCHES OF REVELATION

Black Sea

Byzantium

Sea of Marmara

Dardanelles

Alexandria Troas

Assos

Pergamum · A S I A

Thyatira

Smyrna

Sardis

Philadelphia

Ephesus

Hierapolis

Laodicea · Colosse

Patmos

Attalia

Rhodes

Rhodes

0   50 km.
0   50 miles

◆ Seven Churches of Revelation 1–3

Mediterranean Sea

▲ *Sardis: Gymnasium and palestra of this once very rich city*

Near the center of Akhisar, visible archaeological remains are located in a fenced-off rectangular city block. From inscriptions it is evident that guilds of bakers, bronze smiths, wool workers, potters, linen weavers, and tanners were active in the city. Such guilds often held banquets that included the eating of food offered to idols and participation in immoral sexual acts (cf. Rev 2:20 – 24).

Lydia, who was converted by Paul in Philippi, was a dealer in purple cloth from Thyatira (Acts 16:11 – 15). Evidently she was selling cloth that had been colored with the purple die that was made from the "madder" plant, which was common in the area of Thyatira — higher-quality purple came from the murex shellfish from Phoenicia.

Sardis (1:11; 3:1 – 6)

Sardis, the fifth church addressed by John, was located in the Hermus Valley, about 52 miles east (inland) from Smyrna. During the first millennium BC it was the chief city of the powerful Lydian kingdom. It derived its wealth from the gold that was mined in the nearby Pactolus River, from commerce (connecting eastern kingdoms with the port of Smyrna), and from the manufacture of textiles.

During the first millennium BC Sardis was governed by a series of dynasties, and one of its kings was Gyges, who was referred to as "Gog" in Ezekiel 38 and Revelation 20:8. In 546 BC Cyrus conquered the city, defeating its famed king Croesus (this is probably hinted at in Isaiah 41:1 – 7). In 499 BC the Greeks (Athenians and Ionians) attacked and destroyed the city. In retaliation the Persians recaptured the city and Ionia, and it was from Sardis that first Darius I and then Xerxes I launched attacks on the Greek mainland — eventually burning Athens in 480 BC. The Persians built the "Royal Road" that connected Sardis with Susa, one of the Persian capitals — 1,600 miles to the east-southeast.

By the first century AD Sardis had passed into Roman hands, and although it had approximately 100,000 people, its glory days were behind it. When John admonished the church to "wake up" (Rev 3:2) and asserted that Jesus will "come like a thief" in the night (3:4), he may have been alluding to the two occasions when the enemies of Sardis were able to capture its citadel because of the laxness of its defenders. The reference to true believers being "dressed in white" garments (3:4 – 5) may allude to the famous textile industry of Sardis.

At Sardis, American archaeologists have discovered a huge temple in which, at various times, Artemis, Zeus, and the Roman emperor Antonius Pious (AD 138 – 161) and his wife Faustina were worshiped. A large, but later, synagogue attests to a significant Jewish presence at Sardis.

Philadelphia (1:11; 3:7 – 13)

Philadelphia (modern Alashehir) is the sixth of the seven churches that John addresses. It was founded in the third century BC by one of the Pergamenian kings and named after Attalus II (159 – 138 BC), who maintained "loyalty/love" for his brother, Eumenes II (197 – 159 BC), in spite of Roman pressure — thus the name Philadelphia (lit., "brotherly love").

It is located about 80 miles east of Smyrna (Izmir) in western Turkey. In NT times it was at an important road junction, for a major route coming from the east, which terminated at Smyrna, ran though it. In addition it was on a diagonal northwest – southeast route that ran from Pergamum (111 miles away) to Attalia.

In the ancient sources it was known for housing a number of temples, and in AD 17 it was destroyed by a devastating earthquake. In Revelation 3:12 the believer who "overcomes" is compared to a pillar (stability) in the temple of God, and on him/it three names will be written — compare the names inscribed on the columns of the temple of Zeus at Euromos.

Laodicea (1:11; 3:14 – 27)

Laodicea is the final church addressed in Revelation. The city was founded in the third century BC; by the first century AD it had replaced Colosse and Hierapolis as the chief city of the region.

It was situated at the eastern end of the Meander Valley close to where the smaller Lycus Valley joins it. The caravans coming from the east — from the regions of the Euphrates River and Syria — arrived at

Laodicea. From there they could travel west along the Meander Valley to Ephesus or Miletus, or they could head northwest toward Philadelphia and Sardis and then follow the Hermus Valley to the port of Smyrna. In addition, an important road connecting Pergamum to the northwest and Attalia to the southeast ran through Laodicea.

Besides being a strategically located trading hub Laodicea was also famous for its textiles, which were made out of its famous "raven-black wool," for an eye salve that was produced there, and for its nearby medical center. Laodicea was so wealthy that after the devastating earthquake of AD 60, it paid for its own rebuilding and declined assistance offered by Emperor Nero!

It is possible that Epaphras (Col 1:7; 2:1; 4:12; Philem 23) brought the gospel to the Lycus Valley region, ministering at cities such as Colosse, Laodicea, and Hierapolis. Paul had not only written a letter to the church at Colosse (Col 1:1 – 2), but evidently also to the church at Laodicea (4:16).

The church at Laodicea is the only one of the seven that does not receive any commendation. The church seemed self-satisfied, believing they were rich when in fact they were poor (Rev 3:17). In 3:18 gold, textiles, and eye salve are all mentioned — the very commodities that Laodicea took pride in. The famous reference to the church as being lukewarm, and not hot nor cold (3:15 – 16), may be an allusion to the desirable — for medicinal purposes — hot springs at Hierapolis, 6 miles to the north of Laodicea, and to the cool stream waters

▲ *Laodicea: Calcified syphon that once brought water into the city (Rev 3:15 – 16)*

running by Colosse, 8 miles to the southeast. Laodicea had built an aqueduct to bring water to it from springs 6 miles to the south — the water quality of which was not as useful as the water in Hierapolis or Colosse.

The site of Laodicea is huge, and archaeological excavations have just begun there, but already temples, streets, theaters, an Odeon, a stadium, water installations, etc., have been discovered.

▼ *Laodicea: Excavated street of this proud, self-satisfied city (Rev 3:14 – 22)*

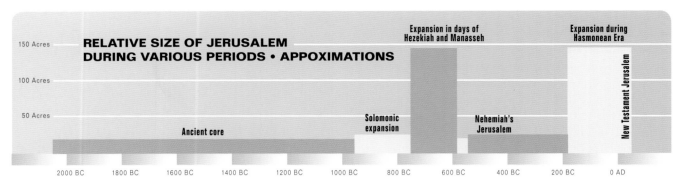

RELATIVE SIZE OF JERUSALEM DURING VARIOUS PERIODS • APPOXIMATIONS

150 Acres — 100 Acres — 50 Acres

Ancient core

Solomonic expansion

Expansion in days of Hezekiah and Manasseh

Nehemiah's Jerusalem

Expansion during Hasmonean Era

New Testament Jerusalem

2000 BC   1800 BC   1600 BC   1400 BC   1200 BC   1000 BC   800 BC   600 BC   400 BC   200 BC   0 AD

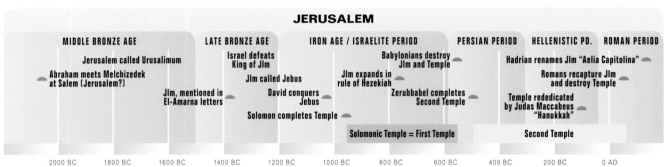

JERUSALEM

MIDDLE BRONZE AGE | LATE BRONZE AGE | IRON AGE / ISRAELITE PERIOD | PERSIAN PERIOD | HELLENISTIC PD. | ROMAN PERIOD

Jerusalem called Urusalimum

Abraham meets Melchizedek at Salem (Jerusalem?)

Jlm, mentioned in El-Amarna letters

Israel defeats King of Jlm

Jlm called Jebus

David conquers Jebus

Solomon completes Temple

Jlm expands in rule of Hezekiah

Babylonians destroy Jlm and Temple

Zerubbabel completes Second Temple

Solomonic Temple = First Temple

Hadrian renames Jlm "Aelia Capitolina"

Romans recapture Jlm and destroy Temple

Temple rededicated by Judas Maccabeus "Hanukkah"

Second Temple

2000 BC   1800 BC   1600 BC   1400 BC   1200 BC   1000 BC   800 BC   600 BC   400 BC   200 BC   0 AD

JERUSALEM

Jerusalem holds a special place in the hearts and minds of Jews, Christians, and Muslims. Of all the cities in the Bible, this is the most prominent one: it is mentioned 667 times in the Old Testament and 139 times in the New. Although today Jerusalem boasts a population of over 750,000 people, its origins were humble.

Jerusalem did not become an important city because of its proximity to any major international highway. In fact, it is far removed from the coastal and the Transjordanian highways. The only route that passed by it was the north–south Ridge Route, and even that ran about 0.5 mile west of the ancient core of the city. A west–east road that connected Gezer on the coastal plain and Jericho in the Jordan Valley via Beth Horon, Ramah, and Geba passed 5.5 miles north of Jerusalem. Although this route was the most natural approach into the hill country, and thus to Jerusalem, whether from the east or west, it was primarily of local significance.

Jerusalem's location in the Hill Country of Judah, at an elevation of 2,500 feet, gave it the benefit of many natural defenses. The Dead Sea, the Rift Valley cliffs, and the wilderness provided protection on the south and east, while the latter two provided security on the northeast as well. It was also difficult to

approach Jerusalem from the west because the hills of the Shephelah and the deep V-shaped valleys carved into the hard limestone of the Judean hills formed a rugged and treacherous landscape. It was somewhat easier to approach Jerusalem from the north or south, along the Ridge Route, but access to the Ridge Route from either the coast or the Rift Valley was difficult. Thus, besides being removed from the main routes of commerce and military expeditions, Jerusalem enjoyed the security of its natural defenses.

▼ *View toward the old ancient core from the south. The golden-colored Dome of the Rock stands where the ancient temples did. Note how the hills surround Jerusalem (Ps 121:1).*

If Jerusalem was not a natural center of commerce because of its location, neither was it situated in the heart of an extraordinarily rich agricultural region. In fact, Jerusalem was perched right on the boundary between the desert and the sown, for east of the Mount of Olives the Judean Desert begins. But Jerusalem itself receives ample supplies of winter rain (approximately 25 in. per year), as do the hills to the west, so that they are able to produce a variety of crops. During early periods, however, agricultural activities must have been limited because of the extensive tree cover on the hills in and around the city.

Beginning in the Bronze Age (ca. 3100 BC) and continuing into later periods, large trees were cut down to provide timber for buildings and ships, while both larger and smaller trees were used to fuel the fires in lime and pottery kilns and to heat houses in the winter months. Areas that had been cleared could be used for agricultural purposes, and on the more level terrain — e.g., in the Valley of Rephaim to the southwest of Jerusalem — grain crops were planted (Isa 17:5). To the north, south, and west of Jerusalem the natural hillside terraces were enhanced as small fields were cleared of boulders, fences and retaining walls were built, and crops were planted. By the time of the United Monarchy (ca. 1000 BC), all three crops in the famous triad of grains, olives, and grapes were being grown in the Jerusalem area, along with other crops. In spite of all this, Jerusalem never really became a large exporter of agricultural produce.

In addition to its unimpressive commercial and agricultural qualities, Jerusalem is not overwhelming from a topographical standpoint. Hills surround its ancient core so that standing on these hills one actually looks down on the city. Roughly speaking, the ancient city can be visualized as sitting on a rise in the bottom of a large bowl, where the rim of the bowl is higher than the rise within it.

Biblical Jerusalem was built on two parallel north – south ridges. The western

▲ *Looking north up the Kidron Valley. The slopes of the City of David are on the left (west) and a corner of the Temple Mount is visible. On the right (east) the Mount of Olives rises out of the Kidron.*

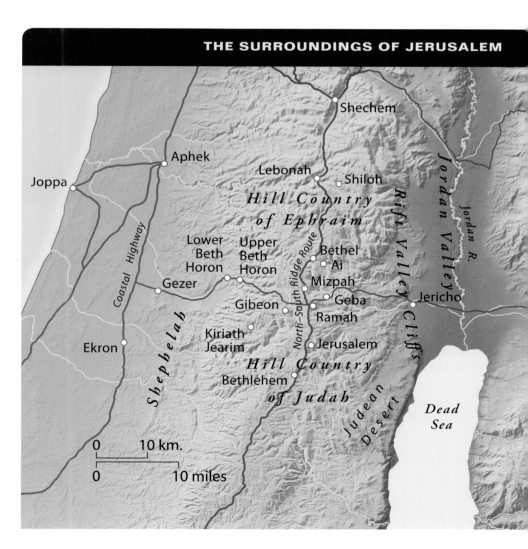

THE SURROUNDINGS OF JERUSALEM

Shechem
Aphek
Lebonah
Shiloh
Joppa
Hill Country of Ephraim
Coastal Highway
Lower Beth Horon
Upper Beth Horon
Bethel
Ai
Mizpah
Gezer
Geba
North–South Ridge Route
Gibeon
Ramah
Jericho
Kiriath Jearim
Jerusalem
Ekron
Shephelah
Hill Country
Bethlehem
of Judah
Judean Desert
Rift Valley Cliffs
Jordan Valley
Jordan R.
Dead Sea

0   10 km.
0   10 miles

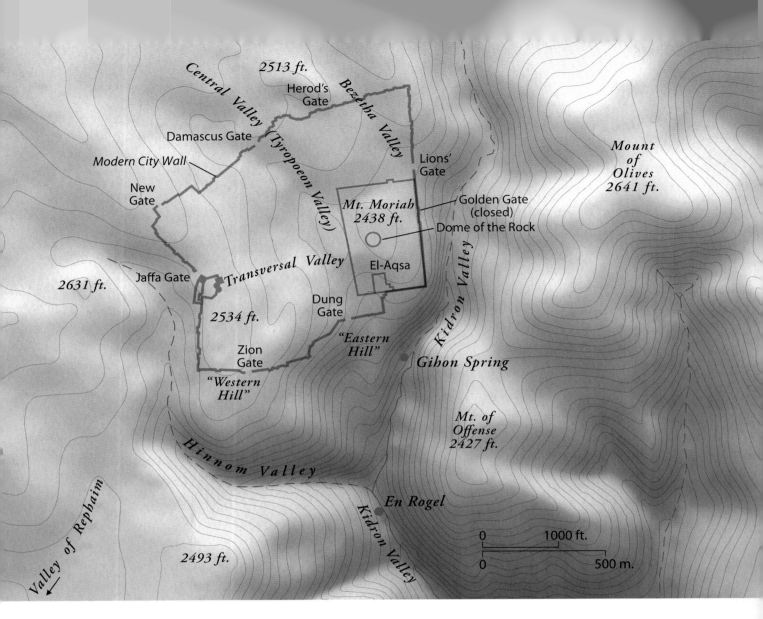

Central Valley (Tyropoeon Valley)

2513 ft.

Herod's Gate

Bezetha Valley

Damascus Gate

Modern City Wall

Lions' Gate

New Gate

Mt. Moriah 2438 ft.

Golden Gate (closed)

Dome of the Rock

El-Aqsa

Mount of Olives 2641 ft.

Transversal Valley

Jaffa Gate

2631 ft.

2534 ft.

Dung Gate

"Eastern Hill"

Kidron Valley

Zion Gate

Gihon Spring

"Western Hill"

Mt. of Offense 2427 ft.

Hinnom Valley

En Rogel

Valley of Rephaim

Kidron Valley

2493 ft.

0   1000 ft.

0   500 m.

ridge, which is the higher and broader of the two, is bounded on the west by the Hinnom Valley. This valley, which begins northwest of the city, runs due south and then turns due east, forming distinctive limits on the western and southern sides of the ridge. The narrower and lower eastern ridge is bounded on the east by the Kidron Valley, which in the Jerusalem area flows basically north to south. After being joined by the Hinnom, the Kidron turns to the southeast and proceeds through the Judean Desert to the Dead Sea. Both the Hinnom and the Kidron are mentioned in the Bible, but the valley between them, which separates the eastern and western ridges, is not. For lack of a better name, geographers often call it the Central Valley, or — following the lead of the Jewish historian Josephus — the Tyropoeon ("Cheesemakers") Valley (*War* 5.4.1 [140]).

On the north, both ridges continue to rise as they veer to the northwest. The major valleys — the Hinnom Valley, the Central Valley, and a tributary of the Kidron — all veer off in that direction as well. The west–east Transversal Valley, which flows into the Central Valley, forms the natural northern boundary of the western ridge, yet in some

periods of Jerusalem's history, the city expanded along that ridge to the northwest. The northern limit of the eastern ridge is not quite as noticeable as it continues rising to the northwest, interrupted only by some minor depressions that flow into the Bezetha Valley. Because of the easier approaches from the north and the northwest, invading armies have often assaulted Jerusalem from a northerly direction.

In many ways the western ridge is the more natural one to settle on, both because it has a relatively large surface area and thus can support more people, and because it is higher and seems to have better natural defenses (higher, steeper slopes) than the eastern ridge. In spite of these features, it was the lower, cigar-shaped, 15-acre southern portion of the eastern ridge that was settled first. The reason why the ancient core of Jerusalem developed on this insignificant, down-in-a-basin hill was that the only good-sized spring in the whole area — the Gihon Spring (photo p. 138) — was located alongside the eastern ridge in the Kidron Valley.

The earliest settlement on the southern portion of the eastern ridge reaches back to the late Chalcolithic and Early Bronze ages, as evidenced by pottery discovered in recent excavations. To date, however, no city wall or buildings have been found that date unequivocally from these periods. From the Middle Bronze I (MB I) period (2200 – 2000 BC) there are few remains, although MB I tombs have been discovered on the Mount of Olives, and some think that there may have been additional MB I tombs on the hill that eventually became the site of the Solomonic temple. According to the chronology adopted in this atlas, it would have been during this period that Abram, returning from rescuing Lot, met Melchizedek, the king of Salem (Gen 14:18; this Salem is evidently to be identified with Zion [Jerusalem], for the two terms occur in synonymous parallelism in Ps 76:2). This meeting took place "in the Valley of Shaveh (that is, the King's Valley)," which in the book of Samuel is said to have been close to Jerusalem (Gen 14:17; 2 Sam 18:18).

Later in his life Abraham took his son Isaac to one of the mountains in the "region of Moriah" to sacrifice him (Gen 22:2). The Chronicler later identifies the mountain where Solomon erected the temple as "Mount Moriah" (2 Chron 3:1). Thus the biblical text places two events in the life of Abraham in close proximity to Jerusalem.

There is no mention in the Bible of Jerusalem during the Middle Bronze II period (2000 – 1550 BC) — unless Abraham should be placed in this period (which is possible). However, Jerusalem is mentioned several times in the Execration texts (see above p. 196). Among the cities, countries, and tribes mentioned, Urusalimum (= Jerusalem; meaning "foundation of the god Shalim" or "city of peace") is men-

tioned several times along with its various rulers. Archaeological excavations have confirmed the existence of Jerusalem during this period (ca. 2000 – 1750 BC). Although excavated building remains are few, 360 feet of a large wall, 20 feet thick, have been uncovered by Kenyon and Shiloh. This wall was apparently built about 1800 BC and continued in use, with rebuilds, until the end of the Judean monarchy (586 BC).

The discovered portions lie about two-thirds of the way down the eastern slope of the eastern ridge. Evidently this site was high enough above the Kidron Valley and far enough away from the foot of the Mount of Olives to be defensible, yet it was far enough down the slope to provide maximum surface area for the city as well as easy access to the Gihon Spring. In spite of the large portions of the wall that have been uncovered, the exact location of the remaining sections on the south, west, and north is conjectural. Pottery finds, or the lack thereof, make it evident, however, that the walled city was confined to the southern portion of the eastern ridge; it covered an area of approximately 15 acres and housed 2,000 – 2,400 people. Basically the city remained this size until it began to expand northward during the days of David and Solomon.

Around 1400 BC the Israelites began their conquest of the land of Canaan. After their initial thrust into the hill country and the capture and destruction of Ai, the Gibeonites made a nonaggression, mutual-defense treaty with Israel (see above, p. 111). When the king of Jerusalem, Adoni-Zedek, heard of this, he realized that his major line of communication with

▼ *Step-stone structure that rises to a height of 80 ft. It evidently supported a monumental building — either Jebusite or Israelite (or both)*

▲ *Looking south at the Dome of the Rock. Many believe that this is where the ancient Israelite temples stood: The Holy of Holies under the dome, and the sacrificial altar near the smaller structure to the left (east) of the dome. Note the cistern head on the left side of the image. There are close to 50 cisterns on the Temple Mount!*

the coast, and hence to Egypt, was in jeopardy. To counter this threat he assembled a coalition of four other Amorite kings and attacked Gibeon (see map p. 110). In compliance with the treaty, Joshua and the Israelites came to the aid of the Gibeonites, and in the ensuing battle the king of Jerusalem was killed. The text does not indicate, however, that Jerusalem itself was captured at this time, although evidently shortly thereafter the tribe of Judah attacked and burned the city (Judg 1:8). But it does not seem that Judah settled in the city, for Jerusalem is mentioned in six of the el-Amarna letters (see above pp. 123 – 125), and the correspondence clearly indicates the non-Israelite character of the city during the reigns of the Egyptian pharaohs Amenhotep III and Akhenaton (ca. 1390 – 1336 BC). In fact, reference is made to the ruler of Jerusalem, Abdi-Heba, a name that does not correspond to the Adoni-Zedek mentioned in the Bible.

At some point during the period of the judges (ca. 1400 – 1050 BC) Jerusalem came under the control of the people called Jebusites and was named Jebus. This status of Jerusalem is reflected in the story of the Levite and his concubine (Judg 19 – 21), where it is called "Jebus ... this city of the Jebusites ... an alien city" (19:11 – 12). In addition, in the description of the boundaries of the tribes of Judah and Benjamin, which date from this period, Jebus is mentioned as a point on the border, and it is specifically noted that it was located within the tribal allotment of Benjamin, not Judah (Josh 15:8; 18:16; but compare 15:63).

But it was the Judahite David who, after being installed as king over the northern tribes as well as over Judah, captured the city in his seventh year (ca. 1003 BC) and made it his capital. In the description of the conquest of the city it is noted that Joab was instrumental in its capture (1 Chron 11:6) and that in some way his use/capture/control of the ṣinnor (NIV "water shaft," 2 Sam 5:8) was a crucial aspect of the assault. Although a number of suggestions have been made regarding the nature of the ṣinnor, it probably was the underground, rock-cut diagonal tunnel (photo p. 139) that led to a large pool that was fed by the waters of the Gihon Spring. Evidently the spring and the pool were each guarded by two large towers, and the whole system may have been several hundred years old by the time Joab entered it! Thus the inhabitants of the city could obtain water without ever having to go outside the city walls, by lowering their buckets down into the pool. If the ṣinnor refers to all, or part of this system, then the biblical reference seems to imply that somehow Joab and his men entered from the outside and ascended into the diagonal tunnel and in turn surprised the Jebusite garrison.

With the capture of Jerusalem David accomplished several important goals. First, he removed a foreign enclave from a border area and thus removed a potentially divisive threat to the Israelite tribes. Second, because of Jerusalem's neutral location — not in the heartland of Judah, like Hebron, and not in the northern part of Israel like Shechem — it was a capital acceptable to both David's own tribe of Judah as well as to the tribes of the north, who had recently acknowledged him as king. By capturing Jerusalem himself, it became his and his descendants' personal property that could not be claimed by his own or any other tribe — it became the royal seat of the Davidic dynasty. In addition, David brought the ark of the covenant from Kiriath Jearim to Jerusalem, thus establishing it as the major worship center for all the Israelite tribes (2 Sam 6:1 – 23; 1 Chron 13:1 – 14).

The city that David captured was small — approximately 15 acres — in size, with a population of 2,000 – 2,400. He evidently took up residence in the old Jebusite fortress called Zion, and from that point on, it, as well as the city as a whole, could be called the "City of David" (e.g., 2 Sam 5:7). Part of the foundation of this pre-Davidic fortress may be the semicircular stepped structure, along with the Late Bronze Age terraces, which Shiloh has discovered on the upper portion of the eastern slope of the eastern ridge. David also engaged in building projects in the city, including the construction of his own palace (2 Sam 5:11). Toward the end of his reign David purchased the threshing floor of Araunah the Jebusite, a site north of and higher than the ancient city core; this is the place where Solomon eventually built the temple (2 Sam 24:18 – 25; 1 Chron 21:18 – 26).

Soon after Solomon's accession to the throne, David died and was buried in the City of David (1 Kings 2:10). Evidently a royal cemetery was established where many of his descendants, up through Hezekiah (d. 686 BC), were buried. The search for the Tombs of the Kings of Judah within the City of David has been long and futile. The rock-hewn cavities excavated by R. Weill had been so quarried away that it is impossible to state with assurance that they were in fact tombs, let alone tombs that date to the time of David. The "tombs" discovered by B. Mazar are from a period later than David's, are not "in the City of David," and do not exhibit any of the fine decorative elements that one would expect in a royal tomb.

Some of the tombs surveyed east of the City of David, in the modern village of Silwan, give some evidence of having been the burial site for members of the upper classes of Jerusalem (e.g., a tomb of a "royal steward" was discovered), but they seem to date to the eighth century BC. North of Jerusalem, on the grounds of the École Biblique (photo p. 171), truly magnificent tombs of the Iron Age have been discovered, but they too are not "in the City of David." Thus, to date, no sure claim for the tomb of David has come to light.

In the fourth year of his reign (966 BC) Solomon began building the temple, a task that took seven years. The building itself was composed of two rooms: the Holy Place, in which the ten candelabra, the table for the bread of the Presence, and the incense altar were placed; and the most sacred, called the Most Holy Place or the Holy of Holies, in which the ark of the covenant was kept. The building faced east and was only 90 feet long, 30 feet wide, and 45 feet high. The whole building was surrounded by courtyards in which were located the sacrificial altar, lavers, etc. The exact location of the temple is not known, although many researchers place it in the immediate vicinity of the existing Muslim shrine called the Dome of the Rock. Given this location, it is possible that the Most Holy Place, or less probably the sacrificial altar, was located on top of the rock that is covered by the Dome of the Rock.

To the south of the temple, but north of the ancient core of Jerusalem, possibly in the vicinity of the present-day El Aqsa mosque, Solomon built his palace and the Palace of the Forest of Lebanon (1 Kings 7:1 – 12). Both of these structures have been compared to similar structures found in Anatolia and northern Syria. The Palace of the Forest of Lebanon was probably a large basilica-type structure that may have served as a ceremonial armory. The palace itself was probably of the *bit-hilanni* style, and thus it would have included the ceremonial colonnaded portico, the throne hall or the Hall of Justice, and then, around the great courtyard, Solomon's quarters, a hall for Pharaoh's daughter, and service areas.

In all probability, other administrative and military buildings were located nearby. It is possible that this royal acropolis, built at least in part on earth-and-stone fill north of the City of David but south of the temple complex, at the narrow neck of the ridge that connected the two areas, was in early times called the Millo (NIV "supporting terraces"; 1 Kings 9:15, 24; 11:27) but later came to be known as the Ophel (the acropolis). Evidently Solomon strengthened the wall of Jerusalem and included the Millo/Ophel, as well as the temple area, within the confines of the wall. Thus the walled city must have expanded in size from 15 acres to approximately 37 acres. Since the newly enclosed areas were not densely populated, Solomonic Jerusalem may have had a population of 4,500 to 5,000 people.

Among the increased population were at least some of the foreign wives whom Solomon married. It was for them that Solomon built a number of pagan shrines "on a hill east of Jerusalem" (1 Kings 11:7 – 8) — probably on the southern portion of the Mount of Olives. The location of these shrines was such that they towered over both the City of David and the temple of the true and living God.

With the secession of the north from the south after Solomon's death (930 BC), Solomon's successors ruled over a much smaller territory consisting of Judah and a portion of Benjamin. Jerusalem remained the seat of the government for the Davidic dynasty, and the Solomonic temple continued to be the focal point for Yahwistic worship.

During the period of the Divided Monarchy (930 – 722 BC), Jerusalem was attacked a number of times: once by the Egyptian pharaoh Shishak during the reign of Rehoboam (925 BC; 1 Kings 14:22 – 28; 2 Chron 12:2 – 4; see pp. 152 – 154 and map p. 153) and once by Hazael of Aram Damascus, during the reign of Joash (ca. 813 BC; 2 Kings 12:17 – 18; 2 Chron 24:17 – 24). But in each instance, lavish gifts, taken from the temple treasury, bought off the aggressors. In the days of Amaziah of Judah, however, Joash of Israel attacked the city and "broke down the wall of Jerusalem from the Ephraim Gate to the Corner Gate — a section about six hundred feet long" (ca. 790 BC; 2 Chron 25:23). Although in this passage and in other places several gates of the city are mentioned, it is difficult to pinpoint their location in the city walls, save perhaps that the "Valley Gate" may be the massive gate discovered by Crowfoot and Fitzgerald (1927 – 28).

Although the biblical text does not refer to any major expansion of Jerusalem during the period of the Divided Monarchy, some building activity is mentioned. We are told that Joash (835 – 796 BC) repaired the temple, probably in response to pagan Athaliah's alterations of the

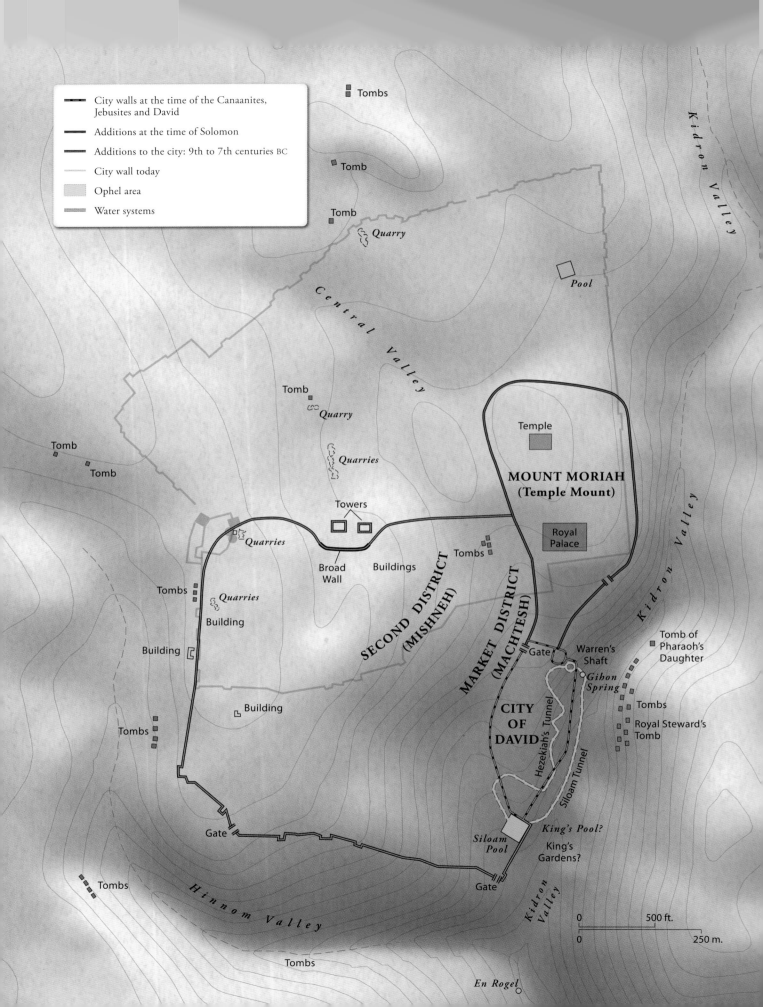

Tombs

Tomb

Tomb

Quarry

City walls at the time of the Canaanites, Jebusites and David

Additions at the time of Solomon

Additions to the city: 9th to 7th centuries BC

City wall today

Ophel area

Water systems

Pool

Central Valley

Kidron Valley

Tomb

Quarry

Quarries

Temple

MOUNT MORIAH
(Temple Mount)

Tomb

Tomb

Towers

Quarries

Broad Wall

Buildings

Tombs

Royal Palace

SECOND DISTRICT (MISHNEH)

Tombs

Quarries

Building

Building

MARKET DISTRICT (MACHTESH)

Gate

Warren's Shaft

Gihon Spring

Tomb of Pharaoh's Daughter

Kidron Valley

Building

CITY OF DAVID

Hezekiah's Tunnel

Siloam Tunnel

Tombs

Royal Steward's Tomb

Tombs

King's Pool?

Siloam Pool

King's Gardens?

Gate

Hinnom Valley

Gate

Kidron Valley

0   500 ft.

0   250 m.

Tombs

En Rogel

structure, but details are lacking (2 Kings 12:4–16; 2 Chron 24:4–14). We are also told that during the eighth century BC "Uzziah built towers in Jerusalem at the Corner Gate, at the Valley Gate and at the angle of the wall" (2 Chron 26:9) as he strengthened the defenses of the city — possibly in response to the growing Assyrian threat. It seems probable that during Uzziah's reign (792–740 BC) and during the reign of his successors, Jerusalem expanded westward so as to include the southern portion of the western ridge. The large increase in the size of Jerusalem at this time was probably due to the fact that settlers from the Northern Kingdom moved south so as to avoid the Assyrian onslaught; they may have thought that Jerusalem would never be taken by a foreign power because the temple of Yahweh was there, and that Yahweh would never allow such an indignity to be perpetrated (Ps 132:13–18).

Soon after the fall of the Northern Kingdom in 722 BC, Hezekiah revolted against his Assyrian overlords (see above, pp. 166–168), and he needed to make defensive preparations for all Jerusalem. Evidently it was during his reign that the suburb that had developed on the southern portion of the western ridge was enclosed by a new wall.

In the excavations in the modern Jewish Quarter of the Old City of Jerusalem, a 210-foot segment of this massive wall, 23 feet thick and in places preserved to a height of 10 feet, was discovered (photo p. 165). Subsequently, additional fragments of this wall have been found in the citadel at the Joppa Gate. Although certainty is not possible, it seems that in Hezekiah's day the whole southern portion of the western ridge was included within the city wall so that the total area of the walled city had swelled to 150 acres and boasted a population of about 25,000.

Since the major water supply of the city, the Gihon Spring, was at some distance from the newly enclosed western suburb and thus was exposed to enemy attack, Hezekiah devised a plan to divert the water to a spot inside the city walls, closer to the western hill. He did this by digging an underground tunnel (photo p. 167) that followed a serpentine path to a point in the Central Valley, which, although it was outside of the old city wall of the City of David, was inside the newly constructed city wall. This diversion of the spring water is mentioned not only in the Bible (2 Kings 20:20; 2 Chron 32:30), but also in a Hebrew inscription that was discovered at the southern end of the 1,750-foot tunnel.

In 701 BC Sennacherib the Assyrian attacked (see above, pp. 167–168). Although he sent some of his army and commanders to Jerusalem to demand its surrender — Sennacherib boasted that he had shut Hezekiah up in Jerusalem like a bird in a cage — he had to retreat when, according to the biblical text, a large portion of his army was destroyed through divine intervention.

As already noted, the eighth century witnessed unprecedented growth in Jerusalem, expanding from 37 acres to about 150 acres. Evidence that the city was so large comes not only from the portions of the city wall that have been discovered but also from noting the location of the cemeteries of the city, which were of course outside the city limits. Iron Age tombs from this period have been discovered on the lower slopes of the Mount of Olives to the east, on the bank of the Hinnom Valley to the south and west, and scattered to the north of Jerusalem. Jerusalem remained basically this size until its destruction in 586 BC.

During the eighth and seventh centuries BC there were both good and bad rulers in Jerusalem. On the negative side were Ahaz and Manasseh, both of whom burned (sacrificed) children in the Valley of Ben Hinnom (2 Chron 28:3; 33:6; cf. 2 Kings 23:10). It was during Ahaz's reign that at least a portion of the temple area was remodeled and a new altar, based on a pagan pattern from Damascus, was built to replace the old one (2 Kings 16:10–18). However, during this period there were also two godly kings, Hezekiah and Josiah, who worked to undo the abominations of their predecessors by taking steps to cleanse and refurbish the temple. It was during such a rebuilding, in the days of Josiah (ca. 622 BC), that the Book of the Law was discovered and in obedience to its commands additional reforms were instituted (2 Kings 22; 2 Chron 34).

But because of the continuing sins of the people and their leaders, God's judgment fell on Jerusalem in 605, in 597, and climactically in 586 BC — the year when the Babylonian king Nebuchadnezzar destroyed both the city and the temple. In the city of David, excavators have found considerable remains of the buildings that were destroyed in this conflagration. In one four-unit house an ostracon mentioning a certain Ahiel was found. In another, extensive burning was evident. And in a third, partially excavated, building, fifty-one bullae (clay lumps used to seal the strings that bound folded papyrus documents) were found. Among the eighty-two legible Hebrew names is "Gemaryahu, the son of Shaphan," a scribe in the court of the Judean king Jehoiakim (cf. Jer 36:9–11). Unfortunately, the papyrus documents that the bullae sealed have perished, but evidently some type of public archive was located here.

Although the temple and major buildings of Jerusalem were destroyed by the Babylonian army, there are indications that some Jews remained in or near Jerusalem. Almost fifty years after the destruction of the city, a large-scale return to Jerusalem began in response to the decree issued by Cyrus in the first year of his reign (539 BC). Led by Sheshbazzar, 49,897 people returned to Jerusalem from Babylon and rebuilt the temple altar and reinstituted sacrificial worship. It was not until the days of the Persian monarch Darius, however, that Jews, under the leadership of Zerubbabel, were able to actually rebuild the temple (520–516 BC; Ezra 6). Although no certain remains of this temple have been found, the Bible indicates that it was a much more modest structure than its Solomonic predecessor (Ezra 3:12).

The second return from Babylon was led by Ezra the scribe (458 BC) and was noted for its spiritual accomplishments. There may have been an attempt to rebuild the city wall at this time (Ezra 4:12), but the actual rebuilding of the walls had to wait another thirteen years until the days of Nehemiah (445 BC). In the book of Nehemiah, portions of three chapters specifically relate to the character of Jerusalem at that time. Chapter 2 describes Nehemiah's night journey around the walls, chapter 3 describes the actual rebuilding of the walls, towers, and gates, and chapter 12 describes the dedication processions around the walls.

It appears that Nehemiah's city was considerably smaller than the one that had been destroyed by the Babylonians. In fact, the city was basically confined to the eastern ridge and was of Solomonic proportions. We should note too that no structures or other artifacts from the Persian or early Hellenistic periods have been found in extensive excavations on the western hill — a clear indication that it was not settled from 586 to ca. 200 BC. In addition, the fact that it took only fifty-two days to refortify the city (Neh 6:15) also indicates that the long line of "Hezekiah's Wall" was not rebuilt at this time. Although the general location and/or geographical sequence of a good number of gates, towers, and buildings mentioned in the book of Nehemiah can be surmised, it is difficult to equate most of them with any known archaeological remains.

From the time of Nehemiah (445 BC) until the beginning of the second century BC, not too much is known about Jerusalem. In 332 BC Alexander the Great conquered the Levant, but his so-called visit to Jerusalem is usually considered to be legendary (see Josephus, *Ant.* 11.8.4 – 5 0321 – 339]). After his death in 323 BC the Ptolemies of Egypt gained control of Palestine and Judah, and it is generally assumed that under their benign rule a priestly aristocracy governed from Jerusalem. However, at the end of the third century BC there was considerable conflict in the area, and Jerusalem suffered at the hands of the Ptolemies. Early in the second century BC the Seleucid king Antiochus III defeated the Ptolemies (198 BC), and the change in rule was welcomed by most of the Jewish population. With his support, repairs were made to the temple and a large pool — possibly the Pool of Bethesda — was constructed (Sir 50:1 – 3).

During the reign of Antiochus IV (175 – 164 BC), however, relations between the Jews and the Seleucids took a decided turn for the worse as Antiochus and his Jewish supporters pressed for a Hellenizing program among all of the Jews. The temple in Jerusalem was desecrated and a statue of Olympian Zeus was set up in its precincts (168 BC). In addition, other Greek structures were erected in Jerusalem, including a gymnasium and a citadel. The citadel, called the "Akra" in Greek, was built on the eastern ridge just south of the temple area and was so tall that it towered over the temple area. Although Judas Maccabeus's forces were able to retake Jerusalem, to purify the temple (164 BC), and to reestablish sacrificial worship, the Seleucid garrison in the Akra remained a thorn in the side of the Jews until Judas's brother Simon (142 – 135 BC) captured and demolished it — even leveling the hill on which it had stood (Josephus, *Ant.* 13.6.7 [215]). To date, no remains of the Akra have been found, and because of its complete destruction and its probable location either under and/or near the southern portion of the present-day Muslim sacred area, it is not likely that any will be found in the near future.

Simon also completed the building of the walls of Jerusalem, a project that his brother Jonathan had begun earlier (1 Macc 10:10 – 11; 13:10). Although specific literary details regarding the exact line of the walls built by the Hasmoneans are not available, Josephus implies — and archaeological excavations confirm — that they were built along the line of Josephus's "First Wall" (see map p. 250). Indeed, not only has Avigad found significant remnants of the Hasmonean wall, which basically followed the east – west line of "Hezekiah's Wall" — thereby fortifying the northern boundary of the western ridge — but additional portions of the Hasmonean wall have been discovered in the present-day citadel, under the north – south wall to the south of the citadel, and also on the upper slope of the southern portion of the western ridge overlooking the Valley of Hinnom. Thus the whole southern portion of the western ridge was once again inside the city after a gap of almost four hundred years. In addition pottery and some remains of modest buildings have been found in excavations in the Jewish Quarter, thus confirming that renewed settlement on the western hill began during the Hasmonean period.

From literary sources it is evident that the Hasmoneans built a palace on the eastern slope of the western ridge, over-

JERUSALEM AT THE TIME OF NEHEMIAH

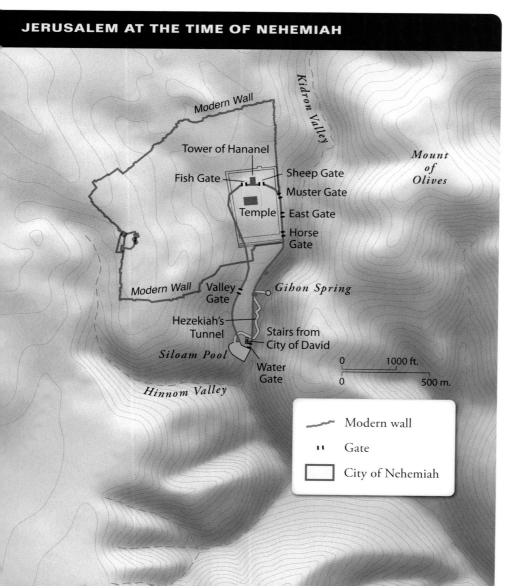

Modern Wall

Kidron Valley

Tower of Hananel

Fish Gate

Sheep Gate

Muster Gate

Temple

East Gate

Mount
of
Olives

Horse
Gate

Modern Wall

Valley
Gate

Gihon Spring

Hezekiah's
Tunnel

Stairs from
City of David

Siloam Pool

Water
Gate

Hinnom Valley

0   1000 ft.

0   500 m.

〜〜〜 Modern wall

ıı Gate

☐ City of Nehemiah

looking the Temple Mount (Josephus, *Ant.* 20.8.11 [189–90]), as well as a bridge that spanned the Central Valley and connected the palace with the temple precincts. This bridge was evidently a predecessor of what today is known as "Wilson's Arch." The only large-scale monuments from the Hellenistic period that are preserved in Jerusalem are several prominent mausoleums located in the Kidron Valley at the foot of the Mount of Olives: the tombs of Beni Hazir and Zechariah (photo p. 191). It is interesting to note that stylistically these tombs combine Hellenistic, Egyptian, and local characteristics; this fact illustrates the various cultural forces that influenced Jewish life during this period.

Jerusalem experienced additional growth and aggrandizement during the long and prosperous reign of Alexander Jannaeus (103–76 BC), but specifics are generally lacking in the historical sources. At the end of the

▲ *Looking east at the northeast corner of the recently discovered "Pool of Siloam" (John 9). Note the series of steps and platforms that lead down to the pool from left to right.*

Hellenistic age the Hasmonean brothers Aristobulus and Hyrcanus II vied with each other for the office of high priest and control of the country. In the end Pompey intervened, and marched on and captured Jerusalem (see above, p. 196). Upon seizing the temple area he entered the Most Holy Place, but he soon allowed the resumption of sacrificial worship. The arrival of Pompey marked the beginning of the long period of control of Jerusalem by Rome and its Byzantine successor, which would last until the time of the Persian and Arab conquests (AD 614 and 638), save for brief periods during the first and second Jewish revolts.

At the beginning of the period of Roman rule Jerusalem experienced great expansion, construction, and beautification under the leadership of the Roman client king, Herod the Great (37–4 BC). Pride of place must certainly go to Herod's refurbishing of the temple and the Temple Mount (photo p. 202). Although he was limited in what he could do to the temple building itself — the divine word and tradition dictated its basic dimensions — he spent over a year and a half beautifying and refurbishing the structure. He did not face similar restrictions when it came to the courts that surrounded the temple, and so he expended great sums on expanding these. He is said to have doubled the size of the platform area so that it reached its present size of 36 acres.

Although no remnants of Herod's temple have been found that can be identified as such with certainty, the huge platform on which its courts were built has survived. The area is now occupied by Muslim structures and is called the Haram esh-Sharif — the Noble Sanctuary. In constructing this large platform, Herod made use of some existing walls, especially on the east, but he expanded the platform to the north, west, and south. Indeed, the western expansion was such that part of the Central Valley was filled in and covered over, and today some twenty-six courses of Herodian stones, founded on bedrock, are still standing. These stones are cut so precisely that no mortar was used in the construction of the wall. A typical stone weighs 2 to 10 tons, while the largest of the known stones

measures 46 x 10 x 10 feet and weighs 415 tons! Almost all the stones, except the very low foundation courses that were covered by dirt, have a flat and finely chiseled boss, and around the edges is a finely dressed narrow margin.

Along the upper perimeter of the huge temple platform Herod built or refurbished a number of covered colonnades. The most famous of these was the southern one, the "Royal Colonnade." It was composed of 162 columns arranged in four rows, forming a long basilica-shaped building. The columns themselves were 27 feet high and 4.6 feet in diameter and were crowned with Corinthian capitals. Although nothing of the colonnade remains today, the appearance of its outer wall can be surmised from the pilaster-recessed design that is evident in the Herodian structure that encloses the Tomb of the Patriarchs in Hebron.

To the south of the Temple Mount large portions of the formal staircase that led up to the Huldah Gates have been found. The outlines of these gates are still visible in the southern wall of the Haram enclosure (the so-called double and triple gates). Although they are now closed, the underground passages that lead up to the top of the mount inside the wall are still preserved. In excavations along the southern portion of the western enclosure wall, portions of the north–south street, a city drain, and most interestingly, the piers that supported a platform and staircase that led south, from a gate in the southern section of the western wall of the Temple Mount into the Central Valley, have been discovered. Among the tumble of debris on the street was a large dressed stone inscribed in Hebrew with the words "For the place of the trumpet blowing." Evidently, this stone had fallen from its position on the southwest pinnacle of the Temple Mount, where it had marked the spot where the priest stood to blow the trumpet to announce to the citizens of Jerusalem the beginning of the Sabbath, New Moon, New Year, etc.

It took Herod almost ten years to complete the major construction on the Temple Mount, but crews were still working on the project

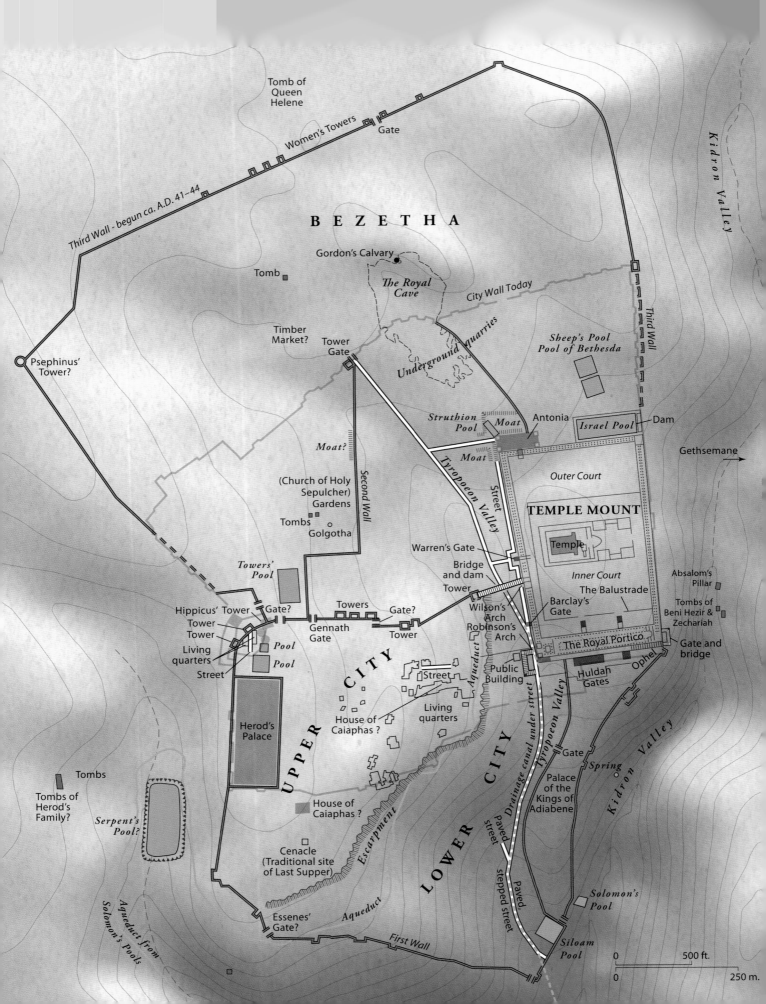

BEZETHA

Tomb of Queen Helene

Women's Towers

Gate

Third Wall - begun ca. A.D. 41–44

Gordon's Calvary

The Royal Cave

City Wall Today

Kidron Valley

Tomb

Sheep's Pool
Pool of Bethesda

Timber Market?

Tower Gate

Underground quarries

Psephinus' Tower?

Struthion Pool

Moat

Antonia

Israel Pool

Dam

Moat?

Moat

Gethsemane

(Church of Holy Sepulcher) Gardens

Second Wall

Tyropoeon Valley

Street

Outer Court

TEMPLE MOUNT

Tombs

Golgotha

Temple

Warren's Gate

Inner Court

Absalom's Pillar

Towers' Pool

Bridge and dam Tower

The Balustrade

Tombs of Beni Hezir & Zechariah

Hippicus' Tower

Gate?

Towers

Gate?

Wilson's Arch

Barclay's Gate

Tower

Tower

Gennath Gate

Tower

Robinson's Arch

The Royal Portico

Gate and bridge

Living quarters

Pool

UPPER

Ophel

Street

Pool

Aqueduct

Huldah Gates

Street

Public Building

CITY

Street

Living quarters

Gate

Tombs

Herod's Palace

House of Caiaphas ?

LOWER

Spring

Tombs of Herod's Family?

Palace of the Kings of Adiabene

Kidron Valley

Serpent's Pool?

House of Caiaphas ?

CITY

Drainage canal under street

Tyropoeon Valley

Paved street

Paved stepped street

Solomon's Pool

Cenacle (Traditional site of Last Supper)

Escarpment

Aqueduct from Solomon's Pools

Essenes' Gate?

Aqueduct

First Wall

Siloam Pool

500 ft.

0

250 m.

during Jesus' lifetime (John 2:20, ca. AD 28) and even on the eve of the first Jewish Revolt, which began in AD 66 (Josephus, *Ant.* 20.9.7 [219]). To the northwest of the temple Herod rebuilt the "Baris" fortress and named it the Antonia, after his friend Mark Antony. This fortress, situated on a rocky scarp, towered over the temple area and housed a garrison whose duty it was to monitor and control the crowds that gathered in the temple precincts.

On the western ridge Herod built a magnificent palace for himself on a site located in the present-day Armenian Quarter of the Old City. Only portions of its substructure have been preserved. To the north of his palace he built a massive defensive complex consisting of three towers named Hippicus (after a friend), Phasael (after his brother), and Mariamne (after his beloved wife, whom he had executed). The massive base of one of these towers (probably Hippicus or Phasael) still remains in the present-day citadel complex just southeast of the Jaffa Gate. In addition, Herod built a second wall that began near these towers — by the Gennath Gate — that ran to the Antonia Fortress, enclosing the northern "Second Quarter" of the city (Josephus *War* 5.4.2 [146]). The exact line of this wall is not known, but it probably ran through the Muristan Quarter of the Old City to the area of the present-day Damascus Gate, and thus it enclosed a portion of the northern reaches of the Central Valley — called the Tyropoeon Valley by Josephus.

Of the location of other Herodian structures, such as the hippodrome, stadium, theater, and amphitheater, little is known. However, excavations in the Jewish Quarter on the eastern slope of the western ridge have yielded significant evidence regarding life in this area during the Herodian period. Large houses have been excavated that contained courtyards, ritual baths, fresco-covered walls, fine pottery, stoneware dishes, huge stone pots, carved stone tables, mosaic floors, etc. The nature of the archaeological finds — among them the discovery of a weight with the name of a priestly family, Kathros — and data from literary sources make it evident that priestly and other aristocratic families lived in this area.

The Jerusalem that Jesus knew during his ministry (ca. AD 26 – 30) was basically the same as Herodian Jerusalem described above. On one of his visits to the city, he healed a thirty-eight-year-old paralyzed invalid at the Pool of Bethesda, north of the Temple Mount near the Sheep Gate (John 5:1 – 14). Portions of a double pool that could have been surrounded by "five covered colonnades" — one on each side and one in the middle separating the two pools — have been discovered just north of the Temple Mount. On another occasion Jesus healed a blind man whom he sent to the Pool of Siloam to wash (John 9; photo p. 249).

Most of the information about Jesus in Jerusalem comes from the period of the last week of his earthly ministry. Jesus evidently spent the nights of this last week with his friends in Bethany, 1.5 miles from Jerusalem on the east side of the Mount of Olives, and he made his triumphal entry into Jerusalem on a donkey that he had mounted in the Bethphage area. After crossing the Mount of Olives he descended

▼ *Garden Tomb located about 260 yards north of the present-day wall of the Old City. Tradition dating back to the 19th century AD.*

▲ *Crusader entrance to the Church of the Holy Sepulcher. This church contains both the site of Calvary and the tomb of Jesus. Tradition dating back to the 4th century AD.*

into the Kidron Valley to shouts of "Hosanna." Although he may have entered the temple complex from the north or east, as has traditionally been supposed, it is equally possible that he approached it from the south, via the monumental staircase that led up to the Huldah Gates. On Monday he again entered the temple area, and this time he drove out the moneychangers who were possibly operating in the Royal Colonnade along the southern perimeter of the Court of the Gentiles. On Tuesday Jesus once again entered the temple complex and evidently later in the day spent time teaching his disciples on the Mount of Olives.

After resting in Bethany on Wednesday, Jesus sent several of his followers into the city to secure a room and prepare a meal so that he could celebrate the Passover with his disciples. In spite of the fact that the structure on the traditional site of the Last Supper (the Cenacle) dates from the Crusader period (at least 1,100 years after the event), it is probable that the site itself, located on the southern portion of the western ridge in a well-to-do section of town, is close to the spot where the meal took place. After the meal, Jesus and his disciples went down to the Garden of Gethsemane (photo p. 214), at the western foot of the Mount of Olives, near the Kidron Valley. There, after praying for a while, he was captured and taken prisoner.

That night he appeared before Caiaphas the high priest, before Pilate the procurator, and before Herod Antipas, the ruler of Galilee,

who was in Jerusalem for the festival. The exact site of each of these interrogations is not known, but it is probable that the residence of Caiaphas was somewhere on the southern portion of the western ridge, and it is possible that Herod Antipas was staying in the old Hasmonean palace on the eastern slope of the western ridge, overlooking the temple. Although Jesus may have appeared before Pilate at the Antonia Fortress, as has traditionally been maintained, it is more probable that as ruler of the country Pilate was residing in Herod's palace and that Jesus was interrogated, humiliated, and condemned there.

According to the gospel accounts, Jesus was led outside the city, crucified, and buried in a nearby tomb belonging to Joseph of Arimathea. In Jerusalem today there are two localities that lay claim to being the place of these events. The first of these is Gordon's Calvary, to the north of the present-day Damascus Gate, with the nearby Garden Tomb. Although this site lies outside the ancient as well as the present-day city wall and is quite amenable to certain types of piety, there is no compelling reason to think that this is either Calvary and/or the tomb — in fact, the tomb may date back to the Iron Age (1000 – 586 BC), and thus could not have been a tomb "in which no one had yet been laid" (Luke 23:53).

More compelling, although still not certain, is the suggestion that the Church of the Holy Sepulcher marks the spot of these dramatic events. It is probable that this site was outside the walled city of Jesus' day and was in fact a burial ground. Ancient Christian traditions, dating back to at least the days of Eusebius (fourth century AD), suggest that the church marks the more probable of the two sites. The New Testament records that after his resurrection Jesus appeared to his disciples for forty days and then, in the presence of his assembled disciples on the Mount of Olives, ascended into heaven.

During the early apostolic period (ca. AD 30 – 41) the church was centered in Jerusalem. A number of events are mentioned in connection with the activities of Peter, John, Stephen, and others as having taken place in Jerusalem, such as house meetings, appearances before the Sanhedrin, and imprisonments, but it is difficult if not impossible to pinpoint the exact location of these events. In the temple precincts a paralyzed man who was sitting at the "Beautiful Gate" (probably the gate that led into the court of women) was healed, and it is apparent that the early Christians often met in Solomon's Colonnade (Acts 3:11; 5:12) — probably the colonnade along the inner side of the eastern enclosure wall of the temple precincts.

Since the days of Herod the Great (37 – 4 BC) the city had been steadily expanding northward and finally, during the reign of Herod Agrippa I (AD 41 – 44), an attempt was made to enclose the northern quarter of the city (Bezetha). This wall ran northwest from the present-day Jaffa Gate to the Tower of Psephinus (probably in or near the present-day Russian Compound) and from there turned east to the Kidron Valley, at which point the wall turned south to join the northeastern corner of the Temple Mount (Josephus, *War* 5.4.2 – 3 [147 – 60]). However, Agrippa became fearful that the Romans might accuse him of rebellion, and he stopped work on the wall. Archaeologists have discovered long sections of the east – west portions of this wall about 0.3 miles north of the present-day Old City wall.

After the death of Agrippa I (AD 44) Roman procurators again ruled Jerusalem directly until the outbreak of the first Jewish revolt (AD 66 – 70). One of the immediate causes of the revolt was Emperor Caligula's attempt to have his statue erected in the temple in Jerusalem. The governor of Syria, Cestius Gallus, had some initial success in subduing the revolt in AD 66 but was not able to capture Jerusalem. The Jews were then able to gain control of the city and most of the remainder of the country. For the next three years the Romans, slowly but surely, subdued the rebels. In the spring of AD 70 the Fifth, Tenth, Twelfth, and Fourteenth legions, and their slave captives — about 80,000 men in all — advanced on Jerusalem.

The Jews attempted to fortify the third, or northern, wall that Agrippa I had begun, but by the end of May the Romans had breached it. A few days later the second wall was also breached and a siege dike was set up around the remainder of the city. The suffering within the city was severe, and in late July the Antonia Fortress was attacked and captured. From there the Romans advanced into the temple precincts, and on the ninth of Ab (August 28) the temple was burned (photo p. 216). By the end of September the lower city (the old City of David) and the upper city (on the southern portion of the western ridge) had been captured.

Although Josephus's figures of 1,000,000 Jews dead and 97,000 captured are probably exaggerated, they do illustrate the magnitude of the defeat. Titus, the Roman general who later became emperor, ordered much of the city to be razed, save the three towers just north of the Herodian palace. These he left standing as mute tribute to the greatness of the city that he had just captured.

THE DISCIPLINES OF HISTORICAL GEOGRAPHY

Judeo-Christian beliefs are deeply rooted in space and time. Rather than elaborating an abstract system of theology, the Bible by and large presents history as a concrete form of theology: what God has done in history shows who he is and what he is like. Consequently, there has always been a keen interest in matters related to biblical history. Not only have people asked questions regarding what happened, why it happened, and who was involved, but they have also been interested in knowing where events took place and where and under what conditions people lived.

This interest in historical and geographical details is already evident in the biblical period. On occasion during that period, the transmitters of oral traditions, the authors and editors of texts, and the copyists of scrolls realized that some of their hearers or readers might not recognize the geographical references in a given historical account, and they therefore inserted the then-contemporary name of a place so that the historical setting of an event would be more intelligible to the hearers and readers. Thus the hearers and readers were informed that Bela (a name they were not familiar with) was the same as the city they knew as Zoar (Gen 14:2), that the Valley of Siddim was the same as the Salt Sea (14:3), En Mishpat as Kadesh (14:7), Kiriath Sepher as Debir (Josh 15:15), Kiriath Arba as Hebron (15:54), Kiriath Baal as Kiriath Jearim (15:60), Luz as Bethel (Judges 1:23), and so on.

Philology: Textual Studies

Although modern research on the subject of historical geography did not begin until the time of Napoleon's invasion of Egypt and part of the Levant in 1798 – 99, biblical geography has been a subject for study since at least the days of Eusebius (fourth century AD). Textual materials, both biblical and extrabiblical, have provided the main impetus as well as much of the raw data for the study of Bible geography. One of the main aims of biblical geographers has been to identify biblical place names and other geographical references with existing cities, towns, villages, antiquity sites, regions, etc. With some of these identifications in hand, one can further attempt to define tribal and clan allotments, administrative districts, and boundaries between states. Furthermore, an understanding of the historical-geographical setting of an event can greatly aid in the overall understanding of the biblical narratives.

Two main sources of textual material are available to the biblical geographer. The first is, of course, the Bible itself, which, although it is by no means mainly a book about geography or history, does include much material of a historical-geographical nature. The second source of textual material is the ancient inscriptions that have been found in the Near East.

The Bible

In his comprehensive book *The Land of the Bible: A Historical Geography*, Yohanan Aharoni notes three classes of biblical documents of geographical interest: historical-geographical descriptions, territorial descriptions, and the records of military expeditions and conquests. The first of these, the historical-geographical descriptions, appear to have been composed primarily for geographical purposes. Among them are the Table of Nations (Gen 10), the list of conquered Canaanite kings (Josh 12), and possibly the descriptions of "the land that remains" (Josh 13:1 – 6; Judg 3:1 – 4: map p. 122), as well as the wilderness itinerary of the Israelites (Num 33).

Of special interest to the historical geographer are the territorial descriptions scattered throughout the historical books of the Bible. These are of two types: boundary descriptions and town lists. In some instances the area belonging to a state or tribe is defined by means of a boundary description. Some of these boundary descriptions are brief: only "terminal points" are given. Examples are the description of the land of Israel as stretching "from Dan [in the north] to Beersheba [in the south]" (e.g., 2 Sam 24:2) and the description of the Promised Land (Exod 23:31), which lists four "terminal points": "from the Red Sea [south(east)], to the Sea of the Philistines [= Mediterranean; northwest], and from the desert [southwest] to the River [= Euphrates; northeast]."

But in other cases the boundary descriptions are much more detailed, and a number of points along a border are mentioned sequentially — in a "dot-to-dot-to-dot" fashion. Examples of this technique include the description of the boundaries of the land of Canaan in Numbers 34:1 – 12 and several of the tribal boundary descriptions in Joshua, such as those of Judah (Josh 15:1 – 12), Benjamin (18:12 – 20), Ephraim (16:1 – 9), Manasseh (17:7 – 11), and Zebulun (19:10 – 16). To be sure, in some instances these descriptions appear to have been abbreviated, but others, such as the description of the border between Judah and Benjamin in the Jerusalem area (Josh 15:7 – 9; 18:15 – 17), remain quite detailed.

For the historical geographer, the "terminal point" descriptions are somewhat useful in determining the perceived boundaries of states and tribes, but they are general. The "sequential" boundary descriptions are often much more useful. They can be plotted on a map, and because the names usually follow a logical geographical sequence, the location of unknown places can be surmised. For example, the exact location of Ekron, a Philistine town, was disputed for many years, but its general location was known because the biblical text placed it on the western portion of the northern boundary of Judah (map p. 114) — specifically, to the west of Beth Shemesh but to the east of Jabneel (Josh 15:10 – 11), two towns whose identifications were generally agreed upon.

In addition, territories were described by drawing up a list of the names of the towns assigned to a given tribe or district. Although the date and function of these town lists have been the subject of much dispute, we are certain that at least one of them, the list of twelve Solomonic districts in 1 Kings 4:7 – 19, was used for administrative purposes (map p. 148). Prominent among these rosters are the town lists of Judah (Josh 15:21 – 63; more than 100 towns are mentioned!), Simeon (Josh 19:2 – 8; 1 Chron 4:28 – 33; 17 towns), and Benjamin (Josh 18:21 – 28; 26 towns), as well as the lists of Levitical towns (Josh 21; 1 Chron 6:54 – 80; about 48 towns), although there are many others.

The town list of Judah is of special interest because it is so extensive and because it is divided into at least nine or ten districts (the Septuagint even includes an eleventh district, around Bethlehem, in Josh 15:59; see the NEB for a translation). Even though the date and original function of the town list of Judah in Joshua 15 has been the subject of debate, it has proved useful in assisting to identify the location of disputed towns. For example, Debir, which some authorities located at Tell Beit Mirsim in the Shephelah, was really located in a southern hill country district (v. 49), while Adullam, which some located in the Judean Hill Country, was actually located in the northern district of the "western foothills" (= Shephelah; v. 35). Thus a list like this can poten-

▲ *Cuneiform tablet from el-Amarna, Egypt (14th century BC). Written documents like this help fill in gaps in our knowledge of ancient Canaan.*

tially help peg at least the general placement of biblical towns whose locations are in dispute.

The third type of biblical material, besides historical-geographical and territorial descriptions, is the records of expeditions and conquests. Thus the routes of military expeditions, such as those of Abijah (2 Chron 13:19), Ben-Hadad (1 Kings 15:20), and Tiglath-Pileser (2 Kings 15:29), can be traced with some certainty. Given the reasonable assumptions that these expeditions followed a more or less sensible geographical progression and that the texts faithfully record these invasions, the identification of certain biblical towns can be confirmed, while in the case of unknown towns, at least their general location can be surmised. For example, although the exact location of Janoah is disputed, it is reasonable, on the basis of 2 Kings 15:29, to place it in the vicinity of Abel Beth Maacah, Kedesh, and Hazor (map p. 157).

Extrabiblical Texts

During the last 100 years, numerous epigraphic finds have been made in Egypt, Mesopotamia, and Syria, and to a lesser extent in Palestine/Israel. By far the most substantial corpus of texts relevant to the historical geography of the Bible has come from Egypt. Major finds from the Middle Bronze II period (2000 – 1550 BC) include the Execration texts (texts invoking a curse), with their listing of geographical, personal, and tribal names in the southern Levant, and the Story of Sinuhe, with its description of the land of Upper Retenu, which also includes incidental geographical and cultural references to the land of Canaan.

From the Late Bronze Age (1550 – 1200 BC; the period of the exodus, conquest, and settlement of the Israelites in Canaan) there is a wealth of information, including a description of the driving of the Hyksos from Egypt back into Canaan. From the Eighteenth Dynasty in Egypt, some of the campaigns of Thutmose III (1479 – 1425 BC), especially the first, are recorded in his annals and elsewhere in great detail (map p. 101; photos pp. 100, 102). Copies of his great topographical list include at least 119 geographical names from the Levant. Also to be mentioned are the descriptions of the campaigns of Amenhotep II (1427 – 1400 BC) into Canaan during the early years of his reign. In addition, from the reigns of Amenhotep III (1390 – 1352 BC) and Akhenaton (1352 – 1336 BC), there is the diplomatic archive, written in Akkadian, that was discovered at el-Amarna. It details Egyptian-Levantine relations as well as the internal struggles in the land of Canaan.

From the Nineteenth Dynasty, the campaigns of Seti I (1294–1279 BC) and Ramses II (1279–1213 BC) are well known, and the stele of Merneptah (1213–1203 BC) even mentions "Israel" as being among other groups living in the southern Levant. These texts are supplemented by Papyrus Anastasi I, which describes the geography of Canaan in some detail, as well as by the story of Wen-Amon and other texts. From later periods, only the account of Shishak's invasion of Israel (925 BC), with its listing of close to 150 settlements in Israel and the Negev, is of major importance (map p. 153; photo p. 152).

Epigraphic finds from Mesopotamia make less of a contribution to the historical geography of the Bible than those from Egypt, although references to Hazor and Laish (= Dan) in the cuneiform tablets found at the Syrian site of Mari shed welcome light on the Middle Bronze II period. From the eighth, seventh, and sixth centuries BC, the records of the expeditions of Assyrian and Babylonian kings (e.g., those of Tiglath-Pileser III, Sargon II, Sennacherib, Esarhaddon, Ashurbanipal, Nebuchadnezzar II) also supply a fair amount of relevant geographical detail regarding Israel and its neighbors. In addition, it is possible that the tablets discovered in northern Syria, at Tell Mardikh (ancient Ebla), will contribute to what until now has been an "epigraphic dark age" in our geographic understanding of Palestine, namely, the third millennium BC.

Epigraphic finds in Israel and Jordan have been scarce. A few cuneiform texts from the Late Bronze Age (1550–1200 BC) have been discovered at Tell el-Hesi, Beth Shemesh, Mount Tabor, Taanach, Hazor, and Aphek, but their contribution to the historical geography of the land has been minimal. From the Iron Age (1200–586 BC), the ostraca (pottery fragments with writing on them) discovered at Samaria shed welcome light on a portion of the tribal inheritance of Manasseh, an area the Bible does not treat in detail (Josh 17). The Moabite Stone provides information about the region east of the Dead Sea during the eighth century BC. From Israel proper, the Arad ostraca contain a few geographical references, as do the ostraca discovered at Lachish. (Translations of many of these texts, as well as many of those from Egypt and Mesopotamia, can be found in *Ancient Near Eastern Texts relating to the Old Testament*.)

Finally, from the Hellenistic and early Roman periods, classical authors (e.g., Herodotus), papyri from Egypt (the Zenon papyri), the Septuagint, the books of the Apocrypha, the Dead Sea Scrolls, Josephus, as well as other epigraphic finds all contribute to our historical geographical understanding of the intertestamental and New Testament periods (= Second Temple period).

One of the goals of textual studies in historical geography is to attempt to write as complete a history as possible of a given place or territory, based on all the available resources. The collection of all references to a given town or territory in the literary sources can be complex, for there may have been several towns that had the same name or a single town or territory may have had more than one name. But even if a similar name for a town is used in various languages, the researcher

▼ *Medeba, Jordan: The Jerusalem portion of a 6th-century map of Palestine. To the left (north) is the Damascus Gate and in the center is the Church of the Holy Sepulcher — upside down. Streets are yellow and white; red-roofed buildings are churches and monasteries.*

must deal with the problem of how a name was transcribed, since the phonetic and spelling systems of the various languages do not completely coincide. Thus a given place-name may occur in Hebrew, Egyptian, and Babylonian documents, but the name may not refer to the same place in all documents, while on occasion apparently different names may well refer to the same place.

A further complication is that some of the documents were written long after the date of the events they record. This means that a geographical name used in such a document or section of a document may be the name that was used at the time of the event or the name of the site that was used at the time of writing. For example, although Genesis records events said to have occurred in the days of Abraham, Isaac, and Jacob, certainly the book was written some time after the recorded events, and, as stated above, the writer in several instances took care to give the new name of a town along with the old one.

Toponymy: Study of Place Names

So far it has been noted that a study of the "primary" ancient sources, biblical and extrabiblical, alerts us to the fact that a certain settlement existed in ancient times, and that often these same sources give us clues as to the general geographical location of the site — in the north or south, in the hills or plains, proximity to other towns and villages, etc. One of the most successful ways of trying to attach the ancient name of a settlement to its correct site on the ground and in turn to its place on a modern map is to see if its name has been preserved through the centuries down to the present time.

At first glance this line of investigation (formally known as the study of toponymy) may seem somewhat futile, given the thousands of years that separate us from the time of the Bible. But the well-watered areas of the land of Israel/Palestine have been inhabited by a continuous chain of peoples who have handed down the name of a given place orally from generation to generation. Thus names like Jerusalem, Hebron, Acco, and Tiberias have been preserved for thousands of years. The preservation of ancient place-names has been helped by the fact that through the ages the languages of the indigenous population groups have all been Semitic. Thus, Canaanite was related to Hebrew, Hebrew in turn to Aramaic, and Aramaic to Arabic; we must recognize, of course, that there were also many linguistic differences between these languages. However, in more remote areas, such as Sinai, there seem to have been significant gaps in the chain of indigenous inhabitants, and thus the ancient geographical names have not been well preserved through the centuries.

Since the preservation of place-names has been primarily an oral process, the modern investigator can either consult local, indigenous sources (usually Arabic-speaking) — who, by the way, are becoming fewer and fewer due to large-scale population movements in the Levant — or he/she can consult written works that have codified the oral traditions at certain points in time. These more recent written sources are "secondary" witnesses to places mentioned in the "primary" sources of the biblical era (but they are, of course, primary sources for their own eras). Major secondary sources include Josephus,

rabbinic literature, the *Onomasticon* of Eusebius, which was translated and expanded upon by Jerome, the pilgrim itineraries from the Byzantine and Crusader periods, the works of Arab geographers, and the work of the Jewish scholar Eshtori ha-Parchi and others.

The modern study of toponymy began with the trip of Edward Robinson and Eli Smith to Palestine in 1838. Robinson was a great biblical scholar who had mastered all of the primary and most of the secondary materials available in his day. His expertise was supplemented by that of his former pupil, Smith, then a missionary in Beirut, who was fluent in many Arabic dialects and who accompanied him on that first trip. It was on that trip and on one made by Robinson alone in 1852 that the names of many biblical places were correctly identified as being preserved by "modern" local toponyms.

Although many individual contributions have subsequently supplemented their pioneering work, it was the great British survey of Palestine between 1871 and 1877 that added substantially to the available toponymic data. As a result of this survey, twenty-six large-scale maps, covering roughly the area from Dan to Beersheba, west of the Jordan River and the Dead Sea, were published, along with a number of volumes describing the topographical, cultural, linguistic, and archaeological findings of the survey. The works of Robinson and Smith and those of the British team remain great written repositories of local indigenous oral toponymic traditions that existed prior to the advent of the modern era.

Modern toponyms may preserve the name of an ancient settlement, but those names may not always be attached to the exact place where the biblical settlement was located. For example, the ancient site of biblical Jericho has been located at Tell es-Sultan, but the name was preserved at the nearby village of er-Rahia. While "er-Rahia" may not look like "Jericho" to a layperson, experts who are familiar with the sounds, scripts, and phonetic laws of Hebrew, Aramaic, and Arabic are able to confidently make such identifications. Thus identifications such as Micmash with the Arab village of Mukhmas, Upper Beth Horon with Beit Ur el-Foqa, and many others, are fairly certain.

Besides the ancient name being preserved at the ancient site itself or in the name of a nearby village, it may also be preserved in the name of a nearby tomb (the name Gezer was preserved in the name of the tomb of Sheikh Jezari on the top of the tell) or water source (biblical Beth Shemesh was preserved in the Arabic toponym of Ain Shems located near the ancient site, and the name of Jabesh [Gilead] has been preserved in the name of the Wadi Yabis, along which it was located).

In addition to the complication that the name may not be preserved exactly at the ancient site, it might also be expected that an intrusive population group and/or language (such as Greek or, to a lesser extent, Latin, during the Hellenistic and Roman periods) might further complicate matters. However, this is not often the case. For example, even though the official name of Beth Shan was changed to Scythopolis during the Hellenistic period, the new name did not stick, for evidently the local population preserved the older Semitic name, and with the departure of Greco-Roman influence the use of a form of the more ancient name again came to the fore. Thus the name had been preserved up until recent times as the name of the small Arab village of Beisan, which was located near the foot of the tell where the biblical town was located.

By contrast, forms of the Greek and Latin names of cities that were established during the Roman period have often been preserved by the local Semitic populace: Tiberias is preserved as Tabariyeh and Caesarea as Qeisarieh. However, in a few rare instances, old Semitic names have been replaced by a Greco-Roman counterpart. For example, biblical Shechem was renamed

Mark Connally, Amman, Archaeological Museum

▲ *Metal bottle found at Tell Siran, Jordan dating to ca. 600 BC mentions "Amminadab king of the Ammonites"*

Neapolis during the Greco-Roman period, and it is that name that is preserved in the name of the modern city Nablus.

The above examples are but a few instances of the complicated process of name transmission and preservation through the ages. Of course, any suggested site identification is greatly enhanced if the ancient name can be traced through a number of the "secondary" sources mentioned above, especially if those sources come from different time periods and place the settlement consistently in the proper geographical area. If this is the case, then one can be reasonably certain that the local populace has preserved the name, and one can assert with a reasonable degree of assurance that the ancient site was situated at or near the locale of the "modern" toponym.

Archaeology

The process of matching a geographical name with a given antiquity site, usually a town, tell, or khirbet, involves the attempt to match the known history of a place (see above) with the archaeological remains found at the proposed site. For example, when places such as Bethel, Ai, Hebron, Gerar, and Shechem are mentioned in the narratives of the lives of Abraham, Isaac, and Jacob, does this mean that all of them or just some of them were cities at the time of the patriarchs? (A gate mentioned in connection with Hebron [Gen 23:18], and a building with a second story in connection with Gerar [26:8], at least imply that these were urban areas.)

On the one hand, if one believes that the text faithfully records the events that took place during the patriarchal era, then one reasonably expects to find archaeological remains from that period at the antiquity sites that are candidates for Gerar, Hebron, etc. On the other hand, if one believes that the purpose of the text is other than historical and that the stories are legendary, then one need not expect to find remains from the period of the recorded event at the site. Thus the writing of a "history" of an ancient site can be a rather complex task, and the result-

ing "history" depends to a large extent on the presuppositions that the researcher brings to the task.

But once a "history" of a city has been written, the researcher attempts to determine which of perhaps several antiquity sites located in the general area that has been proposed as the probable location is in fact the actual location of the ancient city. Once a prime candidate has been chosen, often on the basis of initial surface surveys of several sites (see below), archaeological excavations can begin. It must be noted in passing that although the emphasis in this essay is on the role that archaeology can play in site identification, usually other reasons are more prominent in selecting antiquity sites for archaeological excavation. These include such factors as the site's size, historical importance, and regional significance, and the need for emergency salvage operations in the wake of modern building activities.

Today two major methods are used in excavating a tell. The first can be called the *trench method* (or Wheeler-Kenyon method). The archaeologists select an area of the mound, usually near its edge, where they sink a trench that resembles a thin slice taken from a layer cake. They hope that this trench will in fact intersect all the occupation levels (strata) represented within the mound and also that it will intersect any fortifications that might be preserved along its flank. This "trench" is

▲ *Replica of a Tyrian coin used by Jews to pay the Temple Tax during the Second Temple Period*

▲ *Tell Beth Shan in northern Israel. During the OT period life was focused on the tell, but during the Greco-Roman period the city greatly expanded to the area at the foot of the tell.*

actually a series of 5-by-5-meter squares, set in a line and sometimes arranged in pairs, which have a 1-meter divider (baulk) separating them.

As the excavation progresses, scrupulous records are kept, and drawings, photographs, and measurements are made. Thus the exact provenience of each artifact, be it a pot, a piece of jewelry, or something else, is recorded and it can later be studied in relation to the context in which it was found. In addition, the architectural remains (walls, buildings, floors, fortifications, etc.) are plotted in an effort to reconstruct the layout and defenses of the ancient settlement.

As the excavators progress, they attempt to sort out the various "levels" or "strata" of occupation that are represented in the tell. Some tells, such as Megiddo or Hazor, may have more than twenty strata, while others, such as the lower city at Arad, may have four or even fewer. Once excavated, the finds from a given stratum (pottery, jewelry, etc.) are gathered together into an "assemblage," from which one can draw conclusions regarding the culture of the people who lived at that time. To assist in the sorting out of the strata, the sides (= baulks) of the 5-meter squares are kept perfectly vertical, and thus the archaeologists are able to look back on these as reference points that help them gather additional data on the strata that they have been digging through. Indeed, photographs and/or drawings of these vertical surfaces become part of the permanent record of the excavation.

Why Are There Tells?

The ancients did not set out to build tells. Rather, it often took centuries for tells to develop. The following are some of the more important factors that entered into the complex process of the formation of tells.

1. People preferred to settle close to a source of fresh water—a spring, a well, or, more rarely, a flowing stream.
2. By settling on a hill or a rise near a water source, people could more easily monitor the surrounding landscape and defend themselves...
3. People preferred to live in regions with good agricultural land and/or pasturage.
4. Often people wanted to live close to major and even minor "roads," which may have led to the settlement of some sites.
5. Other sites may have developed because of their religious significance, their proximity to special natural resources, etc.
6. If stone walls, foundation walls, or even just stones from structures of previous inhabitants of a site were available, these could easily be reused in the building of a new settlement. In some areas of the country, the accumulation of mud from mud bricks also significantly contributed to the rise of a tell.

Since there was a limited number of water sources with a limited number of hills near them, and since on these sites building materials from previous settlers were often available, it was here that new settlements were built over old ones, a process that was often repeated many times over. Thus in the end, the distinctive mounds now known as tells were formed.

▲ *OT Jericho: Kathleen Kenyon's cut (trench method) 50 years after her excavations. Even in an arid climate like Jericho (8 in. rain per year) mud brick does deteriorate.*

Once a "trench" has been excavated down to virgin soil or bedrock, there is a high degree of probability that it has intersected all the strata on the tell. Almost always the lower strata are from an earlier date than the upper ones, but how can this relative dating sequence be transformed into an absolute one? Inscriptional evidence mentioning a known and datable historical figure or event, found in a given stratum, can be useful in setting boundaries for the dates of that stratum and of other strata. Unfortunately, not much inscriptional evidence from the biblical period is found in the tells of Israel/Palestine, with the exception of coins from the Hellenistic and Roman periods. In some cases modern technology can help, such as the use of the carbon – 14 method for the dating of organic remains within ± 50 years for the historical periods in question.

Another important factor that enters into the dating process is the principle of "typology." Given that one stratum is older than another, one is able to compare the stylistic development of artifacts (oil lamps, cooking pots, gates, palaces, temple designs, etc.) from early to more recent periods. Over the past fifty years these typologies, particularly pottery types with their various forms, decorations, and textures, have become well established, and various forms have been linked to strata that in turn can be correlated to Egyptian and/or Mesopotamian history where their absolute chronologies have been established with a fair degree of certainty. Thus today, in spite of the lack of epigraphic finds, a given stratum can usually be dated on the basis of the type of pottery it contains.

This trench method of excavation is a valuable technique, for it allows the archaeologist to get an overview of the total history of the tell with relatively little outlay of time and money. In addition, this method leaves large areas of the tell untouched so that future generations of archaeologists, with better techniques and equipment, can return to re-excavate the site to check and improve upon earlier conclusions. However, since only a small area of the tell is opened up, usually only parts of buildings and of other interesting structures may be uncovered. Indeed, it may be that a given stratum is represented elsewhere on the tell but is actually totally absent in the area where the probe trench was sunk, and thus incomplete and inaccurate conclusions may be drawn regarding the occupational history of the site.

A second method of excavation, often used in conjunction with the first, aims at opening up large areas of the tell so as to be able to appreciate overall building designs, town planning, etc. These are worthy goals, but this method is time consuming and costly, and once a given stratum has been completely removed from the mound, it can never be re-excavated. In fact, some of the great controversies of modern archaeology have little chance of resolution because past excavators have completely removed the strata in question from a mound, and their published, or sometimes unpublished, results may lead to all kinds of ambiguity.

Excavation, no matter what method is used, is a costly undertaking. In modern times a technique called *surface survey* has led to some productive results at a comparatively low cost. In this method, surface features (topography, wall outlines, installations, etc.) of a tell are drawn to scale, and pottery sherds (broken pieces) are collected from all over the top, sides, and bottom of the mound, on the assumption that some artifacts from the edge of each stratum will have worked their way to the surface. Often, in an attempt to refine the process, certain areas are selected in advance, and the sherds from each area are kept separate from those found in other areas. This at times is helpful in developing a more accurate history of the occupation of the site.

In either case, an archaeologist, through pottery analysis, can come to some initial conclusions as to when the mound was occupied. It must be remembered, however, that the sherds from a given stratum may not have worked their way to the surface, sides, or foot of the tell, and thus the resulting occupational history may be incomplete. But by using this method, a large number of antiquity sites can be quickly surveyed and a general occupational history of each site can be developed.

Thus for the historical geographer, the excavation or survey of a proposed site can help either to confirm or cast doubt on the proposed identification of that site. If the archaeological data do not conform to the known history of a site, then alternate proposals may be made. For example, for a number of years it was thought that Tell esh-Sheikh Ahmed el-Areini fit the geographical description of the Philistine city of Gath, based on the interpretation of the historical sources that was then current, and it was identified as such on printed maps. However,

when the site was excavated and no Philistine remains were found, the identification was dropped, maps were revised, and alternative suggestions were then made as to the true identification of Philistine Gath.

In other instances, the historical geographer, using written sources, may be almost certain regarding the geographical setting of an ancient settlement and the identification of a tell with that settlement, yet when the archaeological profile of the site in question does not fit with the known history of the site, the geographer may begin to question the thoroughness of the survey or excavation, the accuracy of the historical documents, and/or his or her interpretation of the texts. Thus archaeology, which has many other roles and goals in reconstructing the ancient past, can lend or limit support for a proposed identification of an ancient settlement and in some cases lead the investigators to a reevaluation of their historical sources. But usually at the conclusion of the process of site identification, the researcher is able to say that the proposed identification is either certain, probable, possible, or untenable — with the understanding that even "certain" identifications at times need to be revised in light of new evidence and/or fresh interpretations of old evidence.

Geography

Although the discipline of geography is a broad one, certain aspects of it, such as the investigation of local topography, geology, soil types,

▼ *Beersheba: View of the gate area. Y. Aharoni, the excavator, exposed and rebuilt large areas of the site so that the gate, streets, and buildings are identifiable.*

water sources, and climatic conditions, are useful in studying ancient settlement patterns, the location of routes of travel, and the economic base of a region. Since the early chapters of this book have dealt with how the ancients lived their lives in their immediate environment, these topics will not be treated here, and the reader is referred to the appropriate regional studies in the Geographical Section of this atlas (pp. 15–79).

Conclusion

Although this essay has emphasized the fascinating process of site identification — how an ancient settlement can or cannot be identi-fied with an actual antiquity site that in turn can be represented as a dot on a map in an atlas — the historical geographer is also interested in broader questions, such as historical reconstructions and how the ancients lived life in context. How did their immediate environment influence their lives and how, in turn, did they structure their environment? By following up on such geographical questions, as well as on ecological ones, and with the input from the disciplines of philology, toponymy, and archaeology, we can begin to draw a total regional picture. Along this line, detailed regional investigations have already begun in earnest; they will occupy the attention of historical geographers in the years ahead.

BIBLIOGRAPHY

Abel, F.-M, ed. *Géographie de la Palestine: Par Le P. F.-M. Abel des Freres Precheurs. Tome I. Géographie physique et historique.* Third edition. Paris: Gabalda, 1933.

Abel, F.-M, ed. *Géographie de la Palestine: Par Le P. F.-M. Abel des Freres Precheurs. Tome II. Géographie politique: Les villes.* Third edition. Paris: Gabalda, 1938.

Aharoni, Yohanan. *The Land of the Bible.* Edited and translated by Anson F. Rainey. Revised and enlarged ed. Philadelphia: Westminster, 1979.

Aharoni, Yohanan, and Michael Avi-Yonah. *The Macmillan Bible Atlas.* Edited by Anson F. Rainey and Ze'ev Safrai. Completely revised third ed. New York: Macmillan, 1993.

Atlas of Israel. Amsterdam: Elsevier, 1970.

Avi-Yonah, Michael, ed. *The Holy Land: A Historical Geography from the Persian to Arab Conquests (536 BC to AD 640).* With text revisions and toponymic index by A. F. Rainey. Introduction to the 2002 edition, Y. Tsafrir. Jerusalem: Carta, 2002.

Bahat, Dan, and Chaim T. Rubinstein, eds. *The Illustrated Atlas of Jerusalem.* Translated by Shlomo Ketko. Jerusalem: Carta, 1990.

Baines, J., and J. Malek. *Atlas of Ancient Egypt.* New York: Facts on File, 1982.

Baly, Denis. *The Geography of the Bible.* New and revised ed. New York: Harper & Row, 1974.

Baly, Denis, and A. D. Tushingham. *Atlas of the Biblical World.* New York: World, 1971.

Bimson, John J., ed. *Baker Encyclopedia of Bible Places.* Grand Rapids: Baker, 1995.

Brisco, Thomas V. *Holman Bible Atlas.* Nashville: Broadman & Holman, 1998.

Cornell, T., and J. Matthews. *Atlas of the Roman World.* New York: Facts on File, 1982.

Dorsey, David A. *The Roads and Highways of Ancient Israel.* Baltimore: Johns Hopkins University Press, 1991.

Elitzur, Yoel. *Ancient Place Names in the Holy Land: Preservation and History.* Winona Lake, Ind.: Eisenbrauns, 2004.

Hallo, William W., and K. Lawson Younger Jr., eds. *The Context of Scripture: Volume I: Canonical Compositions from the Biblical World.* Leiden: Brill, 1997.

Hallo, William W., and K. Lawson Younger Jr., eds. *The Context of Scripture: Volume II: Monumental Inscriptions from the Biblical World.* Leiden: Brill, 2000.

Hallo, William W., and K. Lawson Younger Jr., eds. *The Context of Scripture: Volume III: Archival Documents from the Biblical World.* Leiden: Brill, 2002.

Hallo, W. W., and W. K. Simpson. *The Ancient Near East: A History.* Second ed. New York: Harcourt Brace Jovanovich, 1998.

Har-El, Menashe. "The Exodus Route in the Light of Historical – Geographical Research." Pages 373 – 96 in *Geography in Israel.* Edited by D. H. K. Amiran and Y. Ben – Arieh. Jerusalem: Israel National Committee International Geographical Union, 1976.

Hoerth, Alfred J. *Archaeology and the Old Testament.* Grand Rapids: Baker, 1998.

Hoffmeier, James K. *Israel in Egypt: The Evidence for the Authenticity of the Exodus Tradition.* New York: Oxford University Press, 1997.

Kallai, Zecharia. *Historical Geography of the Bible: The Tribal Territories of Israel.* Jerusalem: Magnes, 1986.

Kitchen, Kenneth A. *On the Reliability of the Old Testament.* Grand Rapids: Eerdmans, 2003.

Levi, Peter. *Atlas of the Greek World.* New York: Facts on File, 1982.

MacDonald, Burton. *"East of the Jordan": Territories and Sites of the Hebrew Scriptures.* ASOR Books 6. Edited by Victor Matthews. Boston: The American Schools of Oriental Research, 2000.

Mazar, Amihai, ed. *Archaeology of the Land of the Bible: 10,000 – 586 BCE.* Vol. 1 of The Anchor Bible Reference Library. New York: Doubleday, 1990.

McRay, John. *Archaeology and the New Testament.* Grand Rapids: Baker, 1991.

Mittmann, Siegfried, and Gotz Schmitt, eds. *Tübinger Bibelatlas.* Based on maps from the *Tübinger Atlas des Vorderen Orients (TAVO).* Translated by Keith Myrick. Stuttgart: Deutsche Bibelgesellschaft, 2001.

Monson, James M., and Steven P. Lancaster. *Regions on the Run: Introductory Map Studies in the Land of the Bible.* Third edition. Rockford, Ill.: Biblical Backgrounds, 2009.

Monson, James M., and Steven P. Lancaster. *Geobasics in the Land of the Bible.* Rice Lake, Wisc.: Biblical Backgrounds, 2008.

Murphy-O'Connor, Jerome. *The Holy Land: An Oxford Archaeological Guide from Earliest Times to 1700.* Revised and expanded fifth ed. Oxford: Oxford University Press, 2008.

Notley, R. Steven. *Eusebius, Onomasticon: A Triglot Edition with Notes and Commentary.* Leiden: Brill, 2005.

Orni, E., and E. Efrat, eds. *Geography of Israel*. Revised third ed. Jerusalem: Israel Universities Press, 1971.

Pritchard, J. B., ed. *Ancient Near Eastern Texts relating to the Old Testament*. Third edition. Princeton: Princeton University Press, 1969.

Pritchard, J. B., ed. *Ancient Near East in Pictures relating to the Old Testament*. Second edition. Princeton: Princeton University Press, 1969.

Rainey, Anson F. "The Toponymics of Eretz-Israel." *Bulletin of the American Schools of Oriental Research* 231 (1978): 1 – 17.

Rainey, Anson F., R. Steven Notley, J. Uzziel, I. Shai, and B. Schultz. *The Sacred Bridge: Carta's Atlas of the Biblical World*. Jerusalem: Carta, 2005.

Ramsay, William M., ed. *St. Paul: The Traveler and Roman Citizen*. Edited by Mark W. Wilson. Full-color updated and revised edition. Grand Rapids: Kregel, 2001.

Ritmeyer, Leen. *The Quest: Revealing the Temple Mount in Jerusalem*. Jerusalem: Carta, 2006.

Roaf, Michael. *Cultural Atlas of Mesopotamia and the Ancient Near East*. New York: Facts on File, 1990.

Robinson, E. *Biblical Researches in Palestine and Adjacent Regions: A Journal of Travels in the Years 1838 and 1852*. Reprint of 1856 edition; 3 volumes. Jerusalem: Universitas Booksellers, 1970.

Silva, Moisés, ed. *Zondervan Encyclopedia of the Bible*. Grand Rapids: Zondervan, 2009. Revision of *Zondervan Pictorial Encyclopedia of the Bible*. Edited by Merrill C. Tenney. Grand Rapids: Zondervan, 1976.

Simons, J. *The Geographical and Topographical Texts of the Old Testament*. Leiden: Brill, 1959.

Smith, George Adam. *The Historical Geography of the Holy Land*. Fontana Library edition, 1973, using the twenty-fifth ed. London: Collins, 1931.

Stern, Ephraim, Ayelet Lewinson – Gilboa, and Joseph Aviram, eds. *The New Encyclopedia of Archaeological Excavations in the Holy Land*. 4 volumes. New York: Simon & Schuster, 1993.

Stern, Ephraim, Hillel Geva, Alan Paris, and Joseph Aviram, eds. *The New Encyclopedia of Archaeological Excavations in the Holy Land. Volume 5: Supplementary Volume*. Jerusalem and Washington, D.C.: Israel Exploration Society and Biblical Archaeology Society, 2008.

Thiele, E. R. *The Mysterious Numbers of the Hebrew Kings*. New revised third ed. Grand Rapids: Zondervan, 1983.

Tsafrir, Yoram, Leah Di Segni, and Judith Green. *Tabula Imperii Romani Iudaea Palestina: Eretz Israel in the Hellenistic, Roman and Byzantine Periods*. Jerusalem: The Israel Academy of Sciences and Humanities, 1994.

Wood, Bryant G. "From Ramesses to Shiloh: Archaeological Discoveries Bearing on the Exodus – Judges Period." Pages 256 – 82 in *Giving Sense: Understanding and Using Old Testament Historical Texts*. Edited by David M. Howard and Michael Grisanti. Grand Rapids: Kregel, 2003.

Wright, Paul H. *Greatness Grace & Glory: Carta's Atlas of Biblical Biography*. Jerusalem: Carta, 2008.

www.HolyLandPhotos.org. Web site photos and content provided by Carl Rasmussen.

GLOSSARY

Aeolian	Soil deposited by the action of the wind
Alluvial	Soil/sediment deposited by flowing water
Beer	Well (Hebrew; Arabic: Bir)
Beit	House (Arabic; Hebrew: Bet, Beth)
Beqa	Valley, basin (Arabic; Hebrew: Emeq)
Bet, Beth	House (Hebrew; Arabic: Beit)
Bir	Well (Arabic; Hebrew: Beer)
Cisjordan	Area west of the Jordan Valley, or area west of the Rift Valley
Darb	Path, trail, road (Arabic)
Ein	Spring (Arabic or Ain; Hebrew: En)
Emeq	Valley, basin (Hebrew; Arabic: Beqa)
En	Spring (Hebrew; Arabic: Ein/Ain)
Har	Mountain (Hebrew; Arabic: Jebel)
Horbat	Ruin (Hebrew; Arabic: Khirbet)
Jebel	Mountain (Arabic; Hebrew: Har)
Khirbet	Ruin (Arabic; Hebrew: Horbat)
Kurkar	Fossilized dune sandstone
Levant	The land at the eastern end of the Mediterranean Sea
Loess	Fine aeolian soil found in the biblical Negev
Makhtesh	Mortar- or cigar-shaped depression with steep cliff-like sides (Hebrew)
Mishor	Plain (Hebrew)
Nahal	River, stream, dry riverbed (Hebrew; Arabic: Nahr)
Nahar	River (Hebrew; Arabic: Nahr)
Nahr	River (Arabic; Hebrew: Nahar)
Septuagint	The early Greek translation of the Old Testament
Tel	Antiquity mound (Hebrew; Arabic: Tell)
Tell	Antiquity mound (Arabic; Hebrew: Tel)
Terra rosa	Fertile red-colored soil found in the hill country of Israel, especially suited for growing olives and grapes
Transjordan	Area east of the Jordan Valley, or area east of the Rift Valley
Wadi	River, stream, dry riverbed (Arabic; Hebrew: Nahal)

TIMELINE OF BIBLICAL HISTORY

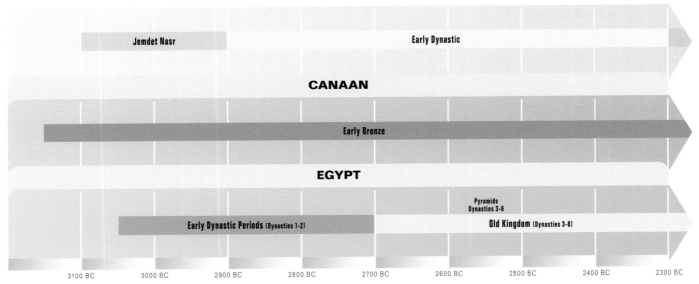

SYRIA/MESOPOTAMIA

Jemdet Nasr
Early Dynastic

CANAAN

Early Bronze

EGYPT

Pyramids
Dynasties 3-6

Early Dynastic Periods (Dynasties 1-2)
Old Kingdom (Dynasties 3-8)

| 3100 BC | 3000 BC | 2900 BC | 2800 BC | 2700 BC | 2600 BC | 2500 BC | 2400 BC | 2300 BC |

● Pre-Patriarchal Period pp. 82–88

Rome

Greece

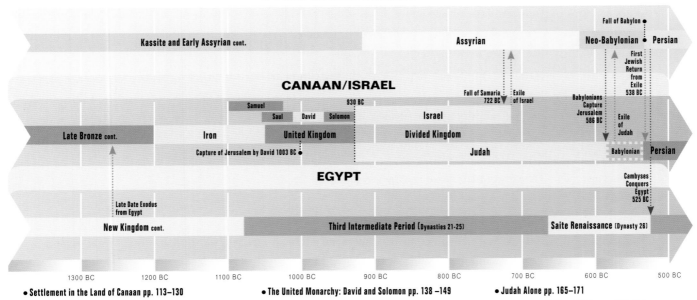

SYRIA/MESOPOTAMIA

Fall of Babylon

Kassite and Early Assyrian cont.
Assyrian
Neo-Babylonian
Persian

First
Jewish
Return
from
Exile
538 BC

CANAAN/ISRAEL

Fall of Samaria
722 BC
Exile
of Israel

Babylonians
Capture
Jerusalem
586 BC

Samuel
930 BC

Saul David Solomon
Israel

Late Bronze cont.
Iron
United Kingdom
Divided Kingdom
Exile
of
Judah

Capture of Jerusalem by David 1003 BC
Judah
Babylonian
Persian

EGYPT

Cambyses
Conquers
Egypt
525 BC

Late Date Exodus
from Egypt

New Kingdom cont.
Third Intermediate Period (Dynasties 21-25)
Saite Renaissance (Dynasty 26)

| 1300 BC | 1200 BC | 1100 BC | 1000 BC | 900 BC | 800 BC | 700 BC | 600 BC | 500 BC |

● Settlement in the Land of Canaan pp. 113–130

● The United Monarchy: David and Solomon pp. 138 –149

● Judah Alone pp. 165–171

● Transition to the Monarchy: Samuel and Saul pp. 131–137

Exile and Return pp. 172–179 ●

● The Divided Kingdom pp. 150–164

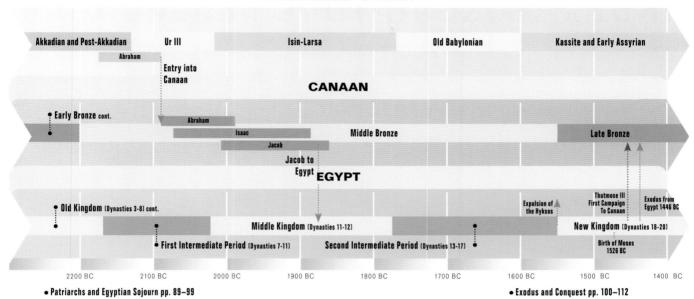

SYRIA/MESOPOTAMIA

| Akkadian and Post-Akkadian | Ur III | Isin-Larsa | Old Babylonian | Kassite and Early Assyrian |

Abraham

Entry into
Canaan

CANAAN

Early Bronze cont.

Abraham

Isaac

Jacob

Middle Bronze

Late Bronze

Jacob to
Egypt

EGYPT

Old Kingdom (Dynasties 3–8) cont.

Expulsion of
the Hyksos

Thutmose III
First Campaign
To Canaan

Exodus from
Egypt 1446 BC

Middle Kingdom (Dynasties 11–12)

New Kingdom (Dynasties 18–20)

First Intermediate Period (Dynasties 7–11)

Second Intermediate Period (Dynasties 13–17)

Birth of Moses
1526 BC

| 2200 BC | 2100 BC | 2000 BC | 1900 BC | 1800 BC | 1700 BC | 1600 BC | 1500 BC | 1400 BC |

- Patriarchs and Egyptian Sojourn pp. 89–99
- Exodus and Conquest pp. 100–112

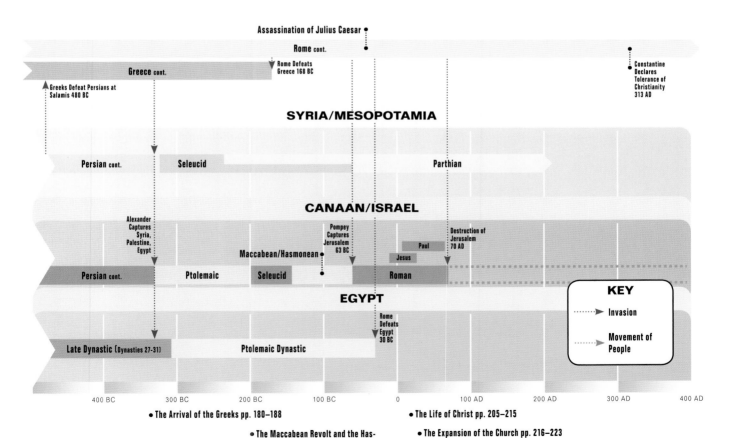

Assassination of Julius Caesar ●

Rome cont.

Rome Defeats
Greece 168 BC

Constantine
Declares
Tolerance of
Christianity
313 AD

Greece cont.

Greeks Defeat Persians at
Salamis 480 BC

SYRIA/MESOPOTAMIA

| Persian cont. | Seleucid | Parthian |

Alexander
Captures
Syria,
Palestine,
Egypt

CANAAN/ISRAEL

Pompey
Captures
Jerusalem
63 BC

Destruction of
Jerusalem
70 AD

Paul

Maccabean/Hasmonean

Jesus

| Persian cont. | Ptolemaic | Seleucid | Roman |

EGYPT

Rome
Defeats
Egypt
30 BC

Late Dynastic (Dynasties 27–31)

Ptolemaic Dynastic

KEY

⋯⋯▶ Invasion

⋯⋯▶ Movement of
People

| 400 BC | 300 BC | 200 BC | 100 BC | 0 | 100 AD | 200 AD | 300 AD | 400 AD |

- The Arrival of the Greeks pp. 180–188
- The Maccabean Revolt and the Hasmonean Dynasty pp. 189–196
- Early Roman Rule in Palestine pp. 197–204
- The Life of Christ pp. 205–215
- The Expansion of the Church pp. 216–223
- The Journeys of Paul pp. 224–234

INDEX OF SCRIPTURE REFERENCES

INDEX OF PERSONS

GEOGRAPHICAL DICTIONARY AND INDEX

The purpose of the Geographical Dictionary is to assist the reader of the Bible in locating places both in this atlas and on modern maps. The Geographical Dictionary lists all places found in the text and on the maps; page numbers in regular type refer to the text, those in bold type to the pages on which specific maps can be found.

Additional information is provided for biblical place names. First, a short description of the significance of a place is given, complete with biblical references. Note that not every reference is given for places mentioned numerous times in the Bible (e.g., Jerusalem, Babylon, Bethel); a concordance such as the *NIV Complete Concordance* or *The Strongest NIV Exhaustive Concordance* lists all occurrences of place names. In addition, many places are treated in more detail in the text (e.g., Jerusalem, which is discussed in a separate chapter, and the regions of the country such as the Bashan, Shephelah, Sea of Galilee, Edom, Moab have special subsections devoted to them), and the reader should consult the appropriate portions of the text.

This description is followed, wherever possible, by an identification of the ancient site with a "modern" place; this material is in italic type. I have attempted to use the most up-to-date site identifications. Some identifications are disputed; in those cases I have cited the identification I think to be the most probably one. (For the process of site identification, see the chapter on "The Disciplines of Historical Geography.")

The modern designation of each site is given in English characters. A simplified system of transliteration from Arabic, Hebrew, Turkish, Greek, etc., into English has been used and is intended to be helpful to the general reader. In the case of sites located in Israel, the Arabic name is usually given first, followed by the Hebrew name. For places in the southern Levant (Israel, Jordan, and sometimes Lebanon and Syria), a six-digit grid reference is provided in parentheses to assist in precisely locating the place on the standard 1:250,000, 1:100,000, and 1:50,000 maps available for those countries (the first three digits serve to locate the site on a north – south axis, while the last three digits locate it on an east – west axis). Also, for quick reference, the approximate distance ("as the crow flies") and direction of the site from a major known ancient place (such as Jerusalem, Joppa, Jericho, Nazareth, Acco, etc.) is given. Abbreviations are used for points of the compass (N = north[ern]; NW = northwest[ern]; NNW = north northwest[ern]; etc.). For sites not in Israel or Jordan, the country of their location is noted. Thus, the reader will be able to quickly locate a place both in this *Atlas* and on a modern map.

For biblical place names, NIV spellings have been used throughout. However, it will generally not be difficult to find place names in the Geographical Dictionary no matter which version of the Bible one is using. In a few cases, readers using other translations of the Bible may find it necessary to look in an NIV translation to check the spelling used here. For example, where the RSV has "Chinnereth" and "Negeb," the NIV uses "Kinnereth" and "Negev." But these instances are relatively few.

Because of the nature of an atlas, the descriptions and discussions of sites are brief. The interested reader should consult the standard dictionaries or encyclopedias, such as *The New Encyclopedia of Archaeological Excavations in the Holy Land* or *The Zondervan Encyclopedia of the Bible* and/or the standard works on historical geography, such as those by Abel, Aharoni, Avi-Yonah, Rainey and Notley, etc., that are listed in the bibliography. Extrabiblical places mentioned in this atlas have not been annotated, but their locations can be found on the maps. The key to the abbreviations used in the Geographical Dictionary is found on p. 12. See also Glossary on p. 265.

Abana – River of Damascus mentioned by Naaman (2 Kgs 5:12) – *Nahr Barada; begins in Anti-Lebanon mountains, flows SE through Damascus, ends in a marsh in the desert.*

Abarim – Mountain range in Transjordan, E of N end of Dead Sea; Mt. Nebo was part of the range (Num 27:12; 33:47, 48; Deut 32:49; Jer 22:20).

Abdon – Levitical town in territory of Asher (Josh 19:28; 21:30; 1 Chron 6:74) – *Kh. Abdeh/ T. Avdon (165272), along N coast of Israel.* 95, 116, 120

Abel – See Abel Beth Maacah (2 Sam 20:18).

Abel Beth Maacah (Abel) – Town in N Israel; Joab pursued Sheba to town (2 Sam 20:14, 15); captured by Ben-Hadad (1 Kgs 15:20; called Abel Maim in 2 Chron 16:4) and later by Tiglath-Pileser III (2 Kgs 15:29). – *Abil el-Qamh/ T. Avel Bet Maakha (204296), N of Huleh Valley, 4 mi. W of Dan.* **32, 34**, 35, 36, **36**, 37, **140**, 143, **143**, 154, **155, 157, 163**, 164, 255

Abel Keramim – Town, captured by Jephthah, either in or on the border of Ammonite territory (Judg 11:33) – *Uncertain; possibly Naur (228142), 8 mi. SW of Amman in Jordan.* 129

Abel Maim – See Abel Beth Maacah.

Abel Meholah – Site along route of Gideon's pursuit of Midianites (Judg 7:22). Town in fifth Solomonic district (along with Beth Shan, et. al.; 1 Kgs 4:12). Home of Elisha the prophet (1 Kgs 19:16) – *T. Abu Sus (203197), 10 mi. SSE of Beth Shan in Jordan Valley.* **38, 115, 126**, 128, **148, 161**

Abel Mizraim – Place where funeral procession of Jacob stopped for 7 days of mourning; also called "threshing floor of Atad" (Gen 50:11). Text says "near the Jordan" which does not fit route from Egypt to Hebron well – *Unknown.*

Abel Shittim (Shittim) – Near Jordan River in plains of Moab, NE of Dead Sea; Israel camped here (Josh 2:1; 3:1; Baalam incident; Num 25:1; 33:49; Micah 6:5) – *T. el-Hammam (214138), 8 mi. NE of NE end of Dead Sea.* 108, **108**

Abila (Beqa). 219, 220

Abila (Decapolis). 33, 35, 187, **187**, 197, **197, 200, 206**, 211, **211**

Abila (Perea). 206, 211, **211, 217, 220, 222**

Abilene – Region/political unit NW of Damascus in Lebanese mountains and Beqa. Lysanias was the tetrarch of Abilene at time of Jesus' birth (Luke 3:1).

Abronah – Stopping place during wilderness wanderings; between Jotbathah and Ezion Geber (Num 33:34, 35) – *Uncertain; possibly Umm Rashrash/ Elath (145884) at N end of Red Sea.*

Abu-Ghosh. 215

Abu Kamal. 76, 77

Abu Rudeis. 69

Abydos. 66, 84, 88, 99, 103

Accaron. 193, 194

Acco – Chief Mediterranean port of Palestine in ancient times. In territory of Asher, but they failed to occupy it (Judg 1:31). Probably under Israelite control during days of David, but ceded to Tyre by Solomon. Later called Ptolemais – *T. el-Fukhkhar/T. Akko (158258), 1 mi. E of Akko/ Acre.* 21, 22, **22**, **23**, 30, 32, 33, 34, 35, 36, 37, 38, **38**, 39, 40, **44, 95, 101, 105, 106, 112**, 116, **116**, 117,

120, 122, 123, 140, 142, 143, 145, 148, 149, **153, 155,** 156, **157, 161, 163,** 168, **170,** 174, **174, 184,** 185, **185, 204,** 257

Acco, Plain of. 22, **36,** 37, **38,** 40, 41, 139, **140,** 149, 201, 205

Achaia – A province of Greece; main city: Corinth. Mentioned 10 times in NT. Achaia and Macedonia were the two major provinces of Greece, and on occasion Achaia could be used to refer to the whole of Greece. **202, 228,** 229, 230, 231, **231,** 232, **233, 234**

Achor, Valley of – Place where Achan was executed (Josh 7:24, 26). On boundary between Judah and Benjamin (Josh 15:7); evidently a wilderness area (Isa 65:10; Hosea 2:15) – *Possibly el-Buqeia, a large broad valley in the Judean Desert, between Jerusalem and the Dead Sea, 10 mi. ESE of Jerusalem.*

Acrabeta. **190,** 192, **193,** 212, 214, 221, **222**

Acshaph – Town assigned to tribe of Asher (Josh 19:25); its king participated in battle of Merom against Joshua (11:1; 12:20). Mentioned in extrabiblical documents as well – *Uncertain; possibly Kh. el-Harbaj/T. Regev (158240), 12 mi. S of Acco.* 111, **112, 116, 123**

Actium. 201, **202**

Aczib (Asher) – Town assigned to tribe of Asher (Josh 19:29) but they were not able to drive out inhabitants (Judg 1:31) – *Ez-Zib/T. Akhziv (159272), 10 mi. N of Acco.* 36, **37, 38, 95,** 115, **115, 122, 139**

Aczib (Judah) – Assigned to tribe of Judah in Shephelah (Josh 15:44); mentioned in a word play by Micah (1:14). Possibly Kezib (Gen 38:5) and Cozeba (1 Chron 4:21) to be identified with Aczib – *Often identified with Kh. T. el-Beida (145116) but see Libnah below.*

Adadah – Town in Negev district of Judah (Josh 15:22) – *Possibly same as Aroer (Negev; 148062) otherwise ID Unknown.* **114**

Adam – Town along Jordan River where water heaped up at time of Israelite crossing (Josh 3:16) – *Tell ed-Damiyeh (201167) on E side of Jordan, ca. 16 mi. NNE of Jericho.* **32, 44,** 45, **58, 60,** 109, **110, 155**

Adamah – One of the fortified cities of Naphtali (Josh 19:36) – *Uncertain; possibly Qarn Hattin/H. Qarne Hittim (193245), 5 mi. W of Tiberias. But see Merom, Waters of.* 116

Adami Nekeb – Town located on the S tribal boundary of Naphtali (Josh 19:33) – *Kh. et-Tell (ed-Damiyeh)/T. Adami (193239), 6 mi. SW of Tiberias.* 116

Adasa. **190,** 192

Addar – Town on S boundary of Judah (Josh 15:3). *Possibly identical to Hazar Addar (Num 34:4).* **114**

Addon – Site in Babylon from which exiles returned (Ezra 2:59; Neh 7:61) – *Unknown.*

Ader. 86, **105**

Adithaim – Town in Shephelah district of Judah (Josh 15:36) – *Unknown.*

Admah – Mentioned in connection with Sodom and Gomorrah (Gen 10:19; 14:2, 8; Deut 29:23). Symbol of God's judgment (Hosea 11:8) – *Unknown; possibly SE of Dead Sea in Rift Valley.* 88, 94

Adora – See Adoraim. **186, 187,** 192, **193,** 195, 197, **198**

Adoraim – Town fortified by Rehoboam (ca. 930 B.C.; 2 Chron 11:9). Called Adora, Dora, or Dor in documents from the intertestamental period – *Dura (152101), ca. 5 mi. ESE of Hebron in Judean Hill Country.* 47, **47, 51,** 54, **151**

Adramyttium – Port city on W coast of Asia Minor (Turkey). In Caesarea, Paul boarded a ship from Adramyttium on the first leg of his journey to Rome (Acts 27:2) – *Karatash near Edremit in W Turkey.* 218, 233, **233**

Adriatic Sea – Paul's ship was "driven across the Adriatic Sea" (Acts 27:27) – *Sea E of Italy, Sicily, and Malta, W of Albania, and Greece.* 217, 229, **223, 234**

Adullam – A town in the Shephelah conquered by Joshua (Josh 12:15) and allotted to Judah (Josh 15:35). Residents intermarried with tribe of Judah (Gen 38:1). David fled to a cave there

(1 Sam 22:1; 2 Sam 23:13; 1 Chron 11:15). Fortified by Rehoboam (2 Chron 11:7). Lamented over by Micah due to pending invasion of Assyrians (1:15). Inhabited after Babylonian Exile (Neh 11:30) – *Kh. esh – Sheikh Madhkur/H. Adullam (150117), 16.5 mi. WSW of Bethlehem in Shephelah.* **51, 97, 114, 134,** 135, **135, 151, 166, 177,** 179, 192, **255**

Adummim, the Pass of – Site mentioned as being on N boundary of Judah (Josh 15:7) and S boundary of Benjamin (Josh 18:17) – *Eusebius places it halfway between Jericho and Jerusalem, possibly the "modern" Talaat ed-Damm ("ascent of blood;" 184136), 8 mi. ENE of Jerusalem.*

Aegean Sea. 17, **83,** 129, **129,** 130, 173, **174,** 176, **181,** 227, **231, 233,** 235, 236, 237

Aelia Capitolina – Hadrainic name for Jerusalem; second century A.D. 223, 240

Aenon – Located near Salim; John was baptizing here (John 3:23) – *Much disputed: (1) NE of modern Nablus in Samaria near sources of W. Farah (182188); (2) 8 mi. SSE of Beth Shan at/near Umm el-Umdan (199199). Ancient sources seem to support number (2).* **212,** 213

Afghanistan. 78, 183

Africa. 17, 20, 24, 27, 41, 52, 64, 65, 67, 69, 70, 84, 175, **202,** 216, 217, 233

Afula. 130

Agrippina (Mount). 207

Ahava (Canal) – In S Babylon (Ezra 8:15, 21, 31). Ahava may be the name of a town or district to/ through which the canal flowed – *Unknown.*

Ahlab – Town in Asher where Canaanites lived (Judg 1:31) – *Problematic; possibly a variant of Mahlab (Josh 19:29 RSV; NIV "in the region of"). If so, then identified with Kh. el-Mahalib (172303), 5 mi. NNE of Tyre along Mediterranean coast.* 116, **116, 122**

Ai (patriarchal) – Abraham pitched his tent between Bethel and Ai (Gen 12:8; 13:3) – *Probably et-Tell (174147), 10 mi. NNE of Jerusalem.* 33, **47, 51,** 49, 89, 91, **91,** 92, 93, **97,** 110, **110,** 115, **258**

Ai (conquest) – Conquered by Joshua (Joshua 7 – 12 passim) – *Usually identified with et-Tell (174147), 10 mi. NNE of Jerusalem however Kh. el – Maqatir (173146) seems to be a more probable location.* 100, 109, 110, **110,** 111, 243

Ai (Transjordan) – Jer 49:3 – *Uncertain.*

Ai/Aiath – Town N of Jerusalem along enemy invasion route (Isa 10:28). After Babylonian exile Jews of Bethel and Ai returned to Judea (Ezra 2:28; Neh 7:32) – *Uncertain; possibly Kh. Haiyan (175145). Some say an alternate form of Ai.* 111

Aija – Town in which Benjamintes settled after exile (Neh 11:31) – *Unknown; possibly identical to Ai or Aiath.* 177

Aijalon (Dan) – Town assigned to tribe of Dan (Josh 19:42) but they failed to take it (Judg 1:35). A Levitical city (Josh 21:24) inhabited by Ephraimites (1 Chron 6:69) and Benjamites (1 Chron 8:13). Philistines fled toward Aijalon (1 Sam 14:31). Fortified by Rehoboam (2 Chron 11:10). Occupied by Philistines during reign of Ahaz (28:18) – *Yalo (152138), 13 mi. NW of Jerusalem in N Shephelah.* **51, 110, 120,** 121, **122, 123, 132, 148, 151, 159,** 160, 165, **166**

Aijalon (Zebulun) – Town in land of Zebulun (N Israel) where Elon the judge was buried (Judg 12:12) – *Unknown.* 129

Aijalon, Valley of – Valley where moon "stood still" as Joshua defeated the Amorite coalition (Josh 10:12) – *Wadi Selman/Emeq Ayyalon; valley named after nearby site of Aijalon.* 47, **47, 51,** 52, 53, 111, 115, 118, **118,** 121, 133, 160, **190,** 191, 192, 215

Ain (Bashan/Golan) – Town along E border of the land of Canaan (Num 34:11) – *Uncertain; possibly Kh. Ayyun (212235), 3 mi. E of S end of Sea of Galilee.* 105

Ain (Judah/Simeon) – Town assigned to both Judah and Simeon (Josh 15:32; 19:7) probably in W Negev. Levitical city (Josh 21:16; 1 Chron 4:32) – *Uncertain; probably Ain and Rimmon should be read together as a compound town name: Ain Rimmon.*

Ain Dara (Syria). 146, 147, 149

Ain Shems, spring of. 257

Akeldama – Field bought with Judas' "blood money," to be used for burials (Acts 1:19) – *Traditional identification along S slope of Valley of Hinnom in Jerusalem.*

Akkad – City mentioned in Gen 10:10. Capital of Akkad, an ancient empire in Mesopotamia. Capital of dynasty of Sargon of Akkad at end of 3rd millennium B.C. Name used in famous phrase "the land of Akkad and Sumer." – *Unknown.* 17, 77, 82, **83, 84,** 85, **90, 93**

Alalakh. 75, **90,** 129, **129**

Alashi(y)a. 84, **93,** 123

Albania. 17, 229

Alema. 192

Alemeth – A Levitical city (Josh 21:18; 1 Chron 6:60) in territory of Benjamin – *Kh. Almit (176136), 4 mi. NE of Jerusalem.*

Aleppo. 17, 18, 33, **72,** 75, **77, 90,** 93, **97,** 160, 169, 171, 174, **182, 225**

Alexandria – City in Egypt along Mediterranean Sea. Established by Alexander (332/1 B.C.). Capital of Egypt, center of commerce and learning. Jews of Alexandria in Jerusalem opposed Stephen (Acts 6:9). Apollos from here (Acts 18:24). Paul sailed on Alexandrian (grain) ship on part of voyage to Rome (Acts 27:6; 28:11). **17, 66,** 181, **181, 182,** 184, **184,** 185, 190, **202, 218,** 231, 233, **233, 234,** 235

Alexandria Troas – See Troas. 237

Alexandrium. **193, 198,** 199, **200,** 201

Al-Kut. 77

Allammelech – Town allotted to tribe of Asher (Josh 19:26) – *Unknown.*

Allenby Bridge. 52, **58, 60,** 61

Allon Bacuth – Burial site of Deborah, Rebekah's nurse (Gen 35:8) – *Unknown; near Bethel.*

Almon – See Alemeth. 115, **120**

Almon Diblathaim – Stopping point of wandering Israelites (Num 33:46, 47) between Dibon Gad and mountains of Abarim. See also Beth Diblathaim – *Uncertain; possibly Kh. Deleilat esh-Sherqiyeh (228116), E of Dead Sea, 21 mi. SE of Amman.*

Aloth – A Solomonic district is described as "in Asher and in Aloth" (1 Kgs 4:16) – *Uncertain; possibly should be read "Bealoth," or even "Zebulun."*

Alush – Stopping point of wandering Israelites, between Dophkah and Rephidim (Num 33:13, 14) – *Unknown.*

Amad – Town allotted to Asher (Josh 19:26) – *Unknown.*

Amalek, city of – Saul attacked it after mustering troops at Telaim (1 Sam 15:5). In Negev of Judah – *Uncertain; possibly T. Masos (146069), 8 mi. E of Beersheba.* 133

Amam – Town in Negev district of Judah (Josh 15:26) – *Unknown.*

Amana, crest of – A mountain (SS 4:8). Probably in Anti-Lebanon range.

Amanus Mountains. 16, **17,** 72, **72,** 75, 78, 84, 85, **97, 181,** 227, **228**

Amastris. 234

Amaw – Home of Balaam the prophet (Num 22:5 RSV; NIV "in his native land") and in extrabiblical texts – *Country W of Euphrates; main center Emar, 50 mi. S of Carchemish.*

Ammah, hill of – Hill where Joab ceased pursuing Abner (2 Sam 2:24) – *Uncertain; E of Gibeon on edge of wilderness.*

Amman. 17, 19, 26, 30, 59, **66,** 211, 258

Ammathus. **193,** 197, **198, 200,** 201, **206,** 211, 212

Ammon – Country of the Ammonites located E of the Jordan River. Capital: Rabbah (modern Amman). **108, 119, 126, 140,** 144, **145, 148,** 153, 155, **161, 163,** 170, **171,** 172, 173, **173, 174,** 177, **177**

Ammonitis. **186, 187,** 188

Amphipolis – Greco-Roman city, capital of first district of Macedonia in NE Greece. Paul passed through it on way to Philippi on second missionary journey (Acts 17:1) – *In NE Greece, 30 mi. SW of Philippi on Strymon (Struma) River.* **228,** 229

Amurru. 72, **93,** 123

Anab – Village in Hill Country of Judah (Josh 15:50) from which Joshua drove out the Anakim

(11:21) – *Kh. Unnab es-Saghir (145091), 12 mi. SW of Hebron.* **114, 166**

Anaharath – Town allotted to Issachar (Josh 19:19) – *Uncertain; possibly T. el-Mukharkhash/T. Rekhesh (194228), 8 mi. SW of S tip of Sea of Galilee.* **105,** 116, **116**

Ananiah – Village to which Benjamites returned after the exile – *El-Azariyeh (174131), 1.5 mi. E of Jerusalem (NT Bethany).* **177**

Anathoth – Levitical city in territory of Benjamin (Josh 21:18; 1 Chron 6:60). Two of David's warriors from there (2 Sam 23:27; 1 Chron 12:3). Abiathar banished there by Solomon (1 Kgs 2:26). Along path of advancing Assyrian invaders (Isa 10:30). Birthplace and place of Jeremiah's activity (Jeremiah passim). Jews lived there after Babylonian exile (Ezra 2:23; Neh 7:27; 11:32) – *Ras el-Kharruba (174135), 2.5 mi. NE of Jerusalem; possibly nearby Der es-Sid is site of Jeremiah's Anathoth.* **47,** 49, **115, 120, 161, 177**

Anatolia. 72, 73, 75, 78, 79, **84,** 85, 102, 129, **129,** 180, 245

Anchialus. 234

Ancyra – See Ankara. **181, 182, 218, 234**

Anem – Levitical town in Issachar (1 Chron 6:73) – Possibly identical to En Gannim (Josh 21:29).

Aner – Levitical town in Manasseh (1 Chron 6:70). Josh 21:25 reads Taanach – *Unknown.*

Anim – Town in Hill Country of Judah (Josh 15:50) – *Kh. Ghuwein et-Tahta/H. Anim (156084), 11 mi. S of Hebron.* **114, 166**

Ankara – See Ancyra. **17, 19**

Antheden. 193, 200, 201

Anti-Lebanon Mountains. 72, **73,** 74, 75

Antioch (Pisidian) – Greco-Roman city in Phrygia, near Pisidia, visited by Paul on all his missionary journeys (Acts passim). Synagogue there is the setting for the longest recorded Pauline sermon (Acts 13:14 – 41) – *In W -central Turkey, NW of Yalvaç.* **225,** 226, 227, **228,** 231, **231**

Antioch (Syrian) – Greco-Roman city in SE Turkey, 15 mi. inland from Mediterranean Sea on Orontes River. Great metropolitan center of over 500,000 people. Christianity spread there after persecution of Stephen; Saul and Barnabas ministered there. "Home base" for Paul on his journeys. Mentioned 17 times in Acts – *Modern Antakya.* **72, 182, 184,** 185, 186, 190, 192, 194, **218,** 224, 225, **225,** 227, **228,** 231, **231,** 232, **233, 234,** 235

Antioch, Plain/Valley of. 75

Antiochenes (See Ptolemais). 187, 187

Antiochia (Dan in north). 187, 188, **193,** 195, **208**

Antiochia (Jerusalem). 187, 187

Antiochia on Chrysorrhoas (Gerasa). 187, 187

Antipatris – City to which Paul was taken by night under military escort from Jerusalem (Acts 23:31). Built by Herod the Great ca. 9 B.C.; named after his father. A military relay station; on border between Judea and Samaria – *Ras el-Ain/Afeq (143168), OT Aphek, ca. 11 mi. NE of Joppa.* **200,** 202, **206, 212, 217,** 219, **219, 220, 222, 233**

Anuathu Borcaeus. 212, 214

Apamea (Syria). 184, 218

Apamea (Turkey). 182, 188, **218, 234**

Aphek (Asher) – Town allotted to Asher (Josh 19:30) who was not able to drive out inhabitants (Judg 1:31) – *Tell Kurdaneh/T. Afeq (160250), 7 mi. SSE of Acco.* 116, **116,** 112

Aphek (Phoenician) – Town on Sidonian-Amorite border (Josh 13:4) – *Afqa (231382), SE of Byblos in Lebanon.*

Aphek (Sharon) – City in SE corner of Sharon Plain at headwaters of Yarkon River. King killed by Joshua (12:18). Philistines mustered troops there for battle of Aphek (1 Sam 4:1) and again for battle of Gilboa (1 Sam 29:1). NT Antipatris – *Ras el-Ain/T. Afeq (143148), 11 mi. ENE of Joppa.* 21, **22, 23, 32,** 43, **44, 51,** 52, **86,** 87, **95, 97, 101,** 105, **108, 110, 115, 130,** 131, **131, 132,** 133, **135,** 136, **142, 155, 159, 163, 166, 170, 177,** 219, 256

Aphek (Upper and Lower) – Site of two Israelite/Aramean battles (1 Kgs 20:26, 30; 2 Kgs 13:17) – *Uncertain; "Lower Aphek" possibly Kh. el-Asheq/En Gev (210243), close to E shore of Sea*

of Galilee. "Upper Aphek" possibly Fiq (216242). **34,** 35, **36, 155, 157**

Aphekah – Town in Hill Country of Judah (Josh 15:53) – *El-Habad (155098), 4 mi. SW of Hebron.* **114, 166**

Apherema. 177, 190, 193, 194

Apollonia – City in Macedonia on Via Egnatia. Paul passed through there on second missionary journey (Acts 17:1) on way from Philippi to Thessalonica – *NE Greece but exact ID Unknown.* **228,** 229

Apollonia (Palestine). 44, **44, 187, 193, 206, 212**

Appius, Forum of – Paul met there by Roman Christians as he travelled to Rome (Acts 28:15) – *On the Appian Way, about 40 mi. S of Rome.* 233, **233**

Apulia. 233

Aqaba – See Elath. **22, 27, 55, 62, 66,** 69, **69**

Aqaba, Gulf of – See Elath, Gulf of. 27, 57, 63, 64, **66,** 68, 69, **69,** 70

Aquitania. 202, 234

Ar (of Moab) – A town or region in Moab near the Arnon River (6 times in OT) – *Uncertain; possibly the same as el-Misna (224076), 15 mi. E of Dead Sea; possibly same as later Areopolis and Rabbath Moab.*

Arab – Town in the Hill Country of Judah (Josh 15:52) – *Possibly Kh. er-Rabiya (153093), 8 mi. SW of Hebron.* **166**

Arabah – OT name for the Rift Valley from the Sea of Galilee to the Red Sea. In OT usually refers to area of the Jordan Valley S of the Sea of Galilee and N of the Dead Sea – *Modern Ghor.* **22, 23,** 25, 49, **55,** 56, 57, **58,** 61, **62,** 63, 64, **69,** 107, **108,** 110, **110, 112, 136, 170,** 171

Arabah, Sea of – Alternate name for the Salt Sea (Deut 3:17; 4:49; Josh 3:16; 12:3; 2 Kgs 14:25) – *Dead Sea.* 49, 109, **110, 155,** 162

Arabah Valley – See Arabah above.

Arabia – Mentioned eight times in NIV; inhabitants ("Arabs") mentioned as well. Can refer to the whole, or part, of the Arabian Peninsula – *Large peninsula in SW Asia, predominantly occupied by Saudi Arabia.* 16, 27, 33, 55, 56, 57, 64, 65, 73, 78, 83, 84, 119, 144, 154, 158, 177, 195, 201, 214, 218, 224

Arabian Desert. 16, **17,** 21, 27, 49, 59, 61, 76, 79, **84, 90, 97, 160, 169,** 171

Arabian Peninsula. 16, **17**

Arabian Sea. 175, 183

Arachosia. 175

Arad – City in Negev of Judah (Judg 1:16; some read Arad for Eder in Josh 15:21). Israelites fought with king of Arad on two occasions (Num 21:1; 33:40) – *T. Arad (162076), 17 mi. S of Hebron, but possibly "Canaanite Arad" was located at T. el-Milh/T. Malhata (152069), 11 mi. E of Beersheba.* **22, 32,** 47, **47,** 49, 54, 55, **55, 69, 84,** 86, **86,** 87, 88, **94,** 106, 107, **108, 135, 145, 153,** 156, **159, 161, 170,** 256, 259

Arad of Jerahmeel. 166

Arad Rabbah. 166

Aradus. 181, 181, 218

Arah – Town belonging to the Sidonians, NW of Israel (Josh 13:4) – *Unknown; RSV has "Mearah."*

Araj – See Bethsaida. 210

Aral Sea. 175, 183

Aram – Used 68 times in the NIV usually to refer to the Aramean peoples (Semitic) and the land(s) in which they lived – *Indefinite geographically; NE of Israel including area around Damascus, Syria, and the upper Tigris-Euphrates Valleys. Used in compound names (e.g., Aram Damascus, Aram Maacah, Aram Zobah, etc.) to refer to small states.* 35, 73, **83,** 84, 90, 125, **140, 146,** 154, **155,** 156, 157, **157,** 160, **160,** 164, 245

Aram Damascus – See Aram. **148,** 154, 160, 164, 245

Aramean kingdom of Damascus – Aramean population in and around Damascus who became subject to David (2 Sam 8:5, 6; 1 Chron 18:6).

Aram Maacah – Alternate name for Maacah (1 Chron 19:6).

Aram Naharaim – Abraham's servant searched for a wife for Isaac here (Gen 24:20). Balaam was from here (Deut 23:4). During period of judges, Cushan-Rishathaim was king (Judg 3:8). Ammonites

hired horsemen and chariots from there to fight against David (1 Chron 19:6 and title of Psalm 60) – *Region bounded by upper Euphrates on W and the Habur River on the E.* **72,** 73, 76, 125, 139

Aram Zobah – Psalm 60 title, see Aram and Zobah.

Ararat – Ark of Noah rested on "the mountains of Ararat" (Gen 8:4). Assassins of Sennacherib escaped to the land of Ararat (2 Kgs 19:37; Isa 37:38). Mentioned in Jer 51:27. Called Urartu is Assyrian inscriptions – *In modern Turkey with center in Lake Van region; where borders of Russia, Iran, Armenia, and Turkey converge.* 16, **17,** 82

Araru. 96

Arava – See Arabah. **22,** 25

Araxes River. 82, 160

Arbata. 192

Arbela (Galilee). 192, **193**

Arbela (Upper Tigris). 175, 182, **182,**

Arbel Cliffs/Pass. 32, 33, 39, 41, 42, **46,** 210

Archelais. 205, **206, 212**

Argob – Name of Og's kingdom (Deut 3:4; 13, 14); mentioned as part of one of the Solomonic districts (1 Kings 4:13). *In, or beside, the Bashan in N Transjordan.* **34,** 33, 35, 148

Aria. 175

Aricia. 218

Ariel – Seldom used name for Jerusalem (Isa 29:1, 2, 7).

Arimathea – Home of Joseph who buried the body of Jesus in his tomb (Matt 27:57; Mark 15:43; Luke 23:51; John 19:38) – *Rentis (151159), 16 mi. E of Joppa.* **190, 193,** 194, 252

Armageddon – Site of assembly of forces of opponents of God's people (Rev 16:16) – *Usually with OT Megiddo (167221) on SW edge of Jezreel Valley, especially if the Greek name is derived from the Hebrew* Har Megiddo *= Mount Megiddo. Alternative interpretations and identifications/ interpretations have been proposed; see the commentaries.* 40

Armenia. 17, 76, 79, 82, 175, **175,** 202

Arnon (Gorge/River) – Mentioned 25 times in the NIV. Traditional N boundary of Moab and S boundary of the Amorites and later the Israelite tribe of Reuben – *Wadi el-Mujib in Transjordan which flows from E to W into the Dead Sea opposite En Gedi.* **22, 23,** 27, **32,** 33, **47, 55, 58,** 62, **62,** 63, **86, 94, 95,** 97, 107, **108, 110,** 118, 119, **119, 120, 122, 126, 135, 140,** 141, **142, 143, 153, 155,** 156, **159,** 162, **177**

Aroer – On boundary between Israelite and Ammonite territories (Josh 13:35) – *Unknown.*

Aroer (Negev) – Village with which David shared Amalekite spoils (1 Sam 30:28). Two of David's might men came from here (1 Chron 11:44) – *Kh. Arareh/Aroer (148062), 12 mi. SE of Beersheba.* **135, 159, 163, 166**

Aroer (Reuben/Gad) – S limit of Sihon's kingdom, allotted to tribe of Reuben (Josh 12:2) but repaired/built by sons of Gad (Num 32:34). Located near Arnon Gorge (Deut 2:36; 3:12). Here Joab began census for David (2 Sam 24:5). Also mentioned in Judges 11:26; 2 Kings 10:33, and Jer 49:19 – *Arair (228097), ca. 14 mi. E of Dead Sea on N bank of Arnon.* **32,** 33, **58,** 62, **86,** 105, **110,** 113, **119,** 129, **140,** 142, **142, 145, 148, 155, 159, 163**

Arpad – A city/province in N Syria conquered by the Assyrians. Always mentioned in connection with Hamath (2 Kings 18:34; 19:13; et al.) – *Tell Erfad, 25 mi. NNW of Aleppo.* **160, 174**

Arqa. 123

Arqad. 84

Arrapkha. 169, 171

Arsinoe. 218

Artacauna. 183

Arubboth – District headquarters for one of Solomon's administrators (1 Kgs 4:10) – *Kh. el-Hammam (162201), 17 mi. NW of Shechem.* 148

Arumah (Galilee). 163

Arumah (Manasseh/Samaria) – Abimelech, son of Gideon, fled there after being driven from Shechem (Judg 9:41) – *Kh. el-Ormah (180172), 5 mi. SE of Shechem.* **44, 115**

Aruna. 32, 33, **38, 44,** 101, **101, 155**

Arvad – Island/city, located 2.5 mi. off the Syrian coast. Famous for its powerful fleets (Ezek 27:8, 11; and extrabiblical references) – *Er-Ruad (229473), ca. 30 mi. N of Tripoli.* **17**, *72*, 74, 75, **83**, 84, **123, 146, 160, 169**

Ascalon (= OT Ashkelon). 186, **186, 193**, 194, **198, 200**, 201, **206, 212, 217**

Ashan – Village first allotted to Judah in Shephelah district but later given to Simeon (Josh 19:7; 1 Chron 4:32). Levitical city (1 Chron 6:59). David and men roamed there (1 Sam 30:30) – *Uncertain; possibly Tell Beit Mirsim (141096), 12.5 mi. SW of Hebron. There are two Ashan's: Judahite in Shephelah; Simeonite in Negev.*

Ashdod – Philistine city located in Philistine Plain 22 mi. S of Joppa, 10 mi. N of Ashkelon. Mentioned 21 times in OT as well as in Acts 8:40 (Azotus). Usually under non-Israelite influence: e.g., Egyptian, Philistine, Assyrian, etc – *Esdud/T. Ashdod (117129), 3 mi. inland from the Mediterranean Sea.* 23, 24, 28, **32, 51**, 52, **95, 105, 110, 122**, 129, **130**, 131, **131, 134, 135, 153, 155**, 158, **159, 160, 163, 166**, 169, **170, 175, 177, 177**, 186

Ashdod – yam. 169, 170

Asher – Israelite tribe that settled in W Galilee – in and east of the Plain of Acco. 37, 39, 115, 116, **116**, 117, **120**, 128, 138, 147, 148, 149, 166, **178**

Ashkelon – Philistine city located in Philistine Plain 32 mi. S of Joppa, 12 mi. NNE of Gaza. Mentioned 13 times in OT. Usually under non-Israelite influence: Egyptian, Philistine, Assyrian, etc – *Asqalan/Ashqelon (107118), on the Mediterranean Sea.* 23, 51, 52, **95**, 96, **105, 110, 114, 122, 123, 124, 126**, 127, 129, 130, **130, 135, 153, 155, 159, 163, 166**, 167, 168, **179**, 177

Ashnah (Judah) – In first Shephelah district of Judah (Josh 15:33) – *Unknown.*

Ashnah (Judah) – In the third Shephelah district of Judah (Josh 15:43) – *Possibly Idna (147107), 8 mi. WNW of Hebron.*

Ashtaroth – Capital city of Og the king of Bashan (Deut 1:4; Josh 9:10; etc.). Captured by Israelites and allotted to half tribe of Manasseh (Josh 13:31). Levitical city (1 Chron 6:71; and Josh 21:27, NIV "Be Eshtarah") – *T. Ashtarah (243244) in Transjordan, 22 mi. E of Sea of Galilee.* 23, 28, **32, 33, 34**, 35, 58, 59, **60**, 88, 94, **94, 101, 106**, 108, **109, 119, 120, 123, 155, 157, 163**

Ashteroth Karnaim – The early name, Ashteroth, is further identified by reference to Karnaim which later replaced it as the chief city in the region (Gen 14:5).

Ashuri. 138

Asia – In NT times a Roman province covering the W third of Turkey, stretching from the Aegean Sea to the central highlands. Mentioned 19 times in NT: Paul established churches there; John wrote Revelation to seven of the churches in the province; Ephesus was evidently the chief city of the province. 199, **202, 225**, 226, 227, **228**, 231, **231**, 232, **233**, 235, 236, 237, **237**

Asia (Minor). 17, 20, 41, 52, 68, 69, 70, 97, 101, **129**, 173, **174**, 176, 180, **181, 182**, 184, 187, 188, 199, 224, 227, 230, 232, 233, 236

Askar. 215

Asochis. 208

Asphaltitis, Lake – See Dead Sea. 50

Asphar, Well of. 190, 192

As – Salt. 30

Asshur – Son of Shem (Gen 10:22; 1 Chron 1:17). Capital city of Assyria (Ezek 27:23, etc.). – *Qalaat Sherqat, 56 mi. S of Mosul/Nineveh on W bank of Tigris River in Iraq.* **17**, 76, **77, 83**, 84, **84, 90, 93, 94, 160**, 169, **171, 175**

Assos – Port city in W Asia perched on hillside overlooking harbor. Paul passed through it on third journey (Acts 20:13, 14) – *Behramkoy in Turkey.* **231**, 232, **237**

Assyria – Country located along upper Tigris River in modern Iraq. Great international power from ca. 900 – 600 B.C. Mentioned over 140 times in OT, took N kingdom, Israel, into captivity in 722 B.C. 13, **17**, 18, 20, 21, 28, 72, 73, 79, 84, 90, **93**, 123, 160, **160**, 166, 168, 169, **169, 171**

Aswan – Town in Upper Egypt on border with Ethiopia (Ezek 29:10; 30:6; RSV: "Syene"). See also Sinim – *Aswan in S Egypt, 550 mi. S of Cairo at first cataract of Nile.* 65, **66**, 173, 175, 182

Atad – Place of mourning for Jacob, near Jordan River (Gen 50:10, 11) – *Unknown.*

Ataroth (Ephraim) – Town in NE Ephraim on border with Manasseh (Josh 16:7) – *Unknown.*

Ataroth Addar – Village on border between Benjamin and Ephraim (Josh 16:5; 18:13) – *Uncertain; possibly Kh. Raddana (169146).* 115

Ataroth (Gad) – Transjordanian town allotted to Gad (Num 32:3, 34). Moabite Stone (10 – 14) mentions that Gad has always dwelt there – *Kh. Attarus, 9 mi. NW of Dibon (213109).* 119

Atbara (River). 66

Athach – Town in S Judah to which David sent booty (1 Sam 30:30) – *Unknown; possibly a scribal error for Ether.*

Atharim – Town in S Judah. Israelites travelled "along the road to Atharim" where they were attacked by the king of Arad (Num 21:1) – *Unknown.*

Athens – City in Greece visited by Paul on second and third journeys (Acts 17:21, 15, 16, 22; 18:1; 1 Thess 3:1). Major cultural and educational center in time of Paul – *5 mi. inland from the Aegean Sea.* **17**, **19**, 175, **176**, **181**, 203, **218**, 228, 229, **229**, 231, 238

Athlit. 44, 44

Athribis. 218

Atlantic Ocean. 202, 234

Atroth Beth Joab – Possibly a village near Bethlehem (1 Chron 2:54) – *Unknown.*

Atroth Shophan – Small village in Transjordan built by Gadites (Num 32:35) – *Uncertain; possibly Rujm Atarus, 1.5 mi. NE of Ataroth, built to protect the latter.*

Attalia – Port city in Pamphylia from which a route led into the interior and from which Paul and Barnabas set sail on their return to Antioch at the end of their first missionary journey (Acts 14:25, 26) – *Antaliya in S Turkey.* **17**, **181**, 225, **225**, 227, **231**, 234, **237**, **237**, 238, 239

Augusta Treverorum. 234

Auja – See Wadi Auja.

Auja (Springs). 58, 61,

Auranitis – See Hauran. 34, 186, **186**, 200, 201, **206**, 210, **212**, 216, **217**, **219**, 220

Avaris. 89, 99, **99**, 102, **103**, 104

Avedat. 32, 55

Aven, Valley of – Valley in Lebanon, between Lebanon and Anti-Lebanon mountains belonging to Aramean kingdom of Damascus (Amos 1:5) – *Uncertain; possibly near modern Baalbek in Lebanon.*

Avith – Town in Edom; home of Hadad son of Bedad, a king of Edom (Gen 36:35; 1 Chron 1:46) – *Unknown.*

Avva – District/province from which Shalmaneser V took people and settled them in Israel (2 Kgs 17:24) – *Unknown. See Ivvah.* 164

Avvim – Town allotted to Benjamin (Josh 18:23) – *Unknown; probably in vicinity of Bethel.*

Ayun Musa (Transjordan) – See Khirbet Ayun Musa. **86**

Ayun Musa (Sinai). 69, 184, 184

Ayyah – Town possessed by Ephraim (1 Chron 7:28) – *Unknown.*

Azekah – Town allotted to Judah located in the Shephelah (Josh 15:35). Joshua pursued Amorite coalition in its direction (10:10, 11); David fought Goliath in the vicinity (1 Sam 17:1). As a city on border between Judah and the Coastal Plain, Rehoboam fortified it (2 Chron 11:9) and Nebuchadnezzar attacked it (Jer 34:7; Lachish Letter 4). Reoccupied by Jews after the exile (Neh 11:30) – *Kh. T. Zakariyeh/T. Azeqa (144123), 15 mi. NW of Hebron.* 47, **51**, 53, 110, 111, **114**, 134, 134, 151, 159, 163, 166, 168, 170, **170, 177**, 179

Azel – Probably a valley just to the south of Jerusalem (Zech 14:5). *Possibly the W. Yasul 1 mi. S of the Hinnom Valley in (modern) Jerusalem.*

Azmaveth – Town where Jews lived after the exile (Ezra 2:24; Neh 12:29) – *Ras Dhukeir (174137) near Hizmeh, ca. 5 mi. NNE of Jerusalem.* 177

Azmon – Town on SW border of land of Canaan (Num 34:4) and Judah (Josh 15:4) E of the Wadi of Egypt (Wadi el-Arish) – *Uncertain; possibly Ain Muweilih (085010).* **106**, 114

Aznoth Tabor – A landmark on SW border of Naphtali – *Uncertain; possibly T. Aznot Tavor (186237), 3 mi. N of Mt. Tabor.* 116

Azor. 94, 105, 130, 166

Azotus – Greco-Roman name from Ashdod (Acts 8:40). See Ashdod. 186, **186, 187**, 188, 192, **193**, 194, **198, 200, 206, 212**, 217, **217, 219**, 220, **220, 222, 234**

Azuru. 168

Baalah (Negev) – Town in Negev district of Judah (Josh 15:29). – *Unknown. Possibly identical to Baalath Beer.*

Baalah, Mount – Mountain on Judean-Benjamite border (Josh 15:11). Probably near Baalah of Judah below.

Baalah/Baalath (of Judah) – Early name for Kiriath Jearim on border of Judah and Benjamin (Josh 15:9, 10) from which David brought ark up to Jerusalem (2 Sam 6:2; 1 Chron 13:6). Fortified by Solomon (2 Chron 8:6). **140**, 144, **145**

Baalath (Dan) – A Danite town (Josh 19:44). – *Possibly el-Mughar (129138). Possibly referred to in 1 Chron 4:33 (N boundary of Simeonites in Shephelah) and in 2 Chron 8:6 and 1 Kgs 9:18 (fortified by Solomon), although latter two references could be to Kiriath Jearim.* **118**

Baalath Beer – Simeonite town in Negev (Josh 19:8 and possibly 1 Chron 4:33). Also called "Ramah in the Negev." – *Unknown; possibly Bir Rekhme/ Beer Yeroham (138043).*

Baalbek. 72, 75

Baal Gad – Town in Valley of Lebanon (Josh 11:17; 12:7) at W foot of Mount Hermon (Josh 13:5). N limit of Joshua's conquest. – *Unknown.*

Baal Hamon – Solomon said to have had a vineyard there (SS 8:11). Possibly only a poetic expression. – *Unknown.*

Baal Hazor – Place near border of Ephraim where Absalom executed Amnon (2 Sam 13:23) – *Jebel Asur/T. Asur (177153), 15 mi NNE of Jerusalem.* **23**, 25, 46, **142**

Baal Hermon – Town on border of Manasseh in Transjordan (1 Chron 5:23); Hivites resided there (Judg 3:3). – *Unknown; probably near Mt. Hermon.*

Baal Meon – Town in Transjordan settled by Reubenites (Num 32:38; 1 Chron 5:8) but at times under Moabite control (Ezek 25:9). Also called Beth Baal Meon (Josh 13:17); Beth Meon (Jer 48:23); and possibly Beon (Num 32:3). – *Main (219120), 23 mi. SW of Amman, 10 mi. E of Dead Sea.*

Baal Perazim – Place where David defeated Philistines soon after becoming king (2 Sam 5:20; 1 Chron 14:11): – *Uncertai; possibly ez-Zuhur (167127), 4 mi. SW of Jerusalem.* **140**

Baal-rosh. 162

Baal Shalishah – Place from which a supporter of Elisha came (2 Kgs 4:42). – *Uncertain; possibly Kh. Marjame (181155), 16 mi NE of Jerusalem.*

Baal Tamar – Town or landmark between Bethel and Gibeah where Israelites lay in wait for pursuing Benjamites (Judg 20:33). – *Unknown.*

Baal Zephon – Site in/near Egypt mentioned in connection with Israelites camping at Pi Hahiroth/ Migdol (Exod 14:2, 9; Num 33:7). – *Unknown although frequently identified with Tahpanhes/ Tell Defnneh near N end of Suez Canal. More probably in area of Bitter Lakes.* 104

Bab edh-Dhra. 47, 50, **55**, 62, 82, **86**, 87, **94**, 95

Babel – Place where tower reaching toward heaven was built (Gen 11:9). – *Unknown; probably in S Mesopotamia. Possibly alternate name for Babylon.* **83**

Babylon – Various forms of "Babylon" are mentioned 237 times in the OT and 11 times in the NT. City was located in S Mesopotamia (Iraq) on the Euphrates River, 50 mi. S of Baghdad. Was the capital of the Neo-Babylonian Empire (ca. 604 – 539 B.C.) Judeans were taken captive by the Babylonian king Nebuchadnezzar in 586 B.C. **17**,

18, 21, 22, 28, 72, 73, 77, **77**, 79, **84**, **90**, **93**, **94**, 123, **146**, **160**, 164, 165, 168, 169, **169**, 170, 171, **171**, 172, **174**, 175, **175**, 179, **182**, 183, **218**, 247, 266, 267

Babylon and Beyond the River. 175, 176

Babylonia – Area in S Mesopotamia ruled from city of Babylon. See Babylon. **17**, 18, 21, 22, 28, 72, 73, 79, 123, **160**, 168, **169**, **171**, 173, **174**, 175, **175**, 186, 266

Baca. 205, **206**, 208

Bactra. 183

Bactria. 175, 179, 182

Baetica. 202

Baghdad. 17, 19, 76, 77

Bahariya Oasis. 66

Bahurim – Site E of Mt. of Olives mentioned several times in events surrounding life of David (2 Sam 3:16; 16:5; 17:18; 19:16; 1 Kgs 2:8; 1 Chron 11:33). – *Uncertain; possibly Ras et-Tumeim (174133), 1.5 mi. NE of Jerusalem.* 142, **143**

Baither – A town allotted to the tribe of Judah but mentioned only in Septuagint of Josh 15:59 (see NEB). – *Kh. el-Yahudi (162126), 7 mi. WSW of Jerusalem.*

Balah – Negev town allotted to Simeon (Josh 19:3). – *Unknown.*

Balik River. 72, 73, 77, **90**, **93**, **97**, 98

Bamoth – Stopping place for Israel in Transjordan, N of Arnon Gorge (Num 21:19, 20). – *Unknown; possibly same as Bamoth Baal.*

Bamoth Baal – Site NE of Dead Sea to which Balak took Balaam to curse Israel (Num 22:41). Allotted to Reuben (Josh 13:17) but claimed by Moabites (Moabite Stone). – *Unknown.*

Barada River. See Abana. 73, 75

Barqay. 94

Barsip. 72

Bascama. 193

Bashan – Mentioned some 60 times in OT. Area E and NE of the Sea of Galilee, bounded by the Rift Valley on the W, Mt. Hermon on the N, Mount Bashan on the E, and Gilead on the S. **22**, **23**, 27, 33, 34, **34**, 35, 38, 39, 40, 41, **58**, 59, **60**, 96, 108, **108**, **109**, **118**, 119, **119**, 125, 148

Bashan, Mount/Mountains of – See Jebel Druze. 33, 34, **34**, 35

Basra. 76, 77

Batanea. 33, **34**, 35, **186**, 187, **187**, 200, 201, **206**, 210, **212**, 216, 217, 219, **219**, **220**

Bealoth – Negev city allotted to Judah (Josh 15:24). – *Unknown; although some suggest Kh. el-Mishash/T. Masos (146069), 8 mi. ESE of Beersheba. Possibly same as Baalath Beer but also see Hormah below.*

Beer (hill country of Ephraim) – Site near Mt. Gerizim to which Jotham fled from Abimelech (Judg 9:21). – *Unknown.*

Beer (Moab) – Site in Moab where Israelites secured well water (Num 21:16). – *Unknown.*

Beer Elim – Moabite town mentioned in oracle of Isaiah (15:8). – *Unknown.*

Beer Lahai Roi – "Well of the Living One that sees me." Place where angel of Yahweh appeared to Hagar (Gen 16:14) and which was frequented by Isaac (Gen 24:62; 25:11). – *Unknown; although Ain Muweileh (085010), 50 mi. SW of Beersheba (see Azmon) has been suggested.* 91, 97

Beer Resisim. 94

Beeroth – One of the four Gibeonite cities that made a treaty with Joshua (Josh 9:17); allotted to Benjamin (Josh 18:25); inhabitants fled to Gittaim (in Saul's day?) and two of their number murdered Ishbosheth (2 Sam 4:2, 3); one of David's elite troops from Beeroth (2 Sam 23:37; 1 Chron 11:39); settled after the exile by Jews (Ezra 2:25; Neh 7:29). – *Kh. el-Burj (167137), 4.5 mi. NW of Jerusalem and/or el-Bira 8 mi. N of Jerusalem.* **110**, 111, **115**, 138, **166**, **177**, 190

Beersheba – Capital of the Negev, mentioned 34 times in OT. Associated with the patriarchs – Abraham, Isaac, and Jacob. Allotted to both Judah (Josh 15:28) and Simeon (Josh 19:2; 1 Chron 4:28). Mentioned as the S boundary of Israel in the phrase "from Dan to Beersheba." Also mentioned in Joab's census (2 Sam 24:7) and the story of Eli-

jah's flight from Jezebel (1 Kgs 19:3). Evidently a shrine/temple was there (Amos 5:5; 8:14) and Jews lived there in the postexilic period (Neh 11:27). – *Tell es-Seba/T. Beer Sheva (134072), 25 mi. SW of Hebron.* 21, 23, 25, **26**, 30, **30**, **32**, 33, 45, 47, **47**, **51**, 54, 55, **55**, **66**, 69, **69**, 87, **91**, 92, 93, **94**, **97**, 114, **114**, **120**, 121, **126**, **130**, 133, **135**, 140, 142, **142**, 145, **153**, 159, **161**, 163, **166**, 169, **170**, 173, **177**, 179, **186**, **187**, **193**, 195, **200**, 201, **206**, 214, 254, 257, 261

Beersheba, Desert/Wilderness of. 161

Be Eshtarah – Levitical city located in Transjordan in allotment of Manasseh (Josh 21:27), same as Ashtaroth in parallel passage (1 Chron 6:71). – See Ashtaroth.

Behistun. 175

Beirut. 17, 19, **72**, 74, 75, **123**, 257

Beisan – See Beth Shan. 258

Beitin. 91, 110, 111

Beit Sahur. 94

Beit Ur el – Foqa – See Beth Horon. 257

Bela – Former name of Zoar (Gen 14:2, 8). – See Zoar. 88, **94**, **94**, 254

Belgica. 202, 234

Bene Berak – Town allotted to tribe of Dan (Josh 19:45) and attacked by Sennacherib the Assyrian (Annals). – *Kheiriyeh/Bene-beraq (133160), 4.5 mi. ESE of Joppa.* 118, **166**, 168

Bene Jaakan – Place of Israelite encampment in Sinai/Negev desert (Num 33:31, 32; Deut 10:6 "wells of the Jaakanites"). – *Unknown.*

Ben Hinnom, Valley of – Mentioned 10 times in OT. Valley to the W and S of OT Jerusalem which joins Kidron just S of Jerusalem. Formed part of the border between Benjamin and Judah (Josh 15:8; 18:16). During later Judean monarchy idolatrous worship took place there (2 Kgs 23:10; 2 Chron 28:3; 33:6; Jeremiah passim) – see Hinnom, Valley of. 247

Beni Hasan. 93, 96, 99

Benjamin – Youngest son of Jacob and the name of the tribe that descended from him. Tribal allotment to the NE, N, and NW of Jerusalem. 45, 46, 47, 49, 113, 114, 115, **115**, 117, 121, 125, 129, 133, 134, 138, 148, 149, 169, 174, 179, 244, 245, 254, 255

Benjamin, Hill Country of. 46, 53, 118, 121, 152, 169, 192

Benjamin Plateau. 46, 47, 132, 133, 134, 199

Beon – Town in Transjordan (Num 32:3). See Baal Meon.

Beqa – See Lebanese Beqa. **23**, 24, 33, **34**, 35, 36, **36**, **72**, 75, **101**, 102, 106, 265

Beracah, Valley of – Valley where Judeans collected booty during days of Jehoshaphat (2 Chron 20:26). – *Uncertain; in or near Judean Desert, possibly Wadi el-Arrub between Bethlehem and Hebron.*

Berea – Large Macedonian city visited by Paul on his second and possibly third missionary journeys (Acts 17:10, 11, 13, 14; 20:4). – *In and around Verria in N Greece, 40 mi. W of Thessalonica.* **218**, **228**, 229, **231**, 232, **234**

Bered – Point mentioned in the story of the flight of Hagar from Sarai (Gen 16:14). – *Unknown; probably in S Israel or NE Sinai.*

Bernice – See Pella. 186, **186**, 187

Berenike. 66

Beroea. 184

Berothai/Berothai – City on N boundary of the land of Canaan (Ezek 47:16, Berothah). Controlled by Hadadezer, king of Hamath, but taken by David (2 Sam 8:8). – *Bereitan (257372), 30 mi. NW of Damascus.* **140**, 141, **178**

Bersabe. 205, **206**, 208

Berytus. 218

Ber-zetha. 190

Besimoth. 222

Besor Ravine – A wadi in SW Israel that David crossed in pursuit of the Amalekites (1 Sam 30:9, 21). – *Wadi Ghazzeh/Nahal Besor.* 21, **23**, **32**, 51, **51**, 52, 54, **55**, **135**, 137

Betah – See Tebah.

Beten – Town allotted to Asher (Josh 19:25). – *Uncertain; possibly Kh. Ibtin/H. Ivtan (160241), in N coastal plain of Israel, 11 mi. SSE of Acco.* 116

Bet Haemeq. 86

Beth Anath – Town assigned to Naphtali (Josh 19:38); inhabitants became forced laborers of Naphtali (Judg 1:33). – *Uncertain; possibly Safed el-Battikh (Lebanon; 190289), 15 mi. SE of Tyre in Upper Galilee.* **36**, 37, **105**, 117, 127

Beth Anoth – Village allotted to Judah in a Hill Country district (Josh 15:59). – *Kh. Beit Anun (162107), 3 mi. NE of Hebron.* **114**, 127, 166

Bethany (near Jerusalem) – Village on road to Jericho (Mark 11:1; Luke 19:29) less than 2 mi. from Jerusalem (John 11:18). Mary, Martha, and Lazarus lived there (John 11) and it seems that Jesus spent the evenings of the week before his crucifixion there. Mentioned 11 times in NT. – *El-Azariya (174131), 1.5 mi. E of Jerusalem, on E side of Mount of Olives.* **212**, 214, **217**, 251, 252

Bethany, on the other side of the Jordan – Place, E of Jordan, where John baptized Jesus (John 1:28) and near which Jesus later ministered (10:40 – 42). – *Uncertain; possibly near Wadi el-Kharar (203138), 6 mi. E of Jericho, E of Jordan, although several other sites have been proposed.* **206**, **212**, 213

Beth Arabah – Village on the boundary between Judah (Josh 15:6) and Benjamin (Josh 18:18). Assigned to both Judah (Josh 15:61) and Benjamin (Josh 18:22). – *Uncertain; possibly Ain el-Gharabah (197139), 4 mi. SE of Jericho.* **114**, **115**, 166

Beth Arbel – Village devastated by Shalman (Hosea 10:14). – *Irbid (229218), 20 mi. NW of Amman in Transjordan.* 60

Beth Ashbea – Town in Judah noted for linen workers (1 Chron 4:21). – *Unknown.*

Beth Aven – Town on N boundary of Benjamin, W of Jericho but E of Luz (Bethel; Josh 18:12) and near Ai (Josh 7:2). Philistines camped at Micmash, E of Beth Aven (1 Sam 13:5; 14:23) and the site was noted for illicit worship (Hosea 4:15; 5:8; 10:5). – *Uncertain; possibly T. Maryam (175141), 7 mi. NE of Jerusalem.* **115**, **132**, 133

Beth Azmaveth – See Azmaveth.

Beth Baal Meon – Town allotted to Reuben (Josh 13:17). See Baal Meon. **119**

Beth Barah – Place in Jordan Valley near which the Ephraimites were encouraged by Gideon to head off the fleeing Midianites (Judg 7:24). – *Unknown. Some emend the text to read "the fords of the Jordan."*

Beth Basi. **190**, 194

Beth Biri – Town in Negev (?) inhabited by Simeonites (1 Chron 4:31). Parallel in Josh 19:6 has "Beth Lebaoth." – *Unknown.*

Beth Car – Israelite pursued Philistines from Mizpah W to a point below Beth Car (1 Sam 7:11). – *Unknown; probably in western Benjamin/Ephraim.* 132

Beth Dagon (Asher) – Boundary town of Asher in N Israel (Josh 19:27). – *Unknown.*

Beth Dagon (Dan area). **166**, 168

Beth Dagon (Judah) – Shephelah town allotted to Judah (Josh 15:41). – *Unknown.*

Beth Diblathaim – Town of Moab "cursed" by Jeremiah (48:22). – See Almon Diblathaim. **163**

Beth Eden – Cursed by Amos along with Damascus and the people of Aram (1:5). – *Aramean principality near the upper Euphrates valley. Possibly equivalent to the "people of Eden (2 Kgs 19:12 = Isa 37:12) and "Eden" (Ezek 27:23). Called Bit-Adini in Assyrian sources.* **160**

Beth Eked – Place where Jehu captured and slaughtered the relatives of the Judean king Ahaziah (2 Kgs 10:12, 14). – *Unknown; probably N of Shechem near Jezreel Valley.*

Bethel (Benjamin) – Town mentioned 71 times in the OT, second in number only to Jerusalem. Was a patriarchal center near which Abram pitched his tent (Gen 12:8; 13:3) and where Jacob had his famous dream of a ladder ascending to heaven (Gen 28:19). Captured by Israelites (Judg 12:16), it was allotted to Benjamin (18:22) but it was situated on the Benjamite-Ephraimite border (Josh 16:1, 2; 18:13). There, Jeroboam built a worship center in which he placed a golden calf (1 Kgs 12:29, 32, 33). As such, Bethel became a target of prophetic rebuke (e.g., Amos 4:4; 5:5, 6). Was resettled by Jews after the Babylonian exile (Ezra 2:28; Neh 7:32; 11:31). – *Beitin (172148), 12 mi. N of Jerusalem. Either Bethel, or a nearby site, was called Luz.* 23, **23**, 25, **32**, 33, **44**, 46, 47, **47**, 49, 91, **91**, 92, 93, **94**, **95**, **97**, **105**, 109, 110, **110**, 111, **115**, 123, **123**, 125, **126**, 127, **130**, **131**, 132, **132**, 150, **153**, 154, **154**, **155**, 161, 163, 166, 170, 177, **190**, 192, **241**, 254, 258

Bethel (Judah) – See Bethuel.

Bethel, hill country of – Hills in Bethel region through which the Ephraimite-Benjamite border ran (Josh 16:2) and where Saul mustered his troops in preparation for battle with Philistines (1 Sam 13:2). **133**

Beth Emek – A town on the boundary of Asher (Josh 19:27). – *T. Mimas/T. Bet Ha-Emeq (164263), 5 mi. NE of Acco.* **116**

Bethennabris. 212, 213

Bether – See Bethther. **114, 166, 222,** 227

Beth Ezel – Town mentioned by Micah as in mourning due to attack (1:11), probably in days of Sennacherib (ca. 701 BC). – *Unknown; probably in Shephelah of Judah.*

Beth Gader – Haraph was the "father" (= ruler, sheikh) of the town (1 Chron 2:51). – *Uncertain; possibly in territory of Judah, possibly the same as Geder.*

Beth Gamul – Town in Moab cursed by Jeremiah (48:23). – *Kh. el-Jemeil (235099), in Transjordan, E of Dead Sea, 33 mi. S of Amman.*

Beth Gilgal – Singers from Beth Gilgal participated in the dedication of the wall of Jerusalem (Neh 12:29). – *Unknown; probably in vicinity of Jerusalem.*

Beth Haggan – Ahaziah, king of Judah, fled from Jehu, "up the road to Beth Haggan" (2 Kgs 9:27) at which time he was wounded. – *Jenin (178207), 11 mi. SE of Megiddo on S edge of Jezreel Valley.* **38**, 39, 40, **44**

Beth Hakkerem – Hill Country town allotted to Judah (Josh 15:59b; LXX). Place associated with Judeans fleeing from Jerusalem from which a signal (fire) could be raised (Jer 6:1). In Nehemiah's day, it was the capital of a district of Judah who's ruler participated in the repair of the Dung Gate of Jerusalem (Neh 3:14). – *Kh. Salih/Ramat Rahel (170127), 3 mi. SSW of Jerusalem, or less probably Ein Karim/En Kerem (165130), 5 mi. W of Jerusalem.* 46, **47**, 166, **177**, 179

Beth Hakkerem Valley. 36, 37, **38**

Beth Haram – Town in Transjordan allotted to tribe of Gad (Josh 13:7; cf. Num 32:36 "Beth Haran"). – *Tell Iktanu (214136), 18 mi. WSW of Amman.* **119**

Beth Hoglah – Town SE of Jericho on border of Judah and Benjamin (Josh 15:6; 18:19) but allotted to Benjamin (18:21). – *Uncertain; possibly Deir Hajlah (197136), 3.5 mi. SE of Jericho.* **114,** 115

Beth Horon (Upper and Lower) – Twin towns located only 2 mi. apart on a ridge guarding the approach to the Hill Country from the Coastal Plain. Mentioned 14 times in OT: Joshua pursued the fleeing Amorites past these towns (Josh 10:9 – 15); they were near Ephraim-Benjamin border, but were in Ephraim (Josh 21:22; 1 Chron 6:68); one, or both, was a Levitical city (Josh 21:22; 1 Chron 6:68); Solomon rebuilt them to protect one of the approaches to Jerusalem from the W (1 Kgs 9:17; 2 Chron 8:5); etc. – *Upper = Beit Ur el-Foqa (160143), Lower = Beit Ur el-Tahta (158144), 12 mi. NW of Jerusalem.* **32, 44,** 47, **47,** 51, 53, 110, 111, **115, 120, 131,** 132, **132,** 133, 139, **140,** 144, **145,** 147, 152, **154, 155, 159,** 166, **177, 190,** 191, 192, **217,** 219, **222, 240,** 241, **257**

Beth Jeshimoth – Last camping place for Israel before crossing the Jordan River into Canaan (on plains of Moab; Num 33:49); on extremity of Sihon the Amorite's territory (Josh 12:3); allotted to Reuben (12:20) but later a town of Moab (Ezek 25:9). – *T. el-Azeimeh (208132), 22 mi. WSW of Amman.* 108, **108, 110,** 119

Beth Lebaoth – Town in Negev allotted to Simeon (Josh 19:6). Possibly same as Lebaoth (Josh

15:32) and Beth Biri (1 Chron 4:31). – *Unknown; in S Israel/Judah.*

Bethlehem (Judah) – Judean town mentioned 40 times in the OT. Naomi and Ruth were from Bethlehem as was David. Most OT references are found in stories concerning David. Fortified by king Rehoboam (2 Chron 11:6), it was settled by Jews during the postexilic period (Ezra 2:21; Neh 7:26). Mentioned 8 times in NT, chiefly in connection with birth of Jesus. – *Beit Lahm (169123), 5 mi. SSW of Jerusalem.* 23, **23,** 25, 33, **47,** 48, 49, 51, **51,** **97,** 114, 114, 123, 126, **129,** 131, 132, 134, **135,** 151, **159,** 163, 166, **177, 190,** 194, **200,** 204, **212,** 214, 217, **217, 241,** 255

Bethlehem (Zebulun) – Town allotted to Zebulun (Josh 19:15) and from which the minor judge Ibzan came (Judg 12:8, 10). – *Beit Lahm/Bet Lehem Hagelilit (168238), 7 mi. WNW of Nazareth.* **116**

Beth Marcaboth – Village in Negev allotted to Simeon (Josh 19:5; 1 Chron 4:31). May be identical to Madmannah which is preceded by Ziklag (Josh 15:31) in the Negev list of Judah, if not, then – *Unknown; in S Judah.*

Beth Meon – See Baal Meon.

Beth Nimrah – Town allotted to Gad that was near good grazing land (Num 32:36; Josh 13:27). Called Nimrah in Num 32:3. – *T. el-Bleibil (210146), in Transjordan, 18 mi. E of Amman.* **119**

Beth Ophrah – Town in/near Shephelah mentioned in oracle of Micah (1:10). – *Uncertain; possibly et-Taiyibeh (153107), 5 mi. NW of Hebron.*

Beth Pazzez – Town allotted to Issachar (Josh 19:21). – *Uncertain; possibly Sheikh Mazghith (199221), 5.5 mi. N of Beth Shan.*

Beth Pelet – Town allotted to Judah, in Negev district (Josh 15:27). After exile, Jews settled there; evidently in Negev (Neh 11:26). – *Uncertain; possibly T. es-Saqti/T. Shoqet (141079), 8 mi. NE of modern Beersheba.* **114,** 166, **177,** 179

Beth Peor – Place in or near the plains of Moab where Israel camped. Setting for giving of laws of Deuteronomy (3:29; 4:46) and place where Moses was buried (Deut 34:6). Allotted to Reuben (Josh 13:20). – *Uncertain; possibly Kh. esh-Sheikh Jayil (215133), 18 mi. WSW of Amman.* **119**

Bethphage – Village on Mt. of Olives, through which Jesus passed on way from Bethany to Jerusalem on day of his triumphal entry into Jerusalem (= Palm Sunday; Matt 21:11; Mark 11:1; Luke 19:29). – *et-Tur (173131), 0.7 mi. E of Jerusalem.* 214, 251

Beth – ramatha. 206, 211

Beth Rehob – Aramean principality N of Israel. Laish/Dan was in, or near, it (Judg 18:28) and the Ammonites hired soldiers from it to fight David's troops in 2 Sam 10:8. – *Probably located in Lebanese Beqa N of Israel.* 139, **140,** 141

Bethsaida – City located on N shore of Sea of Galilee. The disciples Philip, Andrew, and Peter were from the town (John 1:44; 12:21). Jesus performed mighty works there (e.g., healed a blind man [Mark 8:22], and fed 5,000 in vicinity [Matt 14:13; Mark 6:30ff]) yet the town was cursed by him because of unbelief (Matt 11:21; Luke 10:13). – *Possibly there were two Bethsaidas, one on the W (of ancient Jordan River) and the other on the E bank. The one on the W bank would have been a small fishing village "Bethsaida in Galilee" (John 12:21) located at el-Araj (208255), on the shore of the Sea of Galilee – el-Araj is east of the Jordan, but in NT times it may have been W of the Jordan. In contrast, the one on the E bank was a large Gentile city that had been built by Herod Philip, the tetrarch of Gaulanitis. It was also called "Julias" and it was mentioned in extrabiblical literature. – Et-Tell (209257), 1.5 mi. N of lakeshore.* **36,** 42, **43,** 193, **208,** 209, 210, **212**

Bethsaida – Julias – See Bethsaida and Julias. **43, 208,** 210

Bethsaida, Plain of. 43

Beth Shan – Major Canaanite city allotted to Manasseh located in or near Issachar (Josh 17:11; 1 Chron 7:29) but Manasseh was not able to drive

out inhabitants (Josh 17:16; Judg 1:27). Became a Philistine stronghold where bodies of Saul and Jonathan were hung on wall (1 Sam 31:10, 12; 2 Sam 21:12). Was located in 5th Solomonic district (1 Kgs 4:12). In Hellenistic period called Scythopolis and Nysa. – *Tell el-Husn/T. Bet Shean (197212), 15 mi. SSW of Sea of Galilee.* **22, 23,** 25, **30, 32, 33, 34, 38,** 40, **44,** 45, 46, **58, 60,** 61, **86, 94, 95,** 96, **101, 105, 109,** 115, **115,** 116, 121, **122, 123, 124,** 125, **130, 135,** 136, 137, 139, **142, 145,** 148, **148, 155, 157, 163,** 186, **186,** 257, 259

Beth Shemesh (Issachar) – Town in E Lower Galilee on boundary of Issachar. – *Possibly Kh. Sheikh esh-Shamsawi/H. Shemesh (199232) but more probably T. el-Abeidiyeh (202232), 2 mi. S of Sea of Galilee near Jordan River.* **38,** 116

Beth Shemesh (Judah/Dan) – Town in N Shephelah on N boundary of Judah (Josh 15:10) but allotted to Dan (19:41) who was unable to occupy it (Judg 1:35). Designated a Levitical city (Josh 21:16; 1 Chron 6:59) it served as a Judean outpost on their border with the Philistines; Ark of Covenant was returned there (1 Samuel 6); it was in Solomon's second administrative district; fortified by Rehoboam; and was disputed over by both Israel (2 Kgs 14:8 – 14; 2 Chron 25:21) and the Philistines (2 Chron 28:18). – *T. er-Rumeileh/T. Bet Shemesh (147128), 16 mi. W of Jerusalem.* **23,** 24, **32, 47,** 51, 53, **94, 95, 105,** 114, 118, **120, 122,** 130, **130, 131,** 132, **134, 145,** 148, **148,** 158, **159,** 160, 165, **166, 170,** 255, 256, 257

Beth Shemesh (Naphtali) – Town in Upper Galilee allotted to to Naphtali (Josh 19:38); Canaanites continued to live in the town but served as laborers to Naphtalites (Judg 1:33). – *Kh. T. er-Ruweisi/T. Rosh (181271), 17 mi. NE of Acco.* **36,** 37, **116,** 117, **122**

Beth Shittah – Site on route along which Midianites fled before Gideon (Judg 7:22). – *Unknown. Probably SE of Beth Shan in Jordan Valley.* **128**

Beth Tappuah – Town allotted to Judah in the Hill Country (Josh 15:53). – *Taffuh (154105), 3.5 mi. W NW of Hebron.* **114,** 166

Beththter – OT Bether. **222,** 223

Beth Togarmah – N country noted for horses (Ezek 27:14) and mentioned in prophecy against Gog (38:6). – *E Turkey, modern Armenia.* **83**

Bethuel – Negev town allotted to Simeon (1 Chron 4:30 and Josh 19:4 "Bethul"). Probably also 1 Sam 30:27 "Bethel," to which David sent plunder. Parallel Judean list (Josh 15:30) has Kesil. – *Unknown.*

Beth – yerah/Beth Yerah. **38, 43,** 60, **84, 86,** 87, **94**

Beth Zechariah. 190, 192

Beth Zur – Town allotted to Judah in Hill Country (Josh 15:58). Populated by descendants of Caleb (1 Chron 2:45), fortified by Rehoboam (2 Chron 11:7), and administrative center of a half-district in the days of Nehemiah (3:16). Mentioned frequently in the literature of the Intertestamental period. – *Kh. et-Tubeiqeh (159110), 4.5 mi. N of Hebron.* **95, 105, 114, 130, 151, 166,** 173, **177,** 179, **190,** 192, **193,** 194

Bet Netofa Valley. 116

Betogabris. 200, 217, **217,** 222

Betonim – Town allotted to Gad (Josh 13:26). – *Kh. Batneh (217154), 16 mi. NE of Jericho in Transjordan.* **119**

Beyond the Jordan – See Perea. **217**

Beyond the River – See Trans – Euphrates. 73, 141, 175, **175,** 176, 178

Bezek – Town conquered by Simeonites and Judahites (Judg 1:4) and place where Saul mustered troops prior to battle with Ammonites at Jabesh Gilead (1 Sam 11:8). – *Latter at Kh. Ibziq (187197), 13 mi. NE of Shechem; former may be the same or a different, Unknown; site.* **115, 132,** 133

Bezer – One of the three cities of refuge located E of the Jordan River (Deut 4:43; Josh 20:8). Also named as a Levitical city located in Reubenite territory (Josh 21:36; 1 Chron 6:78). – *Possibly Umm el-Amad (235132), 11 mi. S of Amman.* **120,** 121, 156, **163**

Bezer (Golan) – See Bostra. **58, 123**

Bezetha Valley. 242, **242**

Bezetha (Quarter in Jerusalem). 250, 253

Bileam – Levitical city in territory of Manasseh (1 Chron 6:70). – *Probably same as Ibleam.*

Bilhah – Town allotted to Simeon (1 Chron 4:29). Probably same as Balah (Josh 19:3) and Baalah (15:29). – *Unknown; in S Israel.*

Bir el – Abud. 104

Bir el – Mazar. 104

Bir en – Nasb. 69

Bir Mara. 69, 104, 104

Bithron – Joab pursued Abner past Bithron (2 Sam 2:29). – *Unknown; E of Jordan River on way E to Mahanaim.*

Bithynia – A Roman province in N Asia Minor (Turkey) along S coast of Black Sea. On his second missionary journey Paul and Silas were forbidden to enter there (Acts 16:7), yet later Peter addressed the church that had been established there (1 Peter 1:1). **202,** 227, **228, 231, 233**

Bitter Lakes. 69, 104, 104

Biziothiah – Possibly a town in Negev district allotted to Judah (Josh 15:28, NIV), but more probably, following the LXX and Neh 11:27, the passage should be read "Beersheba and its settlements."

Black Sea. 17, 83, 175, 181, 182, 218, 227, 228, 231, 233, 234, 237

Blue Nile. 65, 66

Bohan, Stone of – A marker on the NE border of Judah (Josh 15:6) – SE Benjamin (18:17). – *Unknown; in Jericho region, but closer to Dead Sea.*

Bokim – (weeping) Place where Israel wept in repentance before the Lord (Judg 2:1, 5). – *Uncertain; possibly in Hill Country W of Jericho, near Bethel (cf. Gen 35:8: "oak below Bethel. So it was named Allon Bacuth.").*

Bor Ashan – Place in S Judah where David and men had roamed, and to which he sent plunder (1 Sam 30:30). – *Uncertain; possibly same as Ashan.*

Borsippa. 77, 169, 171

Bosor. 192

Bostra. 192, 206, 211, 234

Bozez – A cliff S of Micmash which Jonathan and his armor bearer climbed in order to attack Philistine garrison (1 Sam 14:4). – *Cliff on N rim of Wadi Suweinit (ca. 176141), 7 mi. NNE of Jerusalem.* **132,** 133

Bozkath – Town in Shephelah district of Judah (Josh 15:39). Home of Jedidah, king Josiah's mother (2 Kgs 22:1). – *Unknown.*

Bozrah (Edom) – City in N Edom located along caravan routes. Early king of Edom from there (Gen 36:33; 1 Chron 1:44) and place was focus of prophecies of judgment (Isa 34:6; 63:1; Jer 49:13, 22; Amos 1:12). – *Buseirah (208016), 25 mi. SE of S tip of Dead Sea in mountains of Edom.* **32,** 33, **55,** 57, **62,** 63, 64, **108, 145, 153, 159,** 163

Bozrah (Moab) – Town in Moab mentioned in judgment prophecy of Jeremiah (48:24). – *Possibly the same as Bezer.*

Britannia. 202

Brook of Besor – See Besor Ravine. 135

Brook of Egypt – See Wadi of Egypt. 140, 164, 169, 69

Bucephalia. 183

Buqeia – see Achor, Valley of. 47, 49

Butastis. 218

Buz – Mentioned in an oracle of Jeremiah (25:23). – *Unknown; probably in the desert to the E of Palestine.*

Byblos – See Gebal. 17, 22, 72, 74, **84,** 86, **90, 93,** 96, **101, 106, 146, 160, 169, 171, 174,** 181, 181

Byzantium. 175, 228, 229, 231, 237

Cabbon – Town allotted to Judah in a Shephelah district (Josh 15:40). – Unknown.

Cabul – Town near/in Plain of Acco allotted to Asher (Josh 19:27). Also name of a district, "Land of Cabul," in the same area; given by Solomon to Hiram, king of Tyre (1 Kgs 9:13). – *Kh. Rosh Zayith (171253), 8 mi. ESE of Acco.* **116, 145,** 148

Cabura. 183

Cadasa. 193, 194, 208, 222

Caesarea/Caesarea Maritima – Mentioned 17 times in the NT – all in Acts. Visited by Philip (8:40) and Saul (9:30) and here Cornelius was converted to Christianity (chapters 10 and 11). It became the Roman capital of Palestine and Herod Agrippa I died there (12:19). Paul passed through it at end of his second (18:22) and third (21:8, 16) missionary journeys. Paul was a prisoner there for several years before being shipped to Rome for trial. Herod the Great (37 – 4 B.C.) had built the city and in NT times it was the main port of Palestine. – *Qaisariye (140212), 31 mi. N of Joppa.* 22, **22, 23,** 34, **38,** 43, **44,** 45, 186, **186, 187, 200,** 201, 202, 203, **206,** 210, **212,** 216, 217, **217,** 219, 219, **220, 222,** 224, **228, 231, 231,** 232, **233, 234,** 258

Caesarea (Cappadocia). 225, 234

Caesarea Philippi – In region of Caesarea Philippi Peter made his great confession that Jesus was the Christ (Matt 16:13 – 20; Mark 8:27 – 30). Located at the foot of Mt. Hermon, formerly called Paneas, it was situated at the headwaters of the Jordan River, and renamed by Philip the tetrarch. – *Banias (215295) on NE slope of Huleh Valley, ca. 50 mi. SW of Damascus. Also called "Neronias" for a brief period of time.* 34, **34,** 35, **36, 38, 43, 206, 208,** 210, 211, **212, 217, 219, 220**

Cairo. 17, 19, 65, **66,** 67, 71, 85

Calah – An important city of the kingdom of Nimrod (Gen 10:11, 12). Frequently mentioned in Assyrian texts. Royal residence of Assyrian kings from Ashurnasirpal II on. – *Nimrud in N Iraq on E bank of Tigris River, N of where the Upper Zab joins it. About 22 mi. S of Nineveh.* 83, **160,** 162, **169,** 171

Caleb Ephrathah – If really a village, Hezron died there (1 Chr 2:24). – *Unknown. RSV, following the LXX translates "after the death of Hezron, Caleb went in to Ephrathah, the wife of Hezron…."*

Callirrhoe. 47, 50, 204, 211

Calneh (S Mesopotamia) – City founded by Nimrod and located in Shinar (S Mesopotamia? Gen 10:10). – *Unknown. But possibly the text is to be translated "all of them in the land of Shinar" (RSV).*

Calneh (Syria) – The Calneh of Amos 6:2 is associated with a northern Hamath and may be identical with Assyrian Kullani. Probably "Calno" (Isa 10:9) and "Canneh" (Ezek 27:23) are the same places as this N "Calneh" with slight spelling variations. – *Kullankoy, 8 mi. NW of Aleppo in Syria.* 160

Calno – See Calneh.

Canaan, (land of) – A geo-political territory mentioned over 75 times in the OT (plus over 75 references to Canaanites). Mentioned in biblical and extrabiblical texts of the second and first millennia B.C. From the Desert of Zin and Kadesh Barnea in the S it stretched to Lebo Hamath in the N, and from the Mediterranean Sea on the W it reached the Dead Sea, the Jordan River, and even E of Damascus on the E (Num 34:1 – 12; Ezek 47:13 – 48:29; see especially p. 106). Mentioned prominently in the Patriarchial, Exodus, and Conquest narratives as well as elsewhere. 13, 18, 21, 59, 68, 75, **83,** 84, 87, 88, 89, 90, 91, 92, **93,** 95, 96, 97, 98, 99, 100, 101, 103, 105, 106, **106,** 107, 108, 111, 112, 113, 115, 116, 118, 119, 121, **122,** 123, 124, **124,** 125, 127, 129, **129,** 130, 131, 142, 243, 254, 255, 256, 266, 267

Cana (of Galilee) – Village where Jesus performed his first (water into wine; John 2:1 – 11) and second (healed official of Capernaum's son; John 4:46 – 54) miracles. It was the home of Nathanael (John 21:2). – *Traditional site Kafr Kana (182239), 4 mi. NE of Nazareth; more probable site Kh. Qana (178247), 8 mi. N of Nazareth.* 36, 38, **38,** 43, 207, **208,** 212

Canatha. 206, 211, **212**

Canneh – See Calneh.

Capercotnei. 217

Capernaum – Village mentioned 16 times in the NT in association with the ministry of Jesus. Jesus' headquarters for much of his public ministry; several of his disciples were from Capernaum; and many miracles were performed there. It was a fishing village, had a toll booth and a synagogue, and was large enough to be the residence of a public official and a Roman centurion. – *Tell Hum (204254), on N shore of Sea of Galilee, 2.5 mi. W of where Jordan enters the sea.* **36,** 38, 42, **43,** 207, **208,** 209, 210, **212, 217, 234**

Capharabis. 222

Capharsalama. 190, 192

Caphartobas. 222

Caphtor(ites) – Place from which the Caphtorites (Deut 2:23) and Philistines (Jer 47:4; Amos 9:7) came. Also mentioned in extrabiblical texts. – *Probably the island of Crete.* 83, 84

Cappadocia – A Roman province in E Asia Minor that served as a buffer on its E frontier. Jews from Cappadocia were present in Jerusalem on day of Pentecost (Acts 2:9) and a church was established there (1 Peter 1:1). *Location: E Turkey.* **175, 202, 218, 225, 228, 231, 233**

Carchemish – Important Hittite-Syrian town situated at an important ford on the W bank of the Euphrates River. Referred to in extrabiblical texts; fought over by the Assyrians, Egyptians, and Babylonians during the second millennium B.C (Isa 10:9; Jer 46:2; 2 Chr 35:20). – *Jerablus, about 60 mi. NE of Aleppo Syria, but actually in Turkey.* 17, **72,** 73, 76, **77,** 78, **84, 93,** 97, 146, 160, 169, **169,** 171

Carem – Village in hill country allotted to Judah. Mentioned in Greek (LXX) text of Josh 15:59 (see NEB). – *Uncertain; possibly same as Beth Hakkerem (Neh 3:14; Jer 6:1).*

Caria. 180, **181, 184,** 185, 227, 228

Carmel (Judah) – Village allotted to tribe of Judah (Josh 15:55). Mentioned in connection with Saul's defeat of the Amalekites (1 Sam 15:12) and figured prominently in the story of David and Abigail – who's husband, Nabal, was from Carmel (1 Samuel 25, passim; 27:3; 30:5; 2 Sam 2:2; 3:3). Davidic warrior from this village (1 Chr 3:1; 2 Sam 23:35). – *Kh. el-Kirmil (162092), 7.5 mi. SE of Hebron on edge of Judean Desert.* 47, **47,** 49, **94, 114,** 133, 135, **135,** 136, 166

Carmel, Mount/Carmel Range – Prominent mountain which interrupts the coastal plain in N Israel. Stretches about 30 mi. from the Mediterranean Sea, SE. NE and W slopes of the triangular shaped range are quite steep. Formed S border of tribe of Asher (Josh 19:26) and was the site of the great contest between Elijah and the prophets of Baal (1 Kings 18). Later, Elisha frequented the region (2 Kgs 2:25; 4:25). Because of its lush tree cover it became a symbol of beauty and fruitfulness (SS 7:5; Isa 35:2; etc.) and its withering became a symbol of destruction and desolation (Amos 1:2; 9:3; etc.). 21, 22, **22, 23, 32,** 33, **36,** 37, **38,** 40, 41, 43, 44, **44,** 45, 47, 59, 88, 101, **101,** 116, **116,** 121, **157, 161,** 162, 169, 175, **193, 200,** 201, 202

Carmel, Shephelah (foothill) of. 38, 44

Carnaim – See Karnaim. 192

Carthage. 234

Casiphia – A village in Babylonian territory from which Ezra summoned temple servants (Ezra 8:17). – *Unknown; probably in S Iraq.*

Caspein. 192

Caspian Sea. 17, 77, 83, 93, 160, 171, 174, 175, 183, 218

Casius, Mount. 72, 75

Caucasus (Mountains). 78, 182

Cauda – Small island in the Mediterranean that the ship Paul was traveling on passed on its lee side (Acts 27:16). – *Gaudhos/Gozzo, ca. 30 mi. S of W Crete.* 233, **233**

Cenchrea – Port at which Paul had his hair cut in fulfillment of a vow before setting sail with Priscilla and Aquila for Ephesus and then Syria on his second journey (Acts 18:18). Phoebe was later commended for her service to the church in Cenchrea (Rom 16:1). Was one of two ports of Corinth; goods (and ships) were transported over the isthmus W to the port of Lechaeum. – *Location: Greece, 7 mi. SE of Corinth on Saronic Gulf.* **228,** 229, 230, 231

Central Mountain Range. 21, **22,** 23, 24, 28, 46, 47, 154

Central Valley (Jerusalem) – See also Tyropoeon. 242, **242, 246,** 247, 249, 251

Chabulon. 205, **206, 208**

Chagar Bazar. 72, 73, 78

Chaldea – A region in S Mesopotamia (modern Iraq) at the head of the Persian Gulf. Chaldeans were a tribe in the area which became powerful in the seventh century B.C. and which eventually established the neo-Babylonian dynasty. Chaldea, Chaldean(s), mentioned 12 times in the NIV translation of the Bible. In some places NIV translates the Hebrew for Chalea/Chaldean(s) as Babylon/Babylonian(s). **90**

Chalcis. **220**

Charax (S Mesopotamia). **182**

Charax (Syrian). 192

Cilicia – A region/province in SE Asia Minor with Tarsus as its leading city. Jews from there disputed with Stephen in Jerusalem (Acts 6:9). Cilicia is mentioned in connection with Paul's home in Tarsus (21:39; 22:3; 23:34) and Paul evangelized the area (15:23, 41; Gal 1:21). On his way to Rome, Paul's boat passed the coast of Cilicia (27:5). *Location: in SE Turkey along shore of Mediterranean Sea – part mountainous and part coastal plain.* 75, **160, 174, 175,** 181, **181, 202, 218,** 219, 224, **225,** 227, **228, 231, 233**

Cilician Gates. **72,** 180, 181, **181, 225,** 226, 227, **228, 231**

Cilician Plain. 75

Cirta. **234**

Cisjordan. 25, 45, 57, 186

Cisjordan Mountains/Range. 45, 46

Cnidus – Paul's ship passed Cnidus on his way to Rome (Acts 27:7). – *City on SW tip of Asia Minor at the W end of a long peninsula between the islands of Cos and Rhodes.* 194, 233, **233**

Cnossus – See Knossos. **234**

Coastal Plain (Levant). 21, 22, **22,** 24, 28, 29, 43, 46, 47, 51, 52, 54, 74, 75, 114, 115, 116, 117, 121, **122,** 125, 127, 132, 133, 134, 139, 149, 151, 157, 158, 186, 191, 192, 199, 217, 240

Coele-Syria. 188, **193**

Colonia/Qaloniya – See Emmaus. 215

Colonia (Germania). **234**

Colosse – A Phrygian city in Roman province of Asia to which Paul wrote an epistle (Col 1:2). Paul probably had not visited the city (Col 2:1) and by his day it had lost some importance to Laodicea, 8 mi. to the NW. – *A Huyuk (tell) located in SW Turkey near Honaz, on the banks of the Aksu River just south of the Lycus Valley.* **234, 237**

Commagene. 199, **202**

Corduba. **234**

Coreae. **212,** 220, **222**

Corinth – Large Roman city strategically located just S of the isthmus that connects mainland Greece with the Peloponnese. Paul visited the city on his second (Acts 18:1 – 18) and third (probably Acts 20:2 – 3) missionary journeys and two of his letters to the church there are preserved. This commercial center was known for its wickedness. **181,** 185, **218,** 227, **228,** 229, **229,** 230, 231, **231,** 232

Corinth, Gulf of (Corinthian Gulf). 230, **229**

Corruption, Hill of – Hill where Solomon had built altars for the pagan deities Ashtoreth, Chemosh, and Molech and which were desecrated by Josiah (2 Kgs 23:13). – *Probably the S most knoll on the Mt. of Olives, .25 mi. E of Jerusalem.*

Corsica. **202**

Cos – Small island on SW coast of Asia Minor (Turkey), which Paul's ship passed on his third journey as he traveled from Ephesus to Rhodes (Acts 21:1).**218, 231,** 232

Cozeba – Village settled by descendants of Judah (1 Chr 4:22). *Possibly the same as Aczib (Judah).*

Craftsmen, Valley of – A low-lying area in the N Philistine Plain where the descendants of Kenaz (1 Chr 4:14, see NIV note) and Benjamin (Neh 11:35) settled. – *A roughly triangular shaped area with the villages of Joppa, Lod, and Aphek situated at the points of the triangle with Ono situated near its center.* 52

Crags of the Wild Goats – Where David hid from Saul (1 Sam 24:1 – 2). – *In the area of En Gedi along the W shore of the Dead Sea.* 136

Crete – A large (160 mi. long) island in the Mediterranean Sea, about 180 mi. S of Athens, Greece. Mentioned 7 times in the NT, people from Crete were in Jerusalem on the day of Pentecost (Acts 2:11). Paul's ship, on its difficult voyage to Rome, passed to the S of Crete in a storm (Acts 27:7 – 16). **17,** 129, **129, 175, 181, 218, 228, 231,** 233, **233,** 234

Crocodile River – See Nahal Tanninim. 43

Crocodilon Polis. 44, **44**

Ctesiphon. **77, 218**

Cumae. **234**

Cun – A town of Hadadezer, the king of Zobah, from which David took a large quantity of bronze (1 Chr 18:8). – *Possibly Ras Baalbek (283406) in Lebanese Beqa, 55 mi. NE of Beirut.* **140,** 141

Cush – A country located immediately to the S of Egypt but N of Ethiopia. Its heartland was between the second and third cataracts of the Nile, although during periods of strength it expanded N to the first and S to the fourth cataracts. On occasion, Cush also seems to be associated with S Arabia (Gen 10:7) and possibly with Mesopotamia as well (Gen 2:13 and 10:8?). Country and people mentioned ca. 50 times in the OT. **66,** 67, 83, **83,** 84, 168

Cushan – Mentioned only in Hab 3:7 in parallel with Midian. May be a region in or near Midian. Compare Moses' "Cushite wife" (Num 12:1).

Cuthah – A city in S Mesopotamia from which Sargon II brought people to settle in N Israel after the N Kingdom had been carried into captivity (2 Kgs 17:24, 30). – *Possibly Tel Ibrahim, ca. 20 mi. NE of Babylon.* **160,** 164, **171**

Cypros. **202,** 201

Cyprus – Third largest island in the Mediterranean Sea, located in its NE corner 45 mi. S of Turkey and 60 mi. W of Syria. Called Kittim (Gen 10:4; Num 24:24; etc.) in OT, which in some cases is translated "Cyprus" (NIV; Isa 23:1; etc.). Barnabas (Acts 4:36), Paul's traveling companion on his first journey was from Cyprus — their first stop on that journey (13:4). Later Barnabas and Mark revisited the island (15:39) and Paul passed it on several of his journeys (21:3; 27:4). **17, 19, 84, 124, 129, 160, 169, 171, 175, 181, 184,** 185, 199, 201, 225, **225,** 226, 227, **228, 231,** 232, **233**

Cyrenaica. **202**

Cyrene – The district capital of the Roman province of Cyrenaica located in N Africa. Simon, who bore Jesus' cross, was from Cyrene (Matt 27:32; Mark 15:21; Luke 23:26), Jews from there were in Jerusalem on the day of Pentecost (Acts 2:10), and Jews from there, living in Jerusalem, disputed with Stephen (6:9). Some of these men from Cyrene were converted and preached the gospel (11:20; 13:1). – *Shakhat, in Libya, E of the Gulf of Sidra, about 100 mi. ENE of Benghazi, a few miles inland from the sea.* **17, 218, 234**

Dabbesheth – Village mentioned in the description of the S border of Zebulun; between Sarid and Jokneam (Josh 19:11). – *T. esh-Shammam/T. Shem (164230), 6 mi. NW of Megiddo.* 116, **116**

Daberath – Town on the border of Zebulun (Josh 19:12) but actually in Issachar (possibly the same as Rabbith; Josh 19:20). Was a Levitical city (Josh 21:28; 1 Chron 6:72). – *Daburiyeh (185233), 5 mi. E of Nazareth at the NW foot of Mt. Tabor.* 116, **116, 120**

Dacia. **202, 234**

Dahshur. **99**

Dalmanutha – Jesus went to the "region" of this village after the feeding of the 4,000 (Mark 8:10). – *Uncertain; may be identical to Magadan/Magdala (Matt 15:39 a parallel text and variant). If so, then Majdal (198247), 3 mi. NW of Tiberias on NW shore of Sea of Galilee.* 42, **43,** 208, **208**

Dalmatia – A district in the S part of Illyricum. Titus traveled there (2 Tim 4:10). – *W Croatia on the Adriatic Sea, around the Dubrovnik region.* **202**

Damascus – Primarily a reference to a major Syrian city, but by extension the term can refer to a geographical region or an Aramean Kingdom. Term is used 43 times in OT and 15 in NT. Bordered on S, W, and N by mountains and fed by two rivers: Biblical Barada and Pharpar. Briefly controlled by David and Solomon, but subsequently a foe of Israel throughout much of OT history. Antioch on the Orontes replaced it as the capital of Syria during the Hellenistic period. Paul was on the road to Damascus when he had his famous vision. – *Esh-Sham/Dimasq (272324), an oasis city at 2000 ft., located E of the Anti-Lebanon mountains in Syria.* **17,** 18, **19,** 21, **23,** 25, **26,** 27, 28, 32, **32,** 33, 34, **34,** 35, 36, **36,** 37, 40, **47, 60, 62,** 64, **66,** 72, **72,** 73, 74, 75, 78, **90,** 91, **93, 94, 97, 101, 106, 109,** 117, 119, **123, 124, 140,** 141, 144, **145, 146, 148,** 149, **153,** 154, 155, **155,** 156, **157,** 158, **160, 161,** 162, **163,** 164, 165, **169,171, 174, 175,** 176, **178, 181, 182, 184,** 185, **186, 193,** 195, 196, **200, 206, 208,** 209, 210, 211, **212, 217,** 218, **218,** 219, **219, 220,** 224, **228, 234,** 247

Damascus Gate (Jerusalem). **242,** 251, 252, 256

Damascus Oasis. 34

Damghan. **175**

Dan – The name of this Israelite tribe was given to the town of Laish/Leshem that they conquered (Judg 18:7; Josh 19:47). As a town name it appears about 24 times in the OT (NIV). Located at the N boundary of Israel, it was used 7 times in the familiar description of Israel's territory as stretching from "Dan to Beersheba" (2 times in reverse order) and it was usually the first city to bear the attack of invaders from the N. There Jeroboam set up a golden calf in a sanctuary he built (1 Kgs 12:29). – *T. el-Qadi/T. Dan (211294), 17 mi. NNE of Hazor at the NE edge of the Huleh Valley near the foot of Mt. Hermon.* 21, **22, 23,** 24, 25, **26,** 27, 29, **30, 32,** 33, **34,** 35, 36, **36,** 37, **38, 43,** 55, **72, 86, 93, 94,** 95, **95,** 96, **97,** 98, **105, 109, 112,** 113, 114, **116,** 117, 118, **118,** 121, **122,** 130, **130, 140,** 142, **142, 145,** 147, **148,** 150, 152, **153,** 154, **155, 157, 161,** 169, **174,** 188, **193,** 195, **208,** 254, 256, 257

Dan Jaan – Place in N Israel mentioned in Joab's census itinerary (2 Sam 24:6). Could be a village in the vicinity of Dan (see above) but possibly the text should be read "to Dan and Ijon (Jaan)" for both are known as towns in the region (cf. 2 Kgs 15:29, where they are mentioned together).

Dannah – Village in a hill country district of Judah. – *Unknown.*

Danube. **175, 182**

Daphne. **208**

Darb el-Gaza. **55, 62, 69,** 107

Darb el-Hagg. **55,** 57, 69, **69,** 104, **104**

Darb es – Sultan – See King's Highway. 107

Dardanelles, Straits of. 180, **181,** 227, **237**

Dathema. 192, **193**

David, City of – Alternate name for Jerusalem. In Luke 2:11 used of Bethlehem, original home of David. 241, 245, 247, **248,** 253

Dead Sea – See Salt Sea. 11, **22, 23,** 24, 25, **26,** 27, **30, 32, 47,** 48, 49, 50, **51, 55,** 55, 56, 57, **58,** 59, 60, 61, 62, **62,** 63, **86,** 87, **91, 94,** 95, **95, 97, 99, 101, 104, 105, 106, 108, 120, 122, 123, 124, 130, 132,** 136, **140,** 141, 142, **142, 143, 145, 148, 153, 154, 161, 163, 166, 170,** 173, **177,** 179, **181, 184,** 186, **186, 187,** 188, **190, 193,** 194, **198,** 199, **200,** 201, 204, **206, 212, 217, 219, 220, 222,** 223, 240, 242, 256, 257

Debeltum. **234**

Debir (Gad) – Gadite town in Gilead (Transjordan; Josh 13:26). – *If same as Lo Debar then Umm ed-Dabar (207219) in Jordan Valley, 10 mi. S of Sea of Galilee.* **119**

Debir (Hill Country of Judah) – Canaanite city formerly known as "Kiriath Sepher/Sannah" (Josh 15:15, 49); conquered by the Israelites (Joshua 10; 11:21; 12:13) and later (?) by Othniel (Josh 15:15; Judg 1:11). Allotted to Judah (Josh 15:49); a Levitical city (Josh 21:15; 1 Chron 6:58). – *Kh. Rabud (151093), 8.5 mi. SSW of Hebron.* 47, **47, 51, 110,** 111, 114, **114, 120, 126, 166,** 254, 255

Debir (Judean boundary point) – A village (?) on the N boundary of Judah near the Valley of Achor (Josh 15:7). – *Unknown.*

Decapolis – A confederation of (usually) ten Greco-Roman (Gentile) cities located primarily E of the Sea of Galilee and the Jordan Valley including: Pella, Dion, Philadelphia, Gadara, Gerasa, Hippos, Damascus, Scythopolis, Raphana, Kanatha, and others. People from the region followed Jesus

(Matt 84 4:25), Jesus healed a demoniac there (Mark 5:20) and he later visited the region on his way to Galilee (Mark 7:31). Parts of the region now in modern Syria, Jordan, and Israel. 42, **58**, 59, **60**, 188, 189, 197, **198**, **200**, 201, **206**, 207, 208, 210, 211, **212**, 213, **217**, **219**, **220**, **222**

Dedad. 84

Dedan(ites) – Name of an individual, a people, a territory, and possibly a settlement. Mentioned primarily in prophetic literature in connection with Arabian (Isa 21:13) places/peoples such as Tema, Buz (Jer 25:23), Edom (Ezek 49:8), Teman (Ezek 25:13), and Sheba (Ezek 38:13, 14). Probably a way station on the incense route from S Arabia to Syria/Palestine. – *El-Ula Oasis in W Arabia.* 83, 84, **174, 175**

Deir Alla. 58, 59

Deir el – Balah. 105

Delphi. 17, 175, 218, 229

Der. 175

Derbe – A Lyconian city visited by Paul on his first (Acts 14:6, 20) and second (16:1) missionary journeys. One of Paul's trusted companions, Gaius, was from there (20:4). – *Kerti Huyuk in Turkey, 14 mi. NNE of Karaman.* 175, 218, 225, 226, 227, **228**, 231, 234

Destruction, City of – A city in Egypt where the "language of Canaan" would be spoken and where Yahweh was worshiped (Isa 19:18). – *Unknown; some read "city of the sun" and identify it with Heliopolis or the nearby site of Tell el-Yehudiyah.*

Dhahab. 69

Diblah – See Riblah.

Dibon (Judah) – Village in Negev settled by Jews after the exile (Neh 11:25). – *Unknown; possibly same as Dimonah.*

Dibon (Moab) – One of the chief cities of Moab. Captured by Israel (Num 21:30), allotted to Reuben (Num 32:3; Josh 13:17), but built by Gad (Num 32:34; also called Dibon Gad [Num 33:45, 46]). Later regained by Moab and mentioned in the Moabite stone and in prophetic oracles (Isaiah 15 [probably same as "Dimon"]; Jeremiah 48). – *Dhiban (224101), 13 mi. E of Dead Sea and 3.5 mi. N of Arnon Gorge.* **22, 23,** 47, 58, **62,** 63, 86, 108, 119, 155, 159, **163,** 179

Dibon Gad – See Dibon.

Dilean – Village in Judean Shephelah (Josh 15:38). – *Unknown.*

Dimnah – Zebulunite village named as a Levitical city. – *Probably the same as Rimmono (1 Chron 6:77) and Rimmon (Josh 19:13).*

Dimon – Variant of Dibon (Moab; Isa 15:9).

Dimona. 55

Dimonah – Negev town allotted to Judah (Josh 15:22). – *Unknown. Possibly Dibon (Judah) is a variant of it.*

Dinhabah – Capital city of Bela, son of Beor, king of Edom (Gen 36:32; 1 Chron 1:43). – *Unknown.*

Dion/Dium. 197, **200**, 206, 211, **212**

Diyala River. 77, **169**, 171

Diyarbakir. 72, 77, 78

Dizahab – A place E of the Jordan, in the Arabah, where Moses delivered (some?) of the messages contained in the book of Deuteronomy (Deut 1:1). – *Unknown; the suggested Dhahab on the Red Sea (104769) does not seem to fit the suggested biblical location.*

Docus. 195

Dophkah – A campsite in Sinai, somewhere between Egypt and Mt. Sinai (Num 33:12, 13). – *Unknown.*

Dor – A port city whose king was defeated by Joshua (12:23) and which was allotted to the tribe of Manasseh (17:11) but possessed by descendants of Ephraim (1 Chron 7:29). The area around Dor was evidently called "Naphoth Dor" (Josh 11:2; 12:23; 17:11) in which Solomon placed a special administrator (1 Kgs 4:11). It was often under non-Israelite control. – *Kh. el-Burj/T. Dor (142224), 15 mi. S of Haifa, 8 mi. N of Caesarea.* **23,** 38, 44, **44,** 88, 95, 105, 112, 115, **115,** 121, **122,** **130,** 148, 153, 155, 157, 163, 170, 186, 186

Dor, Coast/Plain of – See Naphoth Dor. 21, **22,** 38, 43, 149

Dora – OT Dor. **187, 193,** 197, **198, 206,** 212, 217

Dorylaeum. 227, 228

Dothan – A prominent town near which Joseph was sold into slavery by his brothers (Gen 37:17) and in which Elisha was later besieged by the Syrians (2 Kgs 6:13). – *T. Dothan (172202), 11 mi. NNW of Samaria and 5 mi. S of the Jezreel Valley.* 32, 38, **44,** 46, **86, 95, 97,** 105, 115, 161

Dothan Valley. 38, 40, 41, 45, 46, 94

Drehem. 89, 93, 94

Dumah (Arabia, Edom?) – Son of Ishmael (Gen 25:14; 1 Chron 1:30) and founder of an Arab tribe which may have had its center near Dumet ej-Jendel in Arabia, half way between the Persian Gulf and the Gulf of Aqaba/Elath. Possibly the same as Dumah mentioned in an oracle of Isaiah (21:11) although this Dumah may have been located in Edom. – *Uncertain.* **17,** 146, 160, **171, 175**

Dumah (Judah) – A town allotted to Judah (Josh 15:52). – *Kh. ed-Deir Domeh (148093), 10 mi. SW of Hebron in the hill country.* **114**

Dunqul Oasis. 86

Dura Europos. 78

Dura, plain of – A plain located near Babylon where Nebuchadnezzar set up a colossal statue (Dan 3:1). – *Unknown; but probably near Babylon.*

Dur Sharrukin. 77

Dyrrachium. 229

East, land of – Sons of Abraham, save Isaac, were sent to this area (Gen 25:6). – *Uncertain; possibly SE of Israel in Midian/Arabia.*

Eastern Desert/Wilderness. 21, 28, 65, **96,** 106, **106,** 114, 129

Eastern Sea – Name for Dead Sea (Ezek 47:18; Joel 2:20; Zech 14:8). 49, **178**

Ebal, Mount – A 3,080 foot peak (Jebel Islamiyeh) just N of Shechem (176179). There Joshua built an altar and read the curses and blessings of the law (Josh 8:30, 33) as commanded (Deut 11:29; 27:4, 13). **23,** 25, **44,** 45, 46, 91, **91,** 110, 111, **126, 212,** 215, 220, **222**

Ebenezer – There the Israelites lost two battles to the Philistines (1 Sam 4:1; 5:1). This is possibly the same place where Samuel set up a stone to commemorate God's help in a later victory against the Philistines ("between Mizpah and Shen;"1 Sam 7:12). – *Uncertain; possibly Isbat Sarta (146167), 13 mi. E of Joppa.* **44,** 131, **131,** 132

Ebez – One of the towns allotted to the tribe of Issachar (Josh 19:20). – *Uncertain; possibly the unnamed site at grid 197227, 9 mi. N of Beth Shan.*

Ebla. **17,** 18, **72, 82, 84,** 86, 88, **90, 93, 94, 97,** 106, 256

Ecbatana – One of the capitals of the Persian Empire. There Darius found the decree of Cyrus which permitted the temple to be rebuilt in Jerusalem (Ezra 6:2). – *Hamadan in Iran, 175 mi. SW of Tehran.* **17,** 160, 169, 171, 173, **174,** 175, 182, 218

Ecdippa. 208

Ed – Dakhla Oasis. 66, 103

Eden – Evidently a locality in which there was the garden where the first humans were placed (Genesis 2, 3, 4; cf. Ezekiel passim). – *Unknown; possibly in S Mesopotamia or Armenia.* 82, 83

Eden – Ezek 27:23 and 2 Kgs 19:12. See Beth Eden.

Eder – A town allotted to Judah in the Negev district (Josh 15:21). – *Uncertain; possibly a scribal error for Arad.* **114**

Edom – A personal, tribal and geographical name used (in various forms) 116 times in the OT. The geographical location of Edom was S of the Dead Sea, N of the Red Sea, and E of the Rift Valley. On occasion it expanded W into the Negev desert W of the Rift Valley. Edom was usually hostile towards Judah. 28, 49, **55,** 57, **62,** 63, 64, **69,** 97, 103, **106,** 107, **108, 114,** 125, 133, **140,** 144, 158, **159,** 160, 161, 163, 170, **170, 171, 173, 174**

Edom, Mountains of. 22, 25, 27, 56

Edom, Way of. 161

Edrei (Bashan) – A town in the Bashan that was the residence of Og, king of Bashan who was defeated by the Israelites (Num 21:33; Deut 1:4; 3:1, 10; Josh 12:4; 13:12). The town was allotted to the tribe of Manasseh (Josh 13:31). – *Dera (253224),* 60 mi. S of Damascus. **34,** 35, **58,** 108, **109, 119, 155, 157, 206,** 211, **212**

Edrei (Naphtali) – A town allotted to the tribe of Naphtali (Josh 19:37). – *Unknown; probably in Upper Galilee.*

Efrat. 92, 94

Eglaim – A town that was apparently on the border of Moab (Isa 15:8). – *Uncertain; possibly Mazra (201078), on the E shore of the Dead Sea, on the NE section of the Lisan.*

Eglath Shelishiyah – A town of Moab, mentioned in connection with Zoar (Isa 15:5) and Horonaim (Jer 48:34). – *Probably located E, or SE, of the S portion of the Dead Sea.*

Eglon – A village in the Shephelah whose king joined the king of Jerusalem against the invading Israelites and whose city was captured by Joshua (Josh 10 passim; 12:12). The Shephelah town was allotted to the tribe of Judah (Josh 15:39). – *Tell Aitun/T. Eton (143099), 11 mi. WSW of Hebron.* **110,** 111, **114,** 166

Egypt – Major country located in NE Africa. 13, 16, **17,** 18, **19,** 21, 22, 25, 32, 33, 36, **55,** 57, **62,** 64, 65, 66, 67, 68, 69, **69,** 70, 71, 72, 73, 74, 78, 79, 82, **83,** 84, **84,** 85, 86, 88, 89, 90, **91,** 92, 95, 96, **97,** 98, 99, **99,** 100, 101, 102, 103, 104, 104, 105, 106, 108, 123, **124,** 125, 127, **129,** 133, 138, 141, 142, 143, 144, **146,** 149, 150, 151, 152, 154, **159, 160,** 164, 168, 169, 170, 171, **171,** 172, 173, **173,** 174, **174,** 175, **175,** 176, 177, 179, 180, 181, **181, 182,** 184, **184,** 185, 187, 188, 189, 190, 191, 197, 199, 201, **202,** 205, 214, **218,** 231, **233, 234,** 244, 248, 254, 255, 256, 266, 267

Egypt, river of – Alternate designation for Nile (Gen 15:18; Amos 8:8; 9:5).

Egypt, Wadi/Brook of – A prominent wadi that marked a portion of the S boundaries of the land of Canaan (Num 34:5; Ezek 48:28), of Judah (Josh 15:4; 47), and later of Solomon's kingdom (1 Kgs 8:65; 2 Chron 7:8). Mentioned elsewhere in the Bible as a prominent marker (2 Kgs 24:7; Isa 27:12; Ezek 48:28). – *Wadi el-Arish that drains N Sinai and flows into the Mediterranean Sea SW of Gaza, although some identify it with the more northerly Nahal Besor. To be distinguished from the "River of Egypt" = the Nile.* 69, 106, **106, 140,** 164, 169, **178**

Ein (el-)Qudeirat – See Kadesh (Barnea). 69, 104

Ein Faria. 28, 30

Ein Feshkha – Small site on NW corner of Dead Sea 2 mi. SE of Qumran. **23,** 47, 49

Ein Furtaga. 69

Ein Hawwara. 69, 104, 104

Ein Khudra. 69

Ein Qudeirat – See Ein (el-)Qudeirat.

Ein Samiya. 92, 94, 95

Ekallate. 160

Ekron – A city in the coastal plain on the N border of Judah and S border of Dan (Josh 15:11, 46; 19:43). Judah had difficulty taking control of it (LXX of Judg 1:18) for the city was an important Philistine center (Josh 13:3). Mentioned 23 times in the OT and is prominent in the stories of the capture of the ark (1 Sam 5 & 6) and that of David and Goliath (1 Sam 17:52). – *Kh. el-Muqanna/T. Miqne (136131), 20 mi. SE of Joppa.* **23,** 24, **47,** 51, 52, **105, 114,** 118, **118, 122,** 129, **130, 131, 131,** 132, 134, **135, 153, 154, 155, 159,** 161, **163,** 166, **166,** 168, **241, 255**

Elah, Valley of – Valley in which David killed Goliath (1 Sam 17:2, 19; 21:9). – *Wadi es-Sant, ca. 18 mi. WSW of Jerusalem.* **51,** 53, **118,** 134, **134,** 168, 217

Elam – A country located to the NE of the Persian Gulf in modern Iran. In the patriarchal narratives its king, Kedorlaomer, invaded Canaan (Gen 14:1, 9). It later appears in the prophetic books and elsewhere. Eventually it became a satrapy in the Persian Empire and the story of Esther has its setting in Susa, the old capital of Elam. Elamites were present in Jerusalem on the day of Pentecost (Acts 21:9). 79, **83,** 84, **84,** 94, **94,** 160, 169, 171, **174, 218**

Elam (Judea) – Town settled by Jews in postexilic period (Ezra 2:31; Neh 7:34). – *Uncertain;*

possibly *Kh. Beit Alam/H. Bet Elem (145109), 10 mi. WNW of Hebron.* **177**

El-Amarna. 66, 103, 106, 123, 124, 125, 128, 240, 244, 255, 256

El-Arish – See also Wadi el – Arish. 55, 66, 69, **69**, 171

Elath – A place where Israel camped on trek to Canaan (Deut 2:8). From nearby Ezion Geber Solomon dispatched ships on the Red Sea (1 Kgs 9:26; 2 Chron 8:17). Uzziah recaptured the city and rebuilt it (2 Kgs 14:22; 2 Chron 26:1 – 2) but it was lost again in the days' of Ahaz (2 Kgs 16:6). – *Possibly Aqaba (150882), a Jordanian city located at the N end of the Gulf of Aqaba.* 17, 21, **22**, 24, 25, 27, 29, **32**, 55, **55**, 56, 57, **62**, 64, **66**, 68, 69, **69**, 107, **108**, **153**, 158, **160**, 171

Elath, Gulf of – See Aqaba, Gulf of. 27, 55, **55**, 56, 57, 63, 64, **66**, 68, 69, 70, 103, **104**

Elazar. 92, **94**

El-Azariyeh. 214

Elazig. 76

El – Balamun. **103**

El – Baqliya. **103**

El – Bethel – Alternate name for Bethel (Gen 35:6).

Elealah – A village in Transjordan, captured by the Israelites (Num 32:3) and allotted to the tribe of Reuben (Num 32:37) but later recaptured by Moab and mentioned in prophetic oracles as being among the cities of Moab (Isa 15:4; 16:9; Jer 48:34). – *El-Al (228136), 11 mi. SW of Amman.*

Eleasa. 190, 192

Elephantine. 99, 103, 173, 175, 179

Eleutherus River. 184, 185

Elim – An oasis with 12 springs and 70 palm trees where the Israelites camped on their way to Mt. Sinai (Exod 15:27; 16:1; Num 33:9, 10). – *Possibly Ayun Musa (9 mi. SE of Suez; if Jebel Sin Bisher is Mt. Sinai) or Wadi Gharandal (55 mi. SE of Suez; if Jebel Musa).* 104

Elishah – A place in/on the Mediterranean from which blue and purple dyes were secured (Ezek 27:7). – *Probably (a section of?) the island of Cyprus for other ancient near eastern literature refers to Cyprus as Alashia.* 83, 84, 146

El-Jib. 133

El – Kharga Oasis. **66**, **103**

Elkosh – Home of Nahum the prophet (Nahum 1:1). – *Unknown in spite of numerous suggestions.*

El – Lahun. **103**

Ellasar. **94**

El-Lisht – See Lisht. **99**

El-Lubban. 215

Elon – A town allotted to the tribe of Dan (Josh 19:43) located in the second of Solomon's administrative districts (1 Kgs 4:9; = Elon Bethhanan). – *Unknown; but in Coastal Plain, E of Joppa.* 148

Elon Bethhanan – Alternate name for Elon (see above). 148

El Paran – Site near the desert attacked by Kedorlaomer and allies (Gen 14:6). – *Possibly alternate name for Elath.* 94, **94**

El – Qantara. **69**

El-Qubeibeh. 215

Eltekeh – A town allotted to the tribe of Dan (Josh 19:44) that served as a Levitical city (Josh 21:23). – *Uncertain; possibly T. esh-Shallaf/T. Shalaf (128144), 11 mi. SSE of Joppa.* 118, **120**, 166

Eltekeh, Plain of. 168

Eltekon – A town allotted to the tribe of Judah located in the hill country probably just N of Hebron (Josh 15:59).

Eltolad – A town allotted to the tribe of Simeon (Josh 19:4; 1 Chron 4:29 = "Tolad"), within the allotment of Judah (Josh 15:30). – *Unknown but in Negev.*

El – Ula oasis. 84

Elusa. 32

El – Wa'arah. 34

Emar. 72, 84

Emek Keziz – A town allotted to the tribe of Benjamin located in its E portion (Josh 18:21). – *Unknown.*

Emesa. 182

Emmatha. 208

Emmaus – A town to which Cleopas and his companion were traveling as Jesus appeared to them (Luke 24:13). – *Disputed. Best candidate is Emmaus/Nicopolis (Imwas; 149138, 19 mi. W of Jerusalem, thus fitting the reading of 160 stadia [= 20 mi.] of an important NT manuscript). Alternately, Emmaus/Colonia is equally possible (Qalunya; 165133, 3.5 mi. W of Jerusalem, thus fitting the majority of NT manuscripts [60 stadia = 7 mi.] if the NT indicates the round trip distance).* 190, 192, **198**, 199, **200**, 206, 212, 215, **217**, 220, **222**

Enaim – Town near which Tamar enticed Judah (Gen 38:14, 21). – *Probably a variant of Enam.*

Enam – A town in the N Shephelah district of Judah (Josh 15:34). Probably same as Enaim. – *Uncertain; possibly Kh. Beith Ikka (151121), 5 mi. SE of Beth Shemesh.*

En Avedat. 55

En Besor. 86, **86**

En Boqeq. 47, 50

Endor/En Dor – Town "within Issachar" which was assigned to the tribe of Manasseh (Josh 17:11). Nearby, Barak defeated the Canaanites (Psalm 83:10) and there Saul consulted a medium (1 Sam 28:7). – *Kh. Safsafeh/H. Zafzafot (187227), 2.5 mi. due S of Mt. Tabor.* 34, 38, 39, **43**, 115, **116**, 128, **135**

En Eglaim – A town situated on the shore of the Dead Sea where the prophet says fishermen will spread their nets when the water of the sea become fresh (Ezek 47:10). – *Unknown; probably S or SE of the Dead Sea as indicated by Bar Kokhba documents and Eusebius.*

En Gannim (Issachar) – A town allotted to the tribe of Issachar that also served as a Levitical city (Josh 19:21; 21:29; 1 Chron 6:72 [= Anem]). – *Not Kh. Beith Jann (196235) which is too far NW for Issachar, but rather Kh. ed-Dir (200229), 5 mi. SW of the Sea of Galilee.* **116**, **120**

En Gannim (Judah) – A town allotted to the tribe of Judah located in the N Shephelah (Josh 15:34). – *Unknown.*

En Gedi – A town allotted to the tribe of Judah (Josh 15:62) located on the W shore of the Dead Sea (Ezek 47:10) near the edge of the Judean Desert. David, fleeing from Saul, was drawn to its powerful spring (1 Sam 23:29; 24:1). Also called Hazazon Tamar in one passage when invaders from the E mustered their troops (2 Chron 20:2). Was famous for its vineyards (Song of Songs 1:14). – *T. Jurn/T. Goren (187097), 25 mi. SE from Jerusalem.* **22**, **23**, 32, 47, 49, 50, 51, 55, **58**, 86, 114, 135, 136, 155, 158, **159**, 163, 166, 177, 186, 187, 193, 212, 217, 222

En Gedi, Desert of – Where David fled from Saul (1 Sam 24:1). *W of En Gedi.*

En Haddah – A town allotted to the tribe of Issachar (Josh 19:21). – *el-Hadatheh (195231), 6 mi. WSW of the Sea of Galilee.* 116

En Hakkore – A spring from which Samson drank (Judg 15:19). – *Unknown; probably in territory of Judah.*

En Hanaziv. 94

En Harod. 126

En Hazor – A town allotted to the tribe of Naphtali, probably located in Upper Galilee (Josh 19:37). – *Unknown.*

En Mishpat – A place (identified in text as "Kadesh") mentioned in connection with the campaign of the four N kings against Sodom, etc. (Gen 14:7). – *Probably same as Kadesh Barnea.* 94, **94**, 254

Enoch – City built by Cain in honor of son. – *Unknown.*

En Rimmon – Place in Judea settled after the exile (Neh 11:29). – *Unknown; possibly the same as Rimmon that was in the N Negev.* 177, 179

En Rogel – Spring just S of Jerusalem on the boundary of Benjamin and Judah (Josh 15:7; 18:16). There David's spies hid (2 Sam 17:17) and there Adonijah's premature coronation took place (1 Kgs 1:9). – *Bir Ayyub (172130), less than a mile S of Jerusalem in the Kidron Valley.* 114, 115, 143, 242, 246

En Rujum. 104

En Shemesh – A spring on the boundary between Benjamin and Judah, between Jericho and Jerusalem (Josh 15:7; 18:17). – *Ain Hod/"Spring of the Apostles" (175131), 2 mi. E of Jerusalem.*

En Tappuah – A spring near the town of Tappuah, on the boundary between Ephraim and Manasseh (Josh 17:7). – *See Tappuah (Ephraim).*

Ephes Dammim – The place were the Philistines camped in preparation for battle with Saul's forces (1 Sam 17:1) near which Goliath was killed (1 Chron 11:13). – *Uncertain; probably on the S side of the Valley of Elah, ca. 1 mi. SE of Azekah (144123).* 134, **134**

Ephesus – Most important city of Asia Minor (Turkey) located near the shore of the Aegean Sea. A port city at the terminus of the caravan route from the east. Paul briefly visited it on his second journey and resided in the city for almost three years on his third missionary journey. Mentioned 18 times in the NT, it was included among the seven churches of Asia in Revelation (1:11; 2:1). **218**, **228**, 230, 231, **231**, 232, **234**, 235, 236, 237, **237**, 239

Ephraim – The younger son of Joseph and the name of the tribe that descended from him. A territorial name for their tribal allotment N of Benjamin, but S of Manasseh: in rugged hill country. Included in territory were the worship centers of Shiloh and Bethel. **22**, **23**, 24, 43, 44, 45, 46, 48, 53, 61, 91, 114, 115, **115**, 117, **120**, 121, 125, 129, 138, 147, 148, 169, **178**, **212**, 254

Ephraim Gate (Jerusalem). 245

Ephraim (NT) – Town near desert that Jesus retreated to after raising of Lazarus (John 11:45). Probably same as OT Ephron and Ophrah. 214

Ephraim, forest of – An undefined region E of the Jordan River, near Mahanaim and the Jabbok River, in central Gilead, where Absalom was killed (2 Sam 18:6). 143, **143**

Ephraim, Hill country/hills of – Mentioned 32 times in the OT, it refers to the rugged territory, W of the Jordan River, were the tribe of Ephraim settled. At time of conquest it was thickly forested (Josh 17:15 – 18). 25, 33, 43, **44**, 45, 46, 47, 91, 93, **101**, 102, 118, **122**, **123**, 125, 127, 128, 131, **131**, 133, **148**, **241**

Ephrath – Place near which Benjamin was born and Rachel died (Gen 35:16, 19; 48:7). – *Uncertain; seemingly in the territory of Benjamin, but some texts associated it with Bethlehem, S of Jerusalem (see Ephrathah).* 97

Ephrathah – An alternate name for Bethlehem (Micah 5:2; Josh 15:59a; LXX) also mentioned in the story of Ruth (Ruth 4:11), although some have proposed that it is a district (Psalm 132:6) in which Bethlehem, among other villages, was located. There was also a "Caleb Ephrathah" (1 Chron 2:24). – *Either Bethlehem or its immediate vicinity, or a nearby place.*

Ephron – One of the villages N of Jerusalem that Abijah took from Jeroboam (2 Chron 13:19). – *Uncertain; possibly et-Taiyiba (178151), 13 mi. NNE of Jerusalem. Possibly the same as Ophrah (Josh 18:23) and Ephraim (NT).* 154, 192

Ephron, Mount – A hill, or ridge, mentioned in the description of Judah's N boundary (Josh 15:9). – *Uncertain; but W of Jerusalem and E of Beth Shemesh probably in the vicinity of Kiriath Jearim.*

Epirus. 231, 233

Erech – One of several cities founded by Nimrod (Gen 10:10). People from that city were settled in the cities of Samaria by Ashurbanipal (Ezra 4:9). – *Warka, in Iraq, 160 mi. S of Baghdad, E of Euphrates River.* 77, **83**

Eridu. 77

Er – Rahia – See Jericho. 257

Er – Riha. 109

Er – Riqqa. **103**

Esbus – OT Heshbon. 193, **198**, **200**, 201, **212**

Esdraelon, Valley of – OT Jezreel, Valley of. 39, 197, **198**, 199, 205, **206**, 207, **212**, 214, 215

Esek – A well dug by Isaac (Gen 26:20). – *Unknown; located between Beersheba and Gerar.*

Eshan – A Judean town in the hill country (Josh 15:52). – *Unknown.*

Eshcol, Valley of – Valley in the Hebron area where the spies cut a large cluster of grapes (Num 13:23, 24; 32:9; Deut 1:24). – *Uncertain.* 106

Eshnunna. **93, 175**

Eshtaol – Village in the N Shephelah first allotted to Judah but then to the tribe of Dan (Josh 15:33; 19:41). However, some Danites moved N from the area (Judg 18:2, 8, 11). Samson's family was from the area (Judg 13:25; 16:31). – *Ishwa/Eshtaol (1511132), 14 mi. W of Jerusalem.* 24, **114,** 118, **118, 126,** 130, **134, 166**

Eshtemoa/Eshtemoh – A Levitical city located in the hill country of Judah (Josh 15:50; 21:14; 1 Chron 6:57). David sent booty to inhabitants of the town (1 Sam 30:28). – *Es-Semu (156089), 9 mi. SSW of Hebron.* **114, 166, 120**

Es – Safa. 34

Etam (Judah) – A town allotted to the tribe of Judah located in hill country near Bethlehem (Josh 15:59b; LXX) and later fortified by the Judean king Rehoboam (2 Chron 11:6). – *Kh. el-Khokh (166121), 2 mi. SW of Bethlehem.* **114, 151, 166**

Etam (Simeon) – A Simeonite town located in the NW Negev or Shephelah (1 Chron 4:32). – *Unknown.*

Etam, rock of – Place where Samson stayed after battling Philistines (Judg 15:8, 11). – *Unknown; possibly near Etam of Judah.*

Etham – Place of Israel's encampment after leaving Succoth in Egypt (Exod 13:20; Num 3:6–8). – *Uncertain; possibly the name is from the Egyptian word for "fort," and refers to one, or more, of the forts which guarded the E frontier of Egypt in the area of the modern Suez Canal.* 103, 104

Ether – A Judean town in the Shephelah (Josh 15:42). Later assigned to the tribe of Simeon (Josh 19:7). – *Kh. el-Ater/T. Eter (138112), 4 mi. NE of Lachish.* **114**

Ethiopia – See Cush. 67 83, 84, 158, 168, 217

Ethiopian Highlands. 65, **66**

Eth Kazin – Town on NE boundary of Zebulun between Gath Hepher and Rimmon (Josh 19:13). – *Uncertain; possibly Kefr Kenna (182239), 4 mi. NE of Nazareth.*

Et – Taiyiba – See Ophrah (Benjamin). 214

Et – Tell – See Ai. **86,** 87, 89, 91, **94,** 109, 110, **110,** 111

Et – Tell – See Geshur. 34, 141, 210

Et – Tih Desert. 69, **69**

Et – Tur. **69**

Eumenia. **234**

Euphrates River – Major river (1780 mi. long) that begins in Turkey and flows through Syria and Iraq before emptying into the Persian Gulf. Mentioned 18 times in the OT in connection with the Garden of Eden, with the traditional boundaries of the "Promised Land" (Gen 15:8; Deut 1:7; 11:24; Josh 1:4), and in the prophetic literature. Referred to in Rev 9:14 and 16:12. Also called the "great river" (e.g., Deut 1:7) and "the River" (e.g., Gen 31:21). **17,** 18, 28, 33, 72, **72,** 73, 75, 76, 77, **77,** 78, 79, 82, 83, **84,** 85, 90, **90, 93, 94, 97,** 98, 100, 141, 143, **146, 160,** 169, **169, 171, 174,** 175, **175,** 176, **182,** 192, **218,** 238, 254

Exaloth. **208**

Ezel, stone – Place where David was to meet Jonathan (1 Sam 20:19). – *Unknown.*

Ezem – A town allotted to the tribe of Judah in Negev district (Josh 15:29) but later assigned to Simeon (Josh 19:3; 1 Chron 4:29). – *Unknown.*

Ezion Geber – A site at the N end of the Gulf of Elath/Aqaba where the Israelites camped (Num 33:35, 36); evidently located in the Arabah (Deut 2:8). There Solomon built ships to sail to Ophir (1 Kgs 9:26; 2 Chron 8:17) but later Jehoshaphat's and Ahaziah's maritime ventures were not as successful (1 Kgs 22:48; 2 Chron 20:36). – *T. el-Kheleifeh (147884), at N end of Gulf of Elath/Aqaba although the archaeological profile does not completely agree with recorded biblical history. The island anchorage of Jezirat Faraun (135874), 9 mi. SW of Elath, has also been suggested.* 57, **62,** 106, 107, **108, 145, 146, 153,** 158

Fair Havens – A port along the S-central coast of Crete where the pilot and owner of the ship ignored Paul's advise and continued on journey towards Rome – shipwreck ensued (Acts 27:8). – *Kaloi Limenes near Lasea on S coast of Crete; 5 mi. E of Cape Matala.* 233, **233**

Faiyum Oasis. **66**

Farafra Oasis. **66**

Farasha. **99**

Feifa. 50, **86,** 87, 95

Feiran Oasis – See Rephidim and Wadi Feiran. 69, **69**

Fertile Crescent – A modern term that refers to a theoretical schematic arc drawn on a map over the area of major human habitation in the Near East. The territory includes the fertile portions of Iraq, Syria, Lebanon, Israel, and Egypt. **17,** 18, 28, 76, 167

Foothills (western) – See Shephelah. 38, 53, 255

Forum of Appius – A stopping point on the Appian Way, about 40 mi. S of Rome where Paul was met by Christians from Rome (Acts 28:15). 233, **233**

Fureidis. **94**

Gaash – A mountainous region in/near the hill country of Ephraim N of which Joshua was buried at Timnath Serah (Josh 24:30; Judg 2:9). One of David's mighty men was from the ravines of Gaash (2 Sam 23:30; 1 Chron 11:32). – *Unknown; possibly the rugged area ca. 15 mi. NW of Jerusalem.*

Gaba. **200,** 201, **222**

Gabae. **175, 183**

Gabara. **206,** 207, **208**

Gad – Son of Zilpah and Jacob and thus a tribal name. Their tribal allotment was NE of the Dead Sea, and continued up the E side of the Jordan Valley toward the Sea of Galilee. 27, **58, 60,** 62, 118, 119, **119, 120,** 141, 149, 156, 162, **178**

Gadara – City of the Decapolis. See also Gadarenes, region of the – *Umm Qeis (214229) located 6 mi. SE of the Sea of Galilee.* **34,** 35, **38, 43, 58, 60,** 187, **187,** 197, **198, 200,** 201, **206, 208,** 211, **212, 217, 219, 220**

Gadarenes, region of the – The region E or SE of the Sea of Galilee where Jesus healed the demoniacs (Matt 8:28). The parallel texts read "Gerasenes" (Mark 5:1; Luke 8:26, 37). Gadara was a city of the Decapolis – *Umm Qeis (214229) located 6 mi. SE of the Sea of Galilee. Possibly the miracle took place in that region but more probably it occurred near the site of Kursi (Chorsia/Gergesa; 211248) on the E shore of the Sea of Galilee where the tradition of the event was localized by the third century A.D.* 211

Gadora. **193, 206,** 211, **212, 217, 219, 220, 222**

Galaaditis. **187,** 188, **193**

Galatia – A region in the central plateau of modern Turkey that Paul visited on all three of his missionary journeys (Acts 16:6; 18:23; third journey by implication). In Paul's day the term designated a Roman province that included portions of the ancient kingdom of Galatia as well as others. Paul wrote an epistle to the church there and referred to it in 2 Tim 4:10 as did Peter (1 Peter 1:1). See also 1 Cor 16:1. **202, 225,** 226, 227, **228,** 231, **231, 233**

Galeed – See Jegar Sahaduth.

Galilee – A region in northern Israel mentioned six times in the OT and 64 times in the NT – five with reference to the Sea of Galilee; all references being in the Gospels, save four in Acts. The region is bounded on the N by the Litani River, on the W by the Mediterranean Sea, on the S by Mount Carmel, the Jezreel and Harod Valleys, and on the E by the Rift Valley, including the Sea of Galilee. **22,** 23, 24, 25, **36,** 37, 38, **38,** 39, 41, 42, 43, **43, 101,** 102, 116, 117, **122,** 128, **135,** 137, 139, **140,** 164, 186, **186, 187,** 188, 192, **193,** 195, 197, **198,** 199, **200,** 201, 205, **206,** 207, 208, **208,** 210, 211, **212,** 215, 216, **217,** 219, **219,** 220, **220, 222,** 252

Galilee, Lower. **22,** 25, 35, **36,** 37, 38, **38,** 39, 40, 41, 42, **43, 101,** 116, 117, 121, **122,** 201, **206,** 207, **208,** 220, **222**

Galilee of the Gentiles. 37

Galilee, Sea of – The harp-shaped lake in N Israel measuring 13.5 by 7.5 mi. at an elevation of –695 feet. Mentioned 5 times in the Gospels, it was also called the "sea of Tiberias," "the sea," the "Sea of Kinnereth," the "sea of Tiberias," and the "lake of Gennesaret." Much of Jesus' public ministry took place along its northern shore. 11, **22, 23,** 25, **26, 32,** 33, **34,** 35, 36, **36,** 37, **38,** 39, 40, 41, **43, 44,** 56, **58,** 59, **60,** 61, **86, 95, 97, 101,** 102, **105, 108,** 117, **123, 124, 135, 140,** 141, **142, 143, 145, 148, 153,** 154, **155, 161, 163, 170, 181, 184, 186, 187,** 192, **193,** 197, **198, 200,** 202, 205, **206,** 207, 208, **208,** 209, 210, 211, **212, 217, 219,** 220, **220, 222**

Galilee, Upper. **22, 23,** 25, 32, 35, **36,** 37, **38,** 39, **43, 101, 116,** 117, 121, **122,** 194, 201, 205, **206,** 207, **208, 222**

Gallaecia. **202**

Gallia. **202, 234**

Gallim (Benjamin) – A village in Benjamin located along the traditional military approach to Jerusalem from the N (Isa 10:30). Home of Palti, to whom Saul gave his daughter Michal (1 Sam 25:44). – *Uncertain; possibly in the area of Gibeah and Anathoth (Isa 10:30), ca. 4 or 5 mi. N of Jerusalem.*

Gallim (Judah) – A town allotted to the tribe of Judah (Josh 15:59b; LXX only, see NEB) located in the Bethlehem district S of Jerusalem. – *Unknown.*

Gamala. **36, 43, 193,** 218, 220, **220,** 221, **222**

Gammad – Men of this locality aided in defense of Tyre (Ezek 27:11). – *Unknown; probably in vicinity of Tyre; possibly same as Kumidi of el-Amarna letters.*

Gandhara. **175**

Gangites River. 229

Gareb, hill of – Hill in the immediate vicinity of Jerusalem, possibly to the N or W of the city (Jer 31:39), but – *Unknown.*

Gath (Padalla) – Possibly the Gath mentioned in 1 Chron 7:21 from which men of the village had killed the Ephraimites Ezer and Elead. – *Jett (154200), in the Sharon Plain, 12 mi. NW of Samaria, mentioned in extrabiblical sources as Gath Padalla.* **38,** 43, **44, 95, 123**

Gath (Philistine) – One of the five cities of the Philistines (Josh 13:3; 1 Sam 6:17); possibly the chief of those cities for it had a "king" (1 Sam 21:10, 12) while the others had "lords." Mentioned approximately 34 times in the OT. Its most famous resident was Goliath (1 Sam 17:4). – *T. es-Safi/T. Zafit (135123), 25 mi. SSE of Joppa.* **32, 47, 51,** 52, 53, **105, 110, 120, 123,** 124, 129, **130,** 131, **131,** 132, **132,** 134, **134,** 135, **135,** 136, 139, **140, 151, 153, 155,** 158, **159,** 162, **163, 166,** 168, 261

Gath Hepher – A village on the E border of Zebulun in Lower Galilee (Josh 19:13); also home of Jonah the prophet (2 Kgs 14:25). – *Kh. ez-Zurra/T. Gat Hefer (180238), 3 mi. NE of Nazareth.* 116, **116,** 161

Gath Rimmon (Dan) – A town allotted to the tribe of Dan (Josh 19:45; 21:24) but evidently later occupied by Ephraim (1 Chron 6:69). – *Uncertain; but usually identified with T. Jerisheh/T. Gerisa (132166), located in modern Tel Aviv, 4.5 mi. NE of Joppa, although the archaeological profile is problematic.* **118, 120**

Gath Rimmon (Manasseh?) – A town in Manasseh allotted to the Levitical clan of Kohathites (Josh 21:25), but the parallel passage in 1 Chron 6:70 has "Bileam" (= Ibleam) which seems preferable. One Greek version of Josh 21:25 also has "Bileam."

Gaugamela. **182**

Gaul. 216

Gaulanitis. 33, **34, 186, 187,** 188, 197, **198, 200,** 201, **206,** 207, **208,** 210, **212,** 216, **217,** 219, **219, 220**

Gaza – A town in SW Israel on the border between the sown land of Canaan and the desert of N Sinai. Allotted to Judah (Josh 15:47), the Philistines settled there (Josh 13:3; Judg 1:18, LXX "Judah did not take," NIV footnote; 1 Sam 6:17; etc.). Mentioned 22 times in the OT and once in NT (Acts 8:26), it was usually, if not always, under non-Israelite control: the Egyptians used it as a base for expeditions N, while the Assyrians, Babylonians, etc. used it as a base for invasions of Egypt. – *Ghazzeh/Azza (099101), 3.5 mi. inland from the Mediterranean in SW Israel.* **17,** 18, 21,

22, 23, 28, **30,** 32, **32, 47,** 51, **51,** 52, 55, **55,** 56, 57, **62,** 64, **66,** 68, 69, **69,** 88, **93, 97, 99,** 101, **101,** 103, **103, 104, 105, 106, 108, 114, 120, 122, 123, 124, 126,** 129, 130, **130,** 132, **135, 140, 142, 143, 145, 146, 153, 155,** 158, **159, 160, 163,** 164, 166, **166, 169, 170, 171,** 174, 175, 177, **177,** 181, **181, 184,** 186, **186,** 187, **187, 193,** 195, **198, 200,** 201, **206, 212,** 217, **217, 219, 220, 222**

Gazara – OT Gezer. **190, 193**

Geba (Benjamin) – A town allotted to the tribe of Benjamin (Josh 18:24) that served as a Levitical city (Josh 21:17; 1 Chron 6:60; 8:6). Located in an area disputed by Judah and Israel (1 Kgs 15:22) it was near the N border of Judah after fall of northern kingdom (2 Kgs 23:8; some suggest a more N Geba in this instance, but this proposal is not certain). Mentioned 16 times in the OT, it was located along the N approach to Jerusalem (Isa 10:29). After the exile, Jews settled there (Ezra 2:26; Neh 7:30; 11:31; 12:29). – *Jeba (175140), 6 mi. NNE of Jerusalem.* 46, **47, 115, 120, 123, 131, 132,** 133, 139, 154, **154, 166,** 169, **177,** 240, **241**

Geba (Samaria). **212,** 215

Geba Carmel. **94**

Gebal (Phoenicia) – A city N of Israel on Phoenician coast not conquered by Israel, though part of Canaan (Josh 13:5). Solomon secured timber for temple from there (1 Kgs 5:18) and men of Gebal served as "Shipwrights" (Ezek 27:9). – *Jebeil (210391), on Mediterranean coast, 18 mi. NNE of Beirut. Also called Byblos.* 74, **123, 140, 169, 171**

Gebal (Transjordan) – An unidentified geographical region probably SE of the Dead Sea (Ps 83:7).

Gebim – A site along the traditional approach to Jerusalem from the N (Isa 10:31). – *Unknown; but probably only a few mi. N of Jerusalem.*

Geder – A Canaanite city conquered by Joshua (Josh 12:13). – *Unknown; possibly in N central Negev.*

Gederah – A town allotted to the tribe of Judah in the Shephelah (Josh 15:36; 1 Chron 4:23). – *Unknown.*

Gederoth – A town allotted to the tribe of Judah in the Shephelah (Josh 15:41). Conquered by the Philistines in days of Ahaz (2 Chron 28:18). – *Unknown; in Shephelah, possibly in Valley of Elah area.* 160, 165

Gederothaim – According to the NIV, an alternate name for "Gederah," rather than the fifteenth town in a list which is said to contain fourteen villages (Josh 15:36). The LXX reads "Gederah and her sheep pens" which may be preferable.

Gedor (Benjamin) – Two Benjamites from Gedor joined David at Ziklag (1 Chron 12:7). – *Unknown.*

Gedor (Judah) – A town allotted to the tribe of Judah in the hill country near Hebron (Josh 15:58). It is probable that the Gedor's mentioned in 1 Chron 4:4 and 4:18 are towns, but it is not certain if both, or one, of them are identical to this Gedor. Some follow the LXX and read "Gerar" for "Gedor" in 1 Chron 4:39). – *Kh. Jedur (158115), 7.5 mi. NNW of Hebron.* **105,** 114, **114, 166**

Gedor (Simeon) – Possibly alternate form for Gerar (1Chron 4:39; see LXX).

Gedor (Transjordan). **60, 177**

Ge Harashim – A place on the coastal plain where craftsmen settled (1 Chron 4:14). Probably the same as the "Valley of the Craftsmen" where Benjamites settled after the exile (Neh 11:35). – *In Israel, on coastal plain, E of Tel Aviv Joppa. Probably within triangle formed by Lod, Aphek, and Joppa.*

Geliloth (Benjamin/Judah) – A point on the SE boundary of Benjamin near the Pass of Adummim (Josh 18:17). Possibly the same as the Gilgal mentioned in Josh 15:7. – *Unknown.*

Geliloth (near Jordan River) – An unidentified site by the Jordan River near which an altar was built (Josh 22:10 – 11). But RSV translates Geliloth as the "region about the Jordan."

Gennesaret – A village on the NW shore of the Sea of Galilee at, or near which Jesus and his disciples landed (Matt 14:34; Mark 6:53). The nearby plain (1 by 3 mi.) is named after the site. – *Kh. Ureime/T. Kinrot (200252), 6 mi. N of Tiberias.* 42, **193, 208**

Gennesaret, Lake of – Alternate name for the Sea of Galilee (Luke 5:1). See Gennesaret above. 41

Gennesaret, Plain of. 41, 42, **43, 208,** 210

Gerar – A town near Gaza on the SW border of Canaanite territory (Gen 10:19). There Abraham, and later Isaac, encountered two "Philistine" kings, both named Abimelech (Gen 20:1 – 2; Gen 26:1, 6). The herdsmen of Gerar quarreled with those of Isaac over water rights (Gen 26:20, 26). Later the Judean king Asa pursued the Cushite Zerah in this area (2 Chron 14:13, 14). – *T. Abu Hureireh/T. Haror (112087), 17 mi. NW of ancient Beer Sheba.* **32, 51, 55, 91,** 93, **97, 106,** 114, **159, 163, 193,** 258

Gerar, Valley of – See Nahal Gerar. – The valley along which Gerar was located. – *Probably Wadi esh-Sheriah/N. Gerar.*

Gerasa. 187, **187,** 189, **193,** 197, **198, 200, 206,** 211, **212,** 217, 221, **222**

Gerasenes, region of the – The place where Jesus healed the demoniac and where the swine were drowned in the sea according to Mark 5:1 and Luke 8:26, 37. See Gadarenes and Gergesa. 211

Gergesa – *Probaby the site of Kursi (Chorsia/Gergesa; 211248) on E shore of Sea of Galilee where tradition placed it by the third Century AD).* 42, **43, 208,** 211, **212**

Gerizim, Mount – A mountain just S of Mt. Ebal and the city of Shechem from which the blessings were read at the covenant renewal ceremony (Deut 11:29; 27:12; Josh 8:33) and to which Jotham fled from Abimelech (Judg 9:7). Known as place of Samaritan worship referred to as "this mountain" by the woman that Jesus met at the well (John 4:20). – *Jebel et-Tur/Mt. Gerizim (176170), 30 mi. N of Jerusalem.* **23,** 25, **44,** 45, 46, 91, **91, 110,** 111, **126, 177,** 186, **186, 193,** 195, 196, **198, 200, 206, 212,** 215, 220, **222**

Germania Inferior. **234**

Gerrha. **184**

Geruth Kimham – Place near Bethlehem where Johanan stopped before descending into Egypt (Jer 41:17). – *Unknown.*

Geshur – A small kingdom E and NE of the Sea of Galilee that remained independent of Israel (Deut 3:14; Josh 12:5; 13:11, 13) the capital of which was also called "Geshur." David married the daughter of Talmai the king of Geshur and she bore Absalom (2 Sam 3:3). Absalom fled to Geshur after executing Amnon (2 Sam 13:37, 38; 14:23, 32; 15:8). Later Geshur and Aram captured the Israelite territory of Havvoth Jair (1 Chron 2:23). City – *Et-Tell (209257) 1.5 mi. N of the Sea of Galilee. See also Bethsaida.* 33, **34,** 35, 121, **122, 140,** 141, **142, 143, 145, 148, 153, 157**

Geshur(ites) (south) – A people who lived S and SW of Philistia (Josh 13:2). David raided their territory when working for Achish, the king of Gath (1 Sam 27:8). – *Territory in NE Sinai.* 136, 138

Gethsemane – A garden (John 18:1) located on the lower W slope of the Mt. of Olives (Luke 22:39) where Jesus went to pray with his disciples and where he was captured (Matt 26:36; Mark 14:32). – *E of Temple Mount in Jerusalem, on the lower W slope of the Mt. of Olives.* 214, **250,** 252

Gezer – A city in N Shephelah whose king was defeated by Joshua (10:33; 12:12). It was allotted to Ephraim (Josh 16:3) but they failed to drive out the Canaanites living there (Josh 16:10; Judg 1:29). Assigned as a Levitical city (Josh 21:21; 1 Chron 6:67), David drove Philistines from Jerusalem area to it (2 Sam 5:25; 1 Chron 14:16). It was captured by an Egyptian Pharaoh and given as dowry to his daughter who married Solomon (1 Kings 9). – *T. Jezer (Abu Shusheh)/ T. Gezer (142140), 17 mi. SE of Joppa.* 28, **32, 44,** 47, **47, 51,** 53, **86, 95,** 96, **97, 105, 110, 115,** 117, **118, 120,** 121, **122, 123,** 124, **124,** 127, **130, 131,** 132, **132, 135,** 138, 139, **140, 142,** 143, 144, **145, 146,** 147, 149, **153, 154, 155, 159, 163, 177,** 240, **241,** 257

Ghab Valley. 75

Gharandal (Edom). **62,** 64

Gharandal (Sinai). **69,** 104, **104**

Ghita. **99**

Ghor. 25, **58,** 61

Giah – A site in Benjamin passed by Abner as he fled E from Gibeon as he was pursued by Joab and Abishai (2 Sam 2:24). – *Unknown; but in Benjamin, probably E of watershed near edge of wilderness.*

Gibbar – Ancestor of family that returned from exile (Ezra 2:20), but parallel passage (Neh 7:25) has "Gibeon."

Gibbethon – Philistine village in N Philistia assigned to tribe of Dan (Josh 19:44) and designated a Levitical city (Josh 21:23). There, Baashah murdered Nadab (1 Kgs 15:27), and there Omri was proclaimed king of Israel (1 Kgs 16:15). – *T. Melat/T. Malot (137140), 15 mi. SE of Joppa, 3 mi. W of Gezer.* **118, 120, 155, 163**

Gibeah (Ephraim) – Place in hill country of Ephraim where Eleazar, son of Aaron, was buried (Josh 24:33). Possibly Abijah's mother was from this place (2 Chron 13:2). – Unknown.

Gibeah (in/of Benjamin) – Town allotted to the tribe of Benjamin (Josh 18:28). The wicked men of Gibeah abused the Levite's concubine and in turn the city was destroyed as a punishment (Judges 19 and 20); the incident is referred to in Hosea 9:9; 10:9, 10. – *T. el-Ful (172136), 3 mi. N of Jerusalem (site was later called "Gibeah [of Saul]").* – Two or three hundred years later Jonathan launched an attack on the Philistines from "Gibeah in Benjamin" (1 Sam 13:2, 15; 14:16) which was then located 3 mi. to the NE of the old site. – *Jeba (175140). Also mentioned in 2 Sam 23:29; 1 Chron 11:31; Isa 10:29; Hosea 5:8.* 23, **23,** 33, 46, **47, 51, 115, 121,** 123, **123, 126, 131, 132,** 133, **135, 154, 166,** 179

Gibeah (Judah) – A town allotted to the tribe of Judah located in the hill country SE of Hebron (Josh 15:57). – *Unknown.*

Gibeah (of Saul) – Several hundred years after the destruction of Gibeah (in/of Benjamin), Saul rebuilt the site as his capital. In the narratives of 1 Samuel (10:26; 11:4; 14:2; 15:34; 22:6; 23:19; 26:1; and 2 Sam 21:6) it is called Gibeah (of Saul). – *T. el-Ful (172136). See also Gibeah of/in Benjamin.* **132,** 133

Gibeah of God – A hill, near Gibeon, where there was a Philistine outpost (1 Sam 10:5) and where Saul met a band of prophets (10:10). – *Nabi Samwil (167137), 5 mi. NW of Jerusalem. Also called the high place at Gibeon.* **131,** 133

Gibeath. **166**

Gibeath Haaraloth – Site near Gilgal and Jericho where Israelites were circumcised after crossing the Jordan (Josh 5:3). – *Unknown; vicinity of Jericho.*

Gibeon – A major hill country town that headed the Gibeonite (Hivite) league that tricked Joshua into making a treaty with them (Josh 9 – 11). Joshua saved the inhabitants from the attacking Amorite coalition. Allotted to Benjamin (18:25) it served as a Levitical city (21:17). Gibeon was probably Saul's hometown (1 Chron 9:35 – 39). There Joab fought Abner (2 Sam 2 – 3) and later David drove Philistines from vicinity (2 Sam 5:25). Jeremiah cursed a prophet from Gibeon (Jer 28:1) and there Johanan fought with Ishmael (Jer 41:12, 16). Men from the town helped rebuild the walls of Jerusalem after the exile (Neh 3:7 and see also 7:25). – *El-Jib (167139), 6 mi. NW of Jerusalem.* 46, 47, **47, 51,** 53, **94, 95, 105, 110,** 111, **115, 120, 123,** 125, **131,** 132, **132,** 133, **135,** 138, 139, **140, 143, 145, 153, 154, 155, 166, 177,** 179, **241,** 243, **244**

Gibeon, high place at – A worship center near Gibeon where the tabernacle was located during the reigns of David and Solomon (1 Chron 16:39; 21:29; 2 Chron 1:3, 5, 13 and probably also 1 Kgs 3:4, 5; 9:2). – *Nabi Samwil (167137), 5 mi. NW of Jerusalem. Also called Gibeah of God.*

Gibeon, Valley of – Valley in vicinity of Gibeon (el-Jib [167139]) where God fought for his people (Isa 28:21; possible reference to Joshua defeating the Amorites [Josh 10:10 – 12]). – *Unknown.*

Giddi Pass. **69**

Gidom – Defeated Benjamites were pursued to Gidom where several thousand were struck down (Judg 20:45). – *Unknown; probably in NE Benjamin.*

Gihon (river) – One of the four rivers mentioned in connection with the garden of Eden (Gen 2:13). – *Unknown.* 82, 83

Gihon (spring) – The major water source for the city of Jerusalem located just E of the city in the Kidron valley. There Solomon was crowned king (1 Kgs 1:33, 38, 45). Hezekiah blocked its upper outlet (2 Chron 32:30) and repaired the city wall above it (2 Chron 33:14) in anticipation of the Assyrian advance on Jerusalem. – *Ain Umm ed-Deraj/Ain Sitti Maryam.* 138, 143, 167, 242, **242**, 243, 244, **246**, 247, **248**

Gilboa, Mount – Mountain (range) where Saul and three of his sons died in battle against Philistines (1 Sam 28:4; 31:1, 8; 2 Sam 1:6, 21; 21:12; 1 Chron 10:1, 8). – *Jebel Fuquah/Mount Gilboa, 17 mi. SW of the Sea of Galilee overlooking the Jordan, Harod, and Jezreel Valleys.* 38, 40, **44**, 45, 47, 131, **135**, 137, 138

Gilead (Land of) – Primarily a geographical term referring to the mountainous region E of the Jordan, S of the Yarmuk and N of a E-W line drawn at the N end of the Dead Sea and stretching to the edge of the desert on the E (occasionally used as a clan name and personal name). Term use over 97 times in the OT. 21, **22, 23,** 27, 33, **34,** 35, 40, 57, **58,** 59, 60, **60,** 61, 94, 96, 97, **97, 106,** 108, **109,** 115, 119, **119,** 121, **122,** 128, **129,** 138, 141, **142,** 143, **143,** 144, **145,** 148, 152, **161,** 162, **163,** 164, **170,** 176, 177, **178,** 192

Gilead (Sharon Plain). 94

Gilead (Transjordan). 126

Gilgal – A site W of the Jordan River, in the vicinity of Jericho where Israel camped when it entered Canaan (Joshua passim). Mentioned 35 times in the OT, it continued as a worship center during the days of Samuel and Saul (1 Samuel passim) and later became a symbol of wickedness (Hosea and Amos). – *Uncertain; possibly in the Kh. el-Mefjer area (193143), 1.3 mi. NE of OT Jericho.* 109, **110,** 111, 115, **115,** 126, **131,** 132, **132,** 133, **161**

Gilgal (Ephraim) – Elisha and Elijah passed by it on way down to Bethel and on to Jericho (2 Kgs 2:1). Later Elisha returned to the spot (2 Kgs 4:38). – *Unknown.*

Gilgal (Galilee?) – Among the list of kings defeated by Joshua was the "king of Goyim in Gilgal (Josh 11:23; LXX: Galilee). – *Unknown; but probably not the same as the other Gilgals.*

Gilgal (Judah) – A site located on the E portion of the N boundary of Judah in the vicinity of the Pass of Adummim (184136; Josh 15:7), 8 mi. E of Jerusalem. The parallel passage in Josh 18:17 reads "Geliloth." – *Unknown.*

Giloh – A town allotted to the tribe of Judah located in the S hill country (Josh 15:51). Ahithophel, counselor to David and Absalom (2 Sam 15:12; 23:34) was from the town. – *Unknown.*

Gimzo – Town captured by Philistines in days of Ahaz (2 Chron 28:18). – *Jimzu/Gimzo (145148), 15 mi. ESE of Joppa.* 86, **159,** 160, 165, **166**

Gina. 105, 123

Ginae. 212, 214, 215

Gischala. 205, **206,** 208, 212

Gittaim – A village to which people of Beeroth fled (2 Sam 4:3) and in which Jews were living after the exile (Neh 11:33). – *Ras Abu Humeid (140145), 14 mi. SE of Joppa.* 51, 52, **177**

Giza. 66, 84, 85, 86, 103

Goah – A site in the immediate vicinity of Jerusalem (Jer 31:39), but – *Unknown.*

Gob – Place where David's men battled twice with the Philistines (2 Sam 21:18, 19). 1 Chron 20:4 implies that it is another name for Gezer.

Goiim – Tidal king of Goiim participated on the attack of the cities of the plain (Gen 14:1, 9). Tidal may be a Hittite name, but Goiim is unidentified. 94, **94**

Golan – A transjordanian city of the tribe of Manasseh that was allotted to the Levites (Josh 20:8; 1 Chron 6:71). It also served as one of the six cities of refuge (Deut 4:43; Josh 21:27). – *Sahm el-Jolan (238243), 18 mi. E of the Sea of Galilee. In NT times there was a district known as Gaulanitis.* 120, 121

Golan (Heights). 17, 22, 23, 27, 33, 92, **94,** 192

Golgotha – Site (hill?) just outside of NT Jerusalem where Jesus was crucified (Matt 27:33; Mark 15:22; John 19:17. – *Probably, Church of Holy Sepulchre inside of present day old city of Jerusalem.* **250**

Gomorrah – One of the cities of the plain attacked by the kings of the N (Genesis 14) and which was destroyed because of its wickedness (Gen 18 – 19). It, along with Sodom became a symbol of wickedness (23 times in Bible). – *Unknown; possibly one of the Early Bronze III sites, E and SE of the Dead Sea.* 88, **91,** 94, 95

Gophna. 190, 198, **217,** 219, **222**

Gordium. 181, **181, 182**

Gortyna. 175, **218,** 234

Goshen (Egypt) – The region of Egypt where Jacob and his descendants settled (mentioned 15 times in OT). Possibly called "district of Rameses" later in history (Gen 47:11). – *Area of E Nile Delta, near and N of the Wadi Tumilat.* 66, 68, 93, 97, 102

Goshen (Judah) – A town allotted to the tribe of Judah in the S hill country (Josh 15:51). Also the name of a region, possibly in the same area (Josh 10:41; 11:16). – *Unknown.*

Gozan – A city, and also a region, conquered by the Assyrians (1 Kgs 19; Isa 37:12) to which Israelites were deported (1 Kgs 17:6; 18:11; and 1 Chron 5:26). – *T. Halaf, 160 mi. NE of Aleppo (Syria) on a W tributary of the Habor River.* 17, 72, 84, 160, 164, **174**

Granicus River. 180, 181

Great Sea – Used 13 times in NIV for the Mediterranean Sea. 114, 115, 116, 118, 120, 122, 124, 126, 131, 135, 146, 155, 173, 178

Greece – Mentioned 7 times in the NIV translation. – *General area of the modern country of Greece depending upon time frame of text. Paul visited Greece on his second and third missionary journeys.* 16, **17, 19,** 74, **176,** 180, 184, 201, 227, 229, **229,** 266, 267

Gudgodah – A campsite of Israel during wilderness wanderings between Moserah and Jotbatah (Deut 10:7; called "Hor Haggidgad" in Num 33:32, 33). – *Unknown.*

Gur – Spot near Ibleam (177205; S side of Jezreel Valley), where Ahaziah was mortally wounded by Jehu's servants (2 Kgs 9:27). – *Unknown.*

Gur Baal – Uzziah, a Judean king, defeated the Arabs who lived in Gur Baal (2 Chron 26:7). – *Uncertain; to the S or SE of Judah.* 158

Guvrin Valley. 51, 54

Guzana – See also Gozan. **77,** 78

Haarava. 56

Habur/Habor (River) – Deported Israelites were settled in the region of the Habor by the Assyrians (2 Kgs 17:6; 18:11; 1 Chron 5:26). – *Khabur River in NE Syria.* 72, 73, 76, **77,** 84, 90, 93, 97, 98, 160, 164

Hadashah – Town in Judean Shephelah district (Josh 15:37). – *Unknown.*

Hadid – Town resettled by Jews after return from exile (Ezra 2:33; Neh 7:37; 11:34). – *El-Haditheh/T. Hadid (145152), 11 mi. ESE of Joppa on coastal plain.* 86, **177**

Hadrach, land of – Subject of an oracle of Zechariah (9:1). Called "Hatarikka" in Assyrian texts. – *Tell Afis, 28 mi. SW of Aleppo.*

Haeleph – Town in Benjamin (Josh 18:28). – *Unknown; probably W of watershed.*

Hagoshrim. 94

Haifa. 30

Haifa Bay. 37

Hakilah, hill of – Hill in Judean Desert where Saul pursued David (1 Sam 23:18; 26:1). – *Unknown; probably near Ziph.*

Halab – See Aleppo. 75

Halah – Place, in Gozan region, near the Habor River, to which Israelites were exiled by the Assyrians (2 Kgs 17:6; 18:11; 1 Chron 5:26). – *Unknown.* 164

Halak, Mount – Point marking S limit of Joshua's conquests (Josh 11:17; 12:7). – *Uncertain; possibly Jebel Halaq, 28 mi. SE of Beersheba.*

Halhul – Town in hill country district of Judah (Josh 15:58). – *Halhul (160109), 4 mi. N of Hebron.* **166**

Hali – Town allotted to Asher (Josh 19:25). – *Kh. Ras Ali/T. Hali (164241), 11 mi. SSE of Acco.* 116

Halicarnassus. 218

Halys River. 17

Ham – Place in Transjordan where Zuzites lived (Gen 14:5). – *Ham (226213), 40 mi. NNW of Amman in Gilead.*

Ham, land of – Alternate name for Egypt (Ps 105:23, 27; 106:22). 83

Hamakhtesh Hagadol – See also Makhtesh Gadol. **23,** 25, 56, **62**

Hamath (on the Orontes; Land of) – City and state that was located N of the Land of Canaan. Mentioned 24 times in OT. Israelite territory reached to its S boundary during the days of David, Solomon, and Jeroboam II. – *In NW Syria, capital city Hamath (= Hama; 312503), 120 mi. NNE of Damascus.* **17,** 33, **72,** 75, **83,** 84, **84,** 90, 93, **97,** 106, **140,** 141, **143,** 146, **160,** 164, **169,** 171, 174, **175,** 176, 178

Hamath Zobah – City in Aramean state of Zobah captured by Solomon (2 Chron 8:3). – *Unknown; possibly in Beqa in Lebanon.* 143

Hamath, Plain of. 195

Hammamat, Wadi. 65, 68

Hammath (Jordan Valley). 44, **124,** 125

Hammath (Kenite) – Place of origin for some Kenites (1 Chron 2:55). – *Unknown.*

Hammath (Naphtali) – Town allotted to Naphtali (Josh 19:35). – *Hammam Tabariyeh/Hame Teveriya (201241), on W shore of Sea of Galilee.* 36, 38, 42, 43, **43,** 116

Hammath Gader. 43, **43**

Hammon – Levitical city in Naphtali (1 Chron 6:76). – *Uncertain; possibly same as Hammoth Dor.*

Hammon (Asher) – Town allotted to Asher (Josh 19:28). – *Umm el-Awamid (164281), in Lebanon, 14.5 mi. NNE of Acco.* 166

Hammoth Dor – Levitical town in territory of Naphtali (Josh 21:32). Perhaps identical to "Hammath" (Naphtali) of Josh 19:35 and "Hammon" of 1 Chron 6:76.

Hamonah – Symbolic name of town ("horde") in Valley of Hamon Gog (Ezek 39:16).

Hamon Gog, Valley of – Valley, possibly in Transjordan, where army of Gog will be slain and buried (Ezek 39:11, 15). – *Unknown.*

Hanes – Egyptian city mentioned in Isa 30:4. – *Unknown; possibly Heracleopolis Magna, 50 mi. S of Memphis or Heracleopolis Parva in E Delta.* 160

Hanita. 94

Hannathon – Town on NW boundary of Zebulun (Josh 19:14). – *T. el-Bedeiwiyeh/T. Hannaton (174243), 14 mi. SE of Acco.* 32, 33, **36,** 38, 39, 40, **105,** 106, **106,** 116, **116,** 123, **123,** 155, 157, 163

Hapharaim – Town allotted to Issachar (Josh 19:19). – *Affuleh (177223), 6.5 mi. ENE of Megiddo.* 116

Hara – Place to which Assyrians took exiled Israelites (1 Chron 5:26), but text maybe corrupt (cf. 2 Kgs 17:6; 18:11). – *Unknown.*

Haradah – Desert campsite of wandering Israelites (Num 33:24, 25). – *Unknown.*

Haran – Site in N Mesopotamia, located on the Balikh River to which Abram migrated. There Jacob lived with his uncle Laban. Later the place was captured by the Assyrians and it became their last capital. Mentioned 11 times in OT, 2 times in NT. – *Haran, 115 mi. NE of Aleppo in E Turkey, near Turkish-Syrian border.* **17,** 72, 73, **77,** 78, 89, 90, **90,** 91, **93, 94,** 97, 98, **146,** 160, 169, **169,** 171, 174, 175, 182

Harim – People of village participated in first return from Babylonian Exile (Ezra 2:32; Neh 7:35). – *Uncertain; possibly Kh. Hauran/Mezad Hakhlil (144117), 8 mi. NE of Lachish in Shephelah.* 177

Harmon – If text is correct, a place to which Israelites were to be exiled (Amos 4:3). – *Unknown.*

Harmozeia. 183

Harod, spring of – Place where Gideon camped with his men in preparation for battle with Midianites (Judg 7:1). Two of David's mighty men were from a (probably different) town of this name (2 Sam 23:25). – *Ein Jalud/En Harod (183217), 9 mi. WNW of Beth Shan.* **126,** 128

Harod, Valley of. 23, 25, 29, 37, **38,** 39, 40, **44,** 45, **58, 60,** 61, 115, 116, **116,** 135, 137, 139, **140,** 149

Harosheth Haggoyim – Place where Sisera, commander of Jabin's army, resided or ruled over and to which his troops retreated in the face of Deborah and Barak's army (Judg 4:2, 13, 16). – *Uncertain; possibly a reference to the forested region(s) of Galilee or better, to Gentile plantations in the Jezreel Valley in the vicinity of Taanach and Megiddo.*

Har Rahama. 86

Har Ramon. 56

Har Yeroham. 94

Hashmonah – Campsite of Israelites during years of wanderings (Num 33:29, 30). – *Unknown.*

Hatnub. 99

Hatti. 123, 127

Hattin, Horns of – See Adamah and Merom, Water of. **39**

Hattusa. 17, 93

Hauran – An Assyrian district E of the Sea of Galilee mentioned in connection with the boundaries of the Promised Land (Ezek 47:16, 17). – *Territory just W of the Jebel Druze. Called Auranitis during the NT period.* 163, 176, **178,** 186

Hauran, Mount. 34, 38

Havilah – The land through which the Pishon River ran (Gen 2:11). – various proposals, but Unknown. Also known as a region near which the Amalekites settled (Gen 25:18; 1 Sam 15:7). – *Probably NE Sinai or possibly E Negev.* 83, **83,** 133

Havvoth Jair – Settlements in N Gilead which had been part of the kingdom of Og, but which were settled by a descendant of Manasseh named Jair (Josh 13:30; Num 32:41; Deut 3:14). Possibly the judge Jair was from this region (Judg 10:4). Later they became part of a Solomonic district (1 Kgs 4:13) and at some time they were captured by Geshur and Aram (1 Chron 2:23). Also called "settlements of Jair" (Josh 13:30; 1 Kgs 4:13). – *Probably the region of N Gilead between Jabal 'Ajlun and the Yarmuk River.* **34,** 35, **119, 126**

Hazar, Lake. 76

Hazar Addar – Site on S boundary of land of Canaan (Num 34:4). – *Uncertain; possibly Ain Qedeis (100999) or the nearby fort (103000), ca. 51 mi. SW of Beersheba.* **106**

Hazar Enan – Site on the NE border of the land of Canaan (Num 34:9, 10; Ezek 47:17; 48:1). – *Qaryatein (360402), 70 mi. NE of Damascus.* **72,** 106, **178**

Hazar Gad – Village in Negev district allotted to Judah (Josh 15:27). – *Unknown.*

Hazarmaveth. 83

Hazar Shual – Village in Negev allotted to both Judah (Josh 15:28) and Simeon (Josh 19:3; 1 Chron 4:28). Jews settled there after the exile (Neh 11:27). – *Unknown.* 179

Hazar Susah – Negev village allotted to Simeon (Josh 19:5). Name means "corral of the mare." Probably identical to Hazar Susim, "corral of the horses" (1 Chron 4:31). – *Unknown; possibly stables or military stations.*

Hazar Susim – See Hazar Susah.

Hazazon/Hazazon Tamar – Place where kings of N defeated Amorites (Gen 14:7). – *Uncertain; possibly Ein Husb/Hazeva (173024), 27 mi. SSW of Dead Sea. See Tamar.* 94, **94**

Hazazon Tamar (En Gedi) – Place where tribes from the E camped as they invaded Judah (2 Chron 20:2); identified in the text as "En Gedi." – *Probably Ain Jidi/En Gedi (187097), at the mid-point of the W shore of the Dead Sea.* 158

Hazer Hatticon – On N border of Promised Land (Ezek 47:16). Because of position in the list, it seems to be either an alternate form, or scribal error for, Hazar Enan.

Hazeroth – Third campsite after the Israelites left Mt. Sinai. There Miriam and Aaron questioned the propriety of Moses' marriage to a Chushite woman as well as his position as leader (Num 11:35; 12:16; 33:17, 18; Deut 1:1). – *Uncertain;*

possibly Ain Khadra (096814), 40 mi. NE of Jebel Musa (Mt. Sinai?). **104,** 105

Hazi/Haszi. 102, **123**

Hazor (Arabia?) – Probably settlements in Arabia attacked by Nebuchadnezzar (Jer 49:28 – 33). – *Unknown; probably in W or NW Arabia.*

Hazor (Benjamin) – Village where Benjamites settled after the exile (Neh 11:33). – *Uncertain; located in Benjamin Plateau, N of Jerusalem.*

Hazor (Judah 1) – Negev town allotted to Judah (Josh 15:23). – *Unknown; in S Judah.*

Hazor (Judah 2) – Alternate name for Kerioth Hezron (Josh 15:25).

Hazor (Naphtali) – Large Canaanite city conquered by Joshua (11:1, 10, 11, 13; 12:19) and allotted to the tribe of Naphtali (19:36). Had been head of kingdoms in the area. Later its forces fought against Deborah and Barak (Judg 4:2, 17; 1 Sam 12:9). Lying along the major international route through the country, Solomon fortified the city (1 Kgs 9:15), but later it was conquered by the invading Assyrian, Tiglath-Pileser III (2 Kgs 15:29). – *T. el-Qedah/T. Hazor (203269), 9.5 mi. N of Sea of Galilee.* **17,** 18, **32,** 33, **34,** 35, 36, **38, 39,** 42, **43,** 86, 88, **90, 93, 94, 95,** 97, **97,** 98, 100, **101, 105, 106, 109,** 111, 112, **112, 116, 118, 123, 124,** 125, **126,** 127, 130, **140,** 142, 143, **143, 144,** 148, 153, **154, 155, 157, 161, 163,** 164, **170, 193,** 255, 256, 259

Hazorea. 94

Hazor Hadattah – Negev town allotted to Judah (Josh 15:25). – *Unknown.*

Hebron – Town in Hill Country of Judah mentioned 64 times in the OT. There Abraham pitched his tent and built an altar (Gen 13:18). He purchased the cave of Machpelah where the patriarchs and their wives were buried (Gen 23:18; etc.). The Israelite spies passed Hebron on their trip through the land of Canaan (Num 13:22). Hebron was captured by Joshua and given to Caleb (Joshua passim). It was both a Levitical city (Josh 21:11, 13) and a city of refuge (20:7). David ruled from Hebron for 7 and 1/2 years before moving his capital to Jerusalem (2 Samuel passim). It was in Hebron that Absalom began his revolt (2 Sam 15). Later, the city was (re)fortified by Rehoboam (2 Chron 11:10). Its earlier name was Kiriath Arba (Gen 23:2). – *El-Khalil – Jebel er-Rumeideh/Hebron (160103), 20 mi. SSW of Jerusalem.* **22,** 23, **23,** 25, **25, 30, 32,** 33, 46, 47, **47,** 48, 49, **51,** 54, **55, 69, 86, 91,** 92, 93, 94, **94, 95, 97, 101,** 105, 106, **106, 108, 110,** 111, 114, **114, 120,** 121, **122, 124,** 125, **126,** 130, **135,** 138, 139, **139, 140,** 142, **142, 143, 145, 151, 153, 155, 159,** 161, 163, 166, 167, 170, 177, 179, **186, 187,** 190, 192, **193,** 195, **198, 200, 206, 212,** 214, **217, 222,** 244, 249, 254, 257, 258

Helam – Town to which Arameans retreated and where David defeated them (2 Sam 10:16, 17). – *Uncertain; possibly Alma (267240), 35 mi. E of Sea of Galilee.* **140,** 141

Helbah – Town of Asher from which the Canaanites were not driven out (Judg 1:31). – *Uncertain; may be alternate name for Ahlab.* 116

Helbon – Damascus traded wine from Helbon with the city of Tyre (Ezek 27:18). – *Uncertain; possibly the valley of Halbun, 12 mi. N of Damascus.*

Helech – Place from which Tyre hired mercenary soldiers (Ezek 27:11). – *Possibly Cilicia in SE Turkey.*

Heleph – Site at SW corner of boundary of Naphtali (Josh 19:33). – *Uncertain; possibly Kh. Irbadeh/H. Arpad (189236), 3 mi. NE of Mt. Tabor.* **116,** 117

Heliopolis – Major Egyptian religious center mentioned in Ezekiel's lament for Egypt (Ezek 30:17). Also called "On" in NIV (Gen 41:45, 50; 46:20) and possibly identified as the "temple of the sun" in Egypt" in Jer 43:13. – *Tell el-Hisn, near el-Matarieh, 10 mi. NNE of Cairo.* **99,** 103, **104,** 175

Helkath – Town on boundary of Asher (Josh 19:25) that served as a Levitical city (Josh 21:31; but parallel in 1 Chron 6:75 has "Hukok"). – *Uncertain; possibly T. el-Qassis/T. Qashish (160232), at the NW exit of the Jezreel Valley, 17 mi. S of Acco.* **38,** 39, 40, **116, 120**

Helkath Hazzurim – Place where 12 men of Joab fought 12 men of Abner (2 Sam 2:16). – *Unknown; but near Gibeon.*

Hellas (Greece). 181

Hellespont. 227, **228**

Hena – One of a number of cities captured by Sennacherib (2 Kgs 18:34; 19:13; Isa 37:23). – *Uncertain; probably in N Syria.*

Hepher – King of city defeated by Joshua (12:17) and later the city was incorporated into Solomon's third administrative district (1 Kgs 4:10). – *Possibly el-Ifshar/T. Hefer (141197), 25 mi. NNE of Joppa near coast of Mediterranean or more probably T. el –Muhaffar (170205) on the north edge of the Dothan Valley.* 148

Hephzibah – Poetic name for Jerusalem (Isa 62:4).

Heptapegon – See also Tabgha. 42, **43,** 208, 210

Heracleopolis. 93, **99,** 218

Heres, Mount – "Mountain of the sun." A place from which the Danites were not able to drive out the Amorites (Judg 1:35). – *Uncertain; possibly same as, or near, "Ir Shemesh" ("city of the sun;" Josh 19:41) which apparently is the same as "Beth Shemesh" ("city of the sun").*

Heres, Pass of – Place near which Gideon ceased pursuing the defeated Midianites (Judg 8:13). – *Unknown; in Transjordan, probably in/ near desert.*

Hereth, forest of – Place where David hid from Saul after having left Moab (1 Sam 22:5). – *Uncertain; located in Judah, possibly W or SW of Jerusalem.* 135

Hermon, Mount – High mountain in NE Israel that marked the N most limit of Joshua's conquests (11:17; 12:1). Mentioned 14 times in the OT. Snow covered much of the year, tributaries of the Jordan River originate at its base. Also called Senir, Sirion, and Baal Hermon. – *Jebel esh-Sheikh/Jebel eth-Thalj, 27 mi. WSW of Damascus, 9,232 feet high.* 21, **22, 23,** 25, 27, **32,** 33, 34, **34,** 36, **36,** 37, 39, **43, 72,** 75, 108, **109, 112,** 119, **119,** 208, 210, **212**

Hermonit. 210

Hermopolis. 173

Herodian Royal Estates. 201, **217**

Herodium. **47,** 49, **193, 200,** 201, 203, 204

Heshbon – Capital city of Sihon, king of the Amorites, conquered by Israel (Num 21:25 – 34; etc.). Mentioned 38 times in the OT, it was allotted to Reuben but evidently became Gadite. It served as a Levitical city. Later it was recovered by Moab and it is mentioned as Moabite in prophetic oracles (e.g., Isa 15:4; etc.). – *Hesban (226134), 12 mi. SW of Amman, may preserve the name, but the OT site may have been at nearby Jalul where the archaeological profile fits the historical data better in some periods.* **22, 23,** 24, 28, **32,** 33, **58,** 59, 62, **62,** 107, **108, 110,** 113, 119, **119, 120,** 145, **148, 159, 163,** 201

Heshmon – Negev town allotted to Judah (Josh 15:27). – *Unknown.*

Hethlon – A place on N portion of ideal boundary of Israel, between Mediterranean Sea and Lebo Hamath (Ezek 47:15; 48:1). – *Unknown.* **178**

Hezron – Site along S boundary of Judah between Kadesh Barnea and Adar (Josh 15:3). The parallel passage in Num 34:4 had "Hazar Addar." – *Unknown.*

Hierakonpolis. 84

Hierapolis – Paul refers to church in the city (Col 4:13) that he, or Epaphras, may have established while residing in Ephesus. – *In W Turkey, 6 mi. N of Laodicea and 12 mi. NNW of Colosse in Lycus valley.* **237,** 238, 239

Hilen – Levitical city in Hill Country of Judah (1 Chron 6:58). Called Holon in Josh 15:51 and 21:15. – *Unknown.*

Hinnom, Valley of (Ben) – Valley to W and S of Jerusalem along which the border of Judah and Benjamin ran (Josh 15:8; 18:16). In later history it became a place of idolatrous worship (2 Kgs 23:10; 2 Chron 28:3; 33:6; Jer 7:31, 32; 19:2, 6; 32:35). Mentioned as a boundary limit in Neh 11:30. – *Wadi er-Rababi just W and then S of the Old City of Jerusalem. Joins the Kidron just S of the City of David.* 242, **242,** 246, **247,** 248, **248**

Hippus. 34, 35, 42, **43, 193,** 197, **198, 200,** 201, **206, 208,** 211, **212, 217**

Hispania – See Spain. **234**

Hit. 76, 77

Hobah – Place, N of Damascus, to which Abram pursued Amraphel and other kings who had captured Lot (Gen 14:15). – *Unknown.* 95

Holon (Judah) – Town allotted to Judah in S Hill Country (Josh 15:51). Levitical city (Josh 21:15), but parallel passage has Hilen (1 Chron 6:58). – *Unknown.*

Holon (Moab) – Town located on plateau of Moab (Jer 48:21). – *Unknown; E of Dead Sea, N of Arnon Gorge.*

Homs. 72, 73, 74, 75

Hor, Mount (Lebanon) – Mountain on N border of land of Canaan (Num 34:7, 8). – *Uncertain; possibly Ras Shaqqa (213412), 30 mi. NNE of Beirut on Mediterranean coast.* **106**

Hor, Mount (Negev) – Israelite campsite during wilderness wanderings. There Aaron died (Num 20 passim, etc.). Mentioned 10 times in OT. – *Uncertain; possibly Imaret el-Khureisheh (104017), 40 mi. SW of Beer Sheba.* **106,** 107, **108**

Horbat Kufin. 94

Horbat Shivta. 55

Horbat Shovav. 86

Horbat Tov. 86

Horbat Yinon. 86

Horeb – Alternate name for Sinai. 17 times in OT. See Sinai. 104

Horem – Town allotted to Naphtali in Upper Galilee (Josh 19:38). – *Unknown.*

Horesh – Site in the Desert of Ziph, where David hid from Saul and met with Jonathan (1 Sam 23:15 – 16, 18 – 19). – *Uncertain; probably E of Ziph, 5 mi. SE of Hebron.*

Horeshim. 94

Hor Haggidgad – Israelite campsite in wilderness between Bene Jaakan and Jotbathah (Num 33:32, 33). – *Unknown.*

Hormah – Village in Negev allotted to Judah (Josh 15:30) and then to Simeon (19:4; 1 Chron 4:30). There, Israelites had first been defeated by the Canaanites (Num 14:45; Deut 1:44) but later they captured and destroyed the town, which formerly had been called Zephath (Num 21:3; Josh 12:14; Judg 1:17). Later David sent booty to its inhabitants (1 Sam 30:30). – *Uncertain; possibly Kh. el-Meshash/T. Masos (146069), 8 mi. ESE of Beer Sheba (on maps in this Atlas), or T. Khuweilifeh/T. Halif (137087), 9.5 mi. NNE of Beersheba. But see Rimmon.* **114, 166**

Horonaim (Moab) – A town of Moab mentioned in prophetic oracles (Isa 48:3, 5, 34). – *Uncertain; possibly el-Iraq (211055), 8.5 mi. E of S end of Dead Sea.* **62, 159**

Horonaim (N of Jerusalem) – Place mentioned in story of Absalom's killing of Amnon (2 Sam 13:34). – *Unknown.*

Horvat Nahal Nissana. 94

Hosah – Village on N border of Asher (Josh 19:29). – *Uncertain; possibly T. Rashidiyeh (170293), in Lebanon, 3 mi. SSE of Tyre near Mediterranean coast.* **116**

Hukkok – Town on W border of Naphtali (Josh 19:34). – *Uncertain; possibly Kh. el-Jemeijmeh/H. Gamon (175252), 12 mi. ESE of Acco.* **116**

Hukok – A Levitical city in Asher (1 Chron 6:75). – *Alternate form of Helkath (Josh 21:30).*

Huleh (Basin/Lake/Valley). 11, **22, 23,** 25, **34,** 35, 36, **36,** 37, **38,** 41, 42, 139

Humtah – Hill country town allotted to Judah (Josh 15:44). – *Unknown.*

Hurru. 98, 101, 127

Hushah – Evidently a village of Judah (1 Chron 4:4). – *Uncertain; possibly Husan (162124), 4.5 mi. W of Bethlehem.*

Hyrcania (Country). 175, **175**

Hyrcania (Herodian Fortress). 47, 49, **193, 200,** 201

Ibleam – Town allotted to Manasseh beside Issachar (Josh 17:11) from which they were not able to drive out the Canaanites (Judg 1:27). Near it Jehu of Israel killed Ahaziah (king of Judah; 2 Kgs

9:27). The Levitical town of "Bileam" (1 Chron 6:70) is probably identical to Ibleam. – *Kh. Belameh (177205), 12 mi. SE of Megiddo.* **38, 44, 86,** 115, **115, 122,** 139, **140, 155, 157**

Iconium – City visited by Paul on Barnabas on their first missionary journey (Acts 13:51; 14:1, 19, 21; 2 Tim 3:11) and mentioned in passing in Acts 16:2. In all probability Paul also visited it on his second and third journeys as well (16:6 and 18:23). – *Konya, in S central Turkey.* **175, 218, 225,** 226, 227, **228, 231, 234**

Idalah – One of the towns allotted to Zebulun (Josh 19:15). – *Uncertain; possibly Kh. el-Huwarah (167236), 6.5 mi. W of Nazareth.*

Idumea – District S of Judea in time of Jesus. People from there followed Jesus (Mark 3:8). – *Boundaries varied; Dead Sea on E, Beer Sheba on S, Beth Zur in N, and near Marisa on W.* 173, **177,** 185, 186, **186, 187,** 188, **190,** 192, **193,** 195, 196, 197, **198,** 199, **200,** 201, 205, **206, 212,** 214, 216, **219, 220**

Ienysus. 173, 174, 175, 177

Iim – Negev settlement allotted to Judah (Josh 15:29). – *Unknown; in S Judah.*

Ijon – Village in Israel, evidently in Naphtali, taken by both Ben-hadad (1 Kgs 15:20; 2 Chron 16:4) and later by Tiglath-pileser III (2 Kgs 15:29) as they invaded Israel from the N. – *T. ed-Dibbin (205308), 9 mi. NNW of Dan in Lebanon.* **142,** 154, **155, 157, 163,** 164

Illyricum – A Roman province in what is now Albania, Croatia and Bosnia and Herzegovina. Paul evidently visited it on his second and/or third journeys (Rom 15:19). See also Dalmatia. **202,** 232

Imaret el-Khureisheh – See Hor, Mount (Negev). 107

Immer – Babylonian town from which exiles returned (Ezra 2:59; Neh 7:61). – *Unknown.*

Imwas – See Emmaus. 215

India – Country on E boundary of Persian Empire (Esther 1:1; 8:9). – *Area E of Indus River.* 27, 64, 65, 78, **175,** 183, **183,** 195, 236

Indus (River/Valley). 174, 175, **175,** 182, **182,** 184, 185

International Coastal Highway. 28, 102, **108**

Ionia. 84, **175,** 176, 226, 227, 232, 238

Ionian Sea. 181

Iphtah – Town allotted to Judah in a Shephelah district (Josh 15:43). – *Unknown.*

Iphtah El, Valley of – A valley on the boundary between Zebulun (Josh 19:14) and Asher (19:27). – *Uncertain; possibly Wadi el-Malik.* 116

Ipsus. 182, 184

Iran. 13, 17, 19, 76, 78, 183

Iraq. 13, 16, **17,** 18, **19,** 76, 82, 162, 173

Iraq el – Amir. 177, 188

Ir Nahash – Could be a personal name or a geographical name (1 Chron 4:12). – *Uncertain; possibly Kh. en-Nahas (191010), 30 mi. S of Dead Sea in Jordan. Others suggest in Shephelah region of Judah.* **55, 62**

Iron – Town in Upper Galilee allotted to Naphtali (Josh 19:38). – *Yarun (189276), 22 mi. NE of Acco.* **36,** 37, **116, 155, 163**

Irpeel – Town allotted to Benjamin (Josh 18:27). – *Unknown; in area N of Jerusalem.*

Ir Shemesh – Town allotted to Dan (Josh 19:41). Probably alternate form of Beth Shemesh. 118, **118**

Isana. 199

Isfahan. 175

Isin. 89, **93,** 267

Islands of the Sea. 175

Ismailia. 69, 177

Israel – Name for Jacob used numerous times in scripture. Designation for descendants of Jacob and the territory that they lived in, basically from Dan to Beersheba. During the Divided Monarchy (930 – 721 B.C.) it usually refers to the northern kingdom although not exclusively. In post-exilic times it can again designate the whole of God's people. 11, 13, **17,** 18, 19, **20,** 21, 22, 25, 27, 28, 29, 31, 32, 33, 35, 37, 38, 40, 40, 47, 49, 51, 52, 54, 55, 62, 73, 74, 78, 79, 83, 84, 87, 90, 100, 102, 111, 116, 118, 119, 121, **122,** 125, 127, 128, 129, 131, 132, 133,

134, 137, 138, 139, 142, 143, 144, 147, 149, 150, 151, **151, 153,** 154, **154, 155,** 156, 157, **157,** 158, **159,** 160, **160, 161,** 162, 164, 165, 166, 169, 173, 195, 240, 243, 244, 245, 254, 255, 256, 257, 259, 260

Israel Pool. 250

Issachar – One of the tribes of Israel mentioned 42 times in OT and once in NT. Tribe settled in territory SW of Sea of Galilee. 39, 40, 115, 116, **116,** 117, **120,** 125, 128, **148,** 149, **178**

Issus. 181, 181, **182, 225**

Itabyrium. 186, **186, 187**

Italia – See Italy. **202, 234**

Italy – The peninsula (ca. 700 mi. long) which extends S from Europe and which is bounded on the E by the Adriatic, and on the W, by the Tyrrhenian Seas. Mentioned four times in the NT (Acts 18:2; 27:1, 6; Heb 13:24). 16, **218,** 230, 233

Ithlah – Town allotted to Dan (Josh 19:42). – *Unknown.*

Ithnan – Negev town allotted to Judah (Josh 15:23). – *Unknown.*

Iturea – During NT era, a district N of Palestine ruled by Philip the tetrarch (Luke 3:1). – *The Beqa area in Lebanon, just N of Israel, and mountain slopes on E and W of the valley.* 197, **198, 200,** 201

Ivvah – A city/state in N Syria captured by the Assyrians (2 Kgs 18:34; 19:13; Isa 37:13) and from which people were brought to settle in Israel. Called Avva in 2 Kgs 17:24. – *Unknown.*

Iye Abarim – Campsite of wandering Israelites probably E of Rift Valley between Oboth and the Zered Valley (Num 21:11; 33:44). – *Uncertain; possibly el-Medeiyineh (223041), 19 mi. ESE of Dead Sea on N slope of Zered Valley. Called Iyim in Num 33:45.* 107

Iyim – Alternate form of Iye Abarim (Num 33:45).

Jaakanites, wells of – Campsite of Israelites during wilderness wanderings (Deut 10:6) called "Bene Jaakan" in Num 33:32. – *Unknown.*

Jaar, fields of – Possibly a shorten form for Kiriath Jearim (Ps 132:6) where Ark of Covenant rested during part of Saul's reign (1 Sam 7:2).

Jaba. 133

Jabbok (River) – River in Transjordan that begins near Amman, and flows N before turning W and descending to join the Jordan. There Jacob wrestled with God (Gen 32:22). It was the traditional N boundary of the Amorites (Num 21:24; Deut 2:37; Josh 12:2), the boundary between Manasseh to the N and Reuben and Gad to the S (Deut 3:16; Josh 12:2), and the territory in its vicinity was claimed by the Ammonites (Judg 11:13, 22). – *Wadi az-Zarqa in Jordan.* **22, 23,** 25, 27, **32,** 33, **44,** 57, **58,** 59, 60, 61, **86,** 91, **91, 94, 95, 97,** 98, 107, 108, **108, 109, 110,** 119, **119, 120, 122, 123,** 126, 128, **135, 142, 143, 155, 177**

Jabesh – Short form of Jabesh Gilead.

Jabesh (Gilead) – A town in Gilead mentioned in various forms 19 times in the OT. There the surviving men of Benjamin took 400 women as wives (Judges 21) and later Saul (a Benjamite) rescued its inhabitants from an Ammonite attack (1 Sam 11). Later, men of the city rescued the bodies of Saul and Jonathan from the wall of Beth Shan (1 Sam 31; 1 Chron 10) and subsequently, David took their remains from Jabesh and buried them in the land of Benjamin (2 Sam 2). – *T. el-Maqlub (214201) on the N bank of the Jabbok, 7 mi. E of the Jordan River and 13 mi. SE of Beth Shan (preferred in this Atlas). Also Tall al – Meqbereh and Abu al – Kharaz have been suggested.* **38, 44, 60,** 123, **123, 126,** 131, **132,** 133, **135,** 137, **140, 155, 157,** 257

Jabez – A town where scribes lived (1 Chron 2:55), possibly in Bethlehem area. – *Unknown.*

Jabneel Valley – See also Nahal Jabneel. 38, **39, 43,** 116, 117

Jabneel (Judah) – Town on W portion of Judah's N border (Josh 15:11). – *Yebna/Yavne (126141), 13 mi. S of Joppa. Also called "Jabneh" that was captured by Uzziah (2 Chron 26:6). Mentioned frequently in sources of Intertestamental Period, and later, as "Jamnia."* **114,** 255

Jabneel (Naphtali) – Town on S border of Naphtali (Josh 19:33). – *T. en-Naam/T. Yinam (198235),*

3.5 mi. W of S end of Sea of Galilee. **36, 38, 43, 116**

Jabneh – See Jabneel (Judah). **158**

Jackal's Well – A well in vicinity of Jerusalem between the Valley and Dung Gates (Neh 2:13). – *Unknown.*

Jacob's Well – Well near Sychar where Jesus met the Samaritan woman (John 4:5 – 6). – *Bir Yaqub (177179), 0.5 mi. SE of Shechem.* **215**

Jagur – Town in Negev district of Judah (Josh 15:21). – *Unknown.*

Jahaz – Transjordanian town captured from Sihon, king of the Amorites (Num 21:23; Deut 2:32; Judg 11:20). Allotted to Reuben (Josh 13:18), it served as a Levitical city (Josh 21:36; called "Jahzah" in 1 Chron 6:78). Mentioned in prophetic oracles as a town of Moab (Isa 15:4; Jer 48:34). – *Uncertain; possibly Kh. el-Medeiyineh (236110), E of Dead Sea, 11 mi. SE of Medeba.* **107, 108, 110, 119, 120**

Jahzah – Alternate form of Jahaz (1 Chron 6:78).

Jair, settlements of – (Josh 13:30 and 1 Kgs 4:13. See Havvoth Jair. **119**

Jamnia – See OT Jabneel (Judah). **186, 186, 187, 188, 193, 198, 200, 206, 212, 217, 220, 222, 223**

Janim – Judean town in hill country district of Hebron (Josh 15:53). – *Unknown.*

Janoah (Ephraim) – Town on NE boundary of Ephraim (Josh 16:6, 7). – *Kh. Yanun (184173), 6 mi. SE of Shechem.* **44, 115**

Janoah (Galilee) – Town in Upper Galilee captured by Tiglath-Pileser III, king of Assyria (2 Kgs 15:29). – *Yanuh (178296), ca. 6 mi. E of Tyre.* **32, 157, 163, 164**

Japheth. **83, 83**

Japhia – Town on S border of Zebulun (Josh 19:12). – *Possibly Yafa (176232), 1.5 mi. SW of Nazareth, but the archaeological profile does not correspond with the historical record.* **116, 206, 207, 208**

Japhletites, territory of the – Territory on the SW border of Ephraim (Josh 16:3). – *Possibly NE of Beth Horon region.*

Jarkon/Yarkon River – See Me Jarkon and Nahal Yarkon. **11, 22, 23, 32, 33, 43, 44, 44, 51, 51, 52, 118, 118, 131, 177**

Jarmuth (Issachar) – Levitical town in Issachar (Josh 21:29). Called "Ramoth" in 1 Chron 6:73 and "Remeth" in Josh 19:21. – *T. Remet (199221), 6 mi. NNE of Beth Shan.* **120**

Jarmuth (Judah) – Town in Shephelah who's king fought against Joshua (Jos 10:3, 5, 23; 12:11). Assigned to Judah (Josh 15:35), Jews settled here after the return from the Babylonian Exile (Neh 11:29). – *Usually with Kh. el-Yarmuk/T. Yarmut (147124), 17 mi. WSW of Jerusalem but archaeological profile is problematic.* **110, 111, 114, 134, 166, 177, 179**

Jashubi Lehem – Judeans lived in this place (1 Chron 4:22). – *Unknown; text difficult.*

Jattir – Town in hill country allotted to Judah (Josh 15:48) that served as a Levitical city (Josh 21:14; 1 Chron 6:57). David sent spoils to inhabitants of town (1 Sam 30:27). – *Kh. Attir/Yatir (151084), 13 mi. SW of Hebron.* **114, 120, 166**

Javan. **83, 83, 84**

Jaxartes (River). **175, 183**

Jazer – Transjordanian town of Sihon captured by Israelites (Num 21:32) and given to Gad (32:1, 3, 35; Josh 13:25). Served as a Levitical city (Josh 21:29; 1 Chron 6:81) and David stationed soldiers there (1 Chron 26:31). Joab passed through area while taking a census (2 Sam 24:5) but town was taken over by the Moabites later (Isa 16:8, 9; Jer 48:32). – *Uncertain; possibly Kh. es-Sar (228150), 6 mi. W of Amman.* **119, 142**

Jearim, Mount – Mountain (ridge?) on NW boundary of Judah (Josh 15:10) in vicinity of Kesalon – *Kesla (154132), 11 mi. W of Jerusalem.*

Jebel al-Lawz. **104**

Jebel es – Silsilsa. 65, **66**

Jebel Barkal. **66**

Jebel Bishri. **72,** 73

Jebel Druze – See Bashan, Mount, and Harran, Mount. 28, 34, **34,** 72, 106, **106, 109**

Jebel ed – Dabab. 62, **62**

Jebel el – Aziz. **72,** 73

Jebel el – Kabir – See Zalmon, Mount. 128

Jebel el – Maghara. 69

Jebel esh – Sharqia. **72,** 73

Jebel et – Tih. 66, 69

Jebel Geneife. 104, **104**

Jebel Helal. 55, 69, **104**

Jebel Jarmak – See Mount Meron. 210

Jebel/Har Karkom. 55, 69, **104**

Jebel Murr. 103

Jebel Musa – See Sinai (Desert of, Mount). 69, **104, 104,** 105

Jebel Nebi Harun. 107, **108**

Jebel Qaaqir. 94

Jebel Sin Bisher – See Sinai (Desert of, Mount). 69, 104, **104,** 105

Jebel Sinjar. 72, **72,** 73

Jebel Sirbal. 69, **69**

Jebel Umm ed-Daraj. 58, 59, 60

Jebel Yalaq. 69

Jebel Zawiyeh. **72,** 75

Jebus (Jebusite city) – Alternate name for Jerusalem used from the time of Joshua up until the days of David. Mentioned in boundary descriptions of Judah and Benjamin (Josh 15:8; 18:16) and in the city list of Benjamin (18:28). Also in the story of the Levite and his concubine (Judg 19:10, 11) and in story of David's conquest of it (1 Chron 11:4). The Jebusites are mentioned frequently in Scripture. 121, **122,** 123, **123, 132,** 139, **140,** 240, 244, 245, **246**

Jegar Sahaduth – A stone marker set up in Transjordan (Gilead?) by Jacob and Laban (Gen 31:47). Jacob called it "Galeed." – *Unknown.*

Jehoshaphat, Valley of – Valley near Jerusalem where nations are to be judged (Joel 3:2, 12). – *Traditionally with a portion of the Kidron Valley just E of Jerusalem.*

Jehud – Town allotted to tribe of Dan (Josh 19:45). – *El-Yehudiyeh/Yehud (139159), 8.5 mi. E of Joppa in Coastal Plain.* **118**

Jekabzeel – Town in Negev settled by Jews after the exile (Neh 11:25). – *Probably alternate form of Kabzeel.* **177,** 179

Jemdet Nasr. 84, 85, 266

Jenin. 23, 25, 95, 214

Jerahmeelites, towns of the – Settlements of a Judahite clan in the E Negev basin to which David sent booty (1 Sam 30:29) See Negev of Jerahmeel. – *Probably area about 12 mi. E of Beer Sheba in E Negev.*

Jerash. 58, 60, 211

Jericho (NT) – Jesus passed through town on way up to Jerusalem healing the blind (Matt 20:29; Mark 10:46; Luke 18:35) and meeting Zacchaeus along the way (Luke 19:1). Also mentioned in story of Good Samaritan (Luke 10:30). – *Spread out over a large oasis, but Hasmonean and Herodian palaces located at Tulul Abu el-Alayiq (191139) along banks of Wadi Qilt, 9 mi. NNW of the Dead Sea.* **47,** 190, 192, **193,** 195, 196, 197, **198,** 199, **200,** 201, 204, 205, **206, 212,** 213, 214, **217, 219,** 220, **220,** 221, **222,** 240

Jericho (OT) – Town W of Jordan River mentioned 56 times in OT and once in NT (Heb 11:30). Mentioned primarily to identify Israel's campsite in the Plains of Moab and especially in connection with the conquest of Canaan (Joshua passim). Noted as a border city between Ephraim and Benjamin (Josh 16 and 18), it was there that Ehud killed Eglon (called "City of Palms" in Judg 3:13). Also mentioned in the Elijah and Elisha narratives (2 Kings 2). After the exile Jews settled there (Ezra 2:34; Neh 3:2; 7:36). – *T. es-Sultan (192142), 6 mi. W of Jordan River and 10 mi. NNW of Dead Sea alongside a powerful spring that waters the oasis.* 23, 24, 25, 28, **30, 32,** 47, **47,** 48, 49, **58,** 61, **84, 86,** 87, 88, 92, **94, 95,** 100, 105, 108, **108,** 109, 110, **110,** 111, 115, **115,** 125, **126, 131,** 132, **132,** 133, **140,** 154, 161, **163, 170,** 171, **177,** 179, **241,** 257, 260

Jericho, Plain(s) of. 109, 171

Jeruel, Desert/Wilderness of – *An area in the eastern Judean Desert/Wilderness, but W of En Gedi (2 Chron 20:16) near the Nahal Arugot.*

Jerusalem – *el-Quds/Jerusalem (172131).* 11, 13, 17, **19, 22, 23, 23,** 24, 25, **26,** 28, 29, **30, 31, 32,** 33, 37, 46, 47, **47,** 49, **51,** 52, 53, **58, 66, 86,** 93, **93, 95,** 95, 96, **99, 101,** 105, 108, 110, 111, 115, **115, 120,** 121, 123, **123,** 124, **124, 126, 130, 131,** 132, **135,** 138, 139, 140, 141, 142, **142,** 143, 143, 144, 145, 146, 147, **148,** 149, 150, 151, **151, 152, 153,** 154, **154, 155,** 157, 158, **159,** 160, **160, 161,** 162, **163,** 164, 165, 166, **166,** 167, 168, 169, **169,** 170, **170,** 171, **171,** 172, 173, **173,** 174, **174,** 175, **175,** 176, 177, **177, 178,** 179, **181,** 182, **182, 184, 186,** 187, **187,** 188, 189, 190, **190,** 191, 192, **193,** 194, 195, 196, 197, **198,** 199, **200,** 201, 202, 203, 204, **206,** 209, 210, **212,** 213, 214, 215, 216, 217, **217,** 218, 219, **219,** 220, **220,** 221, **222,** 223, 224, 225, 227, **228,** 231, **231,** 232, **233, 234,** 235, 240 – 253, **241, 242, 246, 247, 249,** 255, 256, 257, 266, 267

Jerusalem Plateau. 46, 47

Jeshanah – Town in S Ephraim taken by Abijah (king of Judah) from Jeroboam (king of Israel; 2 Chron 13:19). A gate in Jerusalem was called the Jeshanah Gate (Neh 3:6; 12:39); probably the road from there led to Jeshanah. Possibly "Shen" of 1 Sam 7:12 is a corruption of Jeshanah (as LXX). – *Burj el-Isaneh (174156), 15.5 mi. N of Jerusalem.* 154, **154**

Jeshimon – Desert/wilderness area where David hid from Saul (1 Sam 23:19, 24; 26:1, 3). – *Dry, chalky wilderness E and SE of Ziph between it and the Dead Sea.* 49, 136

Jeshua – Town in S Judah where Jews settled after the exile (Neh 11:26). – *Uncertain; possibly T. es-Saweh/T. Jeshua (149076), 9 mi. ENE of Beersheba.* **177,** 179

Jezirah. 72, 73, 90, 93

Jezreel (Issachar) – Town allotted to tribe of Issachar (Josh 19:18). Saul's troops camped near it in battle with Philistines (1 Sam 29:1 – 11) and later his son, Ish-Bosheth, ruled over area (2 Sam 2:9). It, along with the valley, was included in the fifth Solomonic district (1 Kings 4:12). Ahab had a secondary palace there (1 Kings 18:45) and there Ahab took Naboth's vineyard (1 Kings 21:1). Subsequently it was the site of Jehu's coup in which Joram and Jezebel were killed (2 Kings 9 – 10). – *Zerin/T. Yizreel (181218), 11 mi. WNW of Beth Shan and 8.5 mi. ESE of Megiddo.* **30, 38,** 39, 40, **44,** 116, **116, 135,** 136, 137, 138, **155,** 157, **157,** 158, **161**

Jezreel (Judah) – Town allotted to Judah (Josh 15:56). David's wife Ahinoam was from Jezreel (1 Sam 25:43; 27:3; 30:5; 2 Sam 2:2; 3:2). – *Unknown; but probably in are about 8 – 10 mi. S of Hebron.*

Jezreel, Valley of – Large, triangular-shaped valley in north-central Israel. During period of Joshua (17:16) it was controlled by Canaanites. During days of the judge Gideon, the Midianites, Amalekites and other eastern peoples raided and camped in the area (Judges 6:33). Also mentioned in an oracle of Hosea (1:5). – *The triangular shaped valley that separates the Hill Country of Manasseh from that of Galilee.* 22, 23, **23,** 25, 29, 33, **36,** 37, 38, **38,** 39, 40, 41, **44,** 102, 115, **116, 116,** 121, 124, 127, 128, 139, **140,** 149, 152, 162, 176, 186, 187, 192, **193,** 195, **200,** 201, 202, 207

Jogbehah – Town in Transjordan (Gilead) built up by Gad (Num 32:35). Gideon pursued Midianites in its direction (Judg 8:11). – *El-Jubeihat (231159), 7 mi. NW of Amman.* **119,** 128

Jokdeam – Town in hill country allotted to Judah (Josh 15:56). Possibly the same as "Jorkeam" in 1 Chron 2:44. – *Unknown; probably in area 8 – 10 mi. S of Hebron.*

Jokmeam (east) – Town in fifth Solomonic district (1 Kings 4:12). – *Uncertain; possibly T. es-Samadi (196170), in Jordan Valley, 27 mi. S of Beth Shan.* **32, 44, 148**

Jokmeam (west) – Levitical city, evidently in vicinity of W Ephraim near Gezer and Beth Horon (1 Chron 6:68). – *Unknown.*

Jokneam – Canaanite town conquered by Israel (Josh 12:22). Border of Zebulun extended to the ravine near it (Josh 19:11) and it became a Levitical city (Josh 21:34; although in NIV translation of 1 Chron 6:77, it does not appear in the Hebrew

text there!). – *T. Qeimun/T. Yoqneam (160230), 7 mi. NW of Megiddo on S edge of Jezreel Valley at foot of Mt. Carmel.* **32, 36, 38,** 40, 41, **44, 86, 95,** 101, **101, 105, 112,** 116, **120**
Joktan. 83
Joktheel (Edom) – Amaziah captured Sela and changed its name to Joktheel (2 Kings 14:7) – See Sela. 158, **159**
Joktheel (Judah) – Town in Shephelah allotted to Judah (Josh 15:38). – *Unknown.*
Joppa – Mediterranean seaport evidently allotted to Dan (Josh 19:46). Intermittently served as a Judean port, for logs for the first and second temples were shipped there from Lebanon (2 Chron 2:16; Ezra 3:7) although through out much of the OT period it was in non-Israelite hands. From there Jonah set sail (Jonah 1:3) and it was there that Peter had his vision and received messengers from Cornelius (Acts 9 – 11 passim). Mentioned frequently in extrabiblical literature. – *Yafa/Yafo (126162), just S of modern Tel Aviv.* 22, **22, 23, 26, 30, 32, 44,** 51, **51,** 52, 88, **95, 97, 101, 105, 108, 110, 118, 120, 123, 130, 131, 135, 140, 142, 143, 146,** 147, **148,** 149, **155, 163, 166,** 168, **170, 177, 184,** 186, **186, 187,** 192, **193,** 194, 197, **198,** 199, **200,** 201, **206, 212,** 215, 217, **217, 219,** 220, **220, 222, 234**
Jordan (Country). 11, 13, 16, **17, 19,** 21 – 64
Jordan (River/Valley) – The major river of Israel/Palestine that flows from N to S, into and then out of the Sea of Galilee and then continues S to the Dead Sea. Mentioned 181 times in the OT and 15 times in the NT. 11, **17,** 21, **22,** 23, 25, **26,** 27, **30, 32,** 33, 34, **34,** 35, 36, **36, 38,** 40, 42, **43, 44, 47,** 50, **55, 58,** 59, 60, **60,** 61, **62, 86,** 91, **91,** 93, 94, **94, 95, 97, 101, 105,** 106, **106,** 107, 108, **108,** 109, **109, 110, 112, 114,** 115, **115,** 116, **116,** 118, 119, **119, 120,** 121, **122, 123, 124,** 125, **126,** 128, 129, **130, 131,** 133, **135, 140,** 141, 142, **142, 143, 148,** 149, 152, **153, 154, 155,** 156, **157,** 159, **161, 163, 166, 170,** 172, 176, 177, 177, **178,** 179, **186, 187,** 188, **190, 193,** 196, 197, **198,** 199, **200,** 201, 205, **206, 208,** 209, 210, 211, **212,** 213, **217, 219,** 220, **220, 222,** 240, **241,** 257
Jorkeam – If a place (1 Chron 2:44), and not a personal name, then possibly the same as "Jokdeam" of Josh 15:56. – *Unknown.*
Jotapata. 206, 207, **208,** 220, **222**
Jotbah – Queen mother of Amon was from there (2 Kgs 21:19). – *Jefat/H. Yodefat (176248), 9 mi. NNW of Nazareth.*
Jotbathah – A campsite of the Israelites during the Exodus wanderings (Num 33:33, 34; Deut 10:7). – *Uncertain; possibly Tabeh (139878), 5.5 mi. SW of Elath on W shore of Red Sea.*
Jotbathah (Galilee) – See also Jotapata. **163**
Judah – Descendant of Jacob whose descendants formed on of the 12 tribes. The bulk of Judah's tribal inheritance was the territory between Jerusalem and Beersheba. During the Divided Monarchy, the southern kingdom was called the Kingdom of Judah and later the territory became known as Judea. 21, **22,** 23, **23,** 24, 31, 32, 38, 45, 46, 47, 49, **51,** 52, 55, 113, 114, 115, 117, 118, **120,** 121, **121, 123,** 125, 127, 129, 133, **134,** 136, 137, 138, 139, 142, 143, 147, 150, 151, **151,** 152, **153,** 154, **154, 155,** 157, 158, **159,** 160, **160, 161,** 162, **163,** 164, 165 – 171, **166, 170, 171,** 172, 173, **173,** 175, 176, 177, **177, 178,** 179, 244, 245, 248, 254, 255, 266, 267
Judah, Hill Country/Mountains of. 25, 27, 31, 47, **47,** 48, 49, **51,** 52, 53, 59, **91,** 92, **101,** 102, 113, 121, **122,** 125, 135, 137, 147, **151,** 157, 160, **165,** 166, 173, 179, 192, 240, **241,** 255
Judea – Mentioned 43 times in the NT, it was the Greco-Roman designation for the territory/district/province that succeeded the old Davidic kingdom of Judah. 173, 174, 176, 179, 182, 185, **186,** 187, **187,** 188, 189, 190, **190,** 191, 192, **193,** 194, 195, 196, 197, **198,** 199, **200,** 201, 205, **206,** 211, **212,** 214, 215, 216, 217, **217,** 218, **218,** 219, **219, 220,** 221, **222,** 223, 224, 232
Judean Desert/Wilderness of Judea. 46, 47, **47,** 48, 49, 50, **51,** 135, **135,** 136, 151, 192, 213, **213, 222,** 223, 241, **241,** 242
Julias – See Bethsaida. **43, 208,** 210
Julias (Transjordan). 211, **222**
Juttah – Hill Country city allotted to Judah (Josh 15:55). Served as a Levitical city (21:16; the NIV inserts "Juttah" into 1 Chron 6:59 [following some versions of the LXX] but the MT does not have the name). – *Yatta (158095), 5.5 mi. S of Hebron.* **114, 120, 166**

Kabir River – See Nahr Kabir. 73
Kabzeel – A Negev town allotted to Judah (Josh 15:21). One of David's warriors was from there (2 Sam 23:20; 1 Chron 11:22). – *Uncertain; formerly identified with Kh. Gharreh/T. Ira (148071), 8 mi. E of Beersheba. Also called Jekabzeel.* **114, 166**
Kadesh (Barnea) – Called Kadesh 14 times in OT and Kadesh Barnea 10 times. Mentioned in connection with Abraham (Gen 14:7; 16:14; 20:1). It was located on the S boundary of the land of Canaan (Num 34:4; Josh 15:3). The men that had spied out Canaan returned to the Israelite camp at Kadesh (Num 13:26). Moses' sister, Miriam, died there (Num 20:1) as did Aaron at nearby Mt. Hor (20:22). Israel camped there for many days (Deut 1:46). – *Ain el-Qudeirat (096006), about 50 mi. SW of Beersheba.* **17,** 21, **23, 32,** 54, **55,** 69, **69, 94, 104,** 105, 106, **106,** 107, **108, 114, 153, 159, 178**
Kadesh, Desert of – Mentioned in Psalm 29:8. Since the setting of the Psalm seems to be N of Israel, the Kadesh referred to is probably Kadesh on the Orontes. – *T. Nebi Mind (291444), about 75 mi. N of Damascus in Syria.*
Kadesh (on Orontes River). 33, 72, 75, 101, **101,** 102, 113, **123, 124,** 125, 127
Kain – Hill country town allotted to Judah (Josh 15:57). – *Kh. Bani Dar (164100), 3.5 mi. SE of Hebron.* **114, 166**
Kamon – Transjordanian town where the judge Jair was buried (Judg 10:5). – *Uncertain; possibly Qamm (218221), 11.5 mi. SE of the Sea of Galilee.* **126**
Kanah (Asher) – Town allotted to Asher (Josh 19:28). – *Qana (178290), 7.5 mi. SE of Tyre in Lebanon.* **36,** 37, 116
Kanah (Lower Galilee). 163
Kanah Ravine – A wadi that drained W Ephraim and Manasseh and was a boundary divider for the two tribes (Josh 16:8; 17:9). – *Wadi Qana which flows W out of the Hill Country of Ephraim and Manasseh and joins the Nahal Yarkon a few mi. W of Aphek.* **44,** 45, 46, 114, **115**
Kanish. 17, 93
Kara Kum (Black Desert). 183
Karka – Settlement on SW boundary of Judah (Josh 15:3). – *Uncertain; possibly Ain el-Qeseimeh (089007), about 50 mi. SW of Beersheba in the vicinity of Kadesh Barnea.* **114**
Karkor – Transjordanian site to which Zebah and Zalmunna fled from Gideon (Judg 8:10). – *Unknown.* 128
Karm Abu Girg. 103
Karnaim – Transjordanian city taken by Israel (Amos 6:13). Probably it had replaced Ashtaroth (Karnaim) as the regional center. – *Sheikh Sa'd (247249), 23 mi. E of the Sea of Galilee.* **32,** 33, **34, 35, 94, 109, 153, 155, 157, 163,** 176, **178**
Karnak. 93, 100, 102, 127, 152
Kartah – Levitical town in the territory of Zebulun (Josh 21:34; MT of 1 Chron 6:77 does not have Kartah, see note to NIV translation). – *Unknown.*
Kartan – Levitical town in territory of Naphtali (Josh 21:32). – *Uncertain; possibly the same as "Rakkath" of Josh 19:35 and Kiriathaim (Naphtali; 1 Chron 6:76).* **120**
Karun River. 77
Kashmir. 183
Kattath – A town allotted to Zebulun (Josh 19:15). Possibly the same as "Kitron" of Judg 1:30. – *Unknown.*
Kavalla – See Neapolis (Greece). 227
Kebar River – A canal in ancient Babylon near which Nebuchadnezzar had settled Jewish exiles (Ezek 1:3). It was near this canal that Ezekiel had some of his early visions (1:1; 3:15, 23; 10:15, 20, 22; 43:3). – *Uncertain; possibly same as the ancient "naru kabari" canal which branched off from the Euphrates NW of Babylon and joined it again some 60 mi. to the south.*
Kedar – An Arabian tribe located to the S and SE of Israel mentioned 10 times in OT. 174, 177
Kedemoth – Transjordanian town assigned to tribe of Reuben (Josh 13:18) that served as a Levitical city (Josh 21:37; 1 Chron 6:79). – *Uncertain; possibly Aleiyan (233104), 20 mi. E of Dead Sea, N of Arnon River.* **108, 119, 120**
Kedemoth, Desert of – Wilderness region near Kedemoth from which Moses sent messengers to Sihon, the king of the Amorites (Deut 2:26). 107, **108**
Kedesh (Issachar) – Levitical city in Issachar (1 Chron 6:72). – *Unknown.*
Kedesh (Judah) – Town in Negev district of Judah (Josh 15:23). – *Unknown; some believe it to be identical to Kadesh Barnea but not likely.*
Kedesh (Naphtali, Lower Galilee) – Deborah's general, Barak, was from the place (Judg 4:6, 9) and later he pursued Sisera in the direction of the town (4:11). – *Kh. Qedish/H. Qedesh (202237), 1 mi. W of S end of Sea of Galilee, overlooking the sea.* 43, **126,** 127
Kedesh (Upper Galilee) – Town conquered by Israel (Josh 12:22) and allotted to Naphtali (19:37). Served as a Levitical city (21:32; 1 Chron 6:76) and as a city of refuge (Josh 20:7). Later captured by Tiglah-Pileser III (2 Kgs 15:29). Later called Cadasa. – *T. Qades/T. Qedesh (199279), 17 mi. NNW of Sea of Galilee in Upper Galilee.* **36,** 37, **37, 43, 94, 101, 105, 112, 116, 120,** 121, **155, 157, 163,** 164, **170,** 255
Kefar Hananiya. 205, **206**
Kefar Monash. 86
Kehelathah – Israelite campsite during wilderness wanderings located between Rissah and Mount Shepher (Num 33:22 – 23). – *Unknown.*
Keilah – Shephelah town allotted to Judah (Josh 15:44). David protected its inhabitants from the Philistines but had to flee the area when it became evident that he would be handed over to Saul (1 Sam 23). At the time of the return from the Babylonian exile, Jews settled in Keilah (Neh 3:17, 18). – *Kh. Qila (150113), 18 mi. SW of Jerusalem.* **51, 114, 123,** 124, 134, 135, **135, 166, 177,** 179
Kenath – A Transjordanian town captured by Nobah (Num 32:42) but later lost to Geshur and Aram (1 Chron 2:23). – *Qanawat (302241), 57 mi. E of the Sea of Galilee at the W foot of the Jebel Druze.* **34, 58, 123**
Kephar Ammoni – Town allotted to Benjamin (Josh 18:24). – *Unknown; but N of Jerusalem, possibly on E side of watershed.*
Kephirah – One of four "Gibeonite" (Hivite) cities that tricked Joshua into making a covenant with them (Josh 9:17). Allotted to Benjamin (18:26). Later resettled by Jews returning from Babylonian exile (Ezra 2:25; Neh 7:29). – *Kh. el-Kefireh (160137), 8.5 mi. WNW of Jerusalem.* **110,** 111, **115, 166, 177**
Kerak – See Kir Hareseth. **30**
Kerem Ben Zimra. 94
Kerioth (Moab) – Moabite town mentioned in prophetic oracles (Amos 2:2; Jer 48:24). – *El-Qereiyat (215105), 8 mi. E of Dead Sea.*
Kerioth Hezron – Negev town allotted to Judah (Josh 15:25) also called "Hazor." – *Uncertain; possibly Kh. el-Qaryatein/T.Qeriyot (161083), 12.5 mi. SSE of Hebron.* **114, 166**
Kerith Ravine – Elijah went to this valley for sustenance at the time of prophesied drought (1 Kgs 17:2 – 7). – *Unknown; probably a wadi E of Jordan River, some suggest the Wadi Jabesh.* **161**
Kerman. 175.
Kerub – Jews from there left to return to Judah (Ezra 2:59; Neh 7:61). – *Unknown; in area of Babylonia.*
Kesalon – Town along the W portion of Judah's N boundary (Josh 15:10). – *Kesla/Kesalon (154132), 11 mi. W of Jerusalem.*
Kesil – Town in Negev allotted to Judah (Josh 15:30). Probably same as Simeonite town of Bethul (19:4)/Bethuel (1 Chron 4:30). See Bethuel.

Kestros River. 225

Kesulloth – Town allotted to Issachar (Josh 19:18). Probably the same as Kisloth Tabor mentioned in boundary description of Zebulun (Josh 19:12). – *Iksal (180232), 2 mi. SE of Nazareth.* 116, **116**

Kezib – Town in Judean Shephelah where Shua gave birth to Shelah (Gen 38:5). See Aczib (Judah).

Khanazir. 50, **62**, 82, **86**, **94**, 95

Khashabu. 123

Khirbet Ayun Musa – See Nebo (Moab). **86**

Khirbet el – Mahruq. **86**

Khirbet Ira. 47, **47**

Khirbet Iskander. **86**

Khirbet Qana – See Cana of Galilee. 39, 207

Khirbet Qeiyafa. 53, 134, **134**

Khirbet Rabud – See Debir. 105

Khirbet Uza – See Ramoth Negev.

Kibroth Hattaavah – First site where Israelites camped after leaving Mt. Sinai (Num 33:16, 17). There, they craved the food of Egypt and as a result of the quail which were sent, many Israelites died there as a punishment (11:34, 35; Deut 9:22). Hence the name "graves of desire." – *Uncertain; probably in Sinai Peninsula.* 105

Kibzaim – Levitical city mentioned between Gezer and Beth Horon (Josh 21:22). The parallel list (1 Chron 6:69) has "Jokmeam," with which it may be identical. See Jokmeam (west).

Kidon, threshing floor of – Site between Kiriath Jearim and Jerusalem where Uzzah died because he touched the ark (1 Chron 13:9). Called "the threshing floor of Nacon" in 2 Sam 6:6. – *Unknown.*

Kidron Valley – Valley immediately to the E of Jerusalem. David crossed it as he fled E from his son Absalom (2 Sam 15:23) and Solomon commanded Shimei not to cross it on pain of death (2 Kgs 2:37). The valley was a site where illicit cult objects were destroyed during the religious reforms of Asa, Hezekiah, and Josiah (1 Kgs 15:13; 2 Kgs 23:4, 6, 12; 2 Chron 15:16; 29:16; 30:14). It is also mentioned by Jeremiah (31:40). Jesus and his disciples crossed it as they walked from the upper room to the Garden of Gethsemane (John 18:1). – *The N-S valley E of Jerusalem called from N to S the Wadi ej-Joz, Wadi Sitti Maryam, and Wadi Tantur Farun, Wadi Silwan, and eventually the Wadi en-Nar which flows through the Judean Desert, SE, into the Dead Sea.* 142, 143, 191, 241, 242, **242**, 243, **246**, **248**, 249, **250**, 252, 253

Kilmad – Mentioned after Asshur and other places as having traded with Tyre (Ezek 27:23). – *Unknown; but probably not close to ancient Israel.*

Kinah – Negev town allotted to Judah (Josh 15:22). – *Uncertain; possibly Kh. Taiyib/H. Tov (163081), 19 mi. ENE of Beersheba.* 166

King's Garden – Zedekiah and his troops fled Jerusalem via a gate near this garden (2 Kgs 25:2; Jer 39:4; 52:7). Nehemiah's workers repaired the wall of Jerusalem near this garden (Neh 3:15). – *Uncertain; possibly to the S, E, or SE of Jerusalem in the Kidron Valley. It possibly was watered from the overflow from the pool of Siloam at the S tip of ancient Jerusalem.*

King's Highway – While in Kadesh (Barnea), Israel requested permission from the king of Edom to be able to use this route (Num 20:17). Later, they made a similar request of Sihon, king of the Amorites who ruled from Heshbon (Num 21:22 and possibly Deut 2:27). – *First reference is possibly to the E-W route called the Darb es-Sultan that runs through the Nahal Zin towards Edom. The second reference is possibly to a portion of the N-S Transjordanian route that connects Edom/Arabia with points to the N.* 28, 107, **108**, 156

King's Pool – A pool mentioned in connection with the rebuilding of the wall of Jerusalem (Neh 2:14). Possibly the same as the "Pool of Siloam [Shelah]" in Neh 3:15. – *Uncertain; probably in Kidron Valley near the S or SE point of the City of David.* 46

King's Valley – The Valley of Shaveh, where Abram met king of Sodom after victory over the invading kings, is called the "King's Valley" (Gen 14:17).

Later Absalom erected a pillar in the "King's Valley" as a memorial (2 Sam 18:18). – *Uncertain; possibly in vicinity of Jerusalem if meeting with Melchizedek took place soon after above encounter and if Salem = Jerusalem (see Gen 14).* 95, 243

Kinnereth – Town allotted to Naphtali (Josh 19:35). Ben-Hadad conquered the town, or the area (1 Kgs 15:20). – *Kh. Ureime/T. Kinrot (200252), on NW shore of Sea of Galilee. Called "Gennesaret" in NT.* 33, **36**, **38**, 41, **43**, 95, 105, **112**, 154, **155**, 157

Kinnereth, Sea of – Common name for Sea of Galilee (Num 34:11; Josh 12:3; 13:27). An abbreviated form is evidently used in Deut 3:17 and Josh 11:2, although these references could be to the town. See Galilee, Sea of and Gennesaret, Lake of. 41, **94**, 106, **106**, **109**, 112, **116**, 118, 119, **119**, 120, **122**, **126**, 130, 157

Kios – The ship Paul was on spent a night anchored off Kios (Acts 20:15). – *Greek island, 5 mi. W of central W shore of Asia Minor (Turkey).* 231, 232

Kir – Place to which the Assyrian king Tiglath-pileser III carried captives from Damascus (2 Kgs 16:9). The Arameans originated from Kir (Amos 9:7) and it is mentioned in Amos 1:4 and Isa 22:6. – *Usually thought to be in Mesopotamia, but unknown.*

Kir Hareseth – Evidently the chief city of Moab during period of Monarchy. Attacked by Israel, Judah and Edom (2 Kgs 3:25) and mentioned in prophetic oracles (Isa 16:7, 11; Jer 48:31, 36), often in parallel with Moab. Called "Kir in Moab" in Isa 15:1. – *El-Kerak (217066), 10 mi. E of Dead Sea.* 47, **55**, **62**, 63, 159

Kir in Moab – Isa 15:1. See Kir Hareseth. 33, **140**, **145**, 163

Kiriath – Town in Benjamin (Josh 18:28). See Kiriath Jearim.

Kiriathaim (Naphtali) – Levitical town in Naphtali (1 Chron 6:76). – *Uncertain; possibly same as Kartan in Josh 21:32).*

Kiriathaim (Reuben) – Transjordanian town assigned to Reuben (Num 32:37; Josh 13:19). Mentioned in prophetic oracles against Moab (Jer 48:1, 23; Ezek 25:9). – *Uncertain; possibly Qaryat el-Mekhaiyet (220128), 9 mi. E of Dead Sea.* **94**, 119

Kiriath Arba – Older name (Judg 1:10) for Hebron (Gen 23:2; 35:27; Josh 14:15; 15:13, 54; 20:7; 21:11) but name also used after Babylonian exile (Neh 11:25). See Hebron. **97**, **177**, 179, 254

Kiriath Baal – Town on border of Judah and Benjamin (Josh 15:60; 18:14). See Kiriath Jearim. 254

Kiriath Huzoth – A town in Moab to which Balak took Balaam (Num 22:39). – *Unknown; probably E of N end of Dead Sea.*

Kiriath Jearim – A town in the old Hivite league with which Israel made a treaty (Josh 9:17). Although allotted to Judah (Josh 15:60), it was on the N border of Judah (Josh 15:9) and SW boundary of Benjamin (18:14, 15). The Danites camped in the area (Judg 18:12) and later the ark of God was stored there (1 Sam 6:21, 7:1, 2). David brought the ark from there to Jerusalem (1 Chron 13:5, 6; 2 Chron 1:4). Jeremiah (26:20) mentions a prophet from the town. People from there returned to Judah after the Babylonian exile (Ezra 2:25; Neh 7:29). Also called "Kiriath Baal" (Josh 15:60; 18:14), "Baalah of Judah" (Josh 15:9, 10; 2 Sam 6:2; 1 Chron 13:6), and possibly "Kiriath" (Josh 18:28), "Jaar" (Ps 132:6), as well as "Baalath" (2 Chron 8:6; 1 Kgs 9:18). – *Deir el-Azar/T. Qiryat Yearim (159135); 8.5 mi. W of Jerusalem.* 23, 24, 47, 53, **110**, 111, **114**, 115, 118, 131, **131**, 132, **140**, 142, 144, **154**, **155**, **177**, 241, 245

Kiriath Sannah – Allotted to Judah, an early name for Debir (Josh 15:49). See Debir (Hill Country of Judah).

Kiriath Sepher – Former name of Debir (Judah; Josh 15:15; Judg 1:11). Captured by Othniel (Josh 15:16; Judg 1:12) and settled by Calebites. See Debir (Hill Country of Judah). 254

Kish. 85

Kishion – Town allotted to Issachar (Josh 19:20) that served as a Levitical city (Josh 21:28; the parallel in 1 Chron 6:72 has "Kedesh"). – *El-Khirba/T. Qishyon (187229), 6.5 mi. SE of Nazareth near foot of S slope of Mt. Tabor.* 105, **116**, 120

Kishon River – River that drains the Jezreel Valley from E to W. There Deborah and Barak fought the Canaanites (Judg 4:7, 13; 5:21; Ps 83:9) and Elijah slaughtered the prophets of Baal (1 Kings 18:40). – *Nahr el-Muqatta/Nahal Qishon.* **36**, 37, **38**, 40, 116, **126**, 127, **161**

Kisloth Tabor – Alternate name for Kisloth. Mentioned in boundary description of Zebulun (Josh 19:12). See Kesulloth. **116**

Kitlish – Shephelah town allotted to Judah (Josh 15:40). – *Unknown.*

Kitron – Town in Zebulun from which the Canaanites were not driven out (Judg 1:30). Possibly the same as "Kattath" (Josh 19:15). – *Unknown.* 116

Kittim – One of the descendants of Javan (Gen 10:4; 1 Chron 1:7) but also the ancient name for Cyprus (Num 24:24; Jer 2:4) and a city located on its SE coast. See Cyprus. 83

Kizil Kum Desert. 183

Knossos – See Cnossus. 17

Koa – An Unknown people and/or country (Ezek 23:23), possibly located to the NE of Babylonia.

Kode. 101

Kom Abu Bille. 103

Kom el – Hisn. 99, 103

Korazin – Town visited by Jesus and cursed by him (Matt 11:21; Luke 10:13). – *Kh. Keraze (203257), 2 mi. N of Capernaum and Sea of Galilee.* 42, **43**, **94**, **208**, 210

Kue – Place from which Solomon purchased horses (1 Kgs 10:28; 2 Chron 1:16). – *roughly the same area as Cilicia, i.e., in SE Turkey.* 144, **146**, **169**

Kufrinje. 29

Kumidi. 123

Kurkur Oasis. 66

Kursi – See Gadarenes, region of. **43**, 211, **212**

Laban (Sinai) – Israelite campsite (Deut 1:1). Possibly the same as "Libnah (Sinai)." – *Unknown.*

Lacedaemon. 234

Lachish – Town in Judean Shephelah (Josh 15:39) who's king was defeated by Joshua (Joshua 10; 12:11). It was fortified by Rehoboam (2 Chron 11:9) and the Judean king Amaziah fled there but was assassinated (2 Kgs 14:19; 2 Chron 25:27). The Assyrian monarch Sennacherib attacked and captured the town (2 Kgs 18:14, 17; 19:8; 2 Chron 32:9; Isa 36:2; 37:8; Micah 1:13) as did the Babylonian Nebuchadnezzar (Jer 34:7). Jews returning from the Babylonian exiled settled there (Neh 11:30). Mentioned frequently in extrabiblical documents. – *T. ed-Duweir/T. Lachish (135108), 29 mi. WSW from Jerusalem.* 47, **51**, 54, **86**, 88, **94**, **95**, 105, **110**, 111, **114**, **123**, 124, **130**, **151**, 158, **159**, **163**, **166**, 167, 168, 170, **170**, **177**, 179, 256

Ladder of Tyre – See Rosh HaNiqra. **187**, **188**

Lagash. 89, **94**

Lahmas – Shephelah town allotted to Judah (Josh 15:40). – *Unknown.*

Laish – Older name for Dan (Judg 18:7, 14, 27, 29) also called Leshem (Josh 19:47). See Dan. **93**, **97**, 98, **101**, 105, **109**, 112, 118, **118**, 256

Laishah – Town to N or NNE of Jerusalem along the traditional invasion route from the N (Isa 10:30). – *Unknown.*

Lakkum – Town on S boundary of Naphtali (Josh 19:33). – *Uncertain; possibly Kh. el-Mansurah/Kh. Kush (202233), 1.7 mi. SW of Sea of Galilee.* 116

Lambesis. 234

Laodicea – Greco-Roman city mentioned in letter to church at Colosse (2:1; 4:13, 15, 16) and one of the seven churches of the "province of Asia" mentioned in Revelation (1:11; 3:14). – *Eski Hissar on S bank of Lycus River in SW Turkey, 8 mi. NW of Colosse.* **218**, 231, **231**, **234**, 235, 236, **237**, 238, 239

Larissa. 234

Larsa. 89, **93**, 169, 171, 267

Lasea – The ship Paul was traveling on to Rome passed this port (Acts 27:8). – *Probably the antiquity site located 5 mi. E of Fair Havens on the S – central coast of Crete.* 233

Lasha – Town mentioned in description of boundaries of Canaanite settlement (Gen 10:19). – *Uncertain; some seek in vicinity of Sodom and Gomorrah, i.e., SE of Dead Sea because of context while others suggest Lasha may be a corruption of Laish/Leshem/Dan, in N Israel.*

Lasharon – The king of "Lasharon" was conquered by Joshua (12:18). – *Uncertain; possibly the reading of the LXX, "a king of Aphek of Sharon," is to be preferred? If an independent town, then unidentified.*

Lashon – See Lisan. 49

Latium. 233

Lebanese Beqa. 24, 33, 35

Lebanon – A region to the N of Israel known for its high, often snow-covered, mountains and fertile valleys. Mentioned 70 times in the Bible, its cedar forests were particularly famous. At times it served as a symbol of stability, strength, and fertility. – *Approximately the area of the modern country of Lebanon, excluding the narrow coastal plain (= Phoenicia), but possibly including a portion of W Syria.* 17, 18, **19**, 59, 72 – 75, **72**, 78, 84, **84**, 85, 101, 147, 175, 245

Lebanon, Mount. 72, 84

Lebanon Mountains. 74, 85

Lebaoth – Town in Negev allotted to Judah (Josh 15:32). Possibly the same as "Beth Lebaoth" (Josh 19:6) and "Beth Biri" (1 Chron 4:31). – *Unknown.*

Lebo – See Lebo Hamath. 101, 124

Lebo Hamath – A city located on the S edge of the kingdom of Hamath which is mentioned in the Bible as being on the N boundary of the land of Canaan (Num 13:21; 34:8; Ezek 47:15, 20; 48:1) and which served as the northern most point of "ideal" Israel (Josh 13:5; Judg 3:3; 1 Chron 13:5; 1 Kgs 8:65; 2 Kgs 14:25; 2 Chron 7:8; Amos 6:14). – *Lebweh (277397), in Lebanese Beqa (Valley), 45 mi. N of Damascus.* 33, **72**, **101**, 105, **106**, **106**, 121, **124**, **140**, 142, 143, **146**, 162, **178**

Lebonah – Town in Ephraim S of Shechem but N of Bethel (Judg 21:19). – *El-Lubban (173164), 10 mi. SSW of Shechem.* **32**, **44**, 45, **115**, 123, **123**, 190, **212**, 215, 241

Lebweh – See Lebo Hamath. 106

Lechaeum. 228, 229, 230

Lega, the – See Leja. 34

Lehi – Place where Samson killed 1,000 Philistines (Judg 15:9, 14, 19). – *Uncertain; possibly near Etam of Judah in Hill Country of Judah.*

Leja – See Lega, the. 34, **34**

Leontopolis. 189, **218**

Lesbos. **218**, **231**, 232

Leshem – Older name for Dan (Josh 19:47). Also called Laish. See Dan. 98, 118, **118**

Levant. 73, **84**, 86, 87, 88, 94, 95, 96, 97, 98, 99, 101, 102, 123, 125, 156, **160**, 179, 180, 181, **181**, 182, 186, 248, 254, 255, 256, 257, 265

Libnah (Judah) – One of the Shephelah towns conquered by Joshua (Joshua 10; 12:15) and allotted to Judah (15:42). It served as a Levitical center (Josh 21:13; 1 Chron 6:57). During reign of Jehoram it revolted against Judean rule (2 Kgs 8:22; 2 Chron 21:10). Sennacherib laid siege to the village (2 Kgs 19:8; Isa 37:8) and the mother of the last Judean king, Zedekiah, was from there (2 Kgs 23:32; 24:18; Jer 52:1). – *Disputed, but possibly Kh. T. el-Beida/H. Lavnin (145116), 20 mi. SW of Jerusalem, 8 mi. NE of Lachish or possibly T. Bornat/Burna (138115).* **51**, **110**, 111, **114**, **120**, 158, **159**, **166**, 168

Libnah (Sinai or Negev) – Israelite campsite during wilderness wanderings (Num 33:20). Possibly identical with "Laban" (Deut 1:1). – *Unknown.*

Libya – Mentioned in prophetic oracles (Ezek 30:5; Nahum 3:9). Jews from there were present in Jerusalem on day of Pentecost (Acts 2:10). – *In N Africa, W of Egypt; Libya.* **17**, 65, 84, 127, **175**, **233**, 234

Lisan – See Lashon. 47, 49, 50, 56, 61, 158

Lisht – See el-Lisht. 66, 68, 89

Litani/Litani River. 23, **32**, **33**, 35, **36**, 37, 72, **72**, 74, 75, 112, **112**, 116, **116**, 117, 123

Livias/Julias – See Julias (Transjordan). 211

Lod – Town built by a Benjamite (1 Chron 8:12) to which Jews returned after the Babylonian exile (Ezra 2:33; Neh 7:37; 11:35). Called Lydda in NT era. – *El-Ludd (140151), 11 mi. SE of Joppa in coastal plain.* 51, 86, **177**

Lo Debar – Place of residence of Makir who housed Mephibosheth, son of Jonathan (2 Sam 9:4, 5) and who brought provisions to David when he fled from Absalom (2 Sam 17:27). Probably reconquered by Israel during days of Jeroboam II (Amos 6:13). – *Uncertain; possibly Umm ed-Dabar (207219), 10 mi. SSE of Sea of Galilee, E of Jordan River in the Jordan valley.* **60**, 143, **143**, 155, 157

Lower Egypt – Mentioned in prophetic oracles (Isa 11:11; Jer 44:1). – *Nile delta region, N of Cairo.* 65, **66**, 67, 70, **84**, 85

Luban, Valley of. 45

Lud. 83

Lugdunensis. 202

Lugdunum. 234

Luhith – People from Moab are said to have fled in the direction of this town (Isa 15:5; Jer 48:5). – *Uncertain; possibly Kathrabba (209070), 6 mi. E of the Lissan of the Dead Sea or 6 mi. WNW of Kerak.*

Lusitania. 202

Luz – The name of the Canaanite town where Jacob met with God and renamed it Bethel (Gen 28:19; 35:6; 48:3). The town was located on the boundary between Ephraim and Benjamin (Josh 16:2; 18:13). Bethel, formerly called Luz, was captured by the house of Joseph (Judg 1:23). It is possible that Bethel (originally a worship center?) and Luz (the original town) were nearby each other but distinct (see the note in the NIV to Josh 16:2). See Bethel (Benjamin). 254

Luz (Hittite) – Town founded by survivor of Canaanite Luz in the "land of the Hittites" (Judg 1:26). – *Unknown; possibly somewhere in Lebanon/Syria.*

Lycaonia. 202, **225**, 226

Lycia – A Roman province/region on the W portion of the S coast of Asia Minor (Turkey). Paul changed ships at the port of Myra in Lycia (Acts 27:5) on his voyage to Rome. – *A rugged coastal and inland region in SW Turkey; NE of Rhodes.* 180, **181**, **184**, 185, **202**, 227, **228**, **231**, 232, 233

Lydda – Town visited by Peter where he healed Aeneas the paralytic (Acts 9:32, 35, 38). – See Lod. 190, **193**, 194, **198**, **206**, **212**, 217, **217**, 222, 234

Lydia – There are four references to Lydia and Lydians in the NIV (Isa 66:19; Jer 46:9; Ezek 27:10; 30:5). Lydia, possibly also called "Lud" (Gen 10:13; 1 Chron 1:11), was a country in W Turkey who's most famous king was Croesus. Its capital was at Sardis. However, the references to "Lydia" in the NIV may really be to "Lud" (Hebrew), and refer to an African people/nation. 173, **174**, 175, **175**, 180, **181**, 238

Lystra – A Lycaonian city visited by Paul and Barnabas (Acts 14:6, 8) on their first journey. There they healed a crippled man, but the people turned on them and Paul was stoned and left for dead. Paul revisited Lystra on his second (Acts 16:1) and probably third (18:23) missionary journeys. It seems that Timothy was from Lystra (Acts 16:1 – 5; 2 Tim 3:11). – *A mound N of Turkish village of Hatun Sarai in S central Turkey.* **225**, 226, 227, **228**, **231**, 234

Maacah – A small Aramean (Syrian) kingdom in N Israel whose territory was assigned to the half tribe of Manasseh (Josh 12:5; 13:11) although its inhabitants were not driven out (13:13). In David's day they initially were one of his adversaries (2 Sam 10:6, 8; 1 Chron 19:7) but became his vassal – *The NE portion of the Huleh Valley and the NW slopes of the Golan Heights.* 33, **34**, 121, **122**, 139, **140**, 141, 142

Maan. 55, 62

Maarath – Hill country town allotted to Judah (Josh 15:59). – *Unknown; in the region of Halhul and Beth Zur.*

Mabarot. 94

Macedonia – A Roman province in N Greece mentioned 23 times in the NT. Paul visited it on his

---

second and third missionary journeys. Paul refers to the province and a number of Macedonian Christians in his letters (Romans, 1 and 2 Corinthians, Philippians, 1 Thessalonians, 1 Timothy) – *Mountainous region located in N Greece, S Yugoslavia, and S Albania. Important cities included Philippi, Thessalonica, and Berea.* **17**, 70, 175, **175**, 179, 180, **181**, 182, 184, 186, **202**, 226, 227, **228**, 229, 231, **231**, 232, **233**, **234**

Machaerus. 47, 58, 193, 198, 200, 201, **206**, 211, **212**, 213, 221, **222**

Macbenah – Calebite settlement (1 Chron 2:49) – *Unknown; probably S of Hebron.*

Machpelah – A place near Hebron where there was a field, trees, and a cave. These were purchased by Abraham and the patriarchs (Abraham, Isaac, and Jacob) and their spouses (Sarah, Rebekah, and Leah) were buried there (Gen 23:9, 17, 19; 25:9; 49:30; 50:30) – *Haram el-Khalil in modern Hebron (160103).* 93, 94

Madai. 83, 84

Madaurus. 234

Madmannah – Negev town allotted to Judah (Josh 15:31). Probably the same as "Beth Marcaboth" in Josh 19:15 – *Kh. Tatrit (143084), 9 mi. NE of Beersheba.* 114, 166

Madmen – Town mentioned in oracle against Moab (Jer 48:2) – *Uncertain; possibly Kh. Dimna (217077), 10 mi. E of Dead Sea, in Jordan, and 7.5 mi. N of Kerak.*

Madmenah – Town N of Jerusalem on line of traditional invasion route from the N (Isa 10:31) – *Unknown; probably S of Anathoth.*

Madon – Galilean town whose king, Jobab, assisted Jabin, king of Hazor, in battle against Joshua (Josh 11:1; 12:19) – *Uncertain; possibly same as Merom. Possibly Qarn Hattin/H.Qarne Hittim (193245), 10 mi. ENE of Nazareth.* 111, **112**

Magadan – Jesus and disciples went to "vicinity of Magadan" after the feeding of the 4,000 (Matt 15:39). Parallel text (Mark 8:9) has "Dalmanutha" – *Uncertain; possibly Majdal (198247), 3 mi. NW of Tiberias on NW shore of Sea of Galilee.* **43**, 208, **208**

Magbish – A town settled by Jews after the return from the Babylonian exile (Ezra 2:30) – *Unknown.*

Magdala – Mary, one of Jesus' followers, was from this town (John 19:25; 20:1, 18; and "Magdalene" 9 times in NT) – *Uncertain; possibly same as Magadan/Dalmanutha of the NT and "Taricheae" of Josephus.* 43, **43**, 208, **208**

Maghara. 104

Magnesia. 182, 188, **234**, 236

Magog – A place or a people ruled by "Gog" (Ezek 39:3, 6) – *Uncertain; possibly Magog simply means "land of Gog," although some identify it with the ancient kingdom of Lydia in W Turkey. In Revelation (20:8), Gog and Magog, represent the heathen opponents of the Messiah.* **83**

Mahanaim – Place where Jacob met angel of God on return to Canaan (Gen 32:2). Was located on border between Gad and Manasseh (Josh 13:26, 30) but served as a Levitical city in Gad (Josh 21:38; 1 Chron 6:80). There, Abner set up Saul's son, Ish-bosheth, as king (2 Sam 2:8, 12, 29). David fled to the town when Absalom revolted (2 Sam 17:24, 27; 19:32; 1 Kgs 2:8). Solomon used it as one of his district centers (1 Kgs 4:14) – *Uncertain; possibly edh-dhahab el-Gharbi (214177), 7 mi. E of Jordan River, N of Jabbok (so in this Atlas), or possibly T. er-Reheil (228177), 15.5 mi. E of Jordan River, N of Jabbok.* **58**, 59, 60, **97**, 108, **109**, 119, **119**, 138, **140**, 142, **143**, 148, 155

Mahaneh Dan – The "Camp of Dan." A campsite at which 600 Danites camped before moving N (Judg 18:12) – *Probably to the W of Kiriath Jearim. In Samson's day, it was evidently located between Zorah and Eshtaol (Judg 13:25) – Uncertain; in N Shephelah in Zorah, Eshtaol, Kiriath Jearim area.* 130

Maka. 175

Makaz – One of several towns which formed the second administrative district of Solomon (1 Kgs 4:9) – *Uncertain; but probably in old Danite region in N Shephelah and/or N Philistia.* 148

Maked. 192

Makheloth – Israelite wilderness camp-site between Haradah and Tahath (Num 33:25 – 26) – *Unknown.*

Makhtesh Gadol – See also Hamaktesh Hagadol. **62**

Makhtesh Qatan. 62

Maktesh Ramon. 55, 62

Makkedah – Five Amorite kings hid in cave of Makkedah and were captured by Joshua (Joshua 10). This Shephelah town was captured (Josh 12:16) and allotted to Judah (15:41) – *Kh. el-Kum (146104), 8.5 mi. W of Hebron.* **110,** 111, **114**

Malatha. 193, 200, 201

Malatya. 72

Malatya Mountains. 72, 72, 76

Malta – Island where Paul was shipwrecked on way to Rome (Acts 28:1), called Melita by Greeks and Romans – *Malta, in Mediterranean Sea, 58 mi. S of Sicily. Size: 18 x 9 mi.* 233, **233**

Mampsis. 32

Mamre – Place near Hebron where Abraham and Isaac camped (Gen 13:18; 18:1; 35:27). The "cave of Machpelah" was "near" (Gen 23:17, 19; 25:9; 49:30; 50:13) – *Uncertain; possibly Ramat el-Khalil (160107), 2.5 mi. N of Hebron.* **97**

Manach – Hill Country town allotted to Judah (Josh 15:59b; LXX) – *Possibly same as Manahath.*

Manahath – Town to which some Benjamites from Geba were carried captive (1 Chron 8:6 cf. 2:54). Maybe the same as Manach – *El-Malhah/Mana-hat (167128), 4 mi. SW of Jerusalem.* **166**

Manasseh – Son of Joseph; one of the tribes of Israel. Mentioned over 141 times in OT. Tribe settled in hills S of Jezreel Valley, but N of Ephraim and in Gilead to the E of the Jordan River. **22, 23,** 24, 40, 41, 43, **44,** 45, 46, 48, **58,** 59, **60,** 115, **115,** 116, 118, 119, **119, 120,** 128, 137, 141, 149, 162, 164, 166, 169, **178,** 254, 256

Manasseh, Hill Country of. 16, 25, 41, **44,** 45, 46, 61, 91, 102, 115, 121, **122**

Maon – Town in Hill Country allotted to Judah (Josh 15:55). Nabal, who was antagonistic toward David, was from Maon (1 Sam 25:2). After his death, David married his widow Abigail – *Kh. Main (162090), 8 mi. SSE of Hebron.* 47, **47,** 49, **114,** 135, **135,** 136, **166**

Maon, Desert/Wilderness of – A wilderness area to the E and SE of Maon where David hid from Saul (1 Sam 23:24, 25; 25:1). **135,** 136

Maracanda. 183

Marah – Oasis, about three-days journey from Egypt. There, Moses tossed a piece of wood into the bitter ("Marah") water and it became sweet (Exod 15:23, Num 33:8, 9) – *Uncertain; possibly Bir Mara(h) 10 mi. E of Suez in W central Sinai or alternatively Ain Hawarah, 47 mi. SE of Suez.* **104**

Maralah – Town on W border of Zebulun (Josh 19:11) – *Uncertain; possibly T. el-Ghaltah/T. Reala (166232), on NW edge of Jezreel Valley, 7 mi. NNW of Megiddo, or possibly T. Thorah/T. Shor (166228), 5 mi. NNW of Megiddo.* **116**

Marathon. 175, 176, 180, **181**

Mare Caspium – See Caspian Sea. **202**

Mare Internum – See Mediterranean Sea. **202**

Mareshah – Shephelah town allotted to Judah (Josh 15:44) and fortified by Rehoboam (2 Chron 11:8). Asa, the king of Judah, fought Zerah the Ethiopian in the vicinity (2 Chron 14:9, 10). During the days of Jehoshaphat, Dodavahu of Mareshah prophesied (2 Chron 20:37) and Mareshah is mentioned in the prophecy of Micah (1:15). During the Hellenistic and Roman periods it was an important center known as Marisa – *T. Sandahannah/T. Maresha (140111), 13 mi. NW of Hebron in Shephelah.* 47, **51,** 54, **114,** 151, 158, **159, 166,** 177

Margiana. 175

Mari. 17, 18, **72,** 73, **77,** 78, **84,** 90, **90, 93,** 94, **97,** 98, 106, 256

Marisa – See Mareshah. 54, 185, 186, **186, 187,** 190, 192, **193,** 195

Marmara, Sea of. 227, 228, 231, 237

Maroth – Mentioned in prophecy of Micah as experiencing pain (1:12) – *Uncertain; evidently in Judean Shephelah.*

Masada. 47, 49, 50, 51, **55, 193, 198,** 199, **200,** 201, **206,** 217, 221, **222,** 223

Mashal – Levitical town in Asher (1 Chron 6:74). See Mishal.

Masrekah – Royal city of Edom associated with Samlah (Gen 36:36; 1 Chron 1:47) – *Unknown.*

Massaga. 183

Massah – Place near Horeb and Rephidim where Israelites complained regarding the lack of water. There Moses "struck" the rock and water came forth (Exod 17:7). This incident is referred to in Deut 6:16; 9:22; 33:8 and Psalm 95:8 – *Unknown; somewhere in Sinai.*

Masuate. 163

Mattanah – Israelite campsite on journey to the plains of Moab (Num 21:18, 19) – *Uncertain; probably in Jordan, E of Dead Sea and N of Arnon Gorge.*

Mauretania. 202

Mayan Barukh. 94

Mecca. 27, 57, 66, 69

Meconah – Town settled by Jews after the Babylonian exile (Neh 11:28) – *Uncertain; possibly in the NW Negev or in the Shephelah of Judah.* 179

Medeba – A Moabite town captured from Sihon (Num 21:30) and allotted to Rebuen (Josh 13:9, 16). Joab laid siege to the town and captured it (1 Chron 19:7). Isaiah mentions it in an oracle against Moab (15:2). It was fought over by Moab and Israel on a number of occasions – *Madeba (225142), in Jordan, 20 mi. S of Amman, 12.5 mi. E of Dead Sea.* **22, 47, 58, 62, 63,** 119, 139, **140,** 141, **143,** 156, **159, 163, 193,** 195, **198, 212,** 256

Medes. 84, 169, 173, **174**

Media – Ancient homeland of the Medes located to the SW of the Caspian Sea but NE of the Zagros Mountains. Media/Medes/Mede are mentioned 20 times in the OT, exclusively at the end of the OT period. The Persians took control of the Medes yet their king was called the "king of the Medes and Persians." 160, 164, 169, **169, 171,** 175, **175, 218**

Medina. 57, 66

Mediterranean Sea – See also Great Sea. 16, **17,** 21, **22,** 23, 24, 25, **26,** 28, 29, 31, **32, 34,** 35, **36,** 37, **38,** 40, 43, **44,** 46, 47, 49, 50, 51, **51,** 54, 55, 56, **62,** 64, 65, **66,** 67, 68, 69, **69, 72,** 73, 74, 75, 76, 77, **83, 84,** 85, 86, **86,** 88, **90, 91, 93,** 94, **95,** 96, **97, 99,** 101, 102, 103, 104, 105, 106, 106, **108,** 110, **112,** 113, 114, **115,** 116, **116,** 117, 120, 121, **122, 123,** 124, **129,** 130, **132, 135,** 140, 142, **143,** 144, 148, 153, 154, **155,** 156, **157,** 158, **159, 160, 161,** 163, 164, 166, 168, 169, **170, 171,** 174, **175,** 177, **177,** 181, **181, 182,** 184, 185, 186, **187,** 188, **193,** 195, 197, **198,** 199, **200,** 201, **206, 208,** 209, 210, **212,** 217, **217, 218, 219, 220, 222,** 224, **225, 228, 229,** 231, **231, 233, 233,** 234, **234,** 235, **237,** 254, 265

Megiddo – Major Canaanite city captured by Joshua (12:21) and allotted to Manasseh (17:11; 1 Chron 7:29). However, Manasseh was not able to take possession of the city (Judg 1:27). Evidently it came under Israelite control during the rule of David for it was a district capital of Solomon (1 Kgs 4:12) and he fortified it (9:15). The Judean king Ahaziah died at Megiddo (2 Kgs 9:27) as did the later king Josiah (2 Kgs 23:29, 30; 2 Chron 35:22). It is also mentioned in an oracle of Zechariah (12:11) – *T. el-Mutesellim/T. Megiddo (167221), on SW edge of Jezreel Valley, guarding an important pass through the Carmel Range.* **17, 26, 32,** 33, **34, 36, 38,** 39, 40, 41, **44, 84, 86,** 87, 88, 92, **93, 94, 95,** 97, **97,** 101, **101,** 102, **105, 106, 108,** 112, 115, **115,** 116, **122, 123,** 124, **126,** 127, **130, 135,** 139, **140, 142,** 143, **143, 145, 146, 148, 153, 155, 157, 163,** 169, **169, 170, 171,** 176, **178,** 259

Megiddo, Mountain of – See Armegeddon. 40

Megiddo, waters of – A wadi in the vicinity of Megiddo and Taanach where the Canaanite kings traditionally divided the spoil taken in war (Judg 5:19). 127

Me Jarkon – A place allotted to the tribe of Dan (Josh 19:46). Translation: "the waters of the Jarkon" – *Uncertain and the text is difficult. Reference may be to the Nahr el-Auja/N. Yarkon*

which begins near Aphek and flows into the Mediterranean Sea in N Tel Aviv (adopted in this Atlas) or possibly to W. el – Auja/N. Aijalon. 118, **118**

Melos. 218

Memphis – Capital of Egypt during the Old Kingdom (3 millennium BC) and important city and burial grounds during most periods. Mentioned in prophetic literature of the OT, usually in a negative light (Isa 19:13; Jer 2:16; Jer 46:14; Ezek 30:13, 16). Some exiles from Judah settled there (Jer 44:1) – *Area in and around Mit Rahinah on W bank of Nile, 13 mi. S of Cairo.* **17, 66, 84,** 85, 86, 89, **93, 99, 103, 104, 146, 171,** 173, **174,** 181, **182, 218**

Menahemya. 94

Mephaath – Transjordanian town allotted to Reuben (Josh 13:18) and which served as a Levitical city (Josh 21:37; 1 Chron 6:79). Jeremiah mentions it in an oracle against Moab (48:21) – *Uncertain; possibly T. Jawah (239140), 7 mi. S of Amman in Jordan.* **119,** 120

Merathaim, land of – Symbolic name for Babylonia (Jer 50:21) meaning "double rebellion" and which seems to be a word play on Bablyonian "bitter river."

Meribah – Place, near rock at Horeb (= Sinai?), where Israelites grumbled and Moses struck the rock so that water came out (Exod 17:7; Ps 95:8) – *Unknown; but in Sinai.*

Meribah, waters of – Place near Kadesh Barnea where Moses struck the rock rather than speaking to it as God had commanded (Num 20:13, 24; 27:14; Deut 33:8; Ps 81:7; 106:32). – *In vicinity of Kadesh Barnea.*

Meribah Kadesh – Alternate name for "Meribah, waters of" (Deut 32:51; Ezek 47:19; 48:28). **178**

Merom. 163

Merom, Waters of – Place in (Upper?) Galilee where Joshua defeated the Canaanite forces led by the king of Hazor (Josh 11:5, 7) – *Uncertain; possibly T. el-Khirbeh (190275), in S Lebanon, 19 mi. SE of Tyre, but Qarn Hattin/H.Qarne Hittim (193245), 10 mi. ENE of Nazareth, seems more probable.* 111, **112**

Meron. 94

Meron, Mount. 23, 25, **30, 36,** 37, **43, 208**

Meronoth – A town from which Jadon came to help Nehemiah build the wall of Jerusalem (Neh 3:7). One of David's servants was evidently from this place as well (1 Chron 27:30) – *Uncertain; probably near Gibeah, N of Jerusalem in Benjamin.*

Meroth. 200

Meroz – Village cursed because of failure to assist forces of Deborah against Canaanites (Judg 5:23) – *Uncertain; probably in vicinity of Jezreel Valley.*

Mesad Hasidim. 212

Mesha – On of the limits (town? country? region?) of where the descendants of Joktan lived (Gen 10:30) – *Unknown; possibly in Arabia.*

Meshech – A person and tribal name of a group that settled in the mountainous area SE of the Black Sea. Known in the biblical text as remote and barbaric (Psalm 120:5; Ezek 27:13; 32:26; 38:2; 39:1). **83**

Meshed (Iran). 175

Meshed (Israel). 86

Mesopotamia. 18, 73, 76 – 79, **77,** 82, 84, **84,** 85, 86, 88, 89, 98, 101, 125, 144, 150, 151, 169, 172, 174, **174,** 180, 182, 184, 185, **218,** 224, **233,** 255, 256, 260, 266, 267

Metheg Ammah – "Bridle of the mother city." Taken by David from Philistines (2 Sam 8:1). The parallel passage (1 Chron 18:1) has Gath – *Unknown; if it is not an alternate name for Gath.*

Metulla. 112

Mezad Zohar. 47, 49

Micmash – Village in tribal territory of Benjamin where Philistines mustered their troops against Israel. Jonathan and his armor bearer attacked the Philistine camp there (1 Sam 13 – 14; mentioned 7 times). It was along the traditional northern invasion route to Jerusalem (Isa 10:28). Jews returning from the Babylonian exile settled there (Ezra 2:27; Neh 7:31; 11:31) – *Kh. el-Hara el-*

Fawqa (176142), 7.5 mi. NNE of Jerusalem. 23, **32, 47,** 49, 131, **132,** 133, **154, 166, 177, 190,** 257

Micmethath – A town on the boundary between Ephraim and Manasseh (Josh 16:6; 17:7) – *Uncertain; possibly Kh. Makhneh el-Foqa (175176), 2.5 mi. SSW of Shechem.* **44, 115**

Middin – Town/fort in Wilderness of Judah (Josh 15:61) – *Uncertain; possibly Kh. Abu Tabaq (188127), 10.5 mi. ESE of Jerusalem.* **114, 166**

Midian – Midian was a descendant of Abraham and Keturah and the ancestor of an Arabian tribe that bore his name. Midian/Midianite(s) is mentioned 57 times in the OT and once in the NT (Acts 7:29). This nomadic people seems to have had its center in NW Arabia, E of the Gulf of Aqaba/Elath, but at times entered the Jezreel Valley (story of Gideon), S Transjordan, and maybe even into the Negev and N Sinai. **66,** 69, **69,** 103, 104, **104,** 141

Migdal Eder – Place S of Bethlehem but N of Hebron where Jacob/Israel camped (Gen 35:21) – *Unknown.*

Migdal El – Town in allotted to Naphtali (Josh 19:38) – *Unknown; but probably in Upper Galilee.*

Migdal Gad – Shephelah town allotted to Judah (Josh 15:37) – *Uncertain; possibly Kh. el-Mejdeleh/H. Migdal Gad (140105), 12 mi. W of Hebron.* **114, 166**

Migdol – A town in Lower Egypt where Jews lived (Jer 44:1; 46:14). Ezekiel refers to it as the N(E) extremity of Egypt (Ezek 29:10; 30:6) – *In the vicinity of Tell el-Her, 12.5 mi. NE of Qantar, E of Suez Canal in NW Sinai.* 103, 104, **160, 173**

Migdol (Exodus) – Place (meaning "fort") near which the Israelites camped when fleeing Egypt (Exod 14:2; Num 33:7) – *Uncertain; there were numerous "forts" along Egypt's NE frontier.* 103, 104

Migron – Place to N of Jerusalem where Saul camped with his men, possibly near Gibeah (1 Sam 14:2). Migron is also mentioned as being on the invasion route from the N into Jerusalem, but it seems to be N of Micmash (Isa 10:28) – *Unknown; there may have actually been two Migrons.*

Miletus – A prominent harbor city in SW Asia Minor where Paul, on his third missionary journey, stopped and met with elders from Ephesus, as he made his way to Jerusalem (Acts 20:15, 17; also 2 Tim 4:20) – *Site in SW Turkey; was on a peninsula but now inland. Near town of Balat.* **17, 175,** 176, **231,** 232, 239

Minnith – The judge Jephthah captured 20 cities in its vicinity as he subdued the Ammonites (Judg 11:33). It was also famous for its wheat (Ezek 27:17) – *Unknown; in Transjordan.* 129

Mishal – A town allotted to Asher (Josh 19:26) that also served as a Levitical city (Josh 21:30); called "Mashal" in 1 Chron 6:74) – *Uncertain; possibly T. Kisah/T. Kison (164253), 5 mi. SE of Acco.* **116, 120**

Mishor – See Moab, Tableland of. **22, 23,** 27, **47, 58,** 59, 61, 62, **62,** 119, **119, 122, 126,** 158, **159,** 265

Misrephoth Maim – Joshua pursued defeated Canaanites in this direction (Josh 11:8) and extended Israelite control N to this point (13:6) – *Uncertain; N of Galilee in the vicinity of Sidon; possibly in area of Litani River (Lebanon).* 112, **112**

Mitanni. 101, 102, 123

Mithcah – Israelite campsite during wilderness wanderings between Terah and Hashmonah (Num 33:28, 29) – *Unknown.*

Mitla Pass. 69

Mit Rahina. 99, **103**

Mitylene – Port on island of Lesbos where Paul's ship spent the night as he journeyed to Jerusalem at the end of his 3rd journey (Acts 20:14) – *Mytilene, on E shore of island of Lesbos, Greek island off W coast of Turkey.* **231,** 232

Mitzpe Ramon. 55

Mizar, Mount – A mountain mentioned in connection with Mount Hermon in Psalm 42:6 – *Uncertain; possibly in region of Mt. Hermon, although some think it may be a reference to the "littleness" of Mt. Zion in contrast to the massiveness of Mt. Hermon.*

Mizpah (Benjamin) – Town allotted to Benjamin (Josh 18:26). There the Israelites gathered in preparation for the war against Benjamin on account of the atrocities committed against the Levite's concubine (Judg 20 – 21). It was an important cult center during the days of Samuel (1 Samuel 7 and 10:17). It was fortified by Asa, the Judean king (1 Kgs 15:22; 2 Chron 16:6) and later it became the administrative center of Gedaliah, whom the Babylonians had set up as the governor of defeated Judah (2 Kgs 25:23, 25; Jer 40 – 41). Men from Mizpah helped Nehemiah rebuild the wall of Jerusalem after the exile (Neh 3) – *T. en-Nasbeh (170143), 7.5 mi. NNW of Jerusalem.* 46, **47, 115, 123, 126,** 131, **131, 132, 132,** 133, **154, 166, 170,** 172, **173, 177,** 179, **190,** 192, **241**

Mizpah (Gilead) – Home of the judge Jephthah and place where Israel assembled before battle with the Ammonites (Judg 10:17; 11:11, 29, 34). Also mentioned in Hosea 5:1 – *Uncertain; possibly Kh. Jalad (223169), 15 mi. NW of Amman in Jordan; S of Jabbok, 14 mi. in Gilead.* **126,** 129

Mizpah (Jacob) – Place in Gilead where Jacob and Laban sealed a covenant of friendship (Gen 31:49) – *Unknown; probably near Jabbok River.*

Mizpah (Judah) – Village allotted to Judah (Josh 15:38) – *Unknown but in Judean Shephelah, probably near Lachish.*

Mizpah (Moab) – Place of refuge where David sent his father and mother (1 Sam 22:3) – *Unknown; E of Dead Sea in Moab.*

Mizpah, region/Valley of – Hivites from this region fought against Joshua (11:3) and after their defeat, Joshua pursued them N in this direction (11:8) – *Uncertain; near the foot of Mt. Hermon; possibly in Marj Ayyun region of S Lebanon.* 112, **112**

Mizraim. 83, 84

Moab – A personal, tribal and geographical name used (in various forms) 194 times in the OT. The geographical heartland of Moab was E of the Dead Sea between the Zered and Arnon Rivers. During periods of strength it would expand N into the plateau region N of the Arnon. **22, 23,** 27, **47,** 49, 50, **58,** 61, 62, **62,** 63, 94, 107, 108, **108, 110, 114,** 119, **119,** 125, 133, 135, **135, 140,** 141, 144, **145, 155,** 156, 158, **159, 160, 161, 163,** 170, **170**

Moab, Mountains of. 50

Moab, Plains of. 108, **108,** 110

Moab, Tableland/Plateau of (Mishor) – Area in N Moab stretching from the Arnon Gorge in the S to Heshbon in the N; called the "plateau" in the NIV (Deut 3:10; 4:43; Josh 13:9, 16, 17, 21; 20:8; Jer 48:8, 21). In Hebrew called the "Mishor." 59, 61, **119,** 121, **122, 126,** 129

Moabitis. 186, **187,** 193

Modiin. 190, 191, 194

Moladah – A Negev village allotted to Judah (Josh 15:26) as well as to Simeon (Josh 19:2; 1 Chron 4:28). After the Babylonian exile Jews lived there (Neh 11:26) – *Uncertain; possibly Kh. el-Waten/H. Yittan (142074), 5 mi. E of Beersheba.* **114,** 166, **177,** 179

Monastery of St. Catherine. 69, 104, **104**

Moreh – Place near Shechem where a great oak tree was located. Abram camped there when he first entered Canaan (Gen 12:6) and near there the Israelites were to read the blessings and the curses (Deut 11:30). Possibly the "oak at Shechem" where Jacob hid the teraphim is the same place (Gen 35:4) – *Unknown; near Shechem.* 91

Moreh, Hill/Mount of – Gideon attacked the Midianites who were camped near the hill of Moreh (Judg 7:1) – *Jebel Nabi Dahi/Mt. Moreh, at E end of Jezreel Valley.* 36, **38,** 39, **43, 126,** 128, **135,** 136, **208**

Moresheth – Home of Micah the prophet (Micah 1:1; Jer 26:18) – *See Moresheth Gath.* 54

Moresheth Gath – The full name of the home of Micah the prophet mentioned in Micah 1:14 – *Uncertain; possibly T. el-Judeideh/T. Goded (141115), in Shephelah, 6 mi. NE of Lachish.* **47, 51, 151**

Moriah – Abraham took Isaac to the "region of Moriah" to sacrifice him (Gen 22:2). Later Solomon built the temple on Mount Moriah (2 Chron 3:1) – *The temple mount in Jerusalem now occupied by the Dome of the Rock (172131).* **91,** 93, **242,** 243, **246**

Moserah – Place where Aaron died (Deut 10:6). *Evidently near or identical to Mt. Hor (Num 20 passim).*

Moseroth – Israelite campsite between Hashmonah and Bene Jaakan (Num 33:30, 31) – *Unknown; in Sinai or S Negev. Some suggest identical to Moserah.*

Mozah – Town allotted to Benjamin (Josh 18:26) – *Qalunyah/Mevasseret Ziyyon (165134), 5 mi. WNW of Jerusalem.* **115, 166**

Murat. 82

Mycenae. 17

Myra – City on the S coast of Turkey where Paul and escorts changed to an Alexandrian ship bound for Italy (Acts 27:5) – *Modern Demre.* 233, **233**

Mysia – Paul passed through Mysia on way to Troas on his second journey (Acts 16:7, 8) – *Territory in NW Turkey stretching to Aegean Sea, the Hellespont, and Propontis.* 180, **181,** 227, **228**

Naamah – A town allotted to the tribe of Judah in the Shephelah (Josh 15:41). – *Unknown.*

Naarah – Town on SE border of Ephraim between Ataroth and Jericho (Josh 16:7) called Naaran in 1 Chron 7:28 and said to be in Ephraim. – *T. el-Jisr (190144), 1.8 mi. NW of Jericho.*

Naaran – See Naarah. **115, 154**

Nabatean Kingdom/Nabateans. 28, 57, 64, **184, 193,** 195, 196, **198, 200,** 201, **206, 217,** 218, **219, 220, 222**

Nablus. 30, 258

Nacon, threshing floor of – See Kidon, threshing floor of.

Nafud Desert. 16, 17

Nahal Aijalon – See Aijalon, Valley of and Me Jarkon. **44, 52**

Nahalal – Town allotted to Zebulun (Josh 19:15) which served as a Levitical city (21:35). Called Nahalol in Judg 1:30, where Canaanites lived. – *Unknown; but in or just N of Jezreel Valley.* 116

Nahal Arugot. **47,** 49, 158, **159**

Nahal Beersheba. 55, 201

Nahal Besor – See Besor Ravine. 21, **23, 32,** 51, **51,** 52, 54, **55,** 137

Nahal Gerar – See Gerar, Valley of. 51, **51,** 52, 94

Nahal Harod – See Harod Valley. **38**

Nahal Hebron. 51, 54

Nahaliel – One of last campsites of Israelites N of Arnon Gorge E of Dead Sea but S of Bamoth (Num 21:19). – *Unknown.*

Nahal Jabneel – See Jabneel Valley. **38,** 39, **43**

Nahal Kanah – See Kanah Ravine. 44

Nahal Kesalon. 134

Nahal Kishon – See Kishon River. 38

Nahal Lachish. 51, 134

Nahalol – See Nahalal. 116

Nahal Paran. 55, 56, **62, 104**

Nahal Pattish. 51

Nahal Raqqad – See Raqqad. **32, 33, 34, 35,** 36

Nahal Shechem. 44, 46, 154

Nahal Shiloh. 44, 46

Nahal Shiqma. 51

Nahal Sorek – See Sorek Valley. **51,** 113, **134**

Nahal Tabor. **38,** 39

Nahal Tannanim. 21, **22, 23,** 43, **44**

Nahal Yarkon – See Jarkon River and Me Jarkon. 21, **32,** 51, **52**

Nahal Zin. 23, 25, **32,** 55, 56, **62**

Nahal Zohar. 49

Naharin. 101

Nahariya. 95

Nahor, town of – Could merely be the town in which Abraham's brother Nahor lived (Gen 24:10), but a town named Nahor, near the Habor River in N Syria, is mentioned in the Ebla and Mari texts. – *Unknown.* 76

Nahar Kabir. 72, 73, 74, 75

Nain – Town where Jesus raised the son of a widow from the dead (Luke 7:11). – *Nein (183226), 6 mi. SE of Nazareth on lower N slope of Mt. Moreh.* **38,** 39, **43, 208**

Naioth – A settlement in or near Ramah where Samuel was located and to which David fled and where Saul prophesied (1 Sam 19:18, 19, 22, 23; 20:1). – *Uncertain; near Ramah.* 134

Naphoth – Mentioned in connection with Manasseh's tribal allotment, possibly refering back to "Dor" which is "third in the list" (Josh 17:11). See Naphoth Dor.

Naphoth Dor – Possible translation: "forest in the plain of Dor." Thus descriptive of the region around the coastal city of Dor. King fought against Joshua (11:2; 12:23) and region was the 4th Solomonic district (1 Kgs 4:11). – *Narrow coastal strip E of Dor between Mediterranean Sea and Mt. Carmel.* 148

Naphtali – Son of Jacob and name of tribe; mentioned fifty times in OT and three times in NT. On occasion designates the territory allotted to this tribe to the W and NW of the Sea of Galilee. 37, 39, 42, 116, **116**, 117, **120**, 127, 128, 147, **148**, 164, 169, **178**

Naphtali, hill country of – Designation of the hilly/ mountainous territory of Galilee where the tribe of Naphtali settled and where Kedesh in Galilee was located (Josh 20:7).

Naqb Ishtar. 62, 64

Narbonensis. 202, 234

Nazareth – A small insignificant town (John 1:46) in lower Galilee where Gabriel appeared to Mary (Luke 1:26). Mary and Joseph travelled from there to Bethlehem (Luke 2:4) where Jesus was born. After the flight into Egypt they returned to Nazareth to live (Matt 2:23; Luke 2:4, 39, 51). Jesus eventually left the city for a wider ministry (Matt 4:13; Mark 1:9) but at one point he was rejected by the inhabitants of the town (Luke 4:16). He was called a "prophet from Nazareth in Galilee" (Matt 21:11), and the phrase "Jesus (Christ) of Nazareth" is used 17 times in the NT. – *En-Nasira/Nasrat (178234), today a large Arab city in the S hills of Lower Galilee, 16 mi. W of S tip of Sea of Galilee.* 23, 25, **26**, **36**, 38, **38**, 39, **43**, **94**, 205, **206**, 207, **208**, 212, 217

Neah – Town on the N border of Zebulun between Rimmon and Hannathon (Josh 19:13). – *Possibly Tel el-Wawiyat (178244).*

Neapolis (Greece) – Aegean seaport of Philippi where Paul first set foot on European soil on his second journey (Acts 16:11). Also probably passed through the port twice on his third journey (20:1, 6). – *Kavalla, in N Greece, 10 mi. SE of Philippi.* 227, **228**

Neapolis (Italy). 218, 234

Neapolis (Palestine) – See Nablus and Shechem. **205**, 258

Neara. 206

Neballat – Town near the E edge of the Coastal Plain where Benjamites settled after the Babylonian Exile (Neh 11:34). – *Beit Nabala/H. Nevallat (146154), 13 mi. ESE of Joppa.* 177

Nebi Samwil – See Gibeah of God and Gibeon, high place at. 133

Nebo (of Judah?) – Residents of this town participated in the first return from exile (Ezra 2:29; Neh 7:33) and some later repented of the sin of intermarriage (Ezra 10:43). – *Uncertain; possibly Nuba (153112), 7 mi. NW of Hebron in Shephelah.* 177

Nebo (Moab) – Transjordanian town allotted to the tribe of Reuben (Num 32:3, 38: 33:47; 1 Chron 5:8). Later came under control of the Moabites and mentioned in prophetic oracles concerning them (Isa 15:2; Jer 48:1, 22). – *Kh. Ayun Musa (220131), 9 mi. E of N end of Dead Sea.* 156

Nebo, Mount – "Mount" near Nebo where Moses died (Deut 32:49; 34:1). – *Ras es-Saighah (220130), 10 mi. E of N end of Dead Sea. See also Pisgah.* **23**, 24, **58**, 108, **108**, 110

Negev – The geographical region on the S boundary of Judah somewhat hourglass in shape – placed on its side – with its center at Beersheba. Mentioned 38 times in OT. Some of its subdistricts are also mentioned: Negev of Jerahmeel, Negev of the Kenites, Negev of Judah (1 Sam 27:10), Negev of the Kerethites, and the Negev of Caleb

(1 Sam 30:14). Agriculturally, it was a marginal area because of the low amounts of rainfall in the region (8 to 12 inches). 16, **32**, 33, 47, 51, **47**, **51**, 54, 55, **55**, 56, 57, **69**, **91**, 92, 93, **97**, 102, **106**, 107, **108**, 114, **114**, 136, 137, 144, **145**, 152, **153**, 156, 157, 160, **166**, 179, 214, 256, 265

Negev Basin. 23, **23**, 25, 47, **47**, **51**, **55**, 69

Negev Highlands. 23, **23**, 25, 55, **55**, 56, 92, 94, **94**, 106, 107, **108**, 144

Negev of Caleb – Area of Negev where Calebites lived and which was raided by Amalekites (1 Sam 30:14). – *In Hill Country of Judah, S of Hebron but NE of Beersheba.* **135**, 137

Negev of Jerahmeel – Area of Negev where Jerahmeelites lived and which was defended by David (1 Sam 27:10) – *Probably in E Negev Basin.* **135**, 136

Negev of Judah – Area of Negev defended by David (1 Sam 27:10) and where Joab completed the census (2 Sam 24:7) – *In S Judah, in vicinity of Beersheba.* **135**, 136, 142, **142**

Negev of Kenites – Area of Negev where Kenites lived and which was defended by David (1 Sam 27:10) – *Probably in NE Negev Basin near Arad.* **135**, 136

Negev of the Kerethites – Area in W Negev raided by the Amalekites (1 Sam 30:14). The Kerethites were evidently an Aegean people who settled there (cf. Zeph 2:6). – *Probably S Philistine Plain area and W Negev region to the E and NE of Gaza.* **135**, 137

Nehardea. 218

Neiel – Town allotted to Asher (Josh 19:27). – *Kh. Yanin/H. Yaanin (171255), 8.5 ESE of Acco.* 116

Nephtoah, Waters of – Site on the border between Judah and Benjamin (Josh 15:9; 18:15). – *Lifta/ Me-Neftoah (168133), 3 mi. NW of Jerusalem.* 46, **47**

Netaim – Settlement where royal Judean potters lived (1 Chron 4:23). – *Unknown.*

Netophah – Inhabitants of this town participated in the first return from the Babylonian Exile (Ezra 2:22; Neh 7:26). A number of people are called "Netophathite(s)," 11 times in OT. – *Uncertain; possibly Kh. Bedd Faluh (171141), 8 mi. S of Jerusalem.* 177

Nezib – Town allotted to Judah in a Shephelah district (Josh 15:43). – *Kh. Beit Nesib esh-Sharqiyeh (151110), 7 mi. NW of Hebron.* **114**, 166

Nibshan – Town in wilderness district of Judah (Josh 15:62). – *Uncertain; possibly Kh. el-Maqari (186123), 10 mi. SE of Jerusalem.* **114**, 166

Nicaea. 227, 228

Nicomedia. 234

Nicopolis (Greece) – Paul intended to spend the winter there (Titus 3:12). There are several cities with this name, here the Nicopolis in Epirus seems to be meant. – *Nikopolis, NNE of Preveza in W Greece.* 234

Nicopolis (Palestine) – See Emmaus. **206**, 215

Nile – River beginning in central Africa which flows N through Egypt; 4,145 mi. long. Mentioned 31 times in OT, especially in connection with Israelites in Egypt and in prophetic oracles of Isaiah, Jeremiah, Ezekiel, Amos, Nahum, and Zechariah. 16, **17**, 18, 65, **66**, 67, 68, 70, **84**, 85, 87, 92, **93**, 97, 98, **99**, 100, 101, 102, **103**, **104**, 105, 142, **146**, **160**, **171**, 172, **173**, **174**, 175, **175**, 181, **182**, **184**, 199, 218

Nimrah – Town allotted to Gad (Num 32:3). See Beth Nimrah.

Nimrim, waters of – Place mentioned in oracles against Moab (Isa 15:6; Jer 48:34). – *Uncertain; possibly Wadi Gadira that flows into Gor Numera E of the Dead Sea.*

Nimrod, land of – Land associated with the heroic figure Nimrod (Micah 5:6; cf. Gen 10:8 – 11). – *Evidently area of ancient Babylonia and Assyria, i.e., in modern Iraq.*

Nimrud. 77, 162

Nineveh – Great administrative center in Assyria mentioned 20 times in OT. Jonah prophesied to the inhabitants of the city (Jonah passim) and it figures prominently in the oracles of Nahum (passim) as well as in Zephaniah 2:13. Jesus referred

to its inhabitants at the time of Jonah (Matt 12:41; Luke 11:32). – *The mounds of Quyunjiq and Nebi Yunus, on E bank of Tigris River, opposite modern Mosul, in N Iraq.* **72**, 76, **77**, 78, **83**, **90**, 93, **94**, 146, **160**, 167, 168, 169, **169**, **171**, 174

Nippur. 77, 84, 85, 89, **90**, **94**, **169**, **171**, 173, **174**, 218

Nisibis. 160, 182

No-Amon. 160, 173

Nob – Town to which David fled from Saul (1 Sam 21:1). Evidently the tabernacle, but not the ark, was there at the time. Saul slaughtered the priests that served there in revenge for the assistance that they gave to David (1 Sam 22:9, 11, 19). Attackers of Jerusalem passed it as they approached the city from the N (Isa 10:32). Jews settled there after the Babylonian Exile (Neh 11:32). – *Uncertain; possibly el-Isawiyah (173134), 1.5 mi. NE of Jerusalem.* **135**, **166**, 177

Nobah – Nobah captured Kenath and renamed it after himself (Num 32:42). Mentioned in connection with Gideon's pursuit of the Midianites (Judg 8:11). See Kenath. 128

Noph. 160, 173

Nophah – Town captured by Israelites (Num 21:30) although text is difficult. – *Unknown; but evidently in Moabite Plateau E of the N end of the Dead Sea.*

Northern Kingdom – See Israel. 31, 46, 154, 155, **155**, 157, 160, **163**, 164, 165, 169, 247

North Syrian Plain. 76

Nubia. 27, 28, 50, 62, 63, 64, 65, **66**, 67, 84, 86, 96, 101, 165

Numeira. 47, 50, **55**, **62**, **86**, **94**, 95

Numidia. 202, 234

Nuseiriyeh Mountains. 72, 75

Nuweiba. 69

Nuzi. 17, 84, 90, 93

Oboth – Campsite of Israelites on their trek to Canaan (Num 21:10, 11). Evidently it was between Punon and Iye Abarim (Num 33:43, 44). – *Uncertain; E of S end of Dead Sea, but IDs of Punon and Iye Abarim are also uncertain.*

Oceanus Atlanticus – See Atlantic Ocean. **202**

Odollam – See OT Adullam. **190**, 192

Offense, Mount of. 242

Olives, Mount of – Hill just E of Jerusalem. David crossed it as he fled E from Absalom (2 Sam 15:30). Referred to in oracle of Zechariah (14:4) in connection of appearance of Yahweh. Mentioned frequently in connection with the last week of Jesus' earthly ministry: Jesus passed over it as he triumphantly entered Jerusalem (Matt 21:1; Mark 11:1; Luke 19:29, 37); later it was where he taught his disciples (Matt 24:3; Mark 13:3); and there Jesus took his disciples on the night he was betrayed (Matt 26:30; Mark 14:26; Luke 21:37; 22:39). Forty days after his resurrection Jesus ascended from the mount in the presence of his disciples (Acts 1:12). – *The N-S range, just E of Jerusalem including, from S to N, Jebel Batn el-Hawa, Jebel et-Tur, and Ras Abu Kharnub.* 142, 214, 215, 241, **242**, 243, 245, 247, **248**, 249, 251, 252, 253

Ombos. 103

Omrit, Horvat. 208

On – Joseph married a daughter of Potiphera, a priest of On (Gen 41:45, 50; 46:20). – See Heliopolis. 17, **104**, 160

Ono – Built by a Benjamite (1 Chron 8:12). Jews lived there at the time of the return from the Babylonian Exile (Ezra 2:33; Neh 7:37; 11:35). Sanballat and Geshem attempted to lure Nehemiah to the plain there (Neh 6:2), but he refused to come. – *Kafr Ana/Ono (137159), 7.5 mi. E of Joppa.* **51**, 52, 177

Ono, Plain of – See Ono. 52, 177, **177**

Ophel. 245, **246**, 250

Ophir – A country or region from which David and Solomon obtained fine gold and other exotic products (1 Kgs 9:28; 10:11; 1 Chron 29:4; 2 Chron 8:18; 9:10) but Jehoshaphat's attempt failed (1 Kgs 22:48). The gold of Ophir is also mentioned in the poetic and prophetic literature of the OT (Job 22:24; 28:16; Psalm 45:9; Isa

13:12). – *Uncertain; possibly in the S or SW portion of the Arabian peninsula.* 27, **83**, 158

Ophni – Town allotted to Benjamin (Josh 18:24). – *Unknown; possibly located in Benjamin plateau, but E of watershed.*

Ophrah (Benjamin) – Town allotted to Benjamin (Josh 18:23) towards which a Philistine raiding party advanced (1 Sam 13:17). – *Et-Taiyibeh (178151), 13 mi. NNE of Jerusalem.* **115, 123, 132**, 133, **154, 212,** 214

Ophrah (Manasseh) – Village of Manasseh from which Gideon came and where the angel of the Lord appeared to him and he set up an altar for Yahweh (Judg 6:11, 24). Later, an "ephod" was worshiped at Ophrah (8:27, 32) and Abimelech killed 70 of his brothers there (9:5). – *Kh. Taiyibeh (167213), 5 mi. S of Megiddo.* **126,** 128

Opis. 173, **174**

Orda. 193

Orhesa. 199, **200**

Orontes (River). 72, 75, **93, 97,** 101, **101,** 102, **123,** 124, 125, 156, 175, **184,** 185, 224, **225, 228, 231**

Orthosia. 184

Ostia. 234

Oxus River. 174, **175,** 183

Oxyrrhynchus. 218

Paddan – Shortened form of Paddan Aram (Gen 48:7).

Paddan Aram – Place from which Isaac's wife Rebekah came (Gen 25:20). Later Jacob fled there and lived with his uncle Laban for 20 years. There he married and raised his family and accumulated great wealth (Gen 28:2, 5 – 7; 31:18; 33:18; 35:9, 26; 46:15; 48:7). – *Region in NW Mesopotamia, N of the Euphrates, near the Habur and Balik rivers.* 91, **93, 97,** 98

Pakistan. 183

Palestine. 13, 52, 72, 86, 92, 95, 175, 180, 183, 184, **184,** 185, 187, 188, 191, 195, 197, 199, 201, 205, 216, **217,** 219, 223, 224, **228, 231, 233,** 235, 248, 255, 256, 257, 260

Palmahim. 94

Palms, City of – Evidently another name for Jericho. Mentioned in connection with Moses' viewing the land (Deut 34:3) and the Kenites entering the land with Judah (Judg 1:16). Eglon captured it (Judg 3:13) and much later, Judean prisoners were released there (2 Chron 28:15). – *T. es-Sultan (192142), 14 mi. ENE of Jerusalem in Rift Valley.* 125, **126**

Palmyra – See Tadmor. 73, **218**

Pamphylia – Jews from Pamphylia were in Jerusalem on the day of Pentecost (Acts 2:10). Paul and Barnabas passed through area twice on first missionary journey and there John Mark abandoned them (13:13; 14:24; 15:38). Paul's ship passed by its coast on his journey to Rome (27:5). – *Lowland area along S central coast of Turkey. Was a Roman province in Paul's day bounded by Lycia on SW, Pisidia on N, and Cilicia on E.* 180, **181, 202, 218, 225,** 227, **228, 231, 233**

Paneas – See Caesarea Philippi. 180, **184,** 187, **187,** 188

Pannonia. 202, 234

Paphos – Ancient capital of island of Cyprus. There the governor, Sergius Paulus, was converted by Paul on his first missionary journey (Acts 13:6). – *Baffa, 10 mi. NW of Kouklia on SW edge of Cyprus.* **218,** 225, **225,** 226, **231**

Parah – Town allotted to Benjamin (Josh 18:23). – *Kh. Abu Musarrah (177137), 4.7 mi. NE of Jerusalem.* 115

Paralia. 187, 188

Paran – See Nahal Paran. – Shortened form (Deut 1:1; 1 Kgs 11:18) of Paran, Desert of. 141

Paran, Desert of – Ishmaelites settled in this area (Gen 21:21). Later the Israelites passed through this desert on their way to the land of Canaan (Num 10:12; 12:16; Deut 1:1). From there spies were sent into the land of Canaan and they returned to the Israelite camp at Kadesh (Num 13:3, 26). Hadad the Edomite passed through this desert as he fled to Egypt from Solomon (1 Kgs 11:18). – *Uncertain; seems to refer to a large*

amorphous desert area in central Sinai. Kadesh Barnea and SE Judah (see footnote of NIV LXX implies Hebrew Paran of 1 Sam 25:1) seem to border it on the N and the Desert of Shur on the W. 54, 55, **62,** 97, **104,** 105

Paran, Mount – Poetic name for Mt. Sinai (Deut 33:2; Hab 3:3) that was located in or near the Desert of Paran.

Parium. 234

Paros. 218

Parthia. 175, **175,** 189, 197, 199, **202,** 216, 224, 267

Parvaim – Region from which a special type (reddish?) gold was obtained (2 Chron 3:6). – *Uncertain; possibly in Arabian peninsula.*

Pasargadae. 175, **183**

Pas Dammim – Alternate form for Ephes Dammim where David fought Philistines (2 Sam 23:9; 1 Chron 11:13).

Patala. 175, **183**

Patara – On return from 3rd missionary journey Paul changed ships there (Acts 21:1). – *Located in Lycia, in SW Asia Minor, NE of Rhodes.* 228, **228, 231**

Pathros – Area of Egypt mentioned in prophetic oracles against that country (Ezek 29:14; 30:14). Translated as "Upper Egypt" by NIV in Isa 11:11, and in Jer 44:1, 15. – *Area in Egypt along Nile from Cairo S to Aswan.*

Patmos, island of – Island where John received divine revelation (Rev 1:9). – *Greek island of Patmos, W of SW coast of Turkey in Aegean sea, ca. 35 mi. WSW of Miletus.* 231, 235, **237**

Pau – Town of the Edomite chief Hadad (Gen 36:39; 1 Chron 1:50). – *Unknown; in Edom.*

Pehel – Early name, later NT Pella. **34, 38, 44, 60, 86,** 95, **105, 123,** 124, **124,** 125, 186

Pekod – Name of region in S Babylonia where Aramean tribe of same name was centered. Mentioned in prophetic oracles (Jer 50:21; Ezek 23:23). Probably also a word play on Babylonian "[land of] punishment".

Pella (Macedonia). 181, **182**

Pella (Transjordan) – NT era name, earlier Pehel. **186, 187, 193,** 197, **198, 200, 206,** 211, **212**

Peloponnese. 228, **229, 229,** 233, **233**

Pelusium – Egyptian fortress mentioned in oracle of Ezekiel (30:15, 16). – *Tell Farama, 13 mi. E of Suez Canal, 1.8 mi. from Mediterranean Sea.* 173, **174,** 175, **182, 184, 199,** 202, **218**

Penuel (Peniel) – Place near Jabbok River where Jacob wrestled with a man (angel) of God (Gen 32:30, 31; cf. Hosea 12:4). Inhabitants of Penuel refused to help Gideon against fleeing Midianites and later Gideon punished the town (Judg 8:8, 9, 17). Later, Penuel was fortified by Jeroboam I (1 Kgs 12:25). – *T. edh-Dhahab esh-Sherqiyeh (215176), 8 mi. E of Jordan River.* **58, 60, 97,** 98, **126,** 152, **153, 154, 155,** 157

Peor (Judah) – Town allotted to Judah (Josh 15:59b; LXX). – *Kh. Zakandah (164119), 9.5 mi. SW of Jerusalem.* **114,** 166

Peor (Moab) – A high point in Moab where Balak took Balaam to curse Israel (Num 23:28). Later, the Israelites worshiped the Baal of Peor (Num 25:3, 5, 18; 31:16; Deut 4:3; Josh 22:17; Psalm 106:28). – *Uncertain; possibly near Mt. Nebo, E of N end of Dead Sea.*

Perath – Place mentioned in oracle of Jeremiah where he hid a belt (13:4 – 7). – *Often identified with the Euphrates, but more probably with a valley close to Parah, see there.*

Perazim, Mount – Mentioned in oracle of Isaiah against rulers of Jerusalem (28:21). May be a reference to previous judgment against Philistines at Baal Perazim (2 Sam 5:20; 1 Chron 14:11). See Baal Perazim.

Perea – A Jewish Transjordan district ruled by Herod the Great and then, during the days of Jesus, by Herod Antipas. Stretched from Machaerus in the S to Pella in the N. Not specifically mentioned in NT but note the phrases "region across the Jordan" (Matt 4:25), "on the other side of the Jordan" (Matt 19:1; John 1:28; 3:26), etc. 49, **58,** 59, **186, 187,** 188, **193,** 197, **198, 200,** 201, 205, **206,** 211, **212,** 213, 216, **217, 219,** 220, **220, 222**

Perez Uzzah – Location, W of Jerusalem but E of Kiriath Jearim where Uzzah was struck dead for touching the Ark of the Covenant (2 Sam 6:8; 1 Chron 13:11). – *Unknown.*

Perga – A major city of Pamphylia though which Paul and Barnabas passed on their way into and out of Pamphylia and Pisidia (Acts 13:13, 14; 14:25). – *Ruins near moder Aksu in S Turkey, ca. 8 mi. inland from the Mediterranean coast.* 181, **218,** 225, **225,** 226, **234**

Pergamum – One of the seven churches addressed in the book of Revelation (Rev 1:11; 2:12). – *Bergama in W Turkey, 15 mi. inland (E) from coast of the Aegean Sea; ca. 80 mi. N of Ephesus.* 187, **202,** 231, **231, 234,** 236, 237, **237,** 238, 239

Persepolis. 175, **175,** 179, **183**

Persia – Kingdom mentioned 29 times in OT (2 Chron; Ezra; Esther; Ezek; Daniel), sometimes in connection with Media. – *Heartland in SW Iran, but also used with reference to the extensive Persian Empire.* 72, 78, 79, 172, 173, 174, 175, **175,** 176, 179, 180, 182, **182,** 185, 186, 191, 192, 238, 247, 249

Persian Gulf. 16, **17,** 18, 76, 77, **77,** 78, **83, 84, 93,** 160, 169, 171, 174, **175,** 182, **218**

Pethor – Place from which Balaam the prophet came (Num 22:5; Deut 23:4). – *Uncertain; possibly Tell el-Ahmar, 12 mi. S of Carchemish on W bank of Upper Euphrates river.*

Petra. 23, 28, **32,** 55, 57, **62,** 63, 64, **69, 184, 193,** 195

Phaistos. 17

Pharathon. 190, **192**

Pharpar – One of the rivers of Damascus that Naaman preferred over the Jordan (2 Kgs 5:12). – *Uncertain; possibly the el-Awaj that flows through the plain a few mi. S of Damascus.*

Phasaelis. 200, **212**

Phasis. 175

Philadelphia (Asia Minor) – One of the seven churches addressed in the book of Revelation (Rev 1:11; 3:7). – *Alasehir in W Turkey along banks of Cogmaus river, ca. 75 mi. inland, E of Smyrna.* 231, **236, 237,** 238, 239

Philadelphia (Decapolis) – OT Rabbah of the Ammonites. **184,** 186, **186, 187, 193, 198, 200, 206,** 211, **212, 217, 219,** 220

Philadelphia (Egypt). 218

Philippi – One of the chief cities of Macedonia visited by Paul on his second (Acts 16:12; 20:6; 1 Thess 2:2) and probably third missionary journeys. The letter to the Philippians was addressed to the church there. – *Philippi, in Greece, ca. 9 mi. N of N Aegean coastline near the River Gangites.* 199, **218,** 227, **228,** 229, **231,** 232, **234,** 238

Philistia – See Philistine Plain – Philistia is mentioned only 7 times in OT (Exod 15:14; Psalm 60:8; 83:7; 87:4; Isa 11:4; Joel 3:4; Amos 6:2) although the term Philistine(s) occurs 276 times. Philistia was the Coastal Plain area settled by the Philistines; from the Nahal Besor in the S to the Yarkon River in the N. 18, **51,** 52, 149, 168, 169, **193,** 201

Philistine Plain/ land of the Philistines – See Philistia. 21, **22,** 32, 51, **51,** 52, 53, **83, 91,** 103, 113, 114, 121, **122, 126,** 127, 129, **129,** 130, **130,** 131, **131,** 132, 132, 133, 134, 134, 135, **135,** 136, 137, 138, **140,** 142, **142,** 151, 154, 155, 158, 159, **159,** 160, **161,** 162, 165, 166, **166,** 169, 186, 217, 255, 261

Philistines, Sea of – Exod 21:31. Name for Mediterranean Sea. 254

Philomellum. 234

Philoteria. 187, **193, 198,** 208, 222

Phoenicia – Jesus healed the daughter of a Greek woman from this area (Mark 7:26). Christians came to area due to persecutions (Acts 11:19) and Paul and Barnabas passed through Phoenicia on way to Jerusalem conference (15:3). At Patara, at the end of his third journey, Paul boarded a ship headed for Phoenicia (Acts 21:2). "Phoenicia" of Isa 23:11 should be translated "Canaan" as NIV footnote. – *Narrow region along Mediterranean coast stretching ca. 185 mi. from Mt. Carmel in the S to Mt. Casius in the N. In modern Lebanon and Syria.* 21, 22, **22,** 74, 116, 125, **153, 155,** 156, **157,** 168, 185, **186,** 187, **187,** 188, **193, 198, 200,** 201, 202, **202, 219,** 220, **222,** 225, **228,** 231, 232, **233,** 238

Phoenix – Place where captain of Paul's ship hoped to harbor for winter but due to an intense storm they were blown past and shipwrecked on Malta (Acts 27:12). – *The bay on the W side of Cape Mouros, on the S side of the island of Crete.* 233, **233**

Phrygia – Jews from Phrygia were in Jerusalem on the Day of Pentecost (Acts 2:10). In Hellenistic times a distinct district, during NT times the region was divided between the Roman provinces of Asia and Galatia. Paul visited Galatian Phrygia on all three of his missionary journeys for Antioch and Iconium were located there (Acts 16:6; 18:23). – *Indeterminate area in and near W edge of Anatolian plateau in central Turkey.* 181, **181**, **218**, 225, 226, 227, **228**, 231, **231**

Pi Hahiroth – Place near which the Israelites camped after leaving Etham and after which they crossed the Red (Reed) Sea (Exod 14:2, 9; Num 33:7, 8). – *Uncertain; possibly the low ground near Jebel Geneife, W of the Suez Canal near the Bitter Lakes.* 104

Pirathon – Home and burial place of the judge Abdon (Judg 12:13, 15) as well as the home of Benaiah, one of David's warriors (2 Sam 23:30; 1 Chron 11:31; 27:14). – *Farata (165177), 7 mi. WSW of Shechem in Hill Country of Ephraim.* 44, 115, **126**, 129

Piraeus. 203, 228

Pisgah – Hill top or range near which Israel camped (Num 21:20) and to which Balaam was taken to curse Israel (Num 23:14). From there Moses surveyed the land of Canaan prior to his death (Deut 3:17, 27; 34:1). The area had been ruled by Sihon, king of the Amorites (Deut 4:49; Josh 13:20), but was allotted to Reuben (Josh 13:20). – *Uncertain; probably near Mt. Nebo, ca. 9 mi. E of N end of Dead Sea.*

Pishon – One of the rivers of Eden (Gen 2:11). – *Unknown.* 82, 83

Pisidia – Region through which Paul passed on his first journey as he traveled from Pamphylia to Pisidian Antioch and back (Acts 13:14; 14:24). – *Mountainous district at W end of Taurus range in S central Turkey that was part of province of Galatia in Paul's day.* 225, **225**, 226

Pisidian Antioch – See Antioch (Pisidian). **225**, 226, 227, **228**, 231, **231**

Pithom – One of the two "store cities" built by the Israelites in Egypt (Exod 1:11). – *Uncertain; possibly Tell el-Maskhuta, 9 mi. W of Ismailia, or Tell er-Retabah, 9 mi. W of Tell el-Maskhuta; both in Wadi Tumilat in NE Egypt.* 102, 103, **104**

Plain, Cities of the – Included Sodom, Gomorrah, Admah, Zeboiim, and Bela (Zoar). Lot chose to live in area (Gen 13:12) and God destroyed some of the cities (Gen 19:29). – *Uncertain; possibly SE of the Dead Sea although some argue for a location N of the Dead Sea.*

Plateau – See Moab, Tableland of.

Pompeii. 218

Pontus – Jews from Pontus were in Jerusalem on Pentecost (Acts 2:9). Apollo was from there (Acts 18:2) and Peter addressed his letter to Christians in the area (1 Peter 1:1). – *A Roman province situated along the E portion of the S shore of the Black Sea (in Turkey), bounded by the Halys River on the W, Galatia and Cappadocia on the S, and Armenia on the E. In Paul's day combined with Bithynia into a single province.* 218, 228, 231, 233

Pontus Euxinus – See Black Sea. **202**

Port Said. 66, 68

Poplars, Brook of – Mentioned in oracle against Moab (Isa 15:7). – *On border of Moab, possibly the Zered.*

Prusa. 227, 228

Ptolemais – Greco-Roman name for Acco one of the chief ports of Palestine. Paul stopped there on his return from his third journey (Acts 21:7). – *Acre/ Akko (157258).* 34, 35, 36, 37, 38, 38, 39, 43, 44, 184, 186, **186**, 187, **187**, **193**, 198, 199, **200**, 201, 205, **206**, 207, **208**, 212, 217, 218, 220, **220**, 222, 231, 232, 234

Pumbeditha. 218

Punon – Israelite campsite in wilderness between Zalmonah and Oboth (Num 33:42, 43). – *Uncertain;*

possibly at Feinan (197004), in Jordan, 32 mi. S of Dead Sea. 25, **55**, 56, 57, **62**, 63, 64, **69**, 107, **108**

Pura. 175

Purushkhanda. 85

Put – Descendant of Ham (Gen 10:6; 1 Chron 1:8) but more commonly a country or people mentioned in prophetic oracles (Jer 46:9; Ezek 27:10; 30:5; Nah 3:9; and possibly Isa 66:19, NIV "Libyans" but LXX "Put"). – *Uncertain; but probably all or part of Libya in N Africa.* **83**, 84

Puteoli – Important port S of Rome where Paul landed and stayed there for seven days before proceeding by road N to Rome, ca. 145 mi. distant (Acts 28:13). – *Puzzuoli, W of Naples.* **218**, 233, **233**, 234

Puzzuoli – See Puteoli. 233

Pylos. 17

Qaloniya – See Emmaus. 215

Qantir. 99, 102

Qarat ed – Dahr. **103**

Qarqar. 156, **160**, 162

Qatna. 17, 93, **97**, 123, 174

Qattara. 25, 61

Qedar. 177

Qedem. 96

Qeisarieh – See Caesarea. 258

Qumran. 47, 49, 50, 51, **55**, 194, **212**

Quneitra. 36

Qurna. 76, 77, **77**

Raamah – A personal and tribal name (Gen 10:7; 1 Chron 1:9; Ezek 27:22). – *Uncertain; probably located in Arabia, but exact location disputed.* 83

Rabbah (Judah) – Town allotted to Judah (Josh 15:6). – *Kh. Hamideh (149137), 15 mi. WNW of Jerusalem near the hill country. Mentioned in extrabiblical sources as Rubute.* **114**, 166

Rabbah (of the Ammonites) – Major Transjordanian town where Og's gigantic bed was kept (Deut 3:11). City was excluded from the tribal allotments (Josh 13:25) yet later Joab and David captured it for Israel (2 Sam 11:1; 12:26, 27, 29; 1 Chron 20:1) and its king assisted David as he fled from Absalom (2 Sam 17:27). This capital of the Ammonites is mentioned in later prophetic oracles (Amos 1:14; Jer 49:2, 3; Ezek 21:20; 25:5). In NT times it was called Philadelphia. – *Amman (238151), now capital of Jordan, 24 mi. E of Jordan River.* 22, 23, 28, **32**, 33, **58**, 59, **60**, 63, 105, 108, 110, 119, **119**, 122, 126, **140**, 141, 142, **142**, 143, 145, 148, **153**, 155, 159, 160, 163, 169, 171, 174, 186, **186**

Rabbith – Town allotted to Issachar (Josh 19:20). Possibly a variant of Daberath, a Levitical town in Issachar (Josh 21:28; 1 Chron 6:72) not mentioned in city list of Joshua 19.

Racal – Town in S Judah to which David sent booty taken from Amalekites (1 Sam 30:29). – *Uncertain; possibly a textual error for Carmel that is missing from the list and is near Eshtemoa.*

Raetia Noricum. 202, 234

Rafia – See Raphia. 68, 69

Rakkath – Town allotted to Naphtali situated between Hammath and Kinnereth (Josh 19:35), both on the W shore of the Sea of Galilee. – *Kh. el-Quneitireh/T. Raqqat (199245), 2.5 mi. NW of Tiberias.* 116

Rakkon – Town allotted to Dan (Josh 19:46). – *Unknown; if a town, then in vicinity of Joppa, but Hebrew text is difficult.*

Ramah (Asher) – Town on N border of Asher in vicinity of Tyre and Sidon (Josh 19:29). – *Unknown.*

Ramah (Benjamin) – Town allotted to Benjamin (Josh 18:25). Deborah judged Israel in the vicinity (Judg 4:5) and it was evidently located to the north of Gibeah (Judg 19:13). Was the home of Samuel (1 Sam 1:19; 2:11; also called Ramathaim [1 Sam 1:1]) and from there he went on his circuit (1 Sam 7:15 – 17). Mentioned frequently in connection with the stories of Saul and David (1 Sam passim). Baashah, king of Israel fortified it (1 Kgs 15:17; 2 Chron 16:1) but Asa removed his fortifications (1 Kgs 15:21, 22; 2 Chron 16:5, 6). Mentioned as being along the traditional N inva-

sion route towards Jerusalem (Isa 10:29) as well as in prophetic oracles (Hosea 5:8; Jer 31:15; Matt 2:18). Jeremiah was released by the Babylonians at Ramah (Jer 40:1) and after the Babylonian Exile Jews settled there (Ezra 2:26; Neh 7:30; 11:33). – *Er-Ram (172140), 5 mi. N of Jerusalem.* 23, **23**, **32**, 33, 46, 47, **47**, **51**, 115, 123, **123**, **126**, 127, 131, **131**, **132**, **132**, 133, 134, **135**, 154, **154**, **166**, **177**, 240, **241**

Ramah (Naphtali) – Town allotted to Naphtali on boundary between Lower and Upper Galilee (Josh 19:36). – *Kh. Zeitun er-Rameh (187259), 19 mi. E of Acco.* 36, 38, 116, 117

Ramah in the Negev – Town on the border of Simeon (Josh 19:8). – *Uncertain; possibly Kh. Ghazzah/H. Uza (165068), 20 mi. ESE of Beersheba. Possibly the same as Ramoth Negev (1 Sam 30:27).* 114

Ramathaim – See Ramah (Benjamin). 131

Ramath Lehi – Place near which Samson routed Philistines (Judg 15:17). See Lehi.

Ramath Mizpah – Town allotted to Gad (Josh 13:26). – *Uncertain; possibly Kh. Jelad (223169) NW of Amman.* 119

Rameses – One of the store cities that the Israelites built in Egypt (Exod 1:11). At time of the Exodus they travelled from there to Succoth (Exod 12:37; Num 33:3, 5). – *Tell ed-Daba, ca. 62 mi. NE of Cairo.* 102, 103, **103**, 104

Rameses, district of – Region in NE Nile Delta where Jacob and his descendants settled (Gen 47:11). More commonly called (the land of) Goshen.

Ramoth (Gilead) – Also called Ramoth (2 Kgs 8:29; 2 Chron 22:6) and Ramoth in Gilead (Deut 4:43; Josh 20:8; 21:38; 1 Chron 6:80). A Levitical city of refuge allotted to Gad (Deut 4:43; Josh 20:8; 21:38; 1 Chron 6:80). It served as the headquarters for Ben-Geber in a Solomonic district (1 Kgs 4:13). Here Ahab and Jehoshaphat fought against the Arameans (1 Kgs 22; 2 Chron 18 passim) as did Joram (2 Kgs 8:28; 2 Chron 22:5). There Jehu was anointed king (2 Kgs 9 passim). – *T. Ramith (244210), 36 mi. N of Amman in Jordan.* 17, 23, 28, **30**, **32**, 33, **34**, 40, **58**, 59, **60**, **90**, 91, **109**, 119, **120**, 121, **140**, **142**, **143**, 145, 148, **153**, 155, 156, 157, **157**, 158, **161**, 163

Ramoth (Issachar) – Levitical town in Issachar (1 Chron 6:73). Due to position in list, probably same as Jarmuth (Josh 21:29) and Remeth (Josh 19:21).

Ramoth Negev – Place to which David sent spoils (1 Sam 30:27). – *Uncertain; possibly Kh. Ghazzah/H. Uza (165068), 20 mi. ESE of Beersheba. Possibly the same as Ramah in the Negev (Josh 19:8).* 159

Raphana. 35, **206**, 211, **212**

Raphia. 32, 55, 66, 69, 153, 163, 184, 187, **193**, 197, **198**

Raphon. 192

Raqqad River/Valley – See Nahal Raqqad. 33, 35

Ras Ali. 86

Ras el – Jeifa. **104**

Ras en-Naqb. 62, 63

Ravine of the Poplars – Valley crossed by fleeing Moabites (Isa 15:7). – *Uncertain; possibly Wadi el-Hasa or Sel Esal, E of S end of Dead Sea.*

Recah – Unknown city in Judah (1 Chron 4:12).

Red Sea – 1. Sea that Israel crossed at the time of the Exodus and in which the troops of Pharaoh perished. Mentioned 14 times in the OT and in Acts 7:36 and Heb 11:29. – *Uncertain; possibly in the Bitter Lakes area but there are many other proposals.* 2. Name for Gulf of Suez area from which locust came to the land of Egypt (Exod 10:19) and near which Israel camped before proceeding to the Desert of Sin (Num 33:10, 11). – *Gulf of Suez.* 3. Sea located at SE border of the promise land (Exod 23:31). Israel traveled toward it from Kadesh Barnea as they went around Edom (Num 14:25; 21:4; Deut 1:40; 2:1 and possibly 1:40). Ezion Geber was on its shore (1 Kgs 9:26) as was Edom (Jer 49:21). – *Gulf of Aqabah/Elath. "Red Sea" comes from the LXX translation of the Hebrew "Reed Sea."* 17, 21, 24, 25, 47, 49, 56, 57, **62**, 64, 65, **66**, **83**, 84, **93**, 99, 103, **103**, 104, **104**, 106, 107, **108**, 109, **146**, 147, 149, 158, **160**, **171**, **173**, **174**, **175**, 177, **182**, **202**, **218**, 233, **234**, 254

Reed Sea – See Red Sea. 103, 104

Rehob (Beth Rehob) – Israelite spies traveled N to it (Num 13:21) and troops from there fought against David's army (2 Sam 10:8). Shortened form of Beth Rehob, an Aramean principality N of Israel. 105, 139, **140**, 141

Rehob (Jordan Valley). 44, 96, **105**, 124, 153, 155

Rehob (northern Asher) – Town on N (NE?) border of Asher (Josh 19:28). – *Uncertain; possibly T. el-Balat (177280), 12 mi SE of Tyre in Lebanon.*

Rehob (southern Asher) – Town allotted to Asher (Josh 19:30) that served as a Levitical city (21:31; 1 Chron 6:75) and from which they were not able to drive out the Canaanites (Judg 1:31) – *Uncertain; possibly T. el-Bir el-Gharbi/T. Bira (166256), 5 mi. ESE of Acco.* **38**, **116**, 117, **120**, 122

Rehoboth – Name of well that Isaac dug in the Negev (Gen 26:22). – *Uncertain; possibly Kh. Ruheibe/H. Rehovot (108048), 22 mi. SW of Beersheba, but a site farther N, between Gerar and Beersheba seems more probable.*

Rehoboth (Edom) – Place from which the Edomite ruler Shaul came (Gen 36:37; 1 Chron 1:48). – *Unknown; but in Edom, not along Euphrates as some.*

Rehoboth Ir – Probably not a town, but a reference to the squares or suburbs of Nineveh (Gen 10:11). – *In N Iraq, in or near Nineveh/Calah.*

Rekem – Town allotted to Benjamin (Josh 18:27). – *Unknown; probably W of watershed.*

Rekem (Edom). 153

Remeth – Town allotted to Issachar (Josh 19:21). – *En Hayadid/T. Remet (199221) 9 mi. SSW of Sea of Galilee. Probably the same as Ramoth (1 Chron 6:73) and Jarmuth (Josh 21:29).* 116

Rephaim, Valley of – Valley/plain area to the WSW of Jerusalem on the border between Judah and Benjamin (Josh 15:8; 18:16). On several occasions the Philistines camped in the valley as they attempted to invade Judah/Israel (2 Sam 5:18, 22; 1 Chron 11:15; 14:9). This relatively flat area served as the breadbasket of Jerusalem (Isa 17:5). – *The broad "Baqa" area WSW of ancient Jerusalem (now occupied by several Jerusalem neighborhoods) through which the railroad previously ran W before entering the narrow valley farther W.* **135**, 139, 241, **242**

Rephidim – Campsite of Israelites prior to Mt. Sinai where water was obtained from the rock and where the Amalekites attacked Israel (Exod 17:1, 8; 19:2; Num 33:14, 15). – *Uncertain; some place at the Feiran Oasis in SW Sinai, but location is also dependent upon where Mt. Sinai is placed.*

Resen – A place mentioned in connection with Nineveh and Calah (Gen 10:12). – *Unknown; possibly the reference is to a large water installation associated with the area.*

Retenu (Upper and Lower). 93, 96, 225

Reuben – Eldest son of Jacob and the name of the tribe that descended from him. Their territorial allotment was E and NE of the N end of the Dead Sea. 27, **58**, 62, 113, 118, 119, **119**, **120**, 141, 149, 156, 162, **178**

Rezeph – Sennacherib sent a message to Hezekiah mentioning how he had captured the gods of Rezeph (2 Kgs 19:12; Isa 37:12). – *Uncertain; possibly modern Risafa, 15 mi. S of Euphrates near Sura in Syria, or a Rezeph closer to Jebel Sinjar in N Iraq.* **160**, 169, 171

Rhagae. 183

Rhegium – Paul's ship stopped there for a day on journey from Malta to Puteoli (Acts 28:13). – *Reggio at S tip of Italy, opposite Sicily.* 233, **233**

Rhinocorura. 184, **193**, 195

Rhodes – Inhabitants of Rhodes traded with Tyre (Ezek 27:15; following NIV that follows LXX rather than the Heb. Dedan). Paul's ship stopped there on return voyage of his third journey (Acts 21:1). – *Rhodos a Greek island, the second largest in Aegean Sea (ca. 45 mi. by 22 mi.) ca. 12 mi. off SW coast of Turkey.* 17, 84, **218**, **228**, **231**, 232, **237**

Riblah – City N of Israel on Orontes River, where Pharaoh Neco put Jehoahaz in chains (2 Kgs 23:33). There Nebuchadnezzar, the king of Babylon set

up a major camp in his campaign against the states of the S Levant and Egypt. There Zedekiah, the last Judean king was executed, as were other political and religious leaders (2 Kgs 25:6, 20, 21; Jer 39:5, 6; 52:9, 10, 26, 27). Probably Diblah of Ezek 6:14 refers to the same place. – *Ribleh (296427), in Syria, 65 mi. NNE of Damascus.* 72, **169**, 171, **171**, 174

Riblah – Site mentioned along NE boundary of the land of Canaan (Num 34:11). – *Unknown; probably E or NE of Sea of Galilee.*

Ridge Route. 33, 46, 47, 92, 106, 123, **131**, 132, **132**, 133, 151, 214, 240, **241**

Rift Valley. 21, **22**, 23, **23**, 24, 25, 27, 28, 29, 33, **34**, 35, 37, 39, 41, **43**, **44**, 45, 46, 47, 48, 49, 54, 56, 57, **58**, 59, **60**, 63, 64, 112, 115, 125, 154, 201, 205, 211, 240, **241**, 265

Rimmon – Town in Negev allotted to Judah (Josh 15:32) and Simeon (19:7; 1 Chron 4:32). Zech 14:10 also seems to place it in S Judah. – *Uncertain; possibly Tell Khuweilifeh/Tel Halif (137087), 9.5 mi. NNE of Beersheba.* 166

Rimmon, rock of – Place to which 600 Benjamites fled after being defeated by the rest of Israel (Judg 20:45, 47; 21:31). – *Uncertain; possibly in area of Rammun (178148), 11 mi. NE of Jerusalem.* **115**, 123

Rimmon (Zebulun) – Town on NE border of Zebulun (Josh 19:13) that served as a Levitical city (= Dimnah of Josh 21:35; = Rimmon of 1 Chron 6:77). – *Rummaneh/H. Rimona (179243) or possibly the nearby H. Romah at 177243, 6 mi. NNE of Nazareth.* 116, **116**, 120

Rimmono – Alternate form of Rimmon (Zebulun; 1 Chron 6:77).

Rimmon Perez – Israelite campsite between Rithmah and Libnah (Num 33:19, 20). – *Unknown.*

Rissah – Israelite campsite between Libnah and Kehelathah (Num 33:21, 22). – *Unknown.*

Rithmah – Israelite campsite between Hazeroth and Rimmon Perez (Num 33:18, 19). – *Unknown.*

Rodanim. 83, 84

Rogelim – Home of Barzillai of Gilead who aided David in his flight from Absalom (2 Sam 17:27; 19:31). – *Uncertain; possibly Bersinya (223215), 16. 5 mi. SE of Sea of Galilee, or nearby Dhaharat Soqa.* 143, **143**

Rome – Jews from Rome were present in Jerusalem on Day of Pentecost (Acts 2:10). Claudius had expelled some Jews from the city (Acts 18:2) during his reign. Paul desired to visit the city (Acts 19:21) and wrote an epistle to the Christian community there (Romans 1:7, 15). He finally visited the city, but as a prisoner (Acts 23:11; 25:25; 28:14, 16; 2 Tim 1:17). – *Rome, capital of the Roman Empire, W central Italy.* 187, 188, 189, 191, 192, 195, 196, 197, 199, 201, **202**, 203, 204, 205, 216, **218**, 219, 221, 222, 223, 224, 229, 232, 233, **233**, 234, **234**, 235, 236, 237, 249, 266, 267

Rosh HaNiqra – See Ladder of Tyre. 21, **22**, **23**, **36**, 37, **38**, **86**, 94

Route of the Patriarchs. 33, 92

Royal Road/Way. 175, **175**, 238

Ruba al-Khali. 16

Rubicon. 197, **202**

Rubute (Khirbet Hamida) – See Rabbah (Judah). **123**

Rumah – Home of Zebidah daughter of Pedaiah, the mother of Jehoiakim (2 Kgs 23:36). – *Kh. er-Rumeh/H. Ruma (177243), 6 mi. N of Nazareth.*

Ruweisat el – Ahdar. **104**

Saba. 84

Sabtah – Third son of Cush (Gen 10:7; 1 Chron 1:9). – *Possibly a locality in Arabia or Nubia.* 83

Sabteca – Fifth son of Cush (Gen 10:7; 1 Chron 1:9). – *Possibly a locality in south Arabia.*

Sa el – Hagar. **103**

Safed. 38, 94

Saft el – Hinna. **103**

Sagartia. 175

Sahara Desert. 65, 66

Saint Catherine's Monastery. **69**, 104, **104**

Sais. 175

Sakha. 99

Salamis (Cyprus) – First stop on Paul's first missionary journey where he and Barnabas and John Mark preached in the synagogues (Acts 13:5). – *NE portion of Cyprus, along E coast, just N of Famagusta.* **184**, **218**, 225, **225**

Salamis (Greece). 176, 180, **181**, 267

Salecah – Town on E edge of the Bashan that had been controlled by Og (Deut 3:10). It was allotted to Manasseh (Josh 13:11; cf. 13:30, 31) but Gadites settled the region (1 Chron 5:11). – *Salkhad (311212), in Jordan, ca 62 mi. ESE of S end of Sea of Galilee and 8 mi. S of Jebel Druze.* **34**, 35, **109**

Salem – Home of Melchizedek to whom Abraham paid a tithe (Gen 14:18; Heb 7:1, 2). Associated with Zion in Psalm 76:2. – *Alternate (poetic) form for Jerusalem although other interpretations have been proposed.* 88, **90**, **91**, 94, 95, **97**, 240, 243

Salim – Prominent site used to locate Aenon, where John was baptizing (John 3:23). – *Probably T. er-Radgha/T. Shalem (199200), 7.5 mi. SSE of Scythopolis (OT Beth Shan). See also Aenon.* **212**, 213

Salima Oasis. 66

Salmone – Prominent landmark of the most E portion of Crete. Paul's ship sailed to the S of it passing from Cnidus to Fair Havens (Acts 27:7). – *Cape Sidero, E tip of Crete.* 233

Salona. 234

Salt, City of – Town in the "desert" (= wilderness) district allotted to Judah (Josh 15:62). – *Uncertain; possibly Kh. Qumran (193127), at NW corner of Dead Sea.* **115**, 166

Salt, Valley of – Place where David and his warriors defeated the Edomites (2 Sam 8:13; 1 Chron 18:12 and title of Psalm 60) as did Amaziah later in Judean history (2 Kgs 14:7; 2 Chron 25:11). – *Uncertain; possibly the es-Sebkha region, S of the Dead Sea although the Wadi el-Milh, E of Beersheba has also been suggested.* 47, 56, 94, 95, **140**, 141

Salt Sea – Common biblical name for the "Dead Sea" (Gen 14:3; Num 34:3, 12; Deut 3:17; Josh 3:16; 12:3; 15:2, 5; 18:19). Also called the Sea of the Arabah and the Eastern Sea. See Dead Sea. 25, 49, 50, **69**, **94**, 106, **106**, 108, 109, **110**, 114, **115**, **119**, **120**, 122, **126**, 131, **135**, 136, **155**, 158, **159**, 162, 254

Samaga. 193, 195

Samaria (City) – Capital city of N kingdom of Israel founded by Omri (885 – 874 BC; 1 Kgs 16:24) and developed by subsequent Israelite kings. Mentioned 109 times in the OT and 10 times in the NT. Its capture by the Assyrians ended the northern kingdom of Israel. Subsequently, the district around it became known as Samaria and its inhabitants as Samaritans. It was renamed Sebaste during the days of Herod the Great (37 – 4 B.C.). – *Sebastiyeh (168187), 7 mi. NW of Shechem.* **32**, **38**, **44**, 46, 151, **153**, 154, **155**, 156, **157**, **159**, **160**, **161**, 162, 163, 164, 165, 169, **170**, **171**, 172, **174**, 176, **177**, **181**, 182, **184**, 185, 186, **186**, 187, 191, **193**, 195, 198, **200**, 201, **212**, 214, 217, **217**, **222**, 256, 266

Samaria (district). 31, **163**, 166, 170, 176, 177, **177**, **178**, 185, 186, **186**, 187, 188, **190**, 192, **193**, 194, 199, **200**, 201, 205, **206**, **212**, 213, 214, 215, 216, 217, **217**, **219**, 220, **220**, **222**, 224

Samarra. 76, 77

Samos – Ionian island that Saul's ship passed on return voyage of his third journey (Acts 20:15). – *Greek island of Samos, 10 mi. W of Turkish coast WSW of Ephesus.* **218**, **231**, 232

Samosata. 100

Samothrace – Island in NE Aegean which Paul passed traveling from Asia Minor to Europe on his second (Acts 16:11) and probably third (20:6) missionary journeys. – *Samothrace in the N Aegean Sea.* 227, **228**

San el – Hagar (Tanis). **103**

Sansannah – Town in Negev allotted to Judah (Josh 15:31). – *Kh. esh-Shamsaniyat/H. Sansanna (140083), 8 mi. NE of Beersheba.* **114**, 166

Saqqara. 66, 84, 86, **99**, 103, **104**

Sardinia. 202

Sardis – One of the seven churches addressed in the book of Revelation (1:11; 3:1, 4). Formerly the ancient capital of Lydia; on major highway into interior of Asia. – *Ruins near Turkish village of Sart, 48 mi. E of Smyrna (Izmir).* 173, **174**, 175, **175**, 181, **181**, **182**, **218**, 231, **231**, **234**, 236, 237, **237**, 238, 239

Sarid – Town on S boundary of Zebulun from which the boundary ran W (19:10) and E (19:12). – *T. Shadud (172229), 6 mi. NE of Megiddo on N edge of Jezreel Valley.* 86, 116, **116**

Saronic Gulf. 228, **229**, 230

Saudi Arabia. 13, 16, **17**, 27, 28, 64, 69, 103

Scorpion Pass – Location on SE portion of the boundary of Canaan (Num 34:4) and the territory of Judah (Josh 15:3). Also served as a border point of Amorite territory (Judg 1:36). – *Uncertain; possibly Naqb es-Safa/Maale Aqrabbim (near 162035), 20 mi. SW of S tip of Dead Sea.* 106

Scythopolis – OT Beth Shan. 184, 186, **186**, 187, 192, **193**, 197, **198**, **200**, 206, 211, **212**, 213, **217**, 219, **220**, **222**, 257

Seba – Land associated with Egypt and Cush (Gen 10:7; 1 Chron 1:9; Isa 43:3) as well as with Sheba (Psalm 72:10). – *Uncertain; possibly area of Ethiopia, although some tribal members may have moved E across strait of Bab el-Mandeb and roamed in NW Arabia (cf. Sabeans in Job 1:15; Ezek 23:42; Joel 3:8), but Isa 45:14 associates some Sabeans with Egypt and Cush.* 83

Sebam – Alternate form of Sibmah (Num 32:3).

Sebaste – OT Samaria (City). **200**, 201, 202, **206**, 212, 214, **217**, **219**, **220**, 222

Sebkha. 56, 62

Secacah – Town in desert district allotted to Judah (Josh 15:61). – *Uncertain; possibly Kh. es-Samrah (187125), 10 mi. SE of Jerusalem.* **114**, 166

Secu – Place in Benjamin to which Saul went to gather information regarding Samuel and David (1 Sam 19:22). – *Unknown; probably in Gibeah and Ramah region N of Jerusalem.*

Seir (Mount; Land of) – Name of eponymous ancestor of a group of Horites (Gen 36:20–30) who were displaced by the Edomites (Deut 2:12). Seir is use 38 times in OT, 7 in conjunction with "hill country of," 2 "land of," and 8 "Mount Seir." – *Usually identified with the mountains in Jordan, E of the Rift Valley, stretching from the Zered in the N to Aqaba in the S, but there are numerous biblical texts which indicate that it was also used to refer to territory W of the Rift Valley as well (e.g., Deut 1:2, 44; 33:2; Josh 11:17; 12:7; 1 Chron 4:42, 43; etc.). Sometimes used as a synonym for land of Edom (e.g., Gen 32:3).* 28, **62**, 64, 94, **94**, **97**, 108, 114, 158, **159**, 195

Seir, Mount (Judah) – Geographical marker on N boundary of Judah (Josh 15:10). – *Uncertain; possibly ridge W of Kiriath Jearim and N of Kesalon, ca. 11 mi. W of Jerusalem.*

Seirah – Place to which Ehud escaped after killing Eglon (Judg 3:26). – *Unknown; probably W or NW of Jericho; could be a town or a forested region.* 125

Sela – Edomite fortress captured by Amaziah who renamed it Joktheel (2 Kgs 14:7). Mentioned in oracle concerning Moab (Isa 16:1) as well as in Isa 42:11. The marker of Amorite territory may or may not be the same place (Judg 1:36). – *Uncertain; possibly es-Sela (205020), 23 mi. SE of S tip of Dead Sea in Jordan. Umm el-Bayyara (192971) seems to be too far S for the Moabite associations with this place.* 62, **153**, 158, **159**, 160

Sela Hammahlekoth – Saul abandoned pursuit of David at this place ("rock of parting;" 1 Sam 23:28). – *Unknown; probably E or SE of Maon (23:25).*

Seleucia (Golan). 187, 188

Seleucia (Mesopotamia). 77, 182

Seleucia (Syrian) – Port city of Syrian Antioch (ca. 16 mi. E) from which Paul and Barnabas set sail on their first journey (Acts 13:4). – *Site near Samandag in Turkey, in NE corner of Mediterranean, a few mi. N of present course of Orontes River.* 186, **218**, 225, **225**, **228**, 233

Seleucia Pieria – See Seleucia (Syrian). **225**

Seleucia (Transjordan; see Abila). 187, **187**

Seleucia (Transjordan; see Antiochia). 187, **187**

Semechonitis, Lake. 25, 36, **36**, 197, **198**

Senaah – Some of the inhabitants of this town returned from the Babylonian Exile with Zerubbabel (Ezra 2:35; Neh 7:38). – *Unknown.*

Seneh – A cliff or other geographical feature near the Philistine camp which Jonathan captured (1 Sam 14:4). – *Possibly a cliff on S rim of Wadi Suweinit (ca. 176141), 7 mi. NNW of Jerusalem, near Micmash.* **132**, 133

Senir – Amorite name for Mt. Hermon (Deut 3:9) although some passages seem to distinguish it from Mt. Hermon (1 Chron 5:23; SS 4:8; Ezek 27:5). – *Mt. Hermon in N Israel/SW Syria, or one of its prominent peaks.* 34, **34**

Sennabris. 208

Sephar – Locality on border of lands of Joktan (Gen 10:30). – *Uncertain; probably in S Arabia.*

Sepharad – Obadiah (20) refers to exiles from Jerusalem as being there. – *Uncertain; possibly near the Lower Zab River in NE Iraq, or some suggest Sardis in W Turkey.*

Sepharvaim – People from there were settled in N Israel by the Assyrians (2 Kgs 17:24, 31). The conquest of the city, and its gods, became proverbial (2 Kgs 18:34; 19:13; Isa 36:19; 37:13). – *Unknown; probably in Syria but numerous proposals.* 164

Sepphoris. 38, 39, **43**, 193, 197, **198**, 199, **200**, 205, **206**, 207, 208, **208**, **212**, 219, 220, **220**, 222

Serabit el-Khadem. 69, 70, **99**, 104

Shaalabbin – A town allotted to the tribe of Dan (Josh 19:42). – *Probably a variant of Shaalbim.* 118

Shaalbim – Place where Amorites maintained themselves after the conquest (Judg 1:35). Solomon later placed Ben-Deker in charge of this Danite district (1 Kgs 4:9). – *Selbit/T. Shaalevim (148141), 19 mi. SE of Joppa.* 121, **122**, 148

Shaalim, district of – Area where Saul searched for lost donkeys (1 Sam 9:4). – *Uncertain; possibly in Benjamin, N of Jerusalem.* **131**, 133

Shaaraim (Judah) – A town in Shephelah allotted to Judah (Josh 15:36). Mentioned as a point of reference in story of David and Goliath (1 Sam 17:52). – *Uncertain; possibly Kh. Sairah (152127), in N Shephelah near Azekah, 10.5 mi. NE of Lachish. Recently nearby Kh. Qeiyafa (146122) has been suggested.* **114**, 134, **134**

Shaaraim (Simeon) – Town allotted to Simeon (1 Chron 4:31). – *Unknown; probably in S Shephelah or Negev.*

Shaar HaGay. 28, 30

Shaar Hagolan. 94

Shahazumah – Place on N border of Issachar (Josh 19:22). – *Unknown; probably between Mt. Tabor and Jordan River.*

Shalisha(h) – Area through which Saul passed looking for his donkeys (1 Sam 9:4). – *Uncertain; possibly in vicinity of Baal Shalisha, 16 mi. NE of Jerusalem.* **131**, 133

Shamir (Ephraim) – Home and burial place of the judge Tola in hill country of Ephraim (Judg 10:1, 2). – *Unknown.* **126**, 128

Shamir (Judah) – Hill country town allotted to Judah (Josh 15:48). – *Unknown; probably to S of Hebron.*

Shaphir – Town, probably in Shephelah, mentioned by Micah (1:11). – *Unknown.*

Sharm esh-Sheikh. 66, 68, **69**

Sharon (Plain) – Region of coastal plain N of Yarkon River and S of Mt. Carmel. There cattle were pastured (1 Chron 27:29) in David's day. Its fertility could be a symbol of divine approval (Isa 35:2) and its desolation of disapproval (Isa 33:9). Some inhabitants were influenced by healing of Aeneas in Lydda (Acts 9:35). 21, **22**, 23, 33, **38**, 40, 41, 43, **44**, 45, 46, 51, 52, 124, **135**, 136, 142, **142**, 149, 152, 154, 162, 177, 186, 192

Sharon (Transjordan) – Called a Gadite pastureland in Transjordan (1 Chron 5:16; also Moabite Stone, line 13). – *Unknown; E of Jordan River/Dead Sea.* 62,

Sharuhen – Town in W Negev allotted to Simeon (Josh 19:6) which is called Shilhim in Josh 15:32. – *Probably T. el-Ajjul (093097), 4 mi. SW of Gaza or possibly T. el-Farah (S) (100076).* 99, **99**, 100, **101**, 103, 114, 153

Shatt al-Arab. 77, **77**

Shawbak. 30, 55

Shaveh, Valley of – Place where Abram met king of Sodom after defeating Kedorlaomer (Gen 14:17). See King's Valley. 95, 243

Shaveh Kiriathaim – Transjordanian site where Kedorlaomer defeated the Emites (Gen 14:5). – *Unknown.*

Sheba (Arabia) – Queen of this country visited Solomon (1 Kgs 10; 2 Chron 9; "Queen of the South" Matt 12:42; Luke 11:31). Famous for its merchants, gold, incense, etc. (Job 6:19; Psalm 72:10, 15; Isa 60:6; Jer 6:20; Ezek 27:22, 23; 38:13). – *Uncertain; possibly in SW Arabian peninsula in vicinity of modern Yemen although some suggest the horn of Africa, near modern Djibouti.* 29, 49, **83**, 84, 144, **146**

Sheba (Simeon) – Probably an alternate name for Beersheba (Josh 19:2) but possibly a variant of Shema (Josh 15:26).

Shebarim – Place towards which the inhabitants of Ai chased Israel (Josh 7:5, see footnote). – *Unknown; probably to the E of Ai.*

Shechem – Mentioned close to 60 times in the Bible. Important non-Israelite center that figured prominently in the stories of Abraham, Jacob, and later Abimelech (Judg 9). During the days of Joshua it seems to have served as an Israelite tribal and religious center. Located on the boundary between Manasseh and Ephraim (Josh 17:2, 7) it was designated as a Levitical city as well as a city of refuge (Josh 20:7; 21:21; 1 Chron 6:66, 67). Served briefly as the first capital of the N kingdom but was replaced by Tirzah and then Samaria. – *T. Balatah (176179), 30 mi. N of Jerusalem, on E side of Nablus.* 22, 23, **23**, 26, 32, 33, **38**, **44**, 45, 46, 48, **58**, 59, **60**, 90, **91**, 92, 93, **94**, **95**, 97, **97**, 101, 105, 106, 108, 110, 111, **115**, 120, 121, **122**, 123, **123**, 124, **126**, 128, **132**, 135, 140, **142**, 143, 145, 148, 150, 152, **153**, 154, **155**, **157**, 159, 172, **174**, 177, 193, 198, **212**, 214, 215, **241**, 244, 258

Shem. 83, **83**, 84

Shema – Town in Negev district allotted to Judah (Josh 15:26). – *Unknown.*

Shemer, Hill of – See Samaria. 154

Shen – Samuel set up a marker at Ebenezer, between Mizpah and Shen to commemorate a victory over the Philistines (1 Sam 7:12). – *Uncertain; possibly a variant of Jeshanah.* 132

Shepham – Place on NE boundary of the land of Canaan (Num 34:10, 11). – *Unknown; E or NE of Sea of Galilee.*

Shephelah (of Judah) – Geographical term for the low hills W of the Hill Country of Judah but E of the Philistine Plain (save in Josh 11:2, 16b). Used 20 times in Bible, but NIV translates as "(western) foothills." 24, 28, 38, 47, **47**, 51, 52, 53, 54, **110**, 111, 114, 124, 130, 134, **134**, 135, 142, **142**, 151, 157, 158, 160, 165, 166, 170, 179, 186, 192, 217, 220, 240, **241**, 255

Shephelah (foothill) of Carmel. 38, **38**, 40, 44

Shepher, Mount – Israelite campsite between Kehelathah and Haradah (Num 33:23, 24). – *Unknown.*

Shibah – Name of well dug by Isaac in Negev, later called Beersheba (Gen 26:33).

Shihor (River) – Water course on the E of Egypt which marked SW limits of the land that remained to be conquered (Josh 13:3) as well as the limit of David's influence (1 Chron 13:5). Also mentioned in Isa 23:3 and Jer 2:18. – *NE portion of the Pelusaic branch of the Nile in the E delta of Egypt or the frontier canal in the far E delta.* 142

Shihor Libnath – Point (town? or stream/wadi named after a town) on S boundary of Asher in the Mt. Carmel area (Josh 19:26). – *Uncertain; possibly Libnath in S. T. Abu Huwam (152245), 9 mi. SSW of Acco in modern Haifa.*

Shikkeron – Town on W portion of N boundary of Judah between Ekron on the E and Mount Baalah on the W (Josh 15:11). – *T. el-Ful (132136), 17 mi. SE of Joppa.* 114

Shilhim – Town in Negev allotted to Judah (Josh 15:32). – *Unknown but cf. Josh 19:6.*

Shiloh – Religious center in the Hill Country of Ephraim where some of the tribal allotments were made, the tabernacle was set up, and where Eli and Samuel ministered. Mentioned 32 times in OT, mostly in Joshua, Judges and 1 Samuel, but it also figures prominently in the sermons of Jeremiah (7:12, 14; 26:6, 9). – *Kh. Seilun (177162), 20 mi. NNE of Jerusalem E of ancient road that led from Bethel to Shechem, S of Lebonah (Judg 21:19).* 23, **23, 32, 33, 44,** 46, **51, 95, 105,** 110, 113, 115, **115, 120,** 123, **123, 126,** 131, **131, 132, 161,** 172, **177, 241,** 243, 245

Shimron – Canaanite town whose king fought against Israel (Josh 11:1) and which was allotted to Zebulun (19:15). Possibly "Simeon" in 2 Chron 34:6 refers to this place as well. – *Kh. Sammuniyeh/T. Shimron (170234), 9 mi. NNE of Megiddo.* **36, 38,** 40, **86, 95, 105,** 111, **112,** 116, **116**

Shimron Meron – One of 31 towns vanquished by the Israelites (Josh 12:20), possibly a fuller designation of Shimron, unless the LXX is correct in reading two towns; namely Shimron and Meron.

Shinar – Land that included cities of Erech, Akkad and Calneh (Gen 10:10) and where Babel was located (11:2). Amraphel was king of Shinar (14:1, 9) and Achan stole a "robe from Babylonia" (Josh 7:21; see NIV fn.; Hebrew "Shinar"). Also mentioned in Hebrew text of Isa 11:11; Dan 1:2; and Zech 5:11 (see NIV footnotes). – *Hebrew term for Babylonia, in S Iraq.* **83,** 84, 94

Shion – Town allotted to Issachar (Josh 19:19). – *Uncertain; possibly Kh. Mugheir (183232), 4 mi. SE of Nazareth.*

Shittim – Last campsite of Israel before crossing the Jordan under leadership of Joshua (Num 25:1; Josh 2:1; Micah 6:5). From there spies were sent to Jericho (Josh 3:1). See Abel Shittim. – *T. el-Hammam (214138), in Plains of Moab, E of Jordan River, NE of Dead Sea.*

Shual – A Philistine raiding party turned toward Ophrah in the vicinity of Shual (1 Sam 13:17). – *Uncertain; see Ophrah.*

Shubat-Enlil. 78

Shunem – Town allotted to Issachar (Josh 19:18) near which the Philistines camped in preparation for battle with Saul (1 Sam 28:4). Abishag, who served David (1 Kgs 1 and 2) was from there. A woman of Shunem assisted Elisha (2 Kgs 4). – *Solem (181223), 9 mi. E of Megiddo, at S foot of Mt. Moreh.* **38,** 39, **116, 135,** 136, 137, **153, 161**

Shur – Israelites wandered in the Desert of Shur after crossing the Red Sea (Exod 15:22). In a number of passages it serves as a marker (Gen 16:7; 20:1; 25:18; 1 Sam 15:7; 27:8). – *Uncertain; possibly a line of Egyptian forts near, or E, of the present day Suez Canal.* **55,** 69, **69,** 92, 133, 136

Shushan. 175

Sibmah – Town allotted to Reuben from domain of Sihon (Num 32:38; Josh 13:19) that passed into Moabite hands and was famous for its vineyards (Isa 16:8, 9; Jer 48:32). – *Unknown.*

Sibraim – Site on the N border of Israel between Damascus and Hamath (Ezek 47:16). – *Unknown; N of Damascus.*

Sicily. 233, **233**

Siddim, Valley of – Place where four kings of the N met with the kings of the cities of the plain (Gen 14:3, 8, 10). – *Uncertain; possibly S end of Dead Sea, or the es-Sebkah region to the S of it.* **47,** 50, 94, **94,** 254

Side. 218

Sidon (region) – A principle city of Phoenicia mentioned 34 times in the Bible. Located on the NW border of Israelite settlement (Josh 11:8; 19:28; 2 Sam 24:6) it is frequently mentioned in the Bible in connection with Tyre, especially in the prophetic literature. Jesus visited the region (Matt 15:21; Mark 7:31) and mentioned it in cursing Korazin and Bethsaida (Matt 11:21, 22; Luke 10:13, 14). People of Sidon were present when Herod Agrippa I was struck with a disease (Acts 12:20) and Paul's ship stopped there at the beginning of his journey to Rome (27:3). – *Saida (184329), 24 mi. SSW of Beirut on Lebanese coast.* **17,** 22, **22, 32,** 33, **34,** 37, 40, **72,** 74, **83,** 84, **97, 101,** 112, **112,** 116, **116,** 117, 121, **122, 123, 140,** 142, **142, 153,** 154, **155,** 156, **157, 160, 161,** 162, **163,** 167, 170, **174,** 177, 181, **181, 184,** 185, 186, **186,** 187, **193, 200,** 201, **206,** 210, **212, 218, 219, 220, 228,** 233, **233, 234**

Sile. 101, 103, 103, **103, 124**

Silwan. 245

Simeon – Second son of Jacob, the head of one of the tribes of Israel. Mentioned 43 times in the Bible. Simeon's tribal allotment (Josh 19:2 – 7; 1 Chron 4:28 – 32) was in the S, within the Negev district of Judah (cf. Josh 15:21 – 32). Simeonites seemed to have migrated to various regions in and around Israel (cf. 1 Chron 4:39 – 43; 2 Chron 15:9; 34:6). 114, **114, 120,** 169, **178,** 255

Sin, Desert of – Area that Israel passed through after Elim but before Rephidim (Exod 16:1; 17:1). There manna began to be provided as were quail. – *Uncertain; depends upon location of Mt. Sinai. Probably in W or NW Sinai, E or SE of Suez.*

Sinai (Desert of, Mount, Peninsula) – The term Sinai is used 35 times in the OT (all but 4 in the Pentateuch) and 4 times in the NT. Mount Sinai (17 of the 35 references) seems to be a specific peak located in a larger area called the Desert of Sinai (13 of the 35 references). The latter is sometimes referred to as Horeb (17 times OT), but only once is the mountain specifically referred to as Mount Horeb (Exod 33:6). All the biblical references are associated with Moses and the giving of the law at Sinai (Exodus 19:2; Num 10:12) save for Elijah's flight to Horeb (1 Kgs 19:8). – *Many proposals, but traditionally at Jebel Musa in central portion of S Sinai although Jebel Sin Bisher (W portion of central Sinai) seems to be a viable alternative.* 16, **17,** 18, 32, 33, 47, **55,** 56, 57, 65, **66,** 68, 69, **69,** 70, 83, 86, 88, 92, 100, 101, 103, 104, **104,** 105, 106, 107, 134, 142, 174, 175, 177, 181, **181,** 214, 217, 257

Sinim – In prophetic oracle (Isa 49:12) of restoration used as a figurative reference to the S. – *Aswan, on E bank of Nile in S Egypt. Hebrew word also used in Ezek 29:10 and 30:6 where NIV translates as "Aswan."*

Sinope. 175, 218, 234

Siphmoth – One of a number of villages to which David sent gifts (1 Sam 30:28). – *Unknown; but in S Hill Country of Judah or Negev.*

Sippar. 78, 171, 175

Sirah, well of – Well (or cistern) of Sirah at which Abner received Joab's summons to Hebron (2 Sam 3:26). – *Unknown.*

Sirion – Sidonian name for Mt. Hermon (Deut 3:9) but parallel to "Lebanon" in Psalm 29:6. NIV of Deut 4:48 reads "Mount Siyon." – *Sirion may refer to the whole Anti-Lebanon range of which Mt. Hermon is a part; E Lebanon.* 34, **34**

Sitah – Name of second of three wells dug by Isaac's servants (Gen 26:21). – *Uncertain; probably in one of the valleys between Gerar and Beersheba.*

Siyon, Mount – Alternate name for Mount Hermon (Deut 4:48). Possibly a variant of Sirion.

Smyrna – One of the seven churches addressed in the book of Revelation (1:11; 2:8). Prominent Roman city, famous for worship of emperor. – *Izmir on central portion of W coast of Turkey.* 231, **231, 234,** 236, 237, **237,** 238, 239

Soco – See Socoh.

Socoh (Hill Country of Judah) – Town in hill country allotted to Judah (Josh 15:48; cf. 1 Chron 4:18 "Soco"). – *Kh. Shuweikeh (150090), 10 mi. SW of Hebron.* 114, **166**

Socoh (Sharon Plain) – Town in Solomon's third administrative district (1 Kgs 4:10) associated with Arubboth and land of Hepher. – *Kh. Shuweiket er-Ras (153194) on E edge of Sharon Plain, 26 mi. NNE of Joppa.* **38,** 43, **44, 115,** 148, **155,** 157

Socoh (Shephelah) – Town in N Shephelah district allotted to Judah (Josh 15:35). Near it David fought Goliath (1 Sam 17:1), Rehoboam fortified it for Judah (2 Chron 11:7; "Soco"), but the Philistines took it during reign of Ahaz (2 Chron 28:18; "Soco"). – *Kh. Abbad/H. Sokho (147121), 17 mi. WSW from Jerusalem.* **51,** 53, **114,** 134, **134, 151,** 160, 165, **166,** 167

Sodom – City on SE boundary of Canaan (Gen 10:19) where Lot settled (Gen 13) and which fought against the kings of the N (Gen 14). City was destroyed because of its wickedness (Gen 18, 19), and it, along with Gomorrah, became symbols of wickedness and of cultures that had experienced God's judgment. Mentioned 47 times in Bible. – *Unknown; possibly one of the Early Bronze III sites, E and SE of the Dead Sea in Jordan.* 30, 88, **91,** 94, 95

Sodom, Mount. 47, 49, 50

Sogdiana. 175

Sorek, Valley of – Place where Samson fell in love with Delilah (Judg 6:4). – *Wadi es-Sarar/Nahal Soreq, W of Beth Shemesh (area around Timnah [141132] in NW Shephelah, 21 mi. SE of Joppa).* **51,** 53, 113, 117, **118,** 121, **126,** 130, 131, **131,** 132, 134, 168

Spain – Paul expressed his desire to minister there (Rom 15:24, 28) and some early Christian traditions suggest that he eventually journeyed there. – *Modern Spain.* 16, 234

Sparta. 17, 175, 218

Strato's Tower. 44, **44,** 186, **186,** 187, 193, 198, 200, 201, 202

Subite. 163, 176, **178**

Succoth (Egypt) – Site to which Israel traveled after leaving Rameses but before arriving at Etham (Exod 12:37; 13:20; Num 33:5, 6). – *Uncertain; possibly Tell el-Maskhuta in Wadi Tumilat, 9 mi. W of Ismailia, in NE Egypt.* 103, **104**

Succoth (Transjordan) – Place where Jacob camped after encounter with angel of the Lord (Gen 33:17). Allotted to Gad (Josh 13:27), Gideon punished its leaders after they refused to help him against the Midianites (Judg 8). Solomon cast bronze vessels for the temple in the area (1 Kgs 7:46; 2 Chron 4:17) and the Valley of Succoth is referred to in the Psalms (60:6; 108:7). – *T. Deir Alla (208178), 22 mi. SSE of Beth Shan, E of Jordan near Jabbok.* **44, 58,** 59, **60, 90,** 91, **91, 97, 105,** 119, **126,** 128, **145, 155,** 163

Succoth, Plain of. 59

Suez. 57, **66,** 68, 69, **69,** 101, 103, 104, **104,** 142

Suez, Gulf of. 65, **66,** 68, 69, **69,** 103, 104, **104,** 175

Sumer. 77, **84,** 85, **90,** 93

Sumur. 72, 74, **123,** 160

Suph – Place mentioned to localize the setting of Moses' address to Israel (Deut 1:1). – *Unknown.*

Suphah – Mentioned in the difficult phrase "Waheb in Suphah" (Num 21:14). – *If a place, then in Moab, probably near Arnon Gorge.*

Susa – One of the capitals of the Achaemenian kings. Mentioned in Ezra (4:9) and Daniel (8:2). The city/court is the setting for the story of Esther (mentioned 19 times; king Xerxes) and was where Nehemiah served Artaxerxes I. – *Shush, 65 mi. NE of Tigris River, ca. 200 mi. WSW of Esfahan; in Iran.* **17,** 77, **84,** 169, 171, 174, 175, **175, 183, 218,** 238

Sychar – Town in Samaria where Jesus met woman at well (John 4:5). – *Uncertain; probably at Askar (177180), 1 mi. NE of Shechem if not at Shechem itself.* **212,** 215

Syene – See Aswan. 173

Syracuse – Port city where Paul stayed for 3 days when traveling from Malta to Puteoli on the final leg of voyage to Rome (Acts 28:12). – *Syracuse on SE coast of Sicily.* 233, **233**

Syria – See Aram. – Large Roman province at E end of Mediterranean Sea. Mentioned 8 times in NT; at times in connection with Cilicia (Acts 15:41; Gal 1:21). – *Area included much of modern states of Syria and Lebanon, as well as SE Turkey.* 16, **17,** 18, **19,** 20, 21, 27, 32, 35, 72, 73, 74, 76, 79, 98, 100, 101, 102, 106, 129, 144, 147, 149, **174,** 175, 180, 184, **184,** 185, 186, 188, 192, **193,** 194, 197, 199, **200,** 202, **202,** 205, 206, 212, 216, 217, 219, **219, 220, 222,** 224, **224, 225,** 226, 227, **228,** 231, **231, 233, 234,** 235, 236, 238, 245, 253, 255, 256, 267

Syrian Desert/Plain. 16, 18, **72,** 73, 76

Syrian Gates. 72, 181, **181**

Syro-Arabian Desert. 73, **182**

Syrtis/Gulf of Syrtis – Shallow area of N coast of Libya that sailors on Paul's storm-tossed ship worked to avoid (Acts 27:17). – *Gulf of Sidra on N coast of Libya.* 233, **233**

Taanach – Town conquered by Israelites (Josh 12:21), settled by members of tribe of Manasseh (1 Chron 7:29; Judg 1:27; said to be "within" Issachar [Josh 17:11] but probably "beside" or "on the border of" Issachar is a better translation), and designated as a Levitical city (Josh 17:11). Canaanites lived in town (Judg 1:27) and they traditionally divided war booty in its vicinity (Judg 5:19). Town was in 5th Solomonic administrative district (1 Kgs 4:12). – *T. Tinnik (171214), 4 mi. SE of Megiddo.* 38, 39, 40, 41, **44**, 94, 95, 101, **101**, 105, 115, **115**, 116, 120, 122, 123, 126, 127, **130**, 139, 140, 148, 153, 155, 256

Taanath Shiloh – Town on NE border of Ephraim (Josh 16:6). – *Kh. Tana el-Foqa (185175), 4.5 mi. SE of Shechem.* **115**

Taba. 55, 69

Tabariyeh – See Tiberias. 258

Tabbath – Place towards which Gideon pursued the Midianites (Judg 7:22). – *Unknown; probably E of Jordan River.* 128

Taberah – Place in wilderness where "fire from Lord burned" among Israelites as punishment (Num 11:3; Deut 9:22). – *Unknown.* 105

Tabgha – See Heptapegon. 210

Tabor – Levitical town in Zebulun (1 Chron 6:77). – *Unknown.*

Tabor (Benjamin). 133

Tabor, (Mount) – On NW boundary of Issachar (Josh 19:22). Only in story of Deborah and Barak is Mount Tabor referred to (Judg 4:6, 12, 14). Nearby Zebah and Zalmunna killed Gideon's brothers (Judg 8:18). Mentioned in poetic (Psalm 89:12) and prophetic literature (Jer 46:18; Hosea 5:1). – *Jebel et-Tur/Har Tavor (186232), 4.5 mi. ESE of Nazareth.* 33, **36**, 37, **38**, 39, 40, **43**, 112, 116, **116**, 117, **126**, 127, 128, 186, **208**, 210, **212**, **217**, **256**

Tadmor – Large oasis, caravan stop, in N Syrian desert that Solomon controlled (2 Chron 8:4; cf. 1 Kgs 9:18, NIV Tadmor, but Heb. Tamar). – *Palmyra, 130 mi. NE of Damascus in Syria.* 17, 18, **32**, 72, 73, 74, 75, **77**, 78, **93**, 143, 144, **146**, 160, 169, 171, 174, 175

Tahath – Israelite wilderness campsite after Makheloth but before Terah (Num 33:26, 27). – *Unknown.*

Tahpanhes – Egyptian town mentioned in late prophetic literature (Jer 2:16; Ezek 30:18), especially in connection with Jeremiah's forced exile to Egypt and his prophecy concerning Nebuchadnezzar (Jer 43:7, 8, 9; 44:1; 46:14). – *Tell Defenneh, in NE Egypt, on Pelusaic branch of Nile, 27 mi. SSW of Port Said, 8 mi. W of el-Qantara.* 172, 173

Tahtim Hodshi, region of – A region, or a town of a region, through which Joab passed while taking the census for David (2 Sam 24:6). – *Unknown; seems to be located between Gilead and Dan.*

Tamar – Site mentioned by Ezekiel as being on the SE boundary of a restored Israel (Ezek 47:18, 19; 48:28); also mentioned in Heb. text of 1 Kgs 9:18 as having been built by Solomon. – *Uncertain; possibly Ain Husb/Hazeva (173024), 23 mi. SW of S tip of Dead Sea.* **32**, 55, 56, 57, **62**, 69, 94, **94**, 106, 108, 144, **145**, 153, 158, **159**, 178

Tanis – San el – Hagar. 103, 160

Tappuah (Ephraim) – Town conquered by Israel (Josh 12:17), on the border between Ephraim and Manasseh. Allotted to Ephraim (17:8), but the land around it to Manasseh (16:8; 17:8). – *Sheikh Abu Zarad (172168), 8 mi. SW of Shechem.* 44, 115

Tappuah (Judah) – Town in N Shephelah allotted to Judah (Josh 15:34). – *Unknown.*

Taralah – Town allotted to Benjamin (Josh 18:27). – *Unknown; probably in W Benjamin plateau, NW of Jerusalem.*

Taricheae. 42, **43**, 193, **206**, 207, 208, **208**, 211, 220, **222**

Tarraconensis. 202

Tarshish (Mediterranean) – A place (town? region?) along shores of the Mediterranean Sea (Isa 66:19) W of the Levant to which Jonah attempted to flee (1:3; 4:2) with which the Levantine countries traded (Jer 10:6; Ezek 27:12). – *Uncertain; possibly in Spain, North Africa, Sicily, etc.* **83**

Tarshish (southeast) – Place mentioned in association with Sheba, Seba, and Dedan (Arabian places/tribes; Ps 72:10; Isa 60:9; Ezek 38:13). Solomon sent ships there and Jehoshaphat attempted to do so (1 Kgs 10:22; 2 Chron 20:36, see NIV footnotes in each case). – *Unknown; possibly in Arabia or E Africa.* 158

Tarsus – City where Paul was born (Acts 9:11; 21:39; 22:3) and to which he returned after his conversion (Acts 9:30; 11:25). Major trading and "university city" in Paul's day. – *Tarsus, in S Turkey, in NE corner of Mediterranean Sea, 10 mi. inland from the coast.* 17, 72, 84, **169**, 171, 175, 181, **181**, **218**, 219, 224, **225**, 227, **228**, 231, 234

Taurus Mountains. 17, 78, 84, 85, **90**, 93, **225**, 227, **228**, 231

Tatam – Hill Country town allotted to Judah (Josh 15:59b; in LXX only). – *Unknown.*

Taxila. 175, 183

Tebah – A town of Hadadezer from which David took quantities of bronze (2 Sam 8:8 [Heb. "Betah"]; 1 Chron 18:8 [Heb. "Tibhath"]). Probably "Tibhath" would be a better translation in both instances. – *Unknown; N of Israel.* 141

Tebtynis. 218

Tekoa – Hill country town allotted to Judah (Josh 15:59b; LXX only). One of David's mighty men was from the village (2 Sam 23:26; 1 Chron 11:28; 27:9) and a wise woman from Tekoa was brought to David by Joab to plead for Absalom (2 Sam 14:2, 4, 9). Amos was from the town (1:1) and Jeremiah mentions it in an oracle (6:1). Men of the village helped rebuild the wall of Jerusalem (Neh 3:5, 27). – *Kh. Tequ (170115), 10 mi. SSW of Jerusalem.* 47, 49, **114**, 143, 151, 161, 166, 177, 179, **190**, 192, 194, 199, **200**

Tekoa, Desert/Wilderness of – Wilderness area E of Tekoa into which Jehoshaphat led his troops (2 Chron 20:20).

Tel: see also under Tell.

Tel Agrah. 86

Telaim – Locality in S Judah where Saul mustered his troops prior to battle with Amalekites (1 Sam 15:4). – *Unknown.* 133

Tel Anafa. 86

Tel Aphek – See Aphek (Sharon). **86**, 130

Tel Arad – See Arad. 84, 86

Tel Ashir. 94

Tel Assar – Town that had been conquered by the Assyrians (2 Sam 19:21; Isa 37:12). – *Unknown; possibly in NE Syria or NW Iraq.*

Tel Aviv (Israel). 66

Tel Aviv – Settlement in Babylonia where Ezekiel met Jewish exiles (3:15). – *Unknown; on the Kebar River in S Iraq.*

Tel Bira. 94, 95

Tel Dan – See Dan. 86

Tel Dothan – See Dothan. 86

Teleilat el – Ghassul. 86

Tel Ektenu. 94

Telem – Negev town allotted to Judah (Josh 15:24). Some suggest it is an alternate form of Telaim but Hebrew roots are different. – *Unknown; although Tel Malhata has been recently suggested. But see Arad.*

Tel Erani. 51, 52, 84, 86, **86**

Tel Eton – See Eglon. 105, 130

Tel Gamma – See Yurza and Tel Jemmeh. 105

Tel Gerisa – See Gath Rimmon. 86, 95, 105

Tel Halif – See Rimmon. 86, 86, 94

Tel Haror – See Gerar. 95

Tel Harsha – Jewish settlement in Babylonia, from which Jews returned to Judea (Ezra 2:59; Neh 7:61). – *Unknown; in S Iraq.*

Tel Hefer. 86, 95, 105

Tel Jemmeh – See Tel Gamma. 95

Tel Kabri. 86, 95, 105

Tel Keison. 86

Tel Kison. 95

Tell: see also under Tel.

Tell Abu Huwam – See Shihor Libnath. 38, 105

Tell Atrib. 99

Tell Basta. 99, 103

Tell Beit Mirsim. 51, 92, 94, 95, 105, 255

Tell Defenneh. 69

Tell ed – Dab'a – See Rameses. 93, 99, 102, 103, 104

Tell el – Ajjul – See Sharuhen. 51, 86, 94, 95, 99, 105

Tell el – Farah (North) – See Tirzah. 95

Tell el – Farah (South). 51, 55, 105

Tell el – Farama – See Pelusium. 69, 104

Tell el-Ful – See Gibeah (in/of Benjamin) and Gibeath (of Saul). 133

Tell el – Heir. 69

Tell el-Hesi. 51, 86, 95, 105, 256

Tell el-Maskhuta – See Succoth (Egypt). 99, 103, **103**, 104, 173, 177

Tell el – Yahudiya. 99, 103, 104

Tell er-Retabah – See Pithom. 69, 103, **103**, 104

Tell er – Ruweisa. 95

Tell esh – Sheikh Ahmad el – Areini. 261

Tell es – Sahaba. 99

Tell es – Saidiya – See Zaphon (Gad). 105

Tell es-Samak. 38

Tell es-Safi – See Gath (Philistine). 95

Tell es-Sultan – See Jericho. 95, 109, 257

Tell Mardikh – See Ebla. 86, 256

Tell Marjame – See Baal Shalishah. 95

Tell Muqdam. 103

Tell Nabasha. 103

Tell Umm Hagar. 69

Tell Umm Hamad. 94

Tel Mahfar. 86

Tel Malhata – See Arad and Telem. 47, 86, **86**, 95

Tel Masos – See Amalek, city of, Hormah, and Telem for various proposals. 47, 95, 130, 133

Tel Melah – Jewish settlement in Babylonia, from which Jews returned to Judea (Ezra 2:59; Neh 7:61). – *Unknown; in S Iraq.*

Tel Mevorach. 44, 95, 105

Tel Mikhal. 44, 44, 95

Tel Mor. 95, 105

Tel Nagila. 86, 95

Tel Obeid. 77

Tel Poran. 86

Tel Qadesh. 86, 95

Tel Qashish – See Helkath. 95, 105

Tel Qasile. 44, 94, 130, 145

Tel Qiri. 95

Tel Qishyon – See Kishion. 86

Tel Raqat – See Rakkath. 86

Tel Regev – See Acshaph. 94, 105

Tel Rehov – See Rehob (Jordan Valley). 86, 95

Tel Rekhesh – Anaharath. 86, 95

Tel Ridan. 95

Tel Safit. 86

Tel Sarid – See Sarid. 86

Tel Sera – See Ziklag. 105

Tel Shimron – See Shimron. 86, 95

Tel Shiqmona. 105

Tel Yarmuth – See Jarmuth (Judah). 86, 87

Tel Yinam – See Jabneel (Naphtali). 95

Tel Yosef. 94

Tel Zeror. 95, 105

Tel Zippor. 94, 105

Tema – Desert oasis connected with caravan trade mentioned in Job (6:19) and Isaiah (21:14). Jeremiah seems to locate it in N Arabia (25:23). – *Teima (Tayma), 250 mi. SE of Aqaba/Elath in Saudi Arabia; 200 mi. NNE of Medina.* 17, 146, 160, 171, 174, 175

Teman – Edomite town mentioned in prophetic oracles (Amos 1:12; Ob 9; Hab 3:3; Jer 49:20) and famous for wisdom (Jer 49:7). Eliphaz, one of Job's "comforters," was from Teman (Job six times). – *Uncertain; possibly Tawilan (197971), in S Jordan, 3 mi. E of Petra, 53 mi. S of S end of Dead Sea.* 62, 153, 159

Tepe Gawra. 77

Tephon. 190

Terah – Israelite campsite during wilderness wanderings between Tahath and Mithcah (Num 33:27, 28). – *Unknown.*

Terqa. 72, 84, 93, 160

Zobah – Aramean city-state N of Israel that was defeated by Saul (1 Sam 14:47). Hadadezer, its king, supplied troops to assist the Ammonites against David and Joab, but they were routed and the kingdom eventually fell to David (2 Sam 8, 10 passim; 1 Chron 18, 19 passim; Psalm 60, heading). Also mentioned in connection with the rebel Rezon (1 Kgs 11:23, 24). – *Heartland was the central Beqa Valley region in Lebanon.* 133, 139, **140,** 141, 143

Zobah (Judah) – In tribal list of Judah found in LXX of Josh 15:59b (see NEB). – *Probably Suba/T. Zova (162132).* **166**

Zoheleth, Stone of – Place where Adonijah attempted to initiate his kingship before the death of David (1 Kgs 1:9). – *Near the spring of En Rogel, south of Jerusalem probably in/along the Kidron Valley.*

Zophim, field of – Place to which Balak took Balaam to curse Israel (Num 23:14). – *Uncertain; on top of Pisgah. Meaning: "field of the watchers."*

Zor. 25, 59, 60, 61

Zorah – Town in N Shephelah allotted to Judah (Josh 15:33) as well as to Dan (19:41). Some Danites migrated from the area to Laish in the N (Judg 18:2, 8, 11). Manoah, Samson's father and a Danite, was from Zorah (Judg 13:2). Samson was active in the area (13:25) and was buried in the vicinity (16:31). Rehoboam fortified the town (2 Chron 11:10) and Jews settled there in the postexilic period (Neh 11:29). – *Sarah/T. Zora (148131), 15 mi. W of Jerusalem.* **114,**118, **118, 126,** 130, **134, 151, 166, 177,** 179

Zuph, district of – Area where Saul searched for lost donkeys (1 Sam 9:5). – *Uncertain; probably named after Zuphite clan and located in territory of Benjamin, possibly in area of Ramah.* **131,** 133